Minding the Brain

Also by Georg Northoff

Unlocking the Brain: Volume I Coding; Volume II Consciousness

Neuropsychoanalysis in Practice: Brain, Objects and Self

Minding the Brain

A Guide to Philosophy and Neuroscience

Georg Northoff

© Georg Northoff 2014

All rights reserved. No reproduction, copy or transmission of this publication may be made without written permission.

No portion of this publication may be reproduced, copied or transmitted save with written permission or in accordance with the provisions of the Copyright, Designs and Patents Act 1988, or under the terms of any licence permitting limited copying issued by the Copyright Licensing Agency, Saffron House, 6–10 Kirby Street, London EC1N 8TS.

Any person who does any unauthorized act in relation to this publication may be liable to criminal prosecution and civil claims for damages.

The author has asserted his right to be identified as the author of this work in accordance with the Copyright, Designs and Patents Act 1988.

First published 2014 by
PALGRAVE MACMILLAN

Palgrave Macmillan in the UK is an imprint of Macmillan Publishers Limited, registered in England, company number 785998, of Houndmills, Basingstoke, Hampshire RG21 6XS.

Palgrave Macmillan in the US is a division of St Martin's Press LLC,
175 Fifth Avenue, New York, NY 10010.

Palgrave Macmillan is the global academic imprint of the above companies and has companies and representatives throughout the world.

Palgrave® and Macmillan® are registered trademarks in the United States, the United Kingdom, Europe and other countries.

ISBN 978–0–230–28354–1

This book is printed on paper suitable for recycling and made from fully managed and sustained forest sources. Logging, pulping and manufacturing processes are expected to conform to the environmental regulations of the country of origin.

A catalogue record for this book is available from the British Library.

A catalog record for this book is available from the Library of Congress.

Typeset by MPS Limited, Chennai, India.

Contents

List of Tables and Figures — vii
Preface — ix
Acknowledgements — xi

Introduction — 1

Part I Mind and Brain: From Philosophy through Neuroscience to Neurophilosophy — 21

1. Philosophy and the Mind: Philosophy of Mind and Phenomenology — 23
2. Philosophy and Science: Naturalism — 43
3. Mind, Brain and Science: Psychology and Neuroscience — 69
4. Brain and Philosophy: Neurophilosophy — 91

Part II The Mind–Brain Problem: From Philosophy of Mind to Philosophy of Brain — 125

5. Mental Approaches to the Mind–Brain Problem — 127
6. Physical and Functional Approaches to the Mind–Brain Problem — 155
7. Non-mental and Non-physical Approaches to the Mind–Brain Problem — 183
8. Brain-based Approaches to the Mind–Brain Problem — 212

Part III Philosophy of Psychology and Neuroscience: From Explanation of Mind to Explanation of Brain — 247

9. Philosophy of Psychology: Personal versus Subpersonal Levels of Explanation — 249
10. Philosophy of Psychology: Mind and Meaning — 278
11. Philosophy of Neuroscience: Explanations, Concepts and Observer in Neuroscience — 302
12. Philosophy of Brain: Characterization of the Brain — 329

Part IV Neurophilosophy of Consciousness: From Mind to Consciousness — 351

13. Arguments against the Reduction of Consciousness to the Brain — 353
14. Neural Correlates of Consciousness (NCC) — 382
15. Neural Predispositions of Consciousness (NPC) — 402
16. Conceptual, Phenomenal and Methodological Issues in the Investigation of Consciousness — 426

Part V	Neurophilosophy of Self: From Consciousness to Self	**447**
17	Brain and Self	449
18	Brain and Self-consciousness	471
19	Abnormalities of Self and Brain in Psychiatric Disorders	493
20	Brain and Intersubjectivity	512
	Epilogue: Is the Brain a Door Opener?	535
	References	539
	Index	547

List of Tables and Figures

Table

13.1	Philosophical arguments against the reduction of consciousness to a physical–material basis	358

Figures

I.1	Different disciplines and their domains	12
	I.1a Philosophy of mind	12
	I.1b Neuroscience	12
	I.1c Reductive neurophilosophy as brain-reductive approach	13
	I.1d Non-reductive neurophilosophy as brain-based approach	14
	I.1e Philosophy of brain	15
2.1	Philosophy and naturalism	58
	2.1a Segregation or parallelism between philosophy and science	58
	2.1b Replacement (or incorporation) naturalism (Quine)	64
	2.1c Cooperative naturalism	65
4.1	Different concepts of neurophilosophy	99
	4.1a Reductive neurophilosophy	99
	4.1b Non-reductive neurophilosophy	101
4.2	Concept–fact iterativity	119
7.1	'Inference problem' and 'domain problem' in current mind–brain discussion	196
	7.1a 'Inference problem' as unjustified inferences between different domains	196
	7.1b 'Domain problem' as the 'localization' of the mind–brain problem in different domains	209
8.1	Different concepts of the brain	231
11.1	Relationship between neuroscience and other disciplines	306
	11.1a Neuroscience, philosophy of neuroscience, philosophy of brain and neurophilosophy	306
	11.1b *Incorporation model*: incorporation of other disciplines into neuroscience	309
	11.1c *Contextualization model*: bilateral contextualization between neuroscience and other disciplines	311

12.1	Extrinsic and intrinsic views of the brain	340
14.1	Characterization of consciousness in relation to other functions	396
15.1	Hard problem and neural predispositions, prerequisites and correlates of consciousness	419
20.1	Models of the relationship between intra- and intersubjectivity	530

Preface

Philosophy is an old discipline that is based on and concerned with mind and consciousness in their various facets. Existence and reality as the metaphysics of mind and consciousness is dealt with in what is described as the mind–brain problem. That touches upon how we can know what the mind is, the problem of knowledge as dealt with in epistemology. All this has been threatened by the recent emergence of neuroscience. Neuroscience focuses on the brain and its neural features. More and more, neuroscience also ventures into traditionally philosophical domains when investigating the neural mechanisms that may underlie mental features like consciousness, free will, self, etc.

Does philosophy and its focus on the mind become superfluous? Do we need to replace mind with brain and philosophy as an old discipline with neuroscience as the state of art? So neurophilosophy claims. Roughly, neurophilosophy can be understood as the neuroscientific investigation of traditionally philosophical issues like consciousness, self, mind, free will, etc. Hence, rather than searching for the definition of concepts and arguments for the mind as philosophy does, we may better conduct experiments to figure out the facts of the brain. Focus on neuroscience rather than philosophy! That is what the neurophilosopher tells us.

Is it really so easy? The aim of this book is to show that it is not so easy. We cannot simply replace the mind with the brain and philosophy with neuroscience. Why not? This does not do justice to the features and properties by means of which we characterize the brain. And most important, it does not do justice to the brain itself. The brain itself seems to be more complex than just a bundle of facts we obtain in our experiments in neuroscience. In addition to mere empirical observation, that is, the method of neuroscience, we may also need to investigate the brain in conceptual and argumentative terms, the traditional method of philosophy. We may thus want to use the method of philosophy and apply it to the brain rather than the mind. That may provide us with a more complex picture of the brain which may allow us to account better for both neural and mental features and thus brain and mind.

The present book aims to investigate the brain in this latter way, a conceptual way. This makes it possible to introduce the brain as subject matter into philosophy that so far has focused thematically exclusively on the mind. This makes possible a true neurophilosophy in the literal sense of the term where the brain is introduced as subject matter into philosophy by applying its particular method of investigation. The main title therefore speaks of 'Minding the Brain', which indicates that the brain as such, and more specifically a particular concept of the brain, may be relevant to philosophy. Using the concept 'minding', I suggest that we need to change our opinion of the brain as well as of philosophy in order to make investigation of the brain in philosophy possible.

Such understanding of neurophilosophy has to be distinguished from the current use of the same term. Here the direction is the reverse: rather than from the brain (and neuroscience) to philosophy, the advocates of such a concept of neurophilosophy prefer to move in the reverse direction, e.g. from philosophy and mind to brain and neuroscience. They consequently reduce mind to brain and philosophy to neuroscience, for which reason I speak of reductive neurophilosophy. This is obviously different in my approach. Here the direction is from brain and neuroscience to mind and philosophy. The resulting philosophy or better neurophilosophy is therefore brain-based rather than brain-reductive. For that reason I speak of non-reductive neurophilosophy, and distinguish it from the current rather reductive neurophilosophy.

The main overarching aim of this book is to introduce such non-reductive neurophilosophy to both students at all levels and researchers in the various disciplines. The book targets all those who share a strong interest in philosophical questions revolving around mind and brain. I aim to apply such a non-reductive neurophilosophical approach to traditional philosophical problems like naturalism (Chapter 2), the mind–brain problem (Part II), and issues like explanation and levels as discussed in philosophy of psychology and neuroscience (Part III). Finally, I will demonstrate how such a non-reductive neurophilosophical approach can also be made fruitful for neuroscience in tackling mental features like consciousness (Part IV) and self (Part V).

GEORG NORTHOFF

Acknowledgements

Though many books are written these days, having a book published that aims to break through traditional disciplinary boundaries in a textbook requires the author to overcome several different kinds of challenges. I am therefore happy that this project could be realized. Most importantly, I hope that this book will inspire many students, teachers, and researchers to go further and far beyond the neurophilosophical framework sketched here and to make fruitful the inclusion of the brain as a topic and subject matter in philosophy.

I want to thank Jamie Joseph and his successor Paul Stevens (and his assistant Jenny Hindley) from Palgrave Macmillan for the excellent editorial support in the various stages of this book from its initial idea to its final realization. Suggestions from anonymous reviewers proved very useful along the way. I also want to thank Maia Woolner and Keith Povey for the very helpful editorial work on the English as well as the members of my research group, Eyuep Suzgun, Gabriel Mograbi, Nils-Frederic Wagner, Christine Wiebking, Pengmin Qin, Niall Duncan, Zirui Huang, Zeidy Munoz, Gregory Dumont, as well as Pedro Chavez Lucas Jurkovic and John Atyall (who were especially helpful in the proof readings) for lively discussions and helpful suggestions. Financially, I have to thank the Canada Research Chair program, the Canadian Institute of Health Research (CIHR), and the Michael Smith Foundation who provided me with lavish financial support. Institutionally, I owe gratitude to the Institute of Mental Health Research and the University of Ottawa for providing me with the physical and mental space to realize such a project. Finally, a big thank you goes to John, who copes on a daily basis with an often absent-minded researcher preoccupied by the fascination for the brain and its role in philosophy.

GEORG NORTHOFF

Introduction

Why do we need the brain in philosophy? How can neurophilosophy be non-reductive?

It might be that you are puzzled about the title of this book, *Minding the Brain*. 'No, that does not go together,' you may want to say: 'Isn't philosophy about the mind and its various mental features like consciousness and self?' Rather than minding our brain, we should mind the mind. Traditionally, only the mind can provide a guide to philosophy while the brain shows the road in neuroscience. Hence, the subtitle of the book seems to confuse not only mind and brain but, even worse, philosophy and neuroscience.

Is this yet another of those books where one is forced to read nothing but the latest neuroscientific data explaining the mind? Those sorts of efforts are typical in the recently developed field of neurophilosophy. Roughly, neurophilosophy in the current sense of the term aims to apply neuroscientific methods to the investigation of what were originally philosophical concepts (like mind, self, consciousness, etc.). If you are familiar with neurophilosophy you may now want to ask about the 'non-reductive' qualifier: 'Doesn't all neurophilosophy seek to reduce the mind to the brain by using neuroscientific methods in place of philosophical methods?' The short answer is *not any more*. The approach to neurophilosophy in this book is non-reductive primarily in the methodological (rather than metaphysical) sense.

Now, a longer answer: neurophilosophers so far have not distinguished between the metaphysical question of whether the mind can be reduced to the brain and the more methodological question of whether philosophy of mind should be replaced with neuroscience. To be sure, these matters seem to be tightly related. If the metaphysical claim that the mind is nothing but the brain is accepted, then neuroscience is likely to be championed as an investigative method because it engages with the brain. Similarly, the pace at which neuroscientific research is advancing seems to leave old-fashioned philosophical reflection about the mind behind, which may incline some to accept the superiority of neuroscientific methods over philosophical argumentation in regards to the mind. If one has the impression that neuroscience progresses more quickly than philosophy, then one might infer from this that the mind is nothing but the brain.

Thus, acceptance of either the metaphysical claim (the mind is just the brain) or the methodological claim (neuroscience is more effective than pure philosophy) seems to encourage acceptance of the other. This is not mandatory though, and, as this book will show, there are good reasons to separate the issues. One reason is that neuroscientists do not ignore the mind, and increasingly attempt to shed light on mental/philosophical phenomena such as consciousness, or sense of self. These efforts can be facilitated by traditional philosophical considerations, and conversely may fail to address the philosophical issues as philosophers recognize them if the conceptual side of the work is not sufficiently considered.

Another reason to distinguish between metaphysical and methodological reductionism is that the much sought after reduction of the mind to the brain that has been the vision of neurophilosophy's first wave may be too limited a strategy. It is by no means confirmed that brains all by themselves are sufficient for the phenomenon of mind. The failure to distinguish between metaphysical and methodological reductionism has left neurophilosophy blind to the possibility that phenomena of mind result not just from the brain but from the brain in conjunction with its body and environment. The idea of reducing the mind to *just* the brain is not a necessary feature of neurophilosophy, and only seems to be so because neurophilosophers have been too busy enshrining the neuroscientific method over and against a traditionally philosophical approach to see that neuroscience and neurophilosophy dangle this philosophical loose end.

To sum up, the approach to mental phenomena in this book is non-reductive first and foremost in the methodological sense, meaning that no particular method of investigation will be privileged. Philosophical considerations and empirical methods of neuroscience will both be welcome throughout, rather than either having to be reduced to the other (in a methodological sense). Our approach might also be said to be non-reductive in a subtler, and possibly controversial, way. As we will see, reductionism in the philosophy of mind is taken metaphysically to be more or less synonymous with reduction to the brain. By adding the body and the environment to the store of phenomena that can be appealed to in order to explain mental phenomena, our approach contrasts with mainstream reductivism about the mind.

If non-reductive neurophilosophy as practiced in the book is compared with, say, Descartes' philosophy of mind, it will appear thoroughly reductionist in that we reject the view that minds exist over and above brains, bodies and environments. However, if it is compared with standard neurophilosophy, it can be said to be non-reductive in rejecting the often tacitly presupposed metaphysical orthodoxy of the brain's sufficiency for mental phenomena (a relatively subtle metaphysical adjustment). While in methodological regard, most importantly, non-reductive neurophilosophy does not aim to reduce the practice of philosophy of mind to the practice of neuroscience but rather to profit from, and refine each.

To reinforce this contrast between neurophilosophy as it has been practiced so far, and the approach of this book, we will use the phrase 'reductive neurophilosophy' to refer to extant neurophilosophy, and 'non-reductive neurophilosophy' to refer to the approach just outlined. Our burden is thus to show that neurophilosophy is possible without having to assimilate (and ultimately replace) philosophy to (by) neuroscience. It is to this end that we develop non-reductive neurophilosophy as distinguished from its currently predominating reductive sibling (see Part I). This will allow us to generate what I call a brain-based rather than brain-reductive approach to the mind and the mind-brain problem (see Parts II and III). To see how this can work, and work well, we first need to understand the concept of mind and how it originates in the history of philosophy.

Mind and brain in *past* times: from the presence of mind to philosophy of mind

Philosophy has always had a concern with the mind, and the custom of relating minds to specific organs in the body is typically ancient. For example, the ancient

Egyptians, dating as far back as the third millennium BCE, thought the heart was the locus of our mind. Interestingly, there are old scripts showing that they were aware of the brain. Later, in ancient Greece, there was the medical doctor and philosopher Hippocrates who reasoned that the brain is the origin of mind: 'Men ought to know that from the brain and the brain only arise our pleasures, joys, laughter, and jests as well as sorrows, pain, griefs, and tears... It is the same thing which makes us mad or delirious. Inspires us with dread and fear, whether by night or by day, brings us sleeplessness, inopportune mistakes, aimless anxieties, absent mindedness, and acts that are contrary to habit' (Hippocrates, 'The Sacred Disease').

However, despite Hippocrates' emphasis on the brain, ancient Greek philosophers like Plato and Aristotle associated mental features with a mind or soul. Then throughout the medieval ages and all the way through the sixteenth century, European philosophers reasoned that the mind should be associated with a specific mental substance. This mental substance, thought the French philosopher Rene Descartes (1596–1650), must be distinguished from the physical substance that characterizes a person's body. Despite seeing them as fundamentally distinct sorts of entities, Descartes believed that the mind and body were connected in the brain's pineal gland.

Descartes can be considered the first modern investigator of the relationship between mind and brain. He sought to give a clear definition of mind, and explain how its existence and reality relates to the physical world, including bodies and brains. *This question regarding the existence and reality of mind and body/brain is called the mind–body problem or the mind–brain problem.*

Descartes' interest in explaining how the mind interacts with the world has been shared by philosophers ever since (see Part II). Some even attempted to prove the existence of a mind experimentally. For example, at the beginning of the twentieth century, the physician Dr. Duncan MacDougall (1907) of Boston — in an attempt to provide evidence for the existence of a mind or soul as mental substance — weighed his patients shortly before and immediately after their death. He assumed that the mind or soul leaves the body after death and that as a mental substance the mind should have some kind of weight. This led him to suggest that the body would be lighter after death, since as the soul leaves the body, the weight of the deceased person will decrease.

Indeed, MacDougall observed that two of his patients were lighter after death; this difference in weight he attributed to the mind or soul and its mental substance. In contrast, he did not observe this change in weight in dogs, which were assumed not to have a soul. Based on his experiment, MacDougall assumed the weight of the mind to be 21 grams. To him, this was proof enough that the mind or soul is a mental substance that is distinct from the body.

What does this imply for the relationship between mind and brain? It shows that there is what philosophers call a 'dualism' between the two: mind and brain can be characterized by different kinds of existence, namely the mental and the physical. This position can be traced back to Descartes and is still held by many philosophers today (see Chapter 5). The evaluation of different dualistic theories of the mind–brain relation is a central concern for contemporary 'philosophy of mind'. To put it in a nutshell, philosophy of mind investigates the existence or metaphysical nature of the mind and particular mental features like consciousness, free will, self, etc.

Mind and brain in *present* times: from the elimination of mind to reductive neurophilosophy

Today, we seem to know much more than any of the historical figures discussed in the previous section. We now have much better tools and methods for investigating the brain and the mind. Neuroscience, the discipline that investigates the brain scientifically, provides unprecedented access to the brain and its neuronal states. It also investigates how the brain and its neuronal states relate to mental features like consciousness, the self, free will, emotions, etc. Neuroscience is widely taken to show that the mind is nothing but the brain.

For example, neuroscientists can observe how the brain's neural activity changes in its different regions during experiences of emotion or the exertion of free will. We can inspect the difference between brains in conscious states and unconscious states, or investigate how the brain creates an experience of self. For instance, your feeling of excitement while reading these lines is accompanied with neural activity in specific regions of the brain that are different from those that are activated when you experience boredom. This type of investigation is made possible by functional brain imaging techniques like functional magnetic resonance imaging (fMRI), which create visual representations of neural activity within specific regions of the brain.

In addition to the facts about regional activities within the brain, neuroscience provides insight into the functioning of the brain's cells, the neurons, how they work and how their firing rate changes during mental tasks. This cellular investigation of neural activity is further complemented by additional research about biochemical substances, molecules, and genes in the brain. As we will see in this book, neuroscientists are actively working to explain mental features like self and consciousness by reference to these sorts of features of the brain.

What does this imply for the mind–brain problem? We must assume that the mind neither exists nor is real. There is simply nothing but the brain and its physical features. Mental features like consciousness, emotion, free will, etc., can then be traced back to neural features. This amounts to materialism, or physicalism as it is called in the current mind–brain discussion (see Chapter 6). Despite this position having all the support of the neuroscience's success, it remains a controversial view. While most philosophers agree that bodies, environments, and brains are the entities that comprise the real world, the implications of this materialism/physicalism remain disputed.

There are more or less radical interpretations of materialism/physicalism. Some philosophers acknowledge that the mind has an existence or reality of sorts which can be traced back to the brain and its physical features. Other philosophers go one step further and declare any assumption of a mind to be superfluous, and motivated by illusion; mental concepts like consciousness, free will, etc., are then regarded as illusions that need to be replaced completely by the terms and concepts of neuroscience. For instance, instead of describing emotions in mental terms like sorrow and grief, one could refer to these states by describing activation patterns in certain regions of the brain. This standpoint is called eliminativism in the current mind–brain debate (see Chapter 6).

Eliminativism has serious implications for the conventional view of philosophy as an autonomous discipline. According to eliminativism, any mental features including those underlying philosophy itself and our ability to philosophize must be reduced to (and ultimately eliminated in favour of) the terms of neuroscience. As explained above, the term **neurophilosophy** commonly refers to the work of eliminativist philosophers. *Briefly, neurophilosophy refers to the investigation and application of neuroscientific methods to traditionally philosophical concerns like consciousness, self, free will, etc. (see Chapter 4).*

More specifically, the eliminativist viewpoint recently discussed falls under what I call 'reductive neurophilosophy'. Briefly, 'reductive neurophilosophy' refers to work wherein the mind in general, and mental features in particular, are taken to be completely and exclusively reducible to the brain and its neuronal features (see Chapter 4). Reductive neurophilosophy is therefore really just a part of neuroscience, more specifically its theoretical branch, where traditionally philosophical questions are investigated in the context of neuroscience. More technically, reductive neurophilosophy can be characterized by the search for those neural conditions, the neural correlates that are sufficient to generate mental features. For example, reductive neurophilosophers, who investigate consciousness, are commonly concerned with locating the 'neural correlates of consciousness' (NCC) (see Chapter 14).

Role of the brain: from neural correlates to neural predispositions

The sceptics among you may now want to raise some doubt as to whether mental features like consciousness, the self, free will, etc., can be reduced to and consequently be 'located' in the brain. Not even the most elaborate scanner has ever shown consciousness in the brain. All we see in images of the brain are visualizations of various forms of neural activity in different regions, cells, proteins, etc. This raises the possibility that since consciousness and the other mental features cannot be found within the brain, they cannot be completely reduced to forms of neural activity. In contemporary philosophy of mind, this view is called *non-reductive physicalism*. Most philosophers who are not eliminativists express non-reductive physicalist positions.

Non-reductive physicalists have many different names for the sort of causal relation that holds between mental and physical phenomena (emergentism, supervenience, functionalism, etc. – see Chapter 6), but they are united in the view that when claims about mental phenomena are put in purely neural terms their original content is lost. Non-reductive physicalists' arguments for preserving a fundamental distinction between the mental and the physical casts doubt on the prospect of using neuroscience to shed light on the mind. However, the idea that neuroscience is not capable of shedding such light is also rather implausible, given the abundance of neuroscientific findings that demonstrate the contribution of the brain and its neural features to the generation of mental states.

Let us review the dilemma at hand. We have presented the two main currents in contemporary philosophy of mind (eliminativism and non-reductive physicalism) and found both to be flawed. The eliminativist's vision of a complete reduction of mental features to the neural features of the brain may be empirically implausible because evidence that clearly indicates a location in the brain for mental features is difficult to find. However, non-reductive physicalism, whether it be emergentist, functionalist or based on supervenience does not seem to be empirically implausible either. To deny that neuroscientific claims can be explanatory with respect to the mind clashes with a considerable body of neuroscientific findings.

Thus both eliminativism and non-reductive physicalism seem not to be in sync with actual neuroscientific findings. Since neither eliminativism nor non-reductive physicalism is an attractive option, we need to find another way to make allies of neuroscience and philosophy.

The alternative I present in this book involves a distinction between two different ways in which neural activity could be understood to underlie mental phenomena. As mentioned above, philosophers of mind typically inquire after the neural correlates of mental features, like consciousness. Those who search for neural correlates of mental features are trying to find neural processes that are sufficient for mental phenomena. In contrast, the non-reductive neurophilosophical approach outlined in this book (as inspired by Immanuel Kant and his transcendental method; see Part II), will investigate neural processes as necessary but not sufficient for mental phenomena.

This subtle change allows us to begin investigating the brain's connection to the mind in a new and powerful way. Instead of the neural correlates that are meant to be sufficient for mental phenomena, we will study the neural predispositions of mental features. The neural predispositions of mental features are necessary for the possible (rather than actual) occurrence of those mental features, but they do not bring them about, e.g. their actual occurrence, by themselves. Technically, the concept of neural predisposition refers to the necessary conditions of the possible generation of mental features. This distinguishes neural predispositions from neural correlates, which are sufficient to actually realize and implement mental features (see Chapters 14 and 15 for the application of this distinction in the context of consciousness).

General aim: from neural predispositions to non-reductive neurophilosophy

We can now say more about what non-reductive neurophilosophy is. *Put in a nutshell, non-reductive neurophilosophy considers the brain as relevant to philosophy but, unlike its reductive sibling, does not completely and exclusively reduce mental features to the brain and its neural features (see Chapter 4 for details). Instead, non-reductive neurophilosophy targets the relation of the brain's neural features to the vegetative features of the body and the social features of the environment (see Chapters 8 and 20).*

In addition to its distinction from reductive neurophilosophy, non-reductive neurophilosophy also needs to be distinguished from traditional philosophy and neuroscience. In contrast to traditional philosophy, non-reductive neurophilosophy explicitly considers the relevance of the brain for philosophical questions, and confronts issues that go beyond questions of mind (see Part III). For this, a specific methodological strategy that allows the linking and integration of conceptual and empirical approaches will be developed (see Chapters 2 and 4).

How about neuroscience? Neuroscience focuses on the brain and aims to reveal the various neural mechanisms that operate in it. In non-reductive neurophilosophy, the focus is not so much on the inside of the brain itself but rather on how the brain's neural features relate and link the brain to the world beyond the brain: the body and the environment. Thus non-reductive neurophilosophy strives to put the brain and its neural features into bodily and environmental contexts.

Consciousness in philosophy: the hard problem of the brain

Can neuroscience solve the mind–brain problem? Despite the impressive progress that researchers have made on general and regional neural activity, and even on cellular and molecular-genetic factors, neuroscience has not yet revealed the neural mechanisms underlying mental features like consciousness and self. While there are plenty of suggestions for possible neuronal mechanisms that might make consciousness possible, we have not yet solved the mind–brain problem. We still do not know why there are mental features like consciousness and self, or how they are related to and yielded by a brain that on its own is not conscious and does not possess a self.

Doubts about whether neuroscience promises to explain the mind usually appeal directly to consciousness. What is consciousness? The crucial characteristic of consciousness is that it is subjective, meaning that it is tied to an individual's first-person perspective as opposed to the objective (or third-person perspective) from which science describes the world. For example, while reading these lines, you may be experiencing a sense of boredom. Only you can experience that boredom. Your classmate, sitting just next to you, cannot experience *your* sense of boredom, even though she/he is forced to read the very same book. The way that consciousness seems to always involve the particular subjective experience of an individual seems to suggest that the phenomenon of consciousness is just not the sort of thing that can be discovered in the brain and its neural activity, which can only be observed from the third-person perspective.

The question of how something as subjective as consciousness can be related to the objective brain has been perplexing philosophers since Descartes. Today, eliminativist philosophers and the neuroscientific work they draw on would claim that there is no real problem here. According to this sort of work, which I call reductive neurophilosophy, what seem like subjective states of consciousness will in due time be shown to be nothing but objective neuronal states of the brain.

'No way', you might say. 'I do not experience my brain. Nor do I experience any neuronal state; I experience boredom and the book in front of me.' You are appealing to what philosophers call a *mental state*. In response to the many neuroscientists and philosophers who claim that all the mental states you experience as a conscious being are nothing but the neuronal states of your brain, the more non-reductively minded philosophers tend to ask why it is that nothing in the fancy scans of my brain can show me my particular mental states or what makes them conscious. Those who claim that consciousness can be reduced to neural activity are routinely challenged to explain why all the bright colours of brain scans fail to reveal an individual's subjective experience.

Since consciousness seems to be a paradigmatically subjective phenomenon, while brains seem to be part of the objective world investigated by science, both skeptics and supporters of reducing the mind to the brain treat consciousness as a sort of Holy Grail for the field. Whether neuroscience promises to unlock the secrets of the mind is widely taken to turn on the question of whether a brain-based account of consciousness can be found. Thus, philosophers of mind ask 'Why is there consciousness at all?', or more specifically, 'How is it possible for the objective and non-conscious brain to give rise to something as subjective as consciousness?'

This is what the Australian philosopher David Chalmers described as the 'hard problem'. The hard problem is a metaphysical problem that addresses the following question: Why and how is there consciousness at all rather than none? Reformulated within the context of the brain, the hard problem becomes: How is it possible for the brain, which independently is nonconscious, to yield consciousness? Note that the 'hard problem of the brain' is as metaphysical as the 'hard problem of consciousness': it shifts the brain from the empirical context of neuroscience to the metaphysical context of philosophy (see below).

Metaphorical comparison: hard problem of the heart

Can neuroscience solve the hard problem of the brain? Neuroscientists have provided considerable insight into the brain and how its neuronal states are related to mental states, and various suggestions have been put forward for the neuronal mechanism of consciousness, the so-called neural correlates of consciousness (NCC). However, none of these candidates for the NCC has yielded a definitive answer to the hard problem. We still do not know why and how something as non-conscious as the brain and its neuronal states gives rise to consciousness.

How can we better understand the current shortcomings of neuroscience with regard to consciousness? Let us draw an analogy to the heart. The heart is a pump that circulates blood throughout the whole body. Just as the brain gives rise to consciousness, the heart gives rise to the pumping of blood throughout the body. One can ask what it is about the heart that results in blood-pumping, just as philosophers ask what it is about the brain that results in consciousness. If an explanation of how the heart pumps blood was still being sought, then there would be something like a hard problem of the heart. People would be pondering how something as odd as a heart could do the job of circulating blood throughout the body: 'Why is there pumping rather than none?'

As it is, we already know what it is about the heart that accounts for it being able to pump blood, and it is not simply a matter of this function needing to be performed somewhere in the body. The heart is a muscle that contracts repeatedly and continuously, and it is precisely this activity that results in the circulation of blood throughout the body. The continual contraction of a heart is sufficient in itself to explain how blood is circulated throughout the body. That the heart behaves in this way makes the pumping of blood unavoidable (or necessary). This allows us to clarify what is distinctive about the hard problem of consciousness. While it is widely accepted that the brain does give rise to consciousness, we do not yet know what it is about the brain that makes the generation of consciousness unavoidable (or necessary). Unlike the heart, of which we know both how it pumps blood and why it must, even the most cutting-edge neuroscience has yet to explain what it is about the brain that makes the generation of consciousness not only possible but unavoidable, i.e. necessary.

While the hard problem of the brain remains unresolved at this point in time, the secrets of the heart were discovered some time ago. They are both organs of the body, however, and we will therefore continue to turn to the example of the heart to clarify the sorts of questions asked about brains. We have already clarified that the hard problem of the brain seeks not just to discover what it is about brains that generates consciousness, but rather seeks an explanation of what it is about brains that makes the generation of subjective conscious experience unavoidable in the way that the heart's continuous contraction makes blood-pumping necessary and thus unavoidable. We are not interested just in the functions or purposes of mental phenomena as they appear to common sense, but in the actual mechanisms underlying these phenomena, whether they are purely neural or connected to bodily and/or social phenomena.

Knowledge of the brain: dark spots and the explanatory gap problem

How can we get to work on the hard problem? One clear option would be to simply combine our knowledge about both mental and neuronal states. We can describe mental states in terms of mental concepts like boredom, excitement, etc., and pool these facts with all our knowledge about the brain's neuronal activity. Then we could simply investigate which mental concepts correspond to which neuronal concepts.

The hard problem is not that easy of course. One would soon notice that the concepts used to describe the brain and its neuronal states (neuronal concepts) do not overlap at all with those used to refer to mental states (mental concepts). In other words, there seems to be an explanatory gap between neuronal and mental concepts: neither sort of concept can be used to explain the other.

Let's go into more detail. Mental concepts do not contain any reference to the brain and its neuronal states. You do not experience your brain, let alone the neuronal states of your consciousness. For instance, one can experience excitement due to an upcoming soccer match. One does not, however, experience the specific neuronal activity pattern that is supposed to underlie the experience of excitement. Thus, mental concepts seem unrelated to the brain and its neuronal states.

What about starting with neuronal concepts and trying to link them with mental concepts? We use neuronal concepts like 'the action potential of neurons' or 'degrees of signal changes in specific brain areas' to describe the brain's neural activity on cellular and regional levels. Nothing in these neuronal concepts, however, refers to the mental concepts one uses to describe experience or mental states. Hence, these neuronal concepts do not themselves imply anything about mental concepts.

The result is a gap in our knowledge about neuronal and mental concepts. *This gap between our neuronal and mental explanations is described as the '**explanatory gap problem**' in current philosophy of mind. Rather than being a metaphysical problem about the existence and reality of the mind, the explanatory gap problem is epistemological, referring to a gap in our knowledge of mental and neuronal states.*

Metaphorical comparison: explanatory gap in our knowledge of the heart

How can we better illustrate the explanatory gap problem? Let's invoke the comparison with the heart again. We can discuss the activity of the heart with two different kinds of concepts. First, we have the concepts that describe the heart as an example of a pump; these are engineering concepts like dynamics, velocity, etc. Second, there are those concepts that describe the processes underlying the heart's muscle contractions in physiological or biological terms; examples of these physiological concepts include adenosine, ATP, etc.

How are the physiological and engineering concepts with which we can describe the heart related to each other? At first glance, they are as distinct as the mental and neuronal concepts with which the mind or brain can be described, because nothing about the physiological concepts seems to tell us about the engineering concepts, and vice versa. For example, the biochemical adenosine is a physiological concept that does not contain any reference to velocity, an engineering concept. If scientists still lacked an understanding of how the physiological characteristics of the heart make its engineering characteristics inevitable or necessary, there would be an explanatory gap between our understanding of what the heart is and what it does.

In the case of the mind and the brain, we do not yet have an analogous understanding of how the brain's neuronal features account for the mental features it generates. The explanatory gap refers to a specific dark spot in our current explanations, a lack of knowledge that would explain why mental features must result from the brain's neuronal characteristics. In response to the explanatory gap, non-reductive neurophilosophy makes a methodological move: rather than attempt to analyze concepts of mind in such a way that assimilates them with neuronal concepts (a mind-based metaphysical pursuit), we investigate different ways of thinking about brains to try to discover what neuronal concepts can imply about mental concepts (a brain-based empirical pursuit).

Problems in philosophy: concept(s) of the brain

We have looked at three key problems concerning the mind and brain. Since Descartes, the mind–body problem has challenged philosophers to explain how the mind and the body can both be real and related to each other despite seeming to have different sorts of existence or reality. The more contemporary hard problem singles out consciousness as the aspect of mind for which it is most difficult and thus also most important to provide a physicalist or brain-based explanation. Lastly, the explanatory gap problem describes the missing knowledge that would be needed to connect neuronal and mental concepts.

How can we solve these problems? Philosophy in general — and philosophy of mind in particular — discusses these questions in predominantly conceptual ways. Philosophers of mind define and analyze the various concepts of concern in all sorts of ways in order to get nearer to solutions for these problems. While this has stirred much lively discussion, none of the problems has been solved. And so it has been suggested that the predominantly conceptual methodology of philosophy of mind will not be able to solve these puzzles on its own, however effectively it exposes them.

This claim is common in neuroscience and usually explicit in its recent theoretical-philosophical branch (usually just called neurophilosophy, which we refer to as reductive neurophilosophy). Rather than discussing these problems in a conceptual way, reductive neurophilosophers address these questions empirically, drawing on experimental findings from neuroscience. According to reductive neurophilosophy, the deep philosophical questions about the mind and its mental features should be resolved by careful observation of the brain.

Why, then, consider the brain in the context of philosophy, as suggested by the title and sub-title of this book? If neuroscience can provide all the answers, why should we bother with philosophy at all? The oddness of considering the brain as its own philosophical subject can be further reinforced by the idea that the brain is simply not a subject for philosophy because it is already the subject of neuroscience. Many would say that consciousness, the self, and the mind are subjects of philosophy but that the brain belongs to science. Hence, shifting the brain into the context of philosophy may be a kind of methodological mismatch – similar to, say, studying the Mona Lisa with a spectrometer.

Rather than arguing directly against these vague charges, the whole of this book will serve as counter-evidence to the claim that philosophers should leave the brain to the neuroscientists. We will simply proceed to consider the brain in the context of philosophy and let the accumulation of results speak for themselves. The challenge at hand is to put the brain into the context of philosophy without reducing philosophy to neuroscience, as in reductive neurophilosophy. To avoid this kind of reductionism, we simply need to consider different ways of conceptualizing the brain, just as philosophers have been doing with the mind.

One way of doing this is to generate and evaluate the results of defining the brain according to the different features without which the brain would no longer be a brain. The non-reductive neurophilosophical approach investigates the brain and the features that account for it as a brain, rather than the brain features that correlate with mental features (see Chapters 8 and 12 for details). The advantage of taking the brain to be conceptualizable in multiple ways is that we can fine-tune our brain-based approach according to the problem at hand, rather than treating the brain merely as the mind's empirical counterpart.

Rather than investigating the brain in solely empirical terms, as is done in neuroscience, we will tackle the brain conceptually by proposing different concepts of the brain. This methodological agility is the crucial difference between non-reductive neurophilosophy and its better-known reductive variants. Welcoming the brain into the context of philosophy allows the non-reductive neurophilosopher to investigate the old problems of mind in many new ways. While neuroscientists continue to study actual brains, and philosophers go on pondering the mysteries of mind, we will focus on the concept of the brain, which will enable us to incorporate both the empirical results from neuroscience and the conceptual gains of philosophers.

Investigation of the brain in philosophy: brain in different domains

How can we investigate the brain in philosophy? The usual answer, common to reductive neurophilosophy as well as traditional philosophy, is simply that we can't. The brain belongs to the empirical domain which is the stronghold of the discipline of neuroscience. In contrast, philosophy is traditionally insensitive to the empirical domain and more occupied with the metaphysical and the epistemological domain. Since the brain does not belong to either, the brain is simply not a subject of philosophy. (See the illustrations in Figure 1.).

As discussed, this is regarded as simply unacceptable by the reductive neurophilosopher. Neuroscience shows that the brain must be attributed a central role, and this has inspired some philosophers to give up the mind and focus on the far more observable brain. Thus reductive neurophilosophy, which has dominated the field of

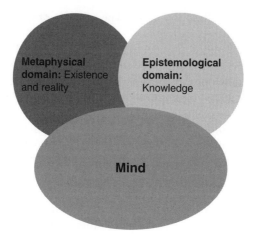

Figure I.1 Different disciplines and their domains
Figure I.1a Philosophy of mind
The figure illustrates different domains in philosophy, neuroscience and neurophilosophy, and how they relate to mind and brain.

(a) The figure shows the traditional association of the mind with the domains of metaphysics and epistemology as it has been (and still is) dealt with in philosophy in general and philosophy of mind in particular.

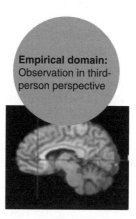

Figure I.1 Different disciplines and their domains
Figure I.1b Neuroscience

(b) The figure shows the association of the brain with the empirical domain as the hallmark feature of neuroscience.

neurophilosophy up to now, is both a methodological endorsement of neuroscience and a methodological indictment of traditional philosophy. It seems that if we focus on the mind, as in traditional philosophy, we must sacrifice accountability to neuroscientific facts, while if we focus on the brain, as in neuroscience, we must sacrifice accountability to the conceptual arguments of philosophers.

To resolve this dispute, we will chart a third way. Direct engagement with the concept of the brain is not a part of either traditional philosophy or reductive

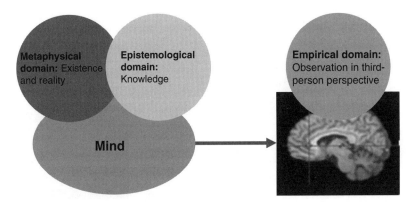

Figure I.1 Different disciplines and their domains
Figure I.1c Reductive neurophilosophy as brain-reductive approach

(c) The figure demonstrates how the epistemological and the metaphysical domains including the mind are reduced to the empirical domain and the brain in neurophilosophy which therefore can be characterized as reductive neurophilosophy.

neurophilosophy, but it is sensitive to influences from both areas. Non-reductive neurophilosophy delineates different concepts of the brain as it appears in different discussions, whether they are metaphysical, epistemological, or empirical. To explain how, we need to clarify the concept of a 'domain'.

What do we mean by the concept of 'domain'? Originally the concept of a 'domain' comes from the Latin terms 'dominium', 'dominus' and 'domus': 'Dominium' describes property, 'dominus' refers to lord, master, or owner, and 'domus' stands for house. These distinct though related meanings are reflected in the various uses of the concept of domain in our current language. *Domain in today's usage can describe a territory governed or ruled; a field of action, thought or influence; a region characterized by specific features; and range of significance. Within the present neurophilosophical context, one can for instance, distinguish between different domains like metaphysical, epistemological and empirical: the metaphysical domain concerns questions of existence and reality, the epistemological domain is about the territory of knowledge, while the empirical domain refers to the field of observation.*

Subject matter and method of investigation: domains and disciplines

The term domain is used in different disciplines like biology, mathematics, computer science and physics. Within the present neurophilosophical context, the concept of domain shall describe the territory, topics, issues, areas, or fields of investigation and thus the questions discussed. This means that we need to distinguish the concept of domain from the one of *discipline* that in addition to the discipline's field also entails a particular methodological strategy (see Chapter 2).

*Roughly, the **concept of domain** concerns the field or area of investigation and thus particular issues or questions as independent of and distinguished from the kind of methodological strategy by means of which these questions are investigated. Put in a slightly distinct way, the concept of domain concerns the subject matter whereas the concept of discipline refers more to the method that is applied to investigate that subject matter.*

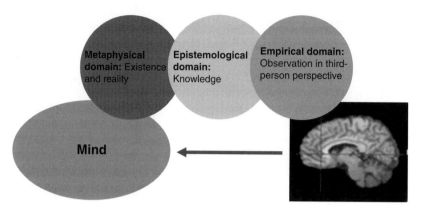

Figure I.1 Different disciplines and their domains
Figure I.1d Non-reductive neurophilosophy as brain-based approach

(d) The figure shows that the brain and its originally empirical domain may serve as a basis or neural predisposition for the metaphysical and the epistemological domains and thus for the mind. This implies non-reductive rather than reductive neurophilosophy.

We thus distinguish between the subject or matter of investigation, the *domain*, and the method of investigation, the *discipline*. More informally, we can describe a particular investigation by distinguishing its matter and its method. The advocate of a traditional discipline like philosophy may want to argue that matter and method of investigation are strongly tied together. For instance, in the history of philosophy the investigation of metaphysical matters, the existence and reality of for instance the mind, has been closely tied to the rational-argumentative method.

This may indeed be *traditionally* so in philosophy. However, the more recent history in especially the second half of the twentieth century shows that such close alignment between matter and method of investigation is no longer a given. Reductive neurophilosophers, for instance, claim that we can investigate metaphysical issues like the existence and reality of the mind in an observational-experimental way rather than using the rational-argumentative method of philosophy. The matter of investigation, the metaphysical domain of existence and reality, becomes here decoupled from its traditional method (and discipline), the rational-argumentative strategy (of philosophy), and then associated with another method, the observational-experimental strategy, and discipline, neuroscience.

It is because the line dividing philosophical and scientific investigations has been blurred like this that we must carefully distinguish the *domain*, the matter of investigation, from the concept of *discipline*, which signifies the method of investigation. *The concept of domain refers to the field, topic or question and thus to the matter of investigation whereas the concept of discipline describes a particular methodological strategy, the method, that is applied to investigate the respective field, the matter.*

We will see over the course of this book that this distinction between matter and method of investigation and thus between domain and discipline will prove central in understanding how the brain can be profitably incorporated into philosophy. Moreover, we will see how different forms of the relationship between domains and disciplines entail different forms of neurophilosophy, reductive and non-reductive (see Chapter 4). Hence, I will use the concept of domain and its distinction from disciplines as one of the guiding ideas in this book. Instead of employing the usual classification of disciplines, I will speak more often of domains – the fields, questions, or matters of investigation. This will make it possible to approach particular domains with different methodological strategies – a hallmark of non-reductive neurophilosophy (see Chapter 4).

Brain in philosophy: from philosophy of mind to philosophy of brain

In short, we will develop what can be called a **'philosophy of the brain'** as distinguished from both philosophy of mind and neuroscience of mind. *Philosophy of the brain refers to the investigation of the brain and how it can be defined and conceptualized in different domains (metaphysical, epistemological and empirical), as well as how we can (and cannot) relate the domain-specific definitions of brain to one another (see Chapters 7, 8, 11, and 12 for details)*. (See Figure 1.1e.)

Unlike neuroscience of mind in particular and neuroscience in general, philosophy of the brain is not restricted to the empirical domain but includes the metaphysical and epistemological domains. Similar to neuroscience though, philosophy of the brain focuses on the brain as its main content or subject matter of investigation. However, unlike neuroscience again, philosophy of the brain applies a multitude of different methods to the investigation of the concept of the brain, combining observational-experimental and conceptual-logical strategies for instance (see Chapters 2 and 4).

The overarching goal of this book is to *demystify* the brain. We do not need to assume special supernatural features like a mental substance to account for what philosophers recognize as mental phenomena. Instead, we can remain within the natural realm of brains, bodies and environments. We will see that a mixture of

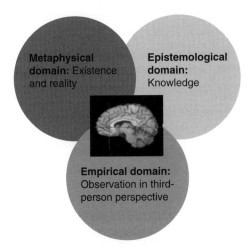

Figure I.1   Different disciplines and their domains
Figure I.1e   Philosophy of brain

(e) The figure illustrates the discussion and investigation of different concepts and definitions of the brain in different domains as the hallmark feature of philosophy of brain.

conceptual and empirical approaches to these physical phenomena can pay greater dividends regarding mental phenomena than the comparatively more limited alternatives of traditional philosophy, reductive neurophilosophy or neuroscience.

Metaphorical comparison: demystification of heart and brain

How can we explain the demystification we're going for in a more illustrative way? Let's invoke the comparison to the heart one last time. The British physiologist William Harvey (1578–1657) sought to discover the 'vital spirits' in the heart and aimed to locate them there in exact detail. He assumed that the heart would produce the 'vital spirits' that keep the organism existent and alive. However, as much as he searched, he did not discover 'vital spirits' nor any other spirits in the heart or elsewhere in the body. Instead of confirming some sort of supernatural spirit, he observed that the heart is a pump that circulates blood throughout the whole body and that it is this that keeps the organism alive. He thus 'demystified' the heart, meaning that he freed concepts of the heart from supernatural features.

So how can we liberate the brain from the tempting assumption of supernatural mental features? To achieve this goal we start with the brain itself, pursuing a brain-based rather than mind-based approach, but as non-reductive neurophilosophers we need not be limited to the brain itself, abstracted from its body and environment. To avoid common forms of reductivism and to develop a more sophisticated way of linking neuroscience and philosophy, it is important to consider the brain in its intrinsic, i.e. necessary or unavoidable relation to the body and the environment. This book will show that mental features are associated more plausibly with the brain's intrinsic (i.e, necessary or unavoidable) relationship to the body and environment than with the brain itself. This openness to the bodily and environmental contexts of brains is another hallmark of non-reductive neurophilosophy. We will see that it is a uniquely fruitful approach when compared with either the eliminativism of reductive neurophilosophy, or the predominantly conceptual techniques of traditional philosophy.

Outline and contents: Part I

Part I of the book is about recent developments in philosophy and neuroscience as well as about the determination of the concept of neurophilosophy. *Chapter 1* gives a brief overview of philosophy of mind and phenomenological philosophy, including their respective methods. This is complemented by *Chapter 2*, which focuses on how philosophy is related to the sciences in general. This touches upon the question of whether or not the mind can be regarded as part of our natural world. Philosophy calls this 'naturalized'. The possibility of such 'naturalization' is a cornerstone for any philosophy that aims to link the mind to the brain and recent neuroscientific findings. *Chapter 3* focuses on neuroscience as a discipline. Here the historical development of neuroscience is briefly described, including the emergence of psychology as a scientific discipline. This is complemented by demonstrating how neuroscience has recently extended its reach to psychology and many other disciplines that focus on the mind. In doing so, neuroscience is focusing not only on the brain itself, but also on the mind and its mental features like consciousness and self. *Chapter 4* discusses the concept of neurophilosophy as it is used currently in the predominantly empirical sense. This includes the investigation of the brain using different philosophical methodological strategies as sketched above. This is supposed to open the door for a more comprehensive concept of neurophilosophy that allows for bilateral traffic of both the mind and the brain between the disciplines of philosophy and neuroscience.

Outline and contents: Parts II and III

Part II of the book focuses on the mind–brain problem as the core nucleus of philosophy of mind. *Chapter 5* discusses mental solutions to the mind–brain problem that consider the mind's existence and reality to be principally different from the existence of brain and body. *Chapter 6* focuses on physical solutions that argue that the mind is merely physical and thus not principally different from the brain. *Chapter 7* presents alternative solutions to the mind–brain problem that defy both mental and physical approaches. Finally, *Chapter 8* discusses the concept of the brain from a philosophical perspective and how that relates to the mind–brain problem.

Part III of the book shifts from philosophy of mind to philosophy of psychology and neuroscience by discussing theoretical and philosophical problems in both disciplines. *Chapter 9* discusses the relationship between different forms of psychology, common sense or folk psychology, and scientific psychology, including their different levels of explanations, personal and subpersonal. This is followed by *Chapter 10*, which addresses the problem of meaning, or semantic content, raising the question of how meaning is assigned to the contents of our mental states.

Chapter 11 shifts to philosophy of neuroscience and discusses different types of concepts used in current neuroscience, including their implications for understanding how the brain yields mental features. Finally, *Chapter 12* aims to characterize the brain independently by determining its intrinsic features, namely those that define the brain as brain and how they are related to previous philosophical characterizations of the mind.

Outline and contents: Part IV

The topic of Part IV is consciousness, the hallmark feature of the mind. Consciousness describes our ability to make subjective experiences in first-person perspective. *Chapter 13* discusses various arguments from the philosophical side against reducing consciousness to the brain and its neuronal states. Among them are the hard problem and the explanatory gap problem as mentioned above.

Chapter 14 focuses on recent neuroscientific suggestions for possible neuronal mechanisms underlying consciousness, the so-called neural correlates of consciousness (NCC). The NCC focus mainly on the neural processing of extrinsic stimuli and how their respective stimulus-induced activity is associated with consciousness. The neuronal, methodological and phenomenal shortcomings of the NCC are discussed and put in the broader context of the hard problem and the explanatory gap problem (and the other problems discussed in Chapter 1).

This is followed by *Chapter 15*, which extends the search for the neuronal mechanisms of consciousness. The focus shifts here from the extrinsic stimulus-induced activity to the brain itself and its intrinsic activity. These intrinsic features, like the brain's intrinsic activity, remain independent of any extrinsic stimuli and are suggested to be necessary for yielding the possibility of consciousness. This amounts to what can be called the 'neural predispositions of consciousness (NPC)'. *Chapter 16* follows up on this and discusses conceptual, phenomenal and methodological issues that need to be considered and refined in the future in order to better understand how the brain yields consciousness rather than nonconsciousness.

Outline and contents: Part V and Epilogue

The fifth and final part discusses the subject of experience, the self, and how it relates to the brain. *Chapter 17* discusses different concepts of self in recent philosophy and how they relate to neuroscientific findings about self-referentiality, or stimuli that refer to the self of a particular person. This is extended in *Chapter 18*, which addresses the question of how the self can be experienced and thus become conscious, or self-consciousness. Here, the respective philosophical issues of accessing the self in consciousness, as well as related neuroscientific findings, are directly compared with each other.

Chapter 19 gathers further empirical support for the assumption of a close linkage between self and the brain's intrinsic activity albeit indirectly. For that purpose, psychiatric disorders like depression and schizophrenia are considered. In these disorders one can find abnormalities in both one's sense of self and in the brain's intrinsic activity.

Chapter 20 extends the focus on the self to its relationship with others. This is known as the 'self–other relationship,' or 'intersubjectivity' as it is described in philosophy. Different philosophical positions about intersubjectivity are described and compared alongside both recent psychological and neuroscientific findings. The psychological focus is here mainly on mind-reading – the ability to understand and read another person's mental states – and empathy, the ability to find resonance in and thus feel another person's emotions. In contrast, the recent neuroscientific results focus on mirror neurons, the neurons that are active when observing another person's actions or emotions. These are discussed and put into the context of the philosophical concept of intersubjectivity.

Finally, the book concludes with a short epilogue: 'Is the brain a door opener?'. Here I summarize the central role of the mind and the mind–brain problem in philosophy and how this can be shifted and enriched by venturing into a novel room, the room of non-reductive neurophilosophy.

Who is this book for? Instruction for students

The primary audience is undergraduate and graduate (and postgraduate) students in neuroscience, philosophy and cognitive science/psychology at different levels, as well as all students who share an interest in the relationship between the mind and the brain. Students with a background in neuroscience may want to start with Parts IV and V about consciousness, self and intersubjectivity. This is where most of the recent neuroscience data are discussed. Students with a background in psychology may be most interested in Part III, the explanation of the mind. Finally, philosophy students may want to start with Parts I and II, those about philosophy and the mind–brain problem.

The book may also be of interest for postgraduate students in cognitive science, neuroscience and/or philosophy if they want to inform themselves about possible research topics and get ideas for their postgraduate research. They may want to focus on the sections dedicated to research themes and ideas for future research. Neuroscientists interested in mental functions like consciousness, self, emotional feelings, etc. may benefit from reading this book too. They may gain an insight into how empirical and conceptual issues are linked with each other and how the determination of their concepts may impact their empirical research and hence their experimental designs.

In order to make this book as accessible as possible, each chapter starts with an overview that sets the context and background for the material to come in that chapter. This is followed by a brief overview of the objectives and the key concepts, which precede the main text. As well as the main thematic sections, the main text also includes some sections on neurophilosophical topics that go into more detail. There are also sections headed 'critical reflection' that discuss the current concepts in a more critical way.

After the main text, toward the end of each chapter, I include a 'take home message', a summary and revision notes for students that include questions about the content of the respective chapter. Finally, I give suggestions for further reading in each chapter. Considering this structure, one may find it useful at first to skip the main text and read the other sections to get a first impression. One may then turn to the relevant sections of the main text.

Finally, it should be noted that I include only the main and most relevant figures in the print version of the book; these figures focus on illustrating how to include the brain in a non-reductive way into philosophy. Many more figures on especially empirical matters as well as other material can be found online on the website related to this book (www.palgrave.com/psychology/northoff).

Who is this book for? Instruction for teachers

The book is divided in five main parts: neurophilosophy, mind–brain problem, explanation of mental features, consciousness, and self. This essentially provides two different options for class material. First, the teacher could take the whole book as the basis for one or two semester terms and work through the parts and chapters subsequently. This would be perfect for an advanced undergraduate class (third or fourth year) or a master class in the first and second years.

Alternatively, each part could be a topic for one semester by itself. In that case, the teacher may want to take the respective part and its chapters provided here as a starting point and pick up some of the secondary literature mentioned. Such a class would then focus either on the development and concept of neurophilosophy (Part I), mind–brain problem (Part II), explanation (Part III), consciousness (Part IV), or self (Part V). This could be of high interest for either Master's degree students or a PhD program as well as for postdoctoral seminars.

I include some imaginary dialogues as well as metaphorical comparisons in the book. Though unusual, this is to put across the sometimes not so easy shift in perspectives to the students in a more illustrative way. The teachers could pick up on that and develop analogous discussions and dialogues in their classes. This is also often proposed in the suggestions for teachers at the end of each chapter. Obviously, these are only suggestions; each teacher may have her or his own ideas and experiences.

In addition to the print version, there is much material to be found on the website related to this book (www.palgrave.com/psychology/northoff). The further reading as suggested in the print version here as well as additional figures, texts and papers can be found online.

Who is this book for? Instruction for researchers

Finally, the book may also be of interest to researchers from both sides, philosophy and neuroscience. The researcher from the philosophical side may find some useful ideas for shifting his/her point of view and framework from a purely philosophical to a rather neurophilosophical one. For that, I include several sections where I directly compare the neurophilosophical view with a more traditional philosophical and neuroscientific one. Moreover, I directly compare reductive and non-reductive neurophilosophical views on topics like the mind–brain problem. This all serves to illustrate how the brain can be conceptualized in different ways and thus become a topic in philosophy itself without reverting to the mind, as has been common practice for so long in philosophy.

In addition, neuroscientific researchers may also be interested. The discussion of different concepts of the brain will open novel ideas for future experimental research. Here the distinction between neural correlates and neural predispositions may prove central; the latter may give a new view on the existing data and thereby generate novel ideas for future experimental paradigms.

In order to bring out the research component, I include two sections that are specifically addressed to the researcher. The 'critical reflections' serve to discuss some problems and issues in the current discussion, while the 'ideas for future research' go beyond the current discussions and make suggestions for a neurophilosophical and more specifically non-reductive approach.

It should also be mentioned, especially to researchers and teachers, that this book is only an introduction to neurophilosophy. As such it cannot serve as a fully exhaustive account of all the issues discussed in philosophy and neuroscience. As such, the reader may want to refer to book editions of actual papers from several authors, as well as the respective neurophilosophical journals.

We should point out that both theorizing about and practicing neurophilosophy go hand in hand here. This makes it clear that the theory and the practice of neurophilosophy cannot be completely separated from each other. What neurophilosophy is and what it should look like might actually only become clear when practicing it. Conversely, practicing neurophilosophy may require some theoretical reflection about its nature, scope and method, and thus what can be described as philosophy of neurophilosophy (see Chapters 4 and 11).

Finally, it should also be noted that this book is accompanied by a special website (www.palgrave.com/psychology/northoff), where additional figures, papers and texts can be found that deepen many of the subjects raised and discussed here. There is also a forum for discussion which may further deepen the research and development of non-reductive neurophilosophy.

Part I
Mind and Brain: From Philosophy through Neuroscience to Neurophilosophy

Philosophy is an old discipline that, in the Western world, dates back to ancient Greece, with Thales of Milet us, 624–546 BC, often being considered one of the first philosophers. At that time philosophy was very much preoccupied with explaining existence, reality, and the nature of the world in particular and the universe in general – these kinds of questions were later subsumed under the term metaphysics. Philosophy was equally concerned with asking epistemological and ethical questions. Epistemological questions concern what we can and cannot know about ourselves and the world, while ethical issues refer to the norms and values of our behaviour.

In the twentieth century the study of philosophy shifted, seeking instead to answer new questions about the nature of the mind: what is the mind's existence and reality when compared to the body and the brain? This is known as the mind–body problem, which can be traced back to Descartes in the seventeenth century and, which is now more commonly described as the mind–brain problem. At the same time, advances were made through the scientific investigation of the brain, and neuroscience yielded much insight into the neuronal mechanisms underlying sensorimotor and cognitive functions.

These simultaneous developments in philosophy and neuroscience led to the increasing convergence of the mind with the brain, ultimately giving birth to neurophilosophy as a field of study interested in the intersection of these problems. The focus of Part I is to provide insight into how neurophilosophy developed on the shoulders of philosophy (Chapters 1 and 2) and neuroscience (Chapter 3), as well as to describe it more precisely as a specific discipline with particular methods and specific topics (Chapter 4). Based on the historical developments, we will also distinguish between reductive and non-reductive forms of neurophilosophy that emphasize different historical lines and traditions in philosophy.

1
Philosophy and the Mind: Philosophy of Mind and Phenomenology

Overview

Philosophy is a broad field of study concerned with many different types of questions. It asks about the fundamental nature of the universe, the existence and reality of the world as a whole, and last, but not least, how we experience that reality. To think and reflect on metaphysics is to ask about reality and existence. To wonder how and what we can (and cannot) know about this reality is to be concerned with epistemology. Philosophy of mind, however, is concerned with the what (or who) that is capable of acquiring knowledge and coming to know something about the world. To ask about the existence and reality of the mind, and how the mind relates to the body and its brain, is to ask a metaphysical question, commonly known as the mind–brain problem. Similarly, 'How can we access and know the mind?' is an epistemological question posed within the context of philosophy of mind. The mind and its metaphysical and epistemological characterization in twentieth-century philosophy are the focus of this chapter.

Objectives

- Understand why philosophy is interested in the mind
- Understand Descartes' metaphysical and methodological dualism of mind and body, and why it is so central
- Explain how philosophy of mind emerged and how it can be defined
- Understand the basic definition of the mind–brain problem
- Discuss the different mental features of mind as put forward in philosophy of mind
- Determine the concepts of phenomena and phenomenology
- Explain the phenomenal hallmark features of consciousness

Key concepts

Dualism, Descartes, phenomena, phenomenology, structure of experience, subjectivity, intentionality, philosophy of mind.

Background 1

Metaphysics and the different domains of philosophy in ancient Greece

The term philosophy means the 'love of wisdom' whose systematic investigation can be traced back to ancient Greece. At that time philosophy dealt with different questions and issues, which will be very briefly explained in the section below.

First, philosophy focused on the basic and fundamental nature of the world and our human place in it: what is the existence and reality of the world and our own selves as human beings? These are questions that are subsumed under the umbrella term *metaphysics. Briefly, metaphysics describes the search for the reality and existence of people, things and the world.*

Ontology, as a sub-branch of metaphysics, is focused on the basic categories of being and existence, and how they are related to each other. While contemporary philosophy especially makes a distinction between ontology and metaphysics, in this book I will consider ontology as a subfield of metaphysics. To keep matters simple and clear, I will therefore use the term metaphysics throughout this textbook.

Second, another area of philosophical investigation concerns the possibility of knowledge: how and what can we know (or not know) about things and the world, given our knowledge capacities and limitations? These questions are addressed in the branch of philosophy known as *'epistemology', which focuses on questions of knowledge.*

Finally, a third domain in philosophy concerns our behavior and more specifically how we should behave: how can we properly and morally behave in relation to others? These problems are discussed in *ethics, the study of norms and how one should behave in order to yield moral behavior.*

In sum, philosophy in ancient Greece already focused on different fields or topics as described by the concept of domains. In addition to the metaphysical domain, the domain of existence and reality, the epistemological domain, the domain of knowledge, and the ethical domain were central in the early philosophical endeavors in the Western world.

Biography 1

Descartes and the mind

Ancient Greek philosophers like Plato (429–347 BC) and Aristotle (384–322 BC) focused mainly on the metaphysics of the world, its existence and reality, and questions pertaining to ethics. This emphasis would drastically change thousands of years later with the philosophical developments of the French philosopher, René Descartes. He lived from 1596 to1650 in various places throughout Europe.

The etymology of Descartes' name can be traced back to the Latin word 'Cartesius', which means 'of the cards'. He did, indeed, reorder the meaning of the philosophical cards: unlike the Greek philosophers who thought of the world as the starting point for philosophical inquiry, Descartes, with his famous sentence – 'I think, therefore I am' – turned inwards, locating the mind as his point of philosophical departure. Thinking is an activity of the mind. 'Being' as referred to in the second part of the sentence ('I am') reflects a metaphysical question. Descartes thus pinpoints the human mind as the very center of philosophical inquiry.

Background 2

Metaphysical dualism between mind and body in Descartes

What is the human mind? This is a metaphysical question that addresses the existence and reality of the mind. Descartes compared the mind to the body. According to him, the body is spatially extended and can be divided into different parts and can therefore be characterized as *'res extensa'*. As such, the body is nothing but a machine that obeys the causal laws of physics. For Descartes, the mind was different. According to this thinking, the mind is neither spatially extended, nor can it be divided into different parts. The mind is characterized by mental features like thinking, consciousness, etc., designating it as *'res cogitans'* rather than as *'res extensa'*.

Descartes concluded that these differences proved the mind and body to be fundamentally different things. He therefore associated their existence and reality with different substances: a mental substance for the mind, and a physical substance for the body. This idea resulted in what is described as 'substance dualism', a 'metaphysical dualism' that describes the mind and the body as two different substances (see Chapter 5 for more details on substance dualism).

As we shall see below, the legacy of Descartes' metaphysical dualism as substance dualism between mind and body has resonated far beyond his lifetime and continues to influence various strands of contemporary thought today (see especially Part II). Due to Descartes, the existence and reality of the mind and its relationship to the one of body and brain, the mind–brain problem, are still major issues in the current discussion in philosophy. This is especially so in philosophy of mind, where the mind–brain problem is the core topic therefore we will discuss in Part II.

Background 3

Methodological dualism between introspection and observation in Descartes

How can we access and know about the mind and the body and their underlying mental and physical substances? According to Descartes, we have direct access to our own mind and its mental states via what is described as introspection. The concept of introspection describes the direct access to our mental states that enables us to become aware of their contents. For instance, while reading these lines you may become aware that you feel completely bored. Knowing you are bored is to have direct access to your mental states and their respective contents from your first-person perspective.

This is different in the case of the body. Unlike with our mental states, we remain unable to directly access our body's states as such from the inside in our first person based experience. Instead, we access the body indirectly via observation from the outside employing a type of third-person perspective.

Descartes' metaphysical dualism between the mind and the body was accompanied by a methodological dualism: in short, he posited two different methods for accessing the mind and the body. More specifically, the body was subject to observation in third-person perspective and was thus subject to scientific investigation. Meanwhile, introspection through first-person perspective would give us access to the mind. What this means is that it is not subject to observation, and thus the same type of scientific study, in third-person perspective.

Why is Descartes' metaphysical and methodological dualism still relevant in the current context of linking philosophy and the brain? What is important is precisely that his dualisms led to a division of labour and disciplines. Following Descartes' metaphysical dualism, his successors concluded that body and brain can be subject to scientific investigation while the mind is not. Instead of being a subject of science, the mind is the domain of philosophy. This implied that the mind was expelled from science while, at the same time, the body and brain were denied access to philosophy. The division of labour was mirrored by a division of disciplines: philosophy and science.

Metaphorical comparison

Dualism between living room and kitchen

How can we illustrate Descartes' metaphysical and methodological dualism more vividly? Let's compare the mind and the body to a living room and a kitchen. A living room is characterized by the presence a sofa, flowers, paintings, books, etc., all objects meant to induce relaxation and provide entertainment. The kitchen is different; here we might find a stove, a microwave, cooking pots and pans, each of which are utilized to cook a proper meal. These various cooking utensils have nothing to do with the activities and atmosphere of the living room in the same way that the tools for relaxing, like the sofa and the paintings, are not meant to be brought into the kitchen.

There is thus a strict dichotomy between each room with regard to the kind of furniture and the kind of action performed in these spaces. This double dichotomy is comparable to Descartes' metaphysical and methodological dualism between the mind and the body. We can imagine the different kinds of furniture in the living room and kitchen as corresponding to the different types of substances: mental and physical (thereby neglecting the aspacial nature of mind which, taken literally, may not correspond to a specific room that is by definition special). Meanwhile, the different kinds of actions – cooking and relaxing – correspond to the different methods of investigating the mind and the body and their related disciplines. In short, introspection versus observation and philosophy versus science.

However, as we all know, kitchens and living rooms are often no longer as strictly separated in the houses of the twenty-first century. The concept of the 'open kitchen' – where the kitchen is open to and, in extreme cases, even an integral part of the living room – is plain enough proof. Most importantly, this direct encounter between spaces and actions changes how we understand and conceive them.

The same intersection happened with the mind and brain in the twentieth century and continues today. Mind and brain and their respective disciplines, philosophy and science, gradually approached each other and became more and more 'open' to one another. In a nutshell, philosophy became aware of the brain's relevance to the mind. Conversely, neuroscience became aware of the mind and its various mental functions like self and consciousness. It equally sought out new ways to investigate the mind scientifically in third-person perspective.

This intersection between mind and brain, and philosophy and neuroscience, is often called 'neurophilosophy'. Before providing you with a more concise definition of what exactly neurophilosophy is, we need to fully understand the developments that led to its emergence. To do this, we first need to explain how philosophy in the twentieth century dealt with the mind. This is the focus of this chapter.

Biography 2

Gilbert Ryle

All metaphysical and epistemological questions and answers are formulated in language, i.e. in sentences. Consequently, in order to fully understand our epistemological and metaphysical assumptions, we have to investigate the features and properties of the sentences we use to express them. To do this is to place language and the linguistic domain at the center of philosophy. This shift happened at the beginning of the twentieth century with the philosophers Gottlob Frege (1848–1925) and Ludwig Wittgenstein (1889–1951).

How is this focus on language related to the search for the nature of the mind? British philosopher Gilbert Ryle (1900–1976) raised this question in his main work, *The Concept of Mind* (1949). During the Second World War, Ryle served Britain by working in intelligence. Eventually, building on his experience of the war, Ryle, in his main work, would deny the mind as the special locus of intelligence. Instead, he considered the concept of the mind as a 'ghost in the machine' of the body.

Philosophy of mind 1

Linguistics of mind as linguistic analysis of mental features

Based on linguistic analysis, Ryle explored our ordinary sentences about mental states, for example our sensations ('I feel an itch'), beliefs ('I believe that neurophilosophy should replace philosophy'), and desires ('I am longing for the chocolate cake rather than this book'). Ryle argued that the linguistic analysis of our mental statements reveals that mental verbs such as believe, perceive, feel, etc. do not imply their ascription to any kind of underlying substance like a mind or a body.

Why, then, are we so inclined to attribute our mental states (as we describe them) to some specific property or substance called mind? Ryle argues that we easily confuse linguistic descriptions of our mental states with something (a mind) to which they can be attributed. This is to confuse the category of linguistic descriptions of mental states with the category of substances.

How can we characterize this confusion in more detail? Ryle argues that there is a confusion of categories or a category mistake, as it is also often called. Mental states are nothing but the behavior as described in the mental verbs and their respective sentences. Our mind thus consists of nothing but observable behavior. In this light, the mind is simply behavior as described in our language. This position is called behaviorism and will be discussed in more detail in Chapter 6.

Behaviorism claims that our mind and its mental states consist in behavior. There is nothing hidden behind the behavior and the mental sentences we use. No mind can ever be found beyond or beneath behavior. Any assumption of such a mind in the gestalt of an underlying mental substance – as Descartes assumed – can thus be considered like a ghost, a ghost that lives in the body. This led Ryle coin his famous characterization of Descartes' concept of the mind as a 'ghost in the machine' of the body.

If the mind does indeed consist in nothing but behavior, the mind is in no way different from the body. Mental states, and thus mental sentences, are consequently simply different ways of describing the body's action and behavior. Since the body's action and behavior can be traced back to its sensorimotor functions, we do not need any kind of special function like cognitive functions to explain mental states. In this light, mental states are bodily states which in turn are sensorimotor states. Here, to assume the existence of cognitive functions – or even a separate mind – amounts to a confusion about our linguistic descriptions of our sensorimotor behavior. In this model, the mind–body problem is thus nothing but a category mistake.

Why are Ryle's concept of the mind and his theory of behaviorism so important? First and foremost, Ryle rejected Descartes' metaphysical dualism between the mind and the brain. And second, it opened up the concept of the mind to linguistic investigation as a predominant methodological strategy in twentieth-century Anglo-American philosophy. Ryle's ideas thus set up the very basis for the linguistic investigation of the mind. This resulted in the development of what is described as 'philosophy of mind' during the second half of the twentieth century in the Anglo-American world. What follows below briefly describes 'philosophy of mind', i.e., what it is and how it characterizes the mind.

Philosophy of mind 2

Philosophy of mind as metaphysics of mind

What is philosophy of mind? Philosophy of mind is the branch of philosophy that focuses on the nature and features of the mind. The nature of the mind concerns the existence and reality of the mind. Is the mind as real and existent as Descartes assumed? Or is the mind an illusion or category mistake, as suggested by Ryle? The mind and its various mental features, for example, consciousness, free will, etc., are the central focus of this branch of philosophy.

What is the mind? And how does the mind stand in relationship to the body? How are the existence and reality of the mind and the body related to each other? This question reflects what is described as the mind–body problem. Since the brain seems to be particularly closely related to the mind and its mental features, the mind–body problem has often been condensed into the mind–brain problem.

The mind–brain problem is a metaphysical problem that raises the question: how are the existence and reality of the mind and its mental features related to those of the brain and its various neuronal states? Different solutions ranging from metaphysical dualism (as in Descartes) to metaphysical monism have been suggested. Different options include: (i) the mind is traced back to the body, leading to what is called physicalism (which in turn can describe different relationships leading from identity through reduction to elimination of the mind with the body; see Chapter 6 for details), (ii) the body is reduced to the mind resulting in idealism, or (iii) both mind and body are traced back to a third entity as described in neutral monism. The different solutions of the mind–brain problem will be discussed in detail in Part II.

What is the mind–brain problem about? It is about the existence and reality of the mind and the brain and how they are related to each other. Since questions of existence and reality are dealt with in metaphysics, the mind–brain problem can be situated in the metaphysical domain.

How does philosophy of mind investigate the mind? This question is one of methodological strategy. Developed in the Anglo-American tradition with its emphasis on language during the twentieth century, philosophy of mind initially strongly relied on the analysis of concepts and sentences (i.e. linguistic analysis). More recently, this methodology has been complemented by considering empirical findings about the brain from neuroscience, and thus joins conceptual, linguistic and empirical methods. What this looks like and how it works will be discussed in further detail (see Chapters 2 and 4).

In sum, philosophy of mind focuses on the existence and reality of the mind and its mental features thus presupposing the metaphysical domain. The mind–brain problem as a metaphysical problem is consequently at the very core of philosophy of mind. Methodologically, philosophy of mind strongly relies on conceptual and linguistic analysis which more recently has been complemented by empirical findings from neuroscience (see especially Parts IV and V on consciousness and self).

Philosophy of mind 3

Mental features of the mind

In addition to the mind's nature, philosophy of mind also investigates the mind's features. What are the features of the mind? The mind is manifest in consciousness, our experience of ourselves and the world. Consciousness, preliminarily defined, describes the subjective experience of ourselves and the world, including its events, objects and people. As such, consciousness is subjective and tied to the first-person perspective.

This characterization of consciousness raises a very specific question: how can something as subjective as consciousness be generated by the brain and its neuronal states, especially when they are considered objective and observable in third-person perspective. Consciousness with its intersection of subjectivity and objectivity poses the definitive question of the mind–brain relationship in a paradigmatic and condensed way. Because consciousness is itself a major topic in current philosophy of mind, we will devote a separate part to it, Part IV.

Besides consciousness, the self can be regarded as another hallmark feature of our mind. Consciousness is experienced by a subject that is manifested linguistically with the terms I, he, or she. Who is the subject of consciousness (and thus of experience)? How can we access and possibly experience this subject itself? These are questions about the self and our sense of self. When asking these types of questions we are ultimately wondering how someone can have a seemingly continuous and coherent self across time, or in other words, personal identity.

These questions about the self, its existence and reality – as well as our experience and consciousness of it – are major topics in current philosophy of mind. Therefore I will devote Chapters 17 and 18 to the self. Questions about the self are closely related to questions about the minds and selves of other persons. How is one person's self related to the selves of other people? In philosophy, this topic is called 'intersubjectivity' and will be discussed in Chapter 20.

In addition to consciousness, self and intersubjectivity, other features of the mind may need to be considered. There is, for instance, the concept of free will, or the experience of free choice in our decision-making. I can decide now to go throw this book into the trashcan. Alternatively, I can decide to keep the book and recommend it to others. This is my free choice. Such free choice presupposes that choices are not fixed – they are indeterminate rather than determinate. This notion, however, contrasts with the supposedly causal nature of the physical processes in our body and brain.

Since they are causal, physical processes seem determined and fixed. They don't allow for any indeterminism or free choice. There thus seems to be a contradiction between our experience and consciousness of indeterminism and free will on the one hand, and the determinism of the physical processes of brain and body on the other. This is discussed under the umbrella term 'free will' in current philosophy of mind.

What about consciousness during sleeping and dreaming? It seems as if consciousness disappears during sleep since we do not experience our sleep as such. However, we experience dreams and their contents during sleep. How does that consciousness during dreams stand in relation to the apparent loss of consciousness during the other stages of sleep and our full-blown consciousness in the state of being awake? In this light, sleep and dreams may be considered paradigmatic test cases for the philosophical investigation of consciousness.

The same applies to various neurological and psychiatric conditions like the 'vegetative state' (complete loss of consciousness) and schizophrenia, where consciousness is abnormally altered. Taken together, dreams, sleep and neurological and psychiatric disorders are increasingly discussed as paradigmatic test cases of consciousness in philosophy of mind. This will be exemplified in more detail in Parts IV and V with regard to consciousness and the self.

Phenomenology 1

Phenomena and consciousness

So far we have focused on philosophy of mind and its relationship to the metaphysics of mind (the mind–brain problem, its mental features, and how they relate to the brain and its neuronal states). Based on the strong consideration of language and linguistic analysis, philosophy of mind was developed mainly in the Anglo-American world during the second half of the twentieth century. At the same time in continental Europe – and especially in Germany and France – there were other philosophical developments that similarly focused on the mind, and especially consciousness. This will be explained in further detail below.

How can we study the mind? One hallmark feature of the mind is what is described as experience and consciousness. While reading the lines of this book you experience a sense of boredom: why do I have to do all this philosophizing about the mind when in fact I want to know more about the brain? Your study partner, sitting beside you, can read the very same lines, but she or he cannot share your individual feeling of boredom. In other words, your own experience is accessible only to you in first-person perspective – by yourself, but not by others and who can only have a third-person perspective of you.

What does our experience in first-person perspective and its specific features tell us about consciousness in general? This is the central question that the German philosopher, Edmund Husserl (1859–1938), asked himself. Husserl studied mathematics, physics and philosophy. He later became a professor of philosophy in southern Germany, in Freiburg, where he was the teacher of many famous philosophers including Martin Heidegger. After 1933, with the rise of fascism in Germany, Husserl was suspended from the university and suffered the humiliation shared by many Jews in Germany at the time. He died in Freiburg in 1938.

Husserl's ultimate aim was to establish philosophy as strict science with a clearly defined methodology. He argued for the need to abstain from any beliefs, preemptions and speculations about the world and instead was in favor of focusing on experience itself and thus on consciousness. He believed that this focus would reveal the things themselves – phenomena – as they are by themselves independent of our preconceptions (hence the name 'phenomenology' for this approach).

By focusing on consciousness itself and what appears to us in it, Husserl shifted the focus of philosophy from metaphysical and epistemological questions to the study of consciousness and how things appear to us in our experience. As such he was interested in the study of phenomena, or the appearance of things in consciousness. For this he is remembered as the founder of the novel philosophical field at the beginning of the twentieth century called *'phenomenology'*.

The term phenomenology refers to the study of phenomena that describes the appearance of things in consciousness as they are by themselves, independent of our preconceptions and presuppositions about their existence and reality. In other words, phenomenology aims to focus on the phenomena themselves as they appear in our consciousness independent of their metaphysical characterization, i.e., their presupposed existence and reality.

Phenomenology 2

Subjectivity and intentionality

How do things appear to us in consciousness? One central hallmark of conscious experience is its subjectivity: any experience is experienced by a subject, for example you or me. This is reflected linguistically in the fact that any description starts with a subject like 'I experience', 'I perceive', 'I imagine', etc... Experience and thus consciousness can be characterized by *subjectivity – by the involvement of a person's specific first-person perspective and its respective point of view* (see Part IV for more details).

Another hallmark of the structure of consciousness is that it is directed toward a particular object. The term 'I' is followed by a verb that stands in connection to an object: 'I perceive the colorful brushstrokes in this painting'. 'Perceive' is the verb that describes a certain activity of the subject in relation to a particular object, the 'painting'. The experience of the person is thus directed towards the 'brushstrokes in the painting'; *this directedness is described as intentionality, the directedness of consciousness towards a specific content*. By means of intentionality, the respective content – the intentional content – becomes semantic content and thus meaningful. Hence, alongside subjectivity, intentionality is considered a main phenomenal feature of consciousness.

In sum, phenomenology is a specific philosophical discipline that investigates how things appear to our consciousness as phenomena independent of our preconceptions of them. As such, phenomenology focuses on the structure of consciousness. The two structural hallmarks of phenomenology are subjectivity and intentionality. Subjectivity details the links between consciousness and first-person perspective, and intentionality describes the directedness of consciousness towards contents.

How do subjectivity and intentionality, as structural features of our experience, relate to consciousness and its various contents? We can, for instance, become conscious of time, space, our body, others and the world. This consciousness is always already characterized by subjectivity and intentionality independent of whether we are conscious of time, space, our body, others, or the world. For example, our consciousness of time is subjective in its first-person perspective and intentional in that it is directed towards the change of time. We do not thus experience physical time in our consciousness but what Edmund Husserl called 'phenomenal time'.

The same holds true for space, our body and the world as a whole. We experience space in relation to ourselves and thus in a subjective way in first-person perspective. We also experience space as homogeneous, rather than segregated. Hence we do not experience physical space in our consciousness, but rather 'phenomenal space'.

It is important to emphasize that the phenomenological approach does not focus on the body, people, events, time, space and the world objectively in third-person perspective. Instead, it considers them in a subjective way in first-person perspective. The reason for this is because phenomenology considers this perspective as a fundamental structural feature of consciousness. In contrast with science, which considers consciousness in a purely objective and non-intentional way, phenomenology attempts to understand and access consciousness through its intentionality and subjectivity.

Critical reflection 1

Body in phenomenology

One may now wonder how phenomenology – the study of the phenomenal features and structures of consciousness – applies to the body and brain as specific contents of our consciousness. The French philosopher Maurice Merleau-Ponty (1908–1961) showed an early interest in Husserl and his phenomenology. Merleau-Ponty taught in the 1950s in Paris at the famous College de France until a car accident in 1961 led to his untimely death. Merleau-Ponty was particularly interested in how we consciously experience the body in first-person perspective. He described the body in our experience as the 'lived body'. This needs to be distinguished from the body as we can observe it in third-person perspective, the 'objective body' (see Chapters 7 and 8 for details).

Thus our bodies can be accessed in two different ways; first, through a non-phenomenal, third-person-based method that utilizes observation of the 'objective body', as in sciences like physiology. Second, we can access our bodies in a first-person, phenomenal way, focusing instead on the experience of the body as appears to us in consciousness, the 'lived body', as is done in phenomenology.

Merleau-Ponty considered the 'lived body' as central for consciousness. More specifically, he regarded the experience of the body as 'lived body' to be fundamental and basic for any subsequent consciousness of other contents like people, things, or events in the world and ultimately for the world as a whole. Without the body and its manifestation in our consciousness as the lived body, we would not able to experience anything at all.

In sum, the lived body anchors us in the world, making it a necessary condition for possible consciousness of the world and its various contents. Therefore some approaches today consider the body and its associated sensorimotor functions as central constituents of consciousness; since the mind and its various mental features are here anchored in the body, these approaches speak of an 'embodied mind' (see Part IV for details). We may thus need to go beyond the brain itself to the body in order to account for the mind. This is the lesson Merleau-Ponty gave us on our way toward understanding consciousness and the mind.

Critical reflection 2

Brain in phenomenology: nothing but a category error?

How about the brain? The brain is not an intentional and meaningful content in our consciousness. Nobody has ever experienced his/her own brain as brain in his/her consciousness. We can experience our own body as lived body and we can experience various other contents like people, objects and events in the world. In contrast, we remain unable to experience our own brain as brain. Unlike in the case of the body, there is no 'lived brain', as one may want to say analogously, but only an 'objective brain'.

The absence of the brain from consciousness implies that we cannot access the brain in first-person perspective in a subjective and intentional, and thus phenomenal, way. Instead, the brain can be accessed only in third-person perspective in an objective and non-intentional and thus non-phenomenal way. The brain is exclusively non-phenomenal as distinguished from the body that is both phenomenal, as in the lived body, and non-phenomenal, as in the objective body. In short, the brain is not subject to the phenomenal domain but belongs exclusively to the empirical domain and hence to science rather than phenomenology.

Traditional philosophy in general and phenomenology in particular cease to be disciplines that are capable of investigating the topic of the brain. The brain is simply not the kind of subject matter that can be investigated in the domains of philosophy, the metaphysical, the epistemological and the phenomenal. Only those phenomena and contents that appear in consciousness and thus the mind and its mental features can become associated with these domains. The brain, in contrast, is not such subject matter and therefore must be associated exclusively with the empirical domain.

Nor can the brain be investigated with the traditional methods of philosophy, the conceptual-logical, the argumentative-rational and the phenomenal methods. Instead, the brain must be investigated by using the observational-experimental approach of science. Accordingly, the brain is thus associated with the empirical domain and the observational-experimental method. Since both domain and method are the hallmark of science, the brain belongs to science in general and neuroscience in particular.

In contrast, any association of the brain and its empirical domain with philosophy and its metaphysical, epistemological and phenomenal domains seems to be impossible. Since these domains are associated with specific methods, any rational-argumentative, conceptual-logical, or phenomenal analysis of the brain seems to be inappropriate too. Hence, any consideration of the brain in the traditional domains of philosophy and its traditional methods must then be considered a category error (see below as well as Chapter 8 for details).

Critical reflection 3

Brain in phenomenology: category error in non-reductive neurophilosophy?

How can we specify the category error? The category error consists here in confusion of domains and methods: the empirical domain of the brain is confused with the metaphysical, epistemological and phenomenal domains. And the observational-experimental approach of science is here confused with the rational-argumentative, conceptual-logical, or phenomenal methods of philosophy. One thus simply confuses mind and brain. We will see in Chapter 2 that this boils down to the confusion between logical and natural worlds: the philosopher is concerned with the logical world and the scientist with the natural world. If one now associates the brain with the domains and methods of philosophy, one confuses the natural world of the brain and science with the logical world of philosophy.

The category error sets up a clear dividing border between philosophy and science in general, and mind and brain in particular. This dividing border is often presupposed in philosophy (and in science) in both past and current discussions. How does the neurophilosopher stand in relation to this category error? Metaphorically speaking, the neurophilosopher aims to enter and cross the border territory that has been set up between brain and mind and thus between neuroscience and philosophy.

How can the neurophilosopher cross the border territory between philosophy and science? Either he/she declares one side of the territory null and void. This is the strategy of the reductive neurophilosopher, who ultimately eliminates the domains and methods of philosophy in favor of those of neuroscience. Alternatively, one may aim to build border stations with border guards where one can cross in both directions. This is the strategy of the non-reductive neurophilosopher.

What does this imply for the investigation of the brain? The reductive neurophilosopher considers the brain exclusively in the empirical domain using the observational-experimental method. He/she is thus following his/her role model, the neuroscientist, and will, like the philosopher, ultimately consider the category error as valid. In contrast, the non-reductive neurophilosopher aims to investigate the brain in different domains including empirical as well as metaphysical, epistemological and phenomenal domains. Moreover, he/she aims to tackle the brain using different methods beyond the observational-experimental approach. This means that he/she will also approach the brain by rational-argumentative, conceptual-logical, or phenomenal methods. The category error here is thus declared as irrelevant and invalid.

There is thus a rather flexible association of the brain with different domains and methods. This flexible characterization is the hallmark of non-reductive neurophilosophy which distinguishes it from reductive neurophilosophy, neuroscience and philosophy (see Chapter 4 for methodological details). This allows, for instance, the investigation of the brain in the phenomenal domain using its method of phenomenal analysis: how does or does not the brain surface in consciousness? And if not why does the brain not surface as distinct content in consciousness? These are the kinds of questions one can now raise. As said already, these are legitimate questions in the framework of non-reductive neurophilosophy whereas they appear as confused and category error from the perspectives of all other disciplines including reductive neurophilosophy, traditional philosophy and neuroscience.

Critical reflection 4

Brain in phenomenology: brain at the junction between disciplines?

The phenomenological approach is thus limited with regard to the brain. The brain is beyond our experience and thus beyond consciousness: if the brain cannot be a content of consciousness, consciousness itself must remain independent of the brain. This is the conclusion many philosophers, including phenomenologists, draw. Therefore, they consider that the brain is irrelevant to consciousness and the mind in particular, and to philosophy in general. Instead of taking the brain as the point of departure, one may rather take the mind as the starting point and thus as the most basic presupposition. This implies a mind-based approach, as we will point out later (see Chapters 7 and 8).

But, in fact, things are not this easy when it comes to the brain. Can we really infer the irrelevance of the brain from our inability to experience our brain in consciousness? That would be to infer from our methodological means, the lack of access to our brain in consciousness, to the role of the brain in consciousness. While we remain unable to experience the brain as brain in consciousness, the brain may nevertheless be central in making consciousness possible. Therefore it could be said that the brain is a necessary condition for possible consciousness (rather than of actual consciousness) (see Part IV for the distinction between actual and possible consciousness).

How can we further substantiate such a claim? Moving beyond consciousness and its phenomenal domain, we might here venture into the empirical domain of observation. One source of evidence comes from patients with changes in their brain and related alterations in consciousness as it can be observed in neurological and psychiatric patients. For instance, patients in a vegetative state (VS) suffer from loss of consciousness while at the same time showing severe changes in their brains (see Part IV for details). Further, neurological patients with selective lesions in their brain show abnormal changes in the contents of their consciousness. Moreover, psychiatric patients suffering from depression or schizophrenia show an abnormal structure in their experience, with altered subjectivity and intentionality (see Part IV for details).

What do these findings tell us about consciousness and its relation to the brain? These findings suggest that the brain must be central for consciousness and its phenomenal features like subjectivity and intentionality. Even though the brain cannot be experienced as such in consciousness, it may nevertheless be necessary to constitute it. This means that the brain may be a necessary condition for the possibility of consciousness; this neural condition will later be described by the term 'neural predisposition of consciousness' (see Introduction and Part IV).

If the brain does indeed predispose (in some as yet unclear way) consciousness and its phenomenal features, the brain is 'located' right at the border between phenomenal and non-phenomenal domains. This has important ramifications; for one, from this vantage point, the study of the brain can therefore be 'located' right at the junction between philosophy and neuroscience, or at the core nucleus of what may be called non-reductive neurophilosophy as distinguished from reductive neurophilosophy (see Chapter 4).

Phenomenology 3

Phenomenology and psychology

First-person-based phenomenal access allows phenomenology to also study how we experience various psychological processes and functions. For instance, how we experience perceptions, thoughts, memories, imagination, emotions, desire, volition, bodily awareness, attention, social interaction and linguistic activity. One may now be puzzled. These are all functions usually associated with psychology as science, rather than with phenomenology as philosophy.

How can we distinguish phenomenology from psychology? Psychology aims to study the mind and consciousness in a scientific way as empirical functions and processes we can observe in third-person perspective (this is called the 'scientific method'; see Chapter 2 for details). This distinguishes psychology from phenomenology. Rather than investigating perception, memory, etc. objectively through third-person perspective, phenomenology aims to reveal how these perceptions, memories, etc. appear to us subjectively from a first-person perspective. In phenomenology, subjectivity and intentionality are considered central features of consciousness.

By shifting the focus from third- to first-person perspective, phenomenology can account for the meaning, and thus the semantic contents, of our experience. Conversely, meaning – or the semantic contents – are lost in psychology because they cannot be captured by mere observation from the third-person perspective (see Part III for a discussion of semantic contents).

What does this imply for the characterization of phenomenology? The thematic field of phenomenology focuses on consciousness itself as well as its structure and features. As such, it aims to reveal what can be described as the phenomenal features of consciousness like intentionality and subjectivity. One may consequently describe the domain of phenomenology as the 'phenomenal domain'. While the term *'domain' refers to thematic field, questions, issues, or problems (see also Chapter 2), the concept of 'phenomenal' stands for 'how things and the world appear to us in consciousness', e.g. in experience in first-person perspective.*

The phenomenal domain must be distinguished from other domains, or thematic fields. First and foremost, the phenomenal domain needs to be distinguished from the empirical domain as presupposed by psychology and other sciences. The empirical domain focuses on the world as we can observe it in third-person perspective. This is well reflected in science in general and psychology in particular. The focus on the world as observed in the third-person perspective distinguishes the empirical domain from the phenomenal domain: instead of observation in third-person perspective, the phenomenal domain focuses on experience in first-person perspective as it appears to us in consciousness. This is what phenomenology is about.

Moreover, the phenomenal domain must be distinguished from the metaphysical domain. The latter focuses on the basic and fundamental nature of reality and existence of beings and the world. This is reflected in metaphysics. In contrast, the phenomenal domain searches for how those beings and the world appear to us in consciousness while avoiding any metaphysical assumptions (and preconceptions) of their underlying reality and existence. That is the claim of phenomenology. Finally, the phenomenal domain must also be distinguished from the epistemological domain: the epistemological domain targets what we can and cannot know about the world including ourselves as it is dealt with in what is described as epistemology. Since it focuses on knowledge, the epistemological domain must be distinguished from the phenomenal domain and its focus on consciousness.

On the whole, phenomenology can be characterized by a distinct domain of investigation (see Introduction for the definition of the concept of domain), the phenomenal domain, and can be characterized by specific phenomenal structures like subjectivity and intentionality. Because of the distinct nature of the phenomenal domain, phenomenology must be distinguished from the sciences like psychology, and other philosophical disciplines like metaphysics and epistemology as well as their respective empirical, epistemological and metaphysical domains.

Phenomenology 4

Experience and introspection

One may now remark that any analysis of experiences presupposes some kind of inner awareness. In order to reveal the phenomenal features of my experience of the perception of the tree standing in front of me, I need to become aware of my own perception as a perception of the tree. This is possible only if I 'look into myself' and 'scan my own experiences' as experiences. This method of becoming aware of one's own internal states (experiential states) is called introspection.

How can we further specify introspection? So far, we have distinguished only between experience and observation. Experience presupposes first-person perspective while observation requires third-person perspective. If I now access and become aware of my own experiences in first-person perspective as experiences in first-person perspective, I take a slightly distinct perspective on them. That perspective has to be distinguished from both first- and third-person perspective: the awareness of my experience can neither be identified with the experience itself in first-person perspective nor its observation in third-person perspective. Several authors therefore associate this type of awareness and introspection with the second-person perspective.

How can we characterize the second-person perspective? *The second-person perspective can be characterized by an awareness of one's own experiences, as inner observation, distinguishable from outer observation in third-person perspective.* In addition, the second-person perspective is also linked to the awareness of others' experiences and how they resonate in one's own experience, thus entailing an explicitly social aspect (see Chapter 20 for details). Here, though, I focus on introspection (rather than the intersubjective and social aspect) as one defining feature of the second-person perspective.

How can we account for the second-person perspective and thus introspection? Phenomenology suggests that the possibility of such awareness of experience is inherent in the experience itself. Becoming aware of experience as experience is thus an integral part of experience. This implies that there is no principal distinction between first- and second-person perspectives with the latter being essentially subsumed in the former.

How can we further illustrate the relationship between first- and second-person perspectives in experience? Let's draw an analogy to a table that consists of legs and a table top. The legs hold the table top so that without them the table top would not amount to a whole table, but rather a plank of wood. Conversely, the legs by themselves do not constitute a table either. Thus legs are an integral part of the table, in the same way that a table top is.

Analogously, phenomenology argues that experience and awareness of experience (introspection) invariably go together. There can be no experience without experience and awareness of that experience. Going back to our metaphor, the legs and table top are intrinsically linked together in the phenomenological conception of consciousness. This is manifest in the lack of distinction between first- and second-person perspectives.

Critical reflection 5

Relation between first- and second-person perspectives

This lack of distinction between first- and second-person perspectives may also be considered differently, as in, for instance, current higher-order cognitive theories of consciousness (HOT). In a nutshell, HOT assumes that experience results from the awareness of particular contents. How can contents become aware as contents? We can become aware of contents as contents by taking their original representation in the processing of the mind (or the brain) and representing that very representation again. This re-representation is described as higher-order or meta-representation. Such higher-order or meta-representation may then elicit experience and thus consciousness (see Part IV for more details on HOT).

How can we link such meta-representation to the concepts of first- and second-person perspectives? Becoming aware of the contents by meta-representing them seems to entail some kind of second-person perspective; the introspection of the contents as contents which in turn makes possible their meta-representation. That meta-representation, in turn, is supposed to elicit the experience of that content as it is associated with the first-person perspective. Hence, the second-person perspective and its associated psychological processes seems to precede here the first-person perspective and its phenomenological features in consciousness.

This is different in phenomenology. Here experience itself forms with the awareness of experience (i.e. introspection) an integral part that is associated with the first-person perspective and its phenomenal features. Phenomenology argues that this first-person approach makes it possible to account for the subjectivity and intentionality of consciousness as its phenomenal hallmark features (see earlier).

How does the phenomenology of consciousness view the psychological approach to consciousness? The phenomenologist claims that neither subjectivity nor intentionality can be captured by mere observation in the third-person-based approach of science in general and psychology in particular. Any psychological approach to consciousness must therefore miss the essence of consciousness and its phenomenal hallmark features, subjectivity and intentionality.

How about complementing or even replacing the first-person-based approach of phenomenology by the third-person-based approach of psychology? Applying the third-person-based approach of psychology to consciousness implies shifting subjectivity and intentionality from their original phenomenal context to a psychological context. That, however, means to 'false positively psychologize' something that is not accessible to psychology and its third-person-based method. This leads to what Husserl himself described as 'psychologism', which in our times is often extended into the realm of neuroscience where it may be designated as 'neuralism' (see Chapters 2 and 4 for more details on such 'neuralism', which may resurface in especially reductive neurophilosophy whereas neuralism may not be virulent in non-reductive neurophilosophy).

Critical reflection 6

Unconscious and consciousness

Phenomenology focuses on consciousness and its phenomenal structures. However, Sigmund Freud (1856–1939), the inventor of psychoanalysis, postulated and found strong evidence in favor of the existence of an unconscious. Freud was born in Příbor, Austria, and lived most of his life in the capital of Austria, Vienna (only in his last years he moved to London, UK, to avoid the occupation of Austria by the Nazis).

What is the unconscious? There are many objects, events and people that are processed by the psyche and ultimately the brain even though we do not experience them in our consciousness. The introduction of the unconscious may be regarded as a threat to the prime focus of phenomenology on consciousness.

Let's imagine that you suffered from a traumatic experience in the past when you were stuck in a glass elevator. That was ten years ago in New York City. You were stuck for two hours in between the 49th and 50th floors until finally a repair man rescued you. While in the elevator, you had a beautiful view of the skyline of Manhattan, but you were unable to enjoy the view because you were constantly afraid that the elevator would crash down and that you would die. Now ten years after that traumatic episode, you live in Shanghai, China, and you still feel your heart rate racing every time the door of an elevator does not open immediately.

How is that possible? It is possible only if the previous experience with the elevator in New York still impacts your current behavior. Even though you do not experience that impact in your consciousness, it nevertheless modulates your consciousness and behavior. This is possible if the previous elevator experience is processed unconsciously. Freud gathered plenty of evidence for such unconscious processing of the previous contents of the person's experiences. Another instance where such unconscious content may surface is in dreams with their sometimes rather bizarre contents that, following Freud, symbolize the unconscious contents.

How does the unconscious stand in relationship to phenomenology and its emphasis on consciousness? Husserl denied that such an unconscious is possible at all. Mental life is conscious life rather than unconscious and conscious life as Freud claimed. If Freud is right (and there is plenty of evidence to support it), we may need to complement the phenomenological approach to consciousness with one that allows us to tap into the unconscious.

Why is all this relevant for our endeavors to understand the mind in general and consciousness in particular? Because we need to distinguish between different concepts, i.e. consciousness, unconsciousness and nonconsciousness: consciousness describes our experience in first-person perspective as in phenomenology. The unconscious contains contents that are not experienced consciously but which can still impact consciousness and, even more importantly, can become conscious by themselves in the 'right' circumstances. As such, the unconscious needs to be distinguished from the nonconscious, or that which can never become conscious at all.

If we now want to understand how consciousness is related to the brain, we need to associate different psychological and/or neuronal mechanisms with the three different concepts. We first need to understand those neuronal mechanisms that allow for the distinction between unconsciousness and consciousness; this is well reflected in what is described as neural correlates of consciousness (NCC) in current neuroscience (see Part IV for details).

And second, we need to investigate those neuronal mechanisms that allow for the distinction between nonconsciousness on the one hand and unconsciousness/consciousness on the other. Conceptually one may characterize those neural mechanisms that allow for the distinction between nonconsciousness and unconsciousness/consciousness as neural predispositions of consciousness (NPC). However, the exact empirical characterization of these neuronal mechanisms remains unclear at this point in time (see Part IV).

Take home message

In the Western world, philosophy is an age-old discipline that grew out of Ancient Greece some 2,000–3,000 years ago. Ancient Greek thinkers were primarily preoccupied with the existence and reality of the world (i.e. metaphysics), whereas twentieth-century philosophers focused much more on the mind and how its existence and reality are related to that of the body and the brain. This relationship is often described as the mind–brain problem and is a metaphysical problem that is discussed in the branch of philosophy called philosophy of mind. Philosophy of mind as a subfield of philosophy was developed primarily in the Anglo-American world where it aligned itself closely with the predominant analysis of concepts and language. In parallel, another approach to the mind, known as phenomenology, was developed in continental Europe. Phenomenology targets how the things themselves appear in consciousness independently of our preconceptions and metaphysical speculations. Both philosophy of mind and phenomenology are important for understanding the current debate about the mind and the brain, since both approaches have been extended and linked to the brain and neuroscience, as will be discussed in later sections of this book. Moreover both approaches raise problems and issues with regard to the mind–brain relationship and are vital to consider for any subsequent study of neurophilosophy.

Summary

Philosophy raises basic and fundamental metaphysical questions about the reality and existence of being and the world. It also raises epistemological questions about the possibility for knowledge of ourselves and the universe. But in order to know anything at all, we need a mind. What is the existence and reality of the mind, and how is it related to the body? By raising these questions, the philosopher René Descartes highlighted the mind as a new center for philosophical inquiry. How can we investigate and approach the mind? Building on the strong linguistic focus of twentieth-century philosophy, philosophy of mind developed as the central branch of the philosophical discipline, especially in the Anglo-American world.

Philosophy of mind investigates the nature of the mind; what the mind's existence and reality is; and how the mind is related to the body. This is described as the mind–body (and more recently as the mind–brain) problem. The mind–brain problem can be 'located' in the metaphysical domain while methodologically it is closely aligned with the predominating linguistic analysis. In addition to the nature of the mind, philosophy of mind also focuses on the various features of the mind like consciousness, self, intersubjectivity, free will, etc.

In addition to philosophy of mind, another branch of philosophy developed particularly in continental Europe, namely phenomenology, also focused on the mind. Phenomenology argued that we have to investigate consciousness and thus our experience as it exists in itself. More specifically, we need to investigate how we experience the meaningful (semantic contents) in consciousness in first-person perspective and how these contents appear to us. Why? What appears to us are phenomena signalling how we experience something, and this experience must, of course, reflect the structure of experience itself. Phenomenology is thus focused on what may be called the phenomenal domain, the domain of experience, or of consciousness, as we can access it in first-person perspective. Rather than investigating functions like perception, memory, etc. in a third-person observational and non-semantic way (as in psychology), phenomenology focuses on the contents of consciousness and their semantic meanings in relation to first-person experience. The same holds true with regard to the body. The body can be observed in third-person perspective as accounted for in the sciences; this is the 'objective body'. At the same time it can also be experienced in first-person perspective, or what is described as the 'lived body'. This is not possible in the case of the brain because it can only be observed in third-person perspective and not experienced as such in a first-person perspective in consciousness. At the same time, however, the brain seems to be essential in constituting consciousness. As such, it may be 'located' at the nexus of phenomenal and non-phenomenal domains. This has important implications for both philosophical and neuroscientific approaches to the investigation of the brain, as we shall see in the subsequent chapters and parts.

Revision notes

- Which domains are investigated in philosophy?
- What methods are applied in philosophy and science?
- Why is the mind a central topic in philosophy? Who was Descartes and what did he claim?
- What is philosophy of mind? How did it emerge?
- What is the mind–brain problem and which domain does it presuppose?
- Name some mental features of consciousness.
- Why is the distinction between unconsciousness and consciousness relevant?
- Define the terms phenomena and phenomenology.
- What is the aim of phenomenology? Why does it focus so much on experience and consciousness?
- Name some typical features of the structure of experience, e.g. phenomenal features.
- Explain the distinction between first- and second-person perspectives and their possible relationships.
- How do the brain and the body differ from each other with regard to experience?

Suggested further reading

- Husserl, E. (1965 [1910]) 'Philosophy as rigorous science', translated in Q. Lauer (ed.), *Phenomenology and the Crisis of Philosophy* (New York: Harper & Row).
- Merleau-Ponty, M. (1962 [1945]) *Phenomenology of Perception* (New York: Humanities Press).
- Petitot, J., Varela, F., Pachoud, B. and Roy, J.-M. (eds) (1999) *Naturalizing Phenomenology. Issues in Contemporary Phenomenology and Cognitive Science* (Stanford, CA: Stanford University Press).
- Searle, John (2004) *Mind: A Brief Introduction* (Oxford, New York: Oxford University Press). *An excellent book that gives a clear and concise overview of philosophy of mind and its various topics.*
- Varela, F. (1996) 'Neurophenomenology: A methodological remedy for the hard problem', *Journal of Consciousness Studies*, 3, 330–49.

2
Philosophy and Science: Naturalism

Overview

The previous chapter focused on how the mind emerged as a central topic in twentieth-century philosophy as reflected in philosophy of mind and phenomenology. Philosophy of mind applies a conceptual and linguistic analysis of the term 'mind' and its features like consciousness, free will, self, etc. In contrast, phenomenology relies on the phenomenal analysis of consciousness as we experience it in first-person perspective. Both conceptual-linguistic and phenomenal analysis must be distinguished from the method applied in science, which is based on observation in third-person perspective. Rather than concepts or consciousness, science investigates the natural world, the world as we can observe it. How can we link both philosophy and science and their respective methods of investigation? These more methodological issues are central if we want to understand how the mind and the brain are related to each other. The mind has traditionally been associated with either conceptual-linguistic or phenomenal analysis in philosophy, while the brain has been investigated by third-person-based observation in science. Hence in order to better understand the relationship between the mind and brain, we need to discuss how their respective methods of investigation relate to each other. That is the focus of this chapter.

Objectives

- Distinguish between domains and methods
- Distinguish between philosophy and science
- Explain the analytic–synthetic distinction
- Account for the difference between a priori and a posteriori
- Understand how epistemology and metaphysics can be naturalized
- Explain the difference between methodological and metaphysical naturalism
- Understand cooperative naturalism and the cross-fertile exchange of domains and methods

Key concepts

Philosophy, science, domain and method, analytic–synthetic distinction, a priori–a posteriori distinction, naturalized epistemology, observation sentences, methodological and metaphysical naturalism, cooperative naturalism, empirical and conceptual plausibility, empirical and transcendental approaches.

Background 1

Philosophy and science

What is philosophy? The term philosophy comes from the Greek concept of 'philosophia', meaning 'love of wisdom'. In the Western world, philosophy originated in ancient Greece with the most important philosophers being Plato (429–347 BC) and Aristotle (384–322 BC). In ancient Greece, philosophy described the investigation of basic and fundamental questions about reality and existence, knowledge, and values and norms. This led to the distinction between metaphysics, epistemology and ethics (the first two of which will be briefly explained below).

We recall from Chapter 1 that metaphysics and epistemology are central areas or domains in philosophy. Metaphysics is concerned with the fundamental existence and reality of being and conceptual grounds (defining concepts) as well as their basic nature. Ontology, as a sub-branch of metaphysics, is focused on the basic categories of being and existence, and how they are related to each other. While contemporary philosophy in particular makes a distinction between ontology and metaphysics, I here subsume the former under the much wider umbrella term metaphysics. To keep matters simple and clear, I will therefore use the term metaphysics throughout this textbook.

In addition to metaphysics, epistemology is regarded as a central field within philosophy. Epistemology is concerned with knowledge. What can we know of ourselves and the world? Are there any limitations to the knowledge of the world including our own selves? What kind of capacities or epistemic abilities (and inabilities) does one need to acquire knowledge of the world? These are the questions dealt with in epistemology.

How can philosophy be distinguished from science? Unlike philosophy, science is rather young. Physics was not really fully developed before the fifteenth and sixteenth centuries, with central figures like Copernicus (1473–1543), Galileo (1564–1642) and Newton (1642–1727). The other scientific disciplines like chemistry, biology, neuroscience, etc. are even younger, dating back (for the most part) to the eighteenth and nineteenth centuries. How can we distinguish science from philosophy? Unlike philosophy, science does not so much address the world in metaphysical and epistemological terms of existence and knowledge but rather in empirical terms of mechanisms and processes.

Background 2

Difference between philosophy and science

Methodology is one major difference between philosophy and science. Philosophy relies (traditionally) on rational and logical argumentation, thereby using concepts and thus language to describe the problem in question. For instance, the philosopher thinks and argues about the nature of objects and events within the world on mainly logical grounds (by making inferences from presuppositions).

In contrast, science is based on observation in third-person perspective of particular processes and mechanisms underlying events or objects in the world, using experimental manipulation. This has been described as the scientific method. Rather than thinking and making logical inferences about the nature of objects and events, the scientist infers from his/her observations their underlying processes and mechanisms and how they change during experimental manipulation.

Let's take the example of the mind. The philosopher thinks about the basic existence and reality of the mind; whether, for instance, it is mental or physical. This addresses the metaphysical question of the existence and reality of the mind and how it is related to the brain and the mind–brain problem (see Part II for details). The philosopher makes different suggestions of how to define the concept of mind, its existence and reality, and from there he/she infers in a logical way how to characterize its relationship to the brain. He/she may also raise the questions: what and how much we can know at all about the mind and its relation to the brain? This provides an epistemological account of the mind–brain problem (see Chapter 7).

The scientist, in contrast, does not consider the mind in a metaphysical or an epistemological context. Instead, he/she is more interested in investigating the mind's various mental faculties like consciousness, self, free will, etc. (see Parts IV and V) in empirical terms. He/she focuses on the processes and mechanisms associated with what is described as consciousness, free will, etc.

In order to understand the processes and mechanisms he/she experimentally manipulates them, observes their changes, and infers from the data how they function and operate. Hence, the 'what' question of philosophy (what is the mind and what can we know about it?) is here replaced by a focus on the 'how': how does the mind and its various features operate and function. Rather than the metaphysical and epistemological domains of philosophy, science operates within the empirical domain, or on that which can be observed.

Let's sketch the difference between the philosophical and scientific methodological strategies with an analogy. You stand in the kitchen and cook a nice meal, cordon bleu with various vegetables. Now your friend, a philosopher, comes into the kitchen and ponders about the food, raising metaphysical questions. Does the cordon bleu really exist? Is it real and existent independent of you cooking it? If so, there must be some kind of physical property or substance underlying the meat. What happens during the process of cooking? Is the presumed existence and reality of the meat still preserved? Or is it lost? And finally he may raise the question: what can he know in principle about the existence and reality of the meat?

You may wonder about all your friend's thoughts: strange, why is he asking all these questions? These are metaphysical and epistemological issues for a philosopher, but they are not really relevant for me as cook. All I need to know is how things function, operate, and are processed. For instance, I do not need to know what meat is, and it is even conceivable that one could be an excellent chef without having any knowledge of where food comes from and what is its nature. Instead, all I need to know is what happens to the meat when it is being cooked rather than the basic existence and being of the meat. I observe the meat and how the heat of the pan changes the consistency and color of the meat. In other words, you apply the scientific method of observation (and operation) rather than the philosophical strategy of making metaphysical and epistemological assumptions and inferences.

Background 3

Are philosophy and science principally different?

How do philosophy and science relate to each other? Before the rise of the sciences in the wake of the scientific revolution in the seventeenth and eighteenth centuries, philosophy was considered one of the most important disciplines. Besides religion, philosophy was assumed to provide a metaphysical and epistemological framework within which science and its focus on the empirical domain took place (see below). Most importantly, there was not yet such a divide between philosophy and science because for many philosophers, like Descartes (and other philosophers after him), both philosophical and scientific issues were not as yet principally distinguished at that time.

This changed with the rise of science as a separate field. Physics with Copernicus, Galileo and Newton provided a major breakthrough in our view of the world that was then no longer considered in religious and philosophical terms but rather in a physical way. This was later complemented by the view of the human being in purely biological terms as, for instance, suggested by Darwin. One of the last vestiges of the 'old' non-scientific, more philosophical view is the mind. However, the concept of the mind is now being taken over more and more by, in particular, neuroscience. This is what this book is about.

To understand this we have to go back to the beginning of science. Why and how did science gain the upper hand over philosophy? The various sciences like physics, chemistry and biology developed the increasing ability to explain the world and its beings in more precise ways than philosophy. The scientific method of observation and experimental manipulation seemed now, in the nineteenth century and even more so in the twentieth century, to provide the prime access to the world in terms of scientific concepts, of which philosophers aimed to provide metaphysical and epistemological analysis. Philosophy was pushed increasingly on to the back burner while science took over the original role of philosophy as the queen of all disciplines.

How do philosophy and science relate to each other today? Can we replace philosophy by science or can both coexist? To address this question, we want briefly to investigate the domains and methods of both philosophy and science.

Are philosophy and science principally different? Different in their respective contents, as well as in the methods they apply to investigate those contents? These are the questions we must address in order to understand how to link philosophy and neuroscience. Then we may need to link their specific contents like the mind and the brain, as well as their respective methods of investigation (conceptual-linguistic versus observational-experimental).

Background 4

Domains versus disciplines

How can we link philosophy and science? To do so, one may want to consider the concept of 'domain'. As already indicated in the Introduction, the concept of a 'domain' comes originally from the Latin terms, 'dominium', 'dominus' and 'domus': 'dominium' describes property, 'dominus' refers to lord, master, or owner, and 'domus' stands for house. These distinct though related meanings are reflected in the various uses of the concept of domain in our current language. Domain in today's usage can describe a territory governed or ruled; a field of action, thought or influence; a region characterized by specific features; and a range of significance.

How can we apply such meanings of the concept of domain to philosophy and science? The discipline of philosophy, for instance, focused originally on metaphysical (existence and being) and epistemological (knowledge) and ethical (morality) questions. This means that it dealt with the metaphysical, epistemological and ethical domains. The sciences, in contrast, are not so much concerned with either metaphysical or epistemological domains (but which, though, they presuppose), but rather with the world and its various processes and mechanisms as we can observe them from the third-person perspective. Thus the domain of scientific inquiry can be identified as the empirical domain.

Though preliminary, the concept of domain also concerns specific contents, including particular questions, problems and issues. Most importantly, the term domain does not contain any specification about the method, or how the respective contents are investigated, be it in a conceptual-logical, phenomenal, or observational-experimental way.

The exclusion of any methodological specification distinguishes the term domain from the term discipline: disciplines like philosophy and the sciences do not only concern particular contents and their respective domains, but are also characterized by a specific methodological strategy of investigation. For instance, philosophy applies the conceptual-linguistic method to investigate questions belonging to the metaphysical and epistemological domains. In contrast, science applies the observational-experimental method to investigate issues belonging to the empirical domain. In short, disciplines are characterized by both content and method, while the concept of domains concerns only content, independent of the respective methodological approach.

Background 5

Naturalism

Why is all this relevant in the present context? We usually take it for granted that specific domains like the metaphysical or epistemological domains use a particular methodological strategy. This is why we associate metaphysical questions with a specific discipline, philosophy. However, this association between domain and method, usually taken for granted, may change as both concepts dissociate from each other and from their traditional combinations.

For instance, a domain like the metaphysical domain may be associated with a different method like the observational-experimental method. The metaphysical domain is then no longer associated with philosophy as a discipline but shifts rather into the territory of science. In other words, there may be flexible and varying constellations between domains and methods, which then can also alter the way in which we define and characterize associated disciplines. Below, we will see that exactly this happens in the relationship between philosophy and science when the metaphysical and epistemological domains become associated with the observational-experimental method of science.

The dissociation between the traditional domain–method constellations may establish continuity between the originally separate disciplines. For instance, the association of the metaphysical and epistemological domains with the observational-experimental method also reorders the relationship between their associated disciplines, philosophy and science. The strict dichotomy between philosophy and science, as based on their seemingly mutually exclusive domains and methods, is then softened and bridged. In other words, some kind of continuity between philosophy and science and their newly ordered domains and methods can be established.

How can we better understand continuity between philosophy and science, including the change in traditional associations between domains and methods? To answer this question we should discuss one of the leading philosophers of the twentieth century, Willard van Orman Quine.

Biography

Willard van Orman Quine and the naturalization of philosophy

Willard van Orman Quine was born in 1908 in the United States and was from very early on interested in mathematics and philosophy. He received his PhD from Harvard University and then traveled to Europe in 1932–33 on a Sheldon fellowship and had the chance to meet many European philosophers. He afterwards maintained a lifelong affiliation with the Philosophy Department at Harvard University, interrupted only by his work for the Navy during the Second World War, where he helped decipher coded messages from German submarines. He later went on to decipher the methodological code of philosophy by ultimately tracing it back to the one of science. Let me detail this idea below.

During the twentieth century, philosophical inquiry was regarded as principally different from science in both content and method. Quine now introduced various arguments (see below) for resolving that principal distinction by showing some continuity between both philosophy and science as disciplines. In a nutshell, he argued that the conceptual-linguistic method of philosophy is basically just an abstraction of the observational-experimental method of science. If true, the traditional philosophical domains (the metaphysical and epistemological domains) can no longer be principally distinguished from the empirical domain of science.

In turn, this makes it possible to apply the observational-experimental method of science to metaphysical and epistemological questions. Since the observational-experimental method presupposes the natural world – the world as we can observe it in third-person perspective – philosophers speak of the 'naturalization' of metaphysics and epistemology or of metaphysical and epistemological domains. However, in order better to understand how such naturalization works, we need to go into more detail about the exact logical features characterizing the methodological strategy of philosophy.

Naturalism 1

Dichotomy between analytic versus synthetic sentences

Why does Quine reject the traditional dichotomy between philosophy and science? One reason is his rejection of the principal distinction between analytic and synthetic sentences in his landmark paper, 'Two dogmas of empiricism' (Quine, 1951). While this paper is rather complex, its main points will nevertheless be briefly discussed in a simplified way below.

What are analytic and synthetic sentences? Analytic sentences are sentences like 'All unmarried men are not married' or 'All bachelors are unmarried'. These sentences are true solely in virtue of the meaning of the terms they contain. 'Of course,' we hear an imaginary philosopher saying, 'the terms "bachelor" and "unmarried" imply that those people are not married. That is pure logic.'

Synthetic sentences, however, are true not only in the virtue of the words they contain, but also by some additional, extra-linguistic facts. For instance, it is not logical to us that 'force equals mass times acceleration', since that is based not only on the words themselves, but also on the additional observation of physical processes. More specifically, the term force does not imply anything about mass, time, or acceleration, which can only be obtained by additional observation that cannot be inferred (or deduced, as the philosopher may say) from the logical structure of the words themselves.

Quine focused on the distinction between analytic and synthetic sentences. Synthetic sentences are subject to observational evidence. We can change them within the same language once our observations tell us more and different things. Each sentence can thus be true or false and we are able to correct them while remaining within the same language. For instance, we may find out in the future that 'force equals mass times acceleration' in a different way than we know and formulate today. Then we can simply change the meaning of the terms mass, force, acceleration, or time and/or combine them in new ways (with or without other terms) without changing the language. How is that possible? This is due to the fact that the meaning of these terms does not only depend on the concepts themselves but also on some additional extra-linguistic factor.

This is different in the case of analytic sentences (sentences like 'All bachelors are unmarried'). These sentences are not subject to observational evidence and justification, they are true by and in themselves, independent of any additional evidence and justification. If we want to change the truth of the sentence, we must redefine its concepts (which, though, as discussed below may run into problems depending on the use, context and equivocity of the terms).

For instance, the term 'bachelor' would then no longer mean 'unmarried'. This implies the need to switch to another language in which the same term (e.g. bachelor), would mean something different. Hence, unlike in synthetic sentences, there is no tolerance for change in the truth of analytic sentences without a concomitant change in the language itself. There is no tolerance for change in analytic sentences, whereas in case of synthetic sentences, change is possible. This is described as the 'principle of tolerance'.

Naturalism 2

Continuum between analytic versus synthetic sentences

We have seen that the principle of tolerance applies only to synthetic sentences but not to analytic sentences. This entails a strict and clear-cut distinction and thus a dichotomy between analytic and synthetic sentences. Quine, however, rejects the principle of tolerance. Why? He argues that there is no principal difference between changes in an analytic or synthetic sentence. In the case of synthetic sentences, we make some small theoretical changes by, for instance, assigning and determining the concepts of mass, time, etc. in a different way. And, more crucially, Quine postulates that is essentially the same thing as when one formulates a new language, as in the case of analytic sentences.

According to Quine, we make theoretical changes in both cases. While such theoretical changes are evident in synthetic sentences, they have been distinguished, until now, from language change in analytic sentences. Quine nevertheless claims that the second case, the language change in analytic sentences, is not principally different from the theoretical change in synthetic sentences.

Why? We make both theoretical and linguistic changes with the same aim, namely better to navigate the world. However, if both linguistic and theoretical changes serve the same aim (e.g. to navigate better in the world), they can no longer be regarded as principally different. The principal distinction between analytic and synthetic sentences thus consequently breaks down. Instead of difference, a continuum replaces their previously mutually exclusive dichotomy.

The failure of the analytic–synthetic distinction has important implications for the relationship between philosophy and science. The distinction between philosophy and science has often been based, especially in the Anglo-American world in the first half of the twentieth century, very much on the distinction between analytic and synthetic sentences. Often philosophy has sought to apply analytic sentences while science relies on synthetic sentences. This is supposed to make possible the distinction between philosophy and science and to thus clearly delineate and demarcate philosophy from science.

Things are not that easy, however. Already the German philosopher Immanuel Kant had argued that philosophy may be able to make synthetic judgments, synthetic a priori judgments, as he called them. Following such synthetic a priori judgment distinguishes philosophy from science, which relies rather on synthetic a posteriori judgments (see below for the distinction between a priori and a posteriori).

While Kant still maintained a clear distinction between analytic and synthetic judgments, Quine was now much more radical and claimed that the dichotomy between analytic and synthetic can no longer be maintained. Instead of a clear-cut dichotomy there is rather a blurry continuum between analytic and synthetic judgments. If, however, the distinction between analytic and synthetic sentences can no longer be maintained, the principal difference between philosophy and science crumbles too. More specifically, the impossibility of principally distinguishing between analytic and synthetic sentences implies a continuum, rather than a principal difference between philosophy and science.

Naturalism 3

Continuum between a priori versus a posteriori knowledge

The continuum is even more apparent in a related distinction, *a priori* versus *a posteriori*. The truth of analytic sentences depends exclusively and completely on the terms themselves, no other extra-linguistic proof such as world experience is needed. As the philosopher says, analytic sentences must be true and can therefore not be false (without any change in language). This can be known by considering the sentence itself, without reverting to any additional evidence outside the sentence itself (like empirical observation). The knowledge about the sentence and its contents thus reflects what philosophers describe as 'a priori knowledge', signifying that the truth or falsity in sentences and concepts can be known without recruiting any additional evidence apart from the sentence itself.

This is different in the case of synthetic sentences. Their truth or falsity depends not only on the concepts and sentence itself, but also on additional evidence as provided by observation of the world. The truth or falsity of synthetic sentences can thus not be known by considering the concepts and sentences in themselves. This means that a priori knowledge remains impossible in the case of synthetic sentences. Instead, due to the recruitment of additional evidence for the truth or falsity of synthetic sentences, the philosophers speak here of 'a posteriori knowledge'.

How now does the distinction between a priori and a posteriori knowledge relate to the distinction between analytic and synthetic sentences? A priori knowledge characterizes analytic sentences while a posteriori knowledge characterizes synthetic sentences. If the distinction between analytic and synthetic sentences breaks down, as described above, it is often supposed to entail the breakdown of the a priori–a posteriori distinction too (see, though, below the counter-argument by the American philosopher Saul Kripke). Hence, the continuum between analytic and synthetic sentences opens up the possibility for a continuum between a priori and a posteriori knowledge.

Accordingly, Quine undermines two traditionally strict dichotomies, the one between analytic and synthetic sentences, as well as the one between a priori and a posteriori knowledge. In the place of these dichotomies, Quine argues for a double continuum between analytic and synthetic as well as between a priori and a posteriori. However, Kripke subsequently claimed the principal difference between the two distinctions; for instance, he argued that analytic a posteriori judgments are possible. This ultimately implies that the breakdown of the former distinction, analytic–synthetic, does not necessarily entail the breakdown of the latter, a priori–a posteriori; see Chapter 6 for details on Kripke's position, which he developed in the context of refuting the identity theory between mind and brain.

Naturalism 4

Continuum between philosophy and science

How does this double continuum relate to philosophy and science? Philosophy has been characterized by a priori knowledge while science is based on a posteriori knowledge. The failure of the a priori–a posteriori distinction consequently necessitates that the principal distinction between philosophy and science also breaks down. Taken together, this means that the philosopher's two essential methodological tools, analytic sentences and a priori knowledge, are no longer principally different from those of the scientist, synthetic sentences and a posteriori knowledge. This has major reverberations for what we consider philosophy, as we will see below.

What is the task of philosophy if methodologically it can no longer be principally distinguished from science? Quine argues that the task and purpose of philosophy consists in (i) analyzing and clarifying the language of science, and (ii) formulating and recommending alternative concepts and languages. These alternative concepts and languages may then be useful for scientists to address and investigate different questions and issues or to see the same question in a different way.

Most importantly, this activity of clarifying concepts and developing alternative ones is not principally different from the activity of the scientist. Why? Because both make theoretical changes according to one and the same overall purpose: to deal with and navigate the world that we live in.

How does this relate to the traditional philosophical field of metaphysics, and questions about existence and reality? Quine argues that metaphysical questions (he speaks of ontological rather than metaphysical) depend on the choice of language. And that choice of language is no longer principally different from the choice of theories in science, since both have the same overall purpose of providing a better understanding of the world.

More specifically, metaphysical (e.g. ontological) assumptions are based on the sum total of our available observation-based theories as structured and paraphrased by the laws of logic. By structuring scientific results and theories in a logical way, they gain simplicity and clarity. Metaphysical assumptions are then nothing but abstractions and are clearer and simpler than the results and theories of the sciences. In short, they are the 'sum total of our scientific knowledge reformulated to fit the framework of logic'. Quine speaks here of what he described as 'regimented theory', namely that scientific results and theories are regimented and condensed into a logical structure.

What does that imply for the relationship between philosophy and science? Rather than being principally different, philosophy is then on a par or continuum with science. There is no principal difference any more between philosophy and science with regard to method: the seemingly different methodological strategy of philosophy and its reliance on analytic sentences and a priori knowledge turns out to be nothing but an abstract variation of the more concrete method used in science, namely synthetic sentences and a posteriori knowledge.

The methodological differences between philosophy and science are consequently no longer mutually exclusive, but rather continuous. Considered methodologically, philosophy and science are thought of as different degrees of abstraction on the same underlying continuum (between synthetic and analytic sentences as well as between a priori and a posteriori knowledge).

Naturalism 5

Epistemology naturalized

How does the proclaimed methodological continuum between philosophy and science impact their respective domains? The methodological continuum between philosophy and science made possible by the breakdown between analytic and synthetic sentences opens the door to new ways of investigating both philosophy and science. For instance, we might investigate the traditionally philosophical domains of metaphysics not only with the conceptual-linguistic approach, but also with the scientific method (observational-experimental strategy). This is exactly what Quine himself suggested, especially with regard to the epistemological domain.

If philosophy is continuous with science, epistemology – the branch of philosophy concerned with our capacities and limitations for knowledge – should also be related to the empirical domain and be investigated by the scientific method. This makes it possible for epistemology to become associated with the natural world, or the world as we can observe it. As such, epistemology can become part of the natural world and thus be naturalized. This topic was addressed by Quine in his paper 'Epistemology naturalized' (Quine, 1969).

Like metaphysics, epistemology is entrenched in language. Thus if we want to understand epistemology, we need to investigate its language. Instead of deploying the conceptual-linguistic approach, however, we should proceed by applying the scientific method of observation-experimentation. Quine focuses on knowledge and more specifically on how we acquire knowledge (knowledge acquisition, as distinguished from knowledge justification) in our language as a paradigmatic example of how to naturalize epistemology and link it to the scientific method. In order to investigate how we acquire knowledge, we thus have to focus on our acquisition of language which in turn requires observation and experimentation.

We have to be careful, though. Here language is not considered by itself, as in the philosophy of language, but rather in terms of its acquisition. Quine investigates how we (and especially children) acquire knowledge via our language. According to Quine, knowledge is based on sensory stimulation, the extrinsic sensory stimuli we receive from the environment. Much of our knowledge is only indirectly and remotely connected to the original sensory stimulation because it is part of an overall system of various sentences that together as a whole serves to cope with and express our sensory experience.

How does a child start to acquire knowledge? A child does not yet possess the complex and sophisticated linguistic competence of adults. Instead, a child's linguistic competence is more basic and elementary. This is reflected linguistically in what Quine calls 'observation sentences', which will be explained in further detail below.

Naturalism 6

Observation sentences: receptual and perceptual similarity

What are observation sentences? Observation sentences are those sentences that are most directly related to sensory stimulation. For instance, the sentences 'I see green' or 'I am warm' are observation sentences where the direct connection to the sensory stimulation is still visible and preserved. Thus they can be considered the most basic form of knowledge. According to Quine, if we want to know how knowledge is yielded, we need to look at how 'observation sentences' are generated.

Quine distinguishes two different criteria for observation sentences to be generated. First, there are individual criteria, and second, social criteria. Let us start with the individual criteria. The stimulus must attach to the receptor in the brain of the respective person. In order to yield knowledge, the stimulus must attach in more or less the same way to the respective receptors across different people. Quine speaks here of 'receptual similarity' which essentially describes the similar stimulation pattern on receptors across different people.

However, mere receptual similarity is not sufficient to yield observation sentences and thus knowledge. In addition to the similarity in receptor stimulation, (receptual similarity), the resulting percepts (what the person perceives) must also be similar between different people. This is what Quine calls 'perceptual similarity' – that which essentially describes similar perceptual response patterns to the same stimulus across different people.

Why do we need perceptual similarity and receptal similarity? The same stimulus may be attached in the same way to the receptor but may nevertheless yield different percepts. For example, the stimulus of a green flower could attach to the 'green receptors' of the visual cortex in the same way across different people. However, one person may perceive that stimulus as green, while another person might perceive it as yellow. Hence, receptal similarity can be accompanied by perceptual dissimilarity (see also analogous arguments in the debate about consciousness: Part IV).

Naturalism 7

Observation sentences: social similarity

In addition to receptal and perceptual similarity as individual criteria, the generation of observation sentences also requires a social criterion. Such social criteria are important because knowledge implies public access and similarity across different people. Hence, Quine argues that the social criteria consist in the fact that the same neural intake yields the same perceptual similarity across different people. In addition to receptal and perceptual similarity, one may therefore speak analogously of 'social similarity'.

Why and how does the same stimulus induce social similarity (perceptual similarity across different people)? Quine assumes that over thousands of years humans have evolved in the same environment which has shaped the brain and thus the brains in different individuals in similar ways. As such, social similarity can ultimately be traced back to a common evolutionary history where different individuals within a particular species shared a common environmental context.

If so, one may be inclined to assume that our brain and its sensory regions are intrinsically social: the neural response pattern of the brain in different individuals may not only reflect receptal and perceptual similarity, but also a common evolutionary background and a shared actual environmental context where both account for social similarity. In other words, the brain and its neural response pattern seems to be intrinsically aligned with the environment.

How is this intrinsic alignment between the brain's neural activity and the environment possible? While Quine discusses social similarity as a criterion for observation sentences as building blocks of knowledge acquisition, he leaves open the exact neuronal – and thus empirical mechanisms – underlying this social similarity. How is it possible for the brain and its neural response to become intrinsically aligned to its respective environmental context? What are the neural mechanisms underlying such neuro-social alignment?

To answer these questions we need to investigate the exact neural mechanisms that allow the brain to align its neural response pattern to the environment and its specific stimulus patterns. Put in a more abstract way, we may say that the mechanisms underlying the intrinsic alignment between brain and environment need to be deciphered. This is highly relevant in order better to understand the brain and its neural mechanisms. At the same time this is especially important in explaining how intersubjectivity is generated as the necessary precondition of any possible knowledge acquisition. This will be discussed in Chapter 20 in full detail.

Critical reflection 1

Observation sentences and internal criteria

Subsequent authors criticized Quine for neglecting internal criteria in his account of observation sentences. More specifically, he neglected the organism's state during the reception of sensory stimuli and thus internal criteria. For instance, water may taste wonderful if one is extremely dehydrated. If, by way of contrast, one had just consumed two litres of water, the same water might taste completely different.

Take another example: if a lion is hungry, it may perceive an antelope as very distinct from its respective surroundings. By contrast, if the lion is not hungry at all (and just had a giraffe for lunch), it may barely perceive the passing antelope, which now more or less blends in with the respective background. Here it is the internal state, and not the sensory stimulus alone, that determines perceptual similarity. The neuroscientist would say that the salience of the stimuli and its objects changes according to the internal state of the organism.

Quine may now want to argue that the internal state of the subject does not change the picture essentially. In addition to the individual and social criteria, one may simply assume an additional criterion for the yielding of observation sentences, namely bodily criteria as internal criteria. Though it may complicate the picture, it does not essentially change the basic account of observation sentences.

How about the brain itself? Does the brain itself provide some internal criterion? At first one may want to think that receptual and (especially) perceptual similarity are internal criteria. However, Quine defines perceptual similarity in a purely behavioral way; it is just a matter of behavioral responses which can be observed. There is no subjective experience (let alone consciousness or awareness) involved. Instead, Quine argues that perceptual similarity consists in behavior that can be explained by the underlying neurophysiology. Since behavior is externally visible, it may be regarded as an external, rather than internal, criterion.

What makes a criterion internal rather than external? As in the case of the body, an internal criterion is a criterion that cannot be observed and accessed from the outside. An internal criterion is only accessible by the respective person him/herself, as is the case with the body. While in the case of the brain, an internal criterion would stem from the brain itself independent of any external criterion (see Chapter 12 for details on this).

Critical reflection 2

Observation sentences and 'experiential similarity'

What if consciousness stemmed from the brain's internal input, as in, for instance, from some intrinsic activity in the brain itself as distinguished from the neural activity induced by external stimuli (extrinsic activity)? The presence of consciousness might then be regarded as an internal criterion that must be distinguished from both receptual and perceptual similarity as purely external criteria (at least in the way Quine conceives them).

What if the early sensory processing that accounts for perceptual similarity is already associated with consciousness due to the brain's intrinsic activity? This would, for instance, empirically be the case if the neural processing in the early sensory cortex (as in, for instance, the primary visual cortex, e.g. V1) already predisposes the association of the stimulus with consciousness. Consciousness would then not be caused by the stimulus itself (which only triggers it), but would be provided internally and stem from the brain itself (its intrinsic activity) (see Parts III and IV for details). The intrinsic activity of the brain and its association with consciousness would then provide an internal criterion.

Most importantly, the association of perceptual processing with consciousness would need to occur by default: the brain could not do otherwise because of its intrinsic activity and its specific neural organization. The brain's neural activity and its organization thus predisposes (makes it necessary; see Part IV for the concept of predisposition) the association of receptual, perceptual and social similarity with experience and thus consciousness.

What does this imply for observation sentences? It implies that one may need to consider the brain's intrinsic activity and its manifestation in experience (consciousness) as an additional criterion. This means that the brain's intrinsic activity and manifestation in consciousness are an internal criterion for observation sentences. Additionally, alongside receptual, perceptual and social similarity, there must also be 'experiential similarity' across different people in response to the same stimulus. Only when there is 'experiential similarity' can different people acquire the same kind of information, which then may be characterized as knowledge.

What does this imply about Quine and his account of observation sentences? It means that he needs to complement his purely behavioral account of perceptual similarity and observation sentences with a phenomenal account (i.e. a subjective-experiential component). Receptual, perceptual and social similarity as external and purely behavioral criteria are then complemented by 'experiential similarity' as internal and rather phenomenal criterion.

Naturalism 8

Naturalization of philosophy

We already discussed the historical background in which Quine and others undermined the traditional dichotomy between philosophy and science. The emergence and increasing success of the sciences like physics, chemistry, biology, etc. in explaining the world made it necessary to readjust the domain and method of philosophy. The concept of the domain contextualizes the kinds of problems, questions, issues or contents that philosophy as a discipline has as its focus.

Philosophy's traditional domains were the metaphysical domain, the epistemological domain, and the normative-ethical domain (for more information on the latter, see Chapter 4). The metaphysical domain describes the existence and reality of being and world, while the epistemological domain is concerned with our knowledge of the world. These domains distinguish philosophy from science in that the latter presupposes an empirical domain where the focus is on the events, people and objects of the world as we can observe them.

What about method? The conceptual-linguistic method, which, as discussed above, presupposes a priori knowledge and analytic sentences, was the preferred way of approaching the questions and problems put forward by the metaphysical and epistemological domains as dealt with in philosophy. In contrast, the scientific method is based on synthetic sentences and a posteriori knowledge. Recall that analytic sentences are those whose meaning and truth can be explained completely and exclusively by reference to the terms themselves, while this remains impossible in the case of synthetic sentences.

How now can we further characterize a priori and a posteriori knowledge? Taken in the traditional sense, a priori knowledge is knowledge that remains independent and prior to any experience and observation of the world, whereas this is not the case in a posteriori knowledge (see, though, Kripke who, claiming for the possibility of analytic a posteriori knowledge that is analytic and nevertheless tied to observation of the world, would not agree with such characterization of a priori and a posteriori knowledge; see Chapter 6 for a discussion of his position).

As we have seen above, Quine undermined the distinction between analytic and synthetic sentences as well as the distinction between a priori and a posteriori knowledge. This implies that the associated principal distinction between philosophy and science can no longer be maintained. Rather than being mutually exclusive, Quine made it possible to consider philosophy and science as existing on a continuum where each discipline employs a different level of abstraction when discussing and dissecting the same problems.

This continuity between philosophy and science implies that both concern and operate within one and the same world: the natural world as we can observe it and as is manifest in synthetic sentences and a posteriori knowledge. This is different from philosophy's original concern with the logical world rather than the natural world. The logical world is the world that can be conceived on purely logical grounds, requiring no observation but only analytic sentences and the postulating of a priori knowledge.

The clear-cut distinction between logical and natural worlds is called into question once the distinctions between a priori and a posteriori knowledge and between analytic and synthetic sentences break down: if a priori knowledge and analytic sentences can no longer be clearly distinguished from synthetic sentences and a posteriori knowledge, the logical world cannot be as categorically separated from the natural world as was once presupposed. Philosophy's shift from the logical world and logical argumentation toward the natural world and empiricism is described as 'naturalism', or the 'naturalization' of philosophy.

Naturalism 9

Metaphysical naturalism: definition

What does the naturalization of philosophy imply for metaphysics and epistemology as the traditional domains of philosophy? Metaphysically, the naturalization of philosophy amounts to what is described as 'metaphysical naturalism', while in the case of epistemology, one can speak of 'naturalized epistemology' (see above). 'Metaphysical naturalism' describes the existence and reality of the world and the universe as fully consisting in natural reality as we can observe it. There is nothing in our world that goes beyond natural reality and its physical and chemical properties. There are no supernatural entities like spirits, ghosts and deities, implying that everything that exists and is real is natural and thus ultimately physical.

Let us use the example of mind. We recall from the first chapter that French philosopher, René Descartes, assumed the reality and existence of two substances: a physical substance that accounts for the body, and a mental substance that underlies the mind. In this model, only the physical substance can be observed and is thus part of the natural world. The mental substance, in contrast, cannot be observed in the natural world and must, therefore, be located in a world outside the natural, physical world. Descartes' assumption of a mental substance may therefore be considered supernatural.

The assumption of the existence and reality of a supernatural mental substance is denied in naturalism because naturalism inherently claims that everything that exists and is real is part of the natural world as we can observe it. The existence and reality of the mind is thus natural rather than supernatural and must therefore be located in and as part of the natural world. More concretely, the naturalization of the mind leads

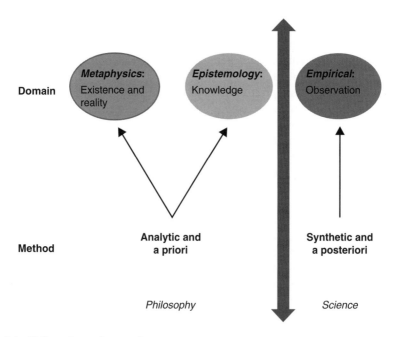

Figure 2.1 Philosophy and naturalism
Figure 2.1a Segregation or parallelism between philosophy and science
The figure illustrates different relationships between philosophy and science on the basis of the distinction between method and domains.

(a) The figure depicts the distinction between the domain, i.e. the area or framework of questions, problems and issues (upper part), and method, i.e. the strategy by means of which the questions, issues and problems are investigated (lower part). Domains include metaphysics, e.g. existence and reality, epistemology, e.g. knowledge, and empirical, e.g. observation, while methods concern the a priori and analytic as well as a posteriori and synthetic sentences. This distinction is considered mutually exclusive and opposite, thereby erecting a principal distinction and segregation between philosophy and science (the vertical arrow).

one to assume that the existence and reality of the mind must ultimately consist in the natural reality and existence of the body and brain as we can observe it. This and other theories about the metaphysics of mind will be discussed in Part II in full detail.

Naturalism 10

Metaphysical naturalism: analytic naturalism

One may now distinguish between different forms of metaphysical naturalism. Let's start with analytic naturalism, or what is also known as a priori physicalism. Analytic naturalism claims that the existences and realities dealt with in higher-level sciences like psychology and biology can be directly inferred from the lower-level sciences like chemistry and physics. No additional observation with a posteriori knowledge and synthetic sentences is needed. Therefore such inference from lower- to higher-level existences and realities is possible on purely logical grounds, relying solely on analytic sentences and a priori knowledge. Hence the name analytic naturalism or a priori physicalism.

Let us give an example. One example often discussed in philosophy of mind is the relationship between pain and C-fibers. Pain is a psychological and thus high-level concept, while the C-fibers mediating pain in the brain are a neuroscientific and thus a lower-level concept. How is pain related to C-fibers? Analytic naturalism claims that we can infer the existence and reality of pain completely from the existence and reality of the C-fibers themselves on purely logical grounds alone. This relatively radical claim will be explained further below.

How can we infer pain from C-fibers? We need to reveal the nature and features of the C-fibers themselves. This will imply and tell us everything there is about pain and how we experience pain. No additional knowledge apart from the nature and features of the C-fibers is required to understand pain and its features. In other words, pain can be understood on the basis of analytic sentences and in a completely a priori way. Considering the mind in general, analytic naturalism would claim that we can infer the mind and its mental features on purely logical grounds (analytic sentences and a priori knowledge) from observing the brain itself and its neuronal states (independent of any mental states).

This example makes it clear that analytic naturalism is a radical thesis. Higher-level descriptions, as in psychology, are not assumed to have a reality and existence independently of lower-level (physical) descriptions. Analytic naturalism claims that what we conceive of as reality and existence in higher-level descriptions is nothing but a conceptual illusion. This implies that there is nothing but the concept of the mind for which there is no corresponding reality or existence, whether physical or mental. One may thus say that the concept of mind (and other higher-level descriptions) signify empty (or illusory) concepts – concepts without any underlying existence and reality.

Naturalism 11

Metaphysical naturalism: synthetic naturalism

A less radical form of naturalism is synthetic-reductive naturalism. In contrast to analytic naturalism, synthetic-reductive naturalism does not opt for a purely logical and conceptual relationship between higher and lower levels of description. Higher levels of description, as in biology and psychology, can be characterized by reality and existence. Most important, their existence and reality is supposed to be distinct from those of the lower levels, as in physics. Unlike in analytic naturalism, the existence and reality of the higher levels is not viewed as illusory.

This means that the biological and psychological concepts that describe those higher levels cannot be reduced to, nor inferred from, the lower-level concepts. In other words, analytic sentences and a priori knowledge are not sufficient. In addition to the merely linguistic-conceptual account we need some additional support from observation to account for the existences and realities of the higher levels. This makes synthetic sentences and a posteriori knowledge necessary, hence the name synthetic naturalism.

Let's go back to our example of pain and C-fibers. In analytic naturalism–pain was completely inferred from C-fibers on purely logical (and conceptual) grounds. This is different in synthetic naturalism. Here pain is assumed to be real and existent rather than denoting a merely empty and illusory concept. To fully understand the reality and existence of pain, however, additional observation, beyond the observation of the C-fibers, is required. We need synthetic sentences and a posteriori knowledge to account for the existence and reality of pain.

How are the existences and realities of the higher levels related to those of the lower levels? There are two different varieties of synthetic naturalism – synthetic-reductive naturalism and synthetic-non-reductive naturalism. Let's start with synthetic-reductive naturalism. The term synthetic-reductive naturalism comes from the fact that the higher-level existences and realities are assumed to be ultimately reduced to and identified with those of the lower levels. When considering pain in this model, it is suggestive that the existence and reality of pain is reduced to and identified completely with the physical features of the C-fibers.

What is the difference between synthetic-reductive naturalism and analytic naturalism? Both claim that the existence and reality of the higher levels consist in the existence and reality of the lower levels. Metaphysically, there seems to be no difference. However, the difference consists in the characterization of the higher level itself. Analytic naturalism claims that the existence and reality of the higher level is an empty and illusory concept like pain. This is denied in synthetic-reductive naturalism. Here the higher level is considered real and existent as, for instance, pain, which though can ultimately be reduced to the existence and reality of the lower level, the C-fibers.

One may also deny that the existence and reality of the higher level can be reduced to and completely identified with the existence and reality of the lower levels. Pain may then be real and existent and, most importantly, its existence and reality cannot be completely identified with the C-fibers.

Critical reflection 3

Metaphysical naturalism and neurophilosophy

Why are these different forms of metaphysical naturalism relevant in our quest to understand the relationship between philosophy and the brain? One central question is how the existence and reality of the mind and the brain are related to each other. The mind was traditionally associated with philosophy and its metaphysical domain. In contrast, the brain is linked to the sciences and its empirical domain. The investigation of the metaphysical relationship between the mind and the brain is consequently confronted with different domains – metaphysical and empirical – that seem to be mutually exclusive.

Thanks to the naturalization of metaphysics (metaphysical naturalism), the metaphysical domain is linked to the empirical domain. Such linkage between metaphysical and empirical domains makes it possible to connect philosophy and science, which so far have been regarded as mutually exclusive. This means that the mind, which up until recently was dealt with exclusively in philosophy, can now enter the empirical domain and science. The mind–brain problem can no longer be considered an exclusively philosophical and thus metaphysical problem, but now must also be considered a scientific and thus empirical one too.

Accordingly, metaphysical naturalism opened the mind up to the natural world, making it possible to study the mind through scientific investigation. This in turn makes it possible to not only link the mind to the brain, but also to connect philosophy and neuroscience. The result is neurophilosophy (see Chapter 4 for details). In this light, the naturalization of philosophy (i.e. metaphysical (and epistemological) naturalism), may be considered a building block for the possible development of neurophilosophy.

However, the impact of metaphysical naturalism goes way beyond being a mere door opener to the living room of neurophilosophy. In addition, its different subsets of analytic and synthetic naturalism also pave different ways within the living room itself. Analytic and synthetic naturalism entail different concepts of neurophilosophy: analytic naturalism leads to a reductive approach to neurophilosophy, reductive neurophilosophy, whereas synthetic naturalism is rather compatible with non-reductive neurophilosophy (see Chapter 4 for details).

More specifically, analytic and synthetic naturalism imply a different stance with regard to the accepted and included domains of investigation. Analytic naturalism exclusively focuses on the physical as it can be observed, which implies that any domain other than the empirical is expelled and excluded. This leads ultimately to domain monism as it characterizes reductive neurophilosophy.

Due to its exclusive focus on the empirical domain in general and the neural states of the brain in particular, reductive neurophilosophy is prone to what can be described as neuralism (see also Chapter 1). Neuralism can be considered the extension of analytic naturalism to the brain and its neural states. More specifically, neuralism in this sense describes the consideration of traditionally philosophical issues like metaphysical, epistemological, phenomenal and ethical questions in exclusively and solely neural terms.

The situation is different in synthetic naturalism (and especially the synthetic-non-reductive versions). Here domains other than the empirical can still be included and linked (without reducing them as in synthetic-non-reductive naturalism) to the empirical domain. There is thus domain pluralism (rather than domain monism) which entails non-reductive (rather than reductive) neurophilosophy (see Chapter 4 for details). Such domain pluralism protects non-reductive neurophilosophy against the charge of neuralism: the focus here is not on considering traditionally philosophical issues like metaphysical, epistemological, phenomenal and ethical questions in exclusively and solely neural terms. Instead the focus is on linking (rather than reducing and eliminating) metaphysical, epistemological, phenomenal and ethical issues and their respective domains to the empirical domain and its neural states. This opens the door for the development of neurophilosophy in general and non-reductive neurophilosophy in particular.

Naturalism 12

Methodological naturalism: definition

In addition to their domains, disciplines can also be characterized by the kinds of methods they apply to investigate their respective subjects (see above and the Introduction for the distinction between domains and disciplines). Philosophy, for instance, traditionally used the conceptual-linguistic method, while modern science has almost always applied the observational-experimental approach. The arguments of Quine showed that the hallmark features of the conceptual-linguistic method – analytic sentences and a priori knowledge – could not be properly distinguished from their counterparts in the scientific method – synthetic sentences and a posteriori knowledge.

What this means is that the philosophical method can no longer be wholly distinguished from the scientific method. If, however, the methodological tools of philosophy can no longer principally be distinguished from the methodological tools of science, both methods must be seen as versions of one and the same strategy, the naturalistic approach. This results in what is described as 'methodological naturalism'.

What exactly do we mean by methodological naturalism? *Methodological naturalism describes the conceptual-logical method of philosophy as not principally different and mutually exclusive from the observational-experimental strategy of science.* In this model, the conceptual-logical method is an abstraction of the observational-experimental strategy. What this means is that both methods are concerned with theories and theoretical assumptions about the natural world, but at different levels – more or less abstract – of description.

The concept 'abstract' refers here to the degree of closeness to the original subject of investigation. Observation and thus the observational-experimental strategy directly target the subject of investigation and are therefore less abstract. In contrast, the conceptual-logical method uses concepts and logical rules to describe the subject of observation in a more indirect way, which is therefore more abstract.

How can methodological naturalism be distinguished from metaphysical naturalism? Metaphysical naturalism concerns a specific domain – the metaphysical domain – that is shifted from the logical world of philosophy to the natural world of science. Hence, a particular type of problem, question, or issue like metaphysical concerns about existence and reality, are now no longer exclusively associated with one discipline, e.g. philosophy, but rather with both philosophy and science. In short, metaphysical naturalism implies a shift in domains from one discipline to another.

Is this different in methodological naturalism? Here, the methods applied in different disciplines like philosophy and science are linked to each other. The conceptual-linguistic method used in philosophy is linked and connected to the observational-experimental method of science. This cross-disciplinary shift in methodological strategies allows each respective discipline to apply new methods to their original domain.

For instance, the metaphysical domain of the mind as investigated on purely conceptual-logical grounds can now also be tackled in an observational-experimental way. In other words, the mind becomes accessible to scientific investigation. This not only opens the door for the scientific investigation of the mind, but also its connection to the neuroscientific investigation of the brain. There is thus a direct methodological interface between mind and brain. This makes possible the linkage and merger between philosophy and neuroscience with the development of non-reductive and reductive forms of neurophilosophy as the scientific investigation of originally philosophical concepts (see Chapter 4 for details).

Naturalism 13

Methodological naturalism: replacement naturalism

Methodological naturalism assumes a continuum between the observational-experimental method of science and the conceptual-logical method of philosophy. What would this methodological continuum between philosophy and science look like? Based on Quine and others, two main suggestions have been put forward: replacement naturalism and cooperative naturalism.

Let us start with replacement naturalism. In the case of replacement naturalism, the method of philosophy – the conceptual-logical strategy – is completely replaced by the method of science – the observational-experimental strategy. The conceptual-logical strategy of philosophy is denied any right and validity in itself because it is seen as yielding nothing but speculation. It is therefore rejected and replaced by the observational-experimental method. This replacement of method is often accompanied by the replacement of domain. For example, the metaphysical and epistemological domains of philosophy are rejected altogether and replaced completely by the empirical domain of science.

Let's use the concept of mind as a specific example. The mind is originally a concept in philosophy where it has been investigated using the conceptual-logical method. Replacement naturalism now claims that we need to abandon the conceptual-logical investigation of the mind because it yields only wild and groundless speculations like the assumption of a mental substance. In its place, we need to investigate the mind in an observational-experimental way, as in psychology and cognitive neuroscience, for example (see Chapter 3).

The replacement of methodological strategies is often accompanied by the abandonment of the domains originally associated with the mind in philosophy (metaphysical and epistemological domains) in favor of the empirical domain of science. The mind is then no longer a subject of philosophy, but one of science with its empirical domain and its observational-experimental method.

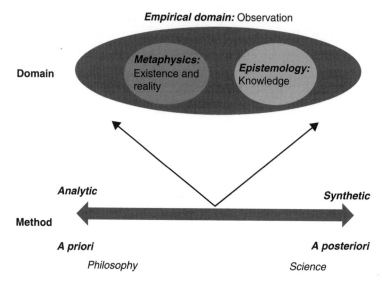

Figure 2.1 Philosophy and naturalism
Figure 2.1b Replacement (or incorporation) naturalism (Quine)

(b) This distinction is resolved in replacement (or incorporation) naturalism as advanced by Quine. He subsumes the metaphysical and epistemological domain into the empirical domain (upper part) while assuming a methodological continuum between analytic and a priori as well as synthetic and a posteriori approaches (lower part). This entails reductive neurophilosophy as distinguished from non-reductive neurophilosophy (see Chapter 4).

This makes it clear that replacement naturalism is quite a radical assumption. It basically leaves no room at all for the conceptual-logical method of philosophy, even within the natural world. Replacement naturalism is even more radical and also replaces the metaphysical and epistemological domains with the empirical domain. In other words, philosophy as a discipline is basically eliminated and replaced completely by science. This is the starting point for the development of especially reductive neurophilosophy which is based on such replacement naturalism (see Chapter 4).

Naturalism 14

Methodological naturalism: cooperative naturalism

What is the alternative to replacement naturalism? This is what is described as 'cooperative naturalism'. *Cooperative naturalism assumes cooperation and thus interaction between philosophical and scientific methodological strategies.* Unlike replacement naturalism, cooperative naturalism does not aim to completely replace the conceptual-logical method of philosophy by the observational-experimental method of science. Nor does it claim that there is exclusively one valid method, the observational-experimental.

Instead, cooperative naturalism considers both methods – conceptual-logical and observational-experimental – to be on a continuum. As claimed by Quine, this makes their cooperation possible. Let us discuss the example of the mind. In the case of replacement naturalism, we assumed the mind to be solely and exclusively

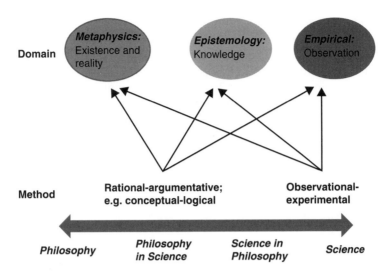

Figure 2.1 Philosophy and naturalism
Figure 2.1c Cooperative naturalism

(c) This is different in cooperative naturalism. Here the different methods are applied to all domains (see additional arrows in middle part) thereby constituting a methodological continuum between science in philosophy and philosophy in science (see lower part). This entails non-reductive neurophilosophy rather than reductive neurophilosophy (see Chapter 4).

investigated by the observational-experimental method of science. In this model, philosophy of mind is replaced by the sciences of mind – for example, psychology and, more recently, cognitive science and cognitive neuroscience.

This is obviously different in cooperative naturalism. In cooperative naturalism the door is opened to both the observational-experimental method of science and the observational-experimental approach of philosophy. In short, science of mind is complemented by philosophy of mind.

What remains unclear is exactly how both science of mind and philosophy of mind can cooperate with each other. More specifically, we need to devise a methodological strategy with rules and criteria to delineate how both methods – observational-experimental and conceptual-logical – can be combined and linked in a valid and reliable way. This is essential for any neurophilosophy that is truly trans- or cross-disciplinary and relies on cooperative naturalism rather than replacement naturalism (see Chapter 4 for more detailed discussion).

In addition to cross-disciplinary cooperation between philosophy of mind and science of mind, cooperative naturalism also makes another scenario possible. In the case of the mind, it opens the door from philosophy to science. However, cooperative naturalism may also raise the opposite scenario – opening the door from science to philosophy – in the case of the brain.

The concept of the brain is originally a neuroscientific concept that is associated with the observational-experimental method and empirical domain of science. The postulated continuum and cooperation between philosophical and scientific methods opens up the possibility of applying the conceptual-linguistic approach of philosophy to the investigation of the brain.

The conceptual-logical investigation of the brain makes it possible to raise metaphysical and epistemological questions about the brain that have thus far only been associated with the mind. This in turn opens the door to developing a 'philosophy of brain' that complements the 'philosophy of mind' (see Chapter 1), the 'science of mind', e.g. psychology (see Chapter 3), and the 'science of brain', e.g. neuroscience (see Chapter 3). Hence, unlike replacement naturalism, cooperative naturalism opens the doors in both directions, from philosophy to science in the case of the mind, (i.e. science of mind), as well as from science to philosophy in the case of the brain (i.e. philosophy of brain; see Chapters 8 and 12). This, as we will see in Chapter 4, opens the door for the development of non-reductive (rather than reductive) neurophilosophy which is based on such cooperative naturalism.

Critical reflection 4

Methodological naturalism and neurophilosophy

Why is the distinction between replacement and cooperative naturalism relevant in our search for the relationship between philosophy and the brain? Both forms of naturalism entail different concepts of neurophilosophy. In the case of replacement naturalism, neurophilosophy can be characterized by the predominance of the observational-experimental method of science over the conceptual-logical approach of philosophy. The relationship between neuroscience and philosophy is here tilted unilaterally toward neuroscience, allowing only for unilateral exchange from philosophy to science.

This is different in the case of cooperative naturalism. Here, both methods – conceptual-logical and observational-experimental – stand side-by-side with their exact linkages and combinations to be developed. The relationship between philosophy and science is consequently considered more bilaterally, making exchange possible in both directions, from philosophy to science and from science to philosophy. This has major ramifications for the concept of neurophilosophy: neurophilosophy can then be based on the inclusion of several methodological strategies, that is, methodological pluralism, which implies non-reductive (rather than reductive) neurophilosophy (see Chapter 4 for details).

Such methodological pluralism in non-reductive neurophilosophy remains impossible, however, once one presupposes replacement rather than cooperative naturalism. In that case, any methodological strategy other than the observational-experimental method is replaced and thus eliminated. This means that methodological pluralism is here replaced by methodological monism. In turn, this entails a reductive neurophilosophy rather than non-reductive neurophilosophy (see Chapter 4 for details).

Take home message

How are philosophy and science related to each other? Philosophy can be characterized as a discipline that asks about the fundamental nature of being and the world and how we can know about it. As such, it focuses on metaphysical issues concerning reality and existence and epistemological questions concerning knowledge of the world. In order to investigate metaphysical and epistemological domains (the fields of their respective problems), philosophy applies a specific methodological strategy that is based on conceptual-logical analysis. This distinguishes philosophy from the sciences that focus on the empirical domain (the observable world) rather than the metaphysical and epistemological domain. The empirical domain is investigated using the scientific method which employs an observational-experimental strategy. Taken together, the difference in both domain and method lead to the segregation and principal distinction between philosophy and science.

This principal distinction between philosophy and science was called into question during the twentieth century with the rise of the sciences. This concerned both methodology and thematic fields. The American philosopher, Quine was instrumental in demonstrating that the philosophical methodology of conceptual-logical investigation is just an abstraction of the observational-experimental approach in science. This opened the door for linking philosophy and science and also cast a doubt on the mutually exclusive distinction of their thematic fields. On the basis of such a methodological continuity, concepts like that of the mind that were originally associated with philosophy could now be investigated in science using the observational-experimental approach. This opened the door to addressing the mind and the mind–brain problem in a scientific, rather than philosophical, way. In turn, this provided the basis for developing the concept of neurophilosophy in general as well as different approaches, reductive and non-reductive, as will be discussed in Chapter 4.

Summary

The previous chapter focused on the mind as it was discussed during the twentieth century in both philosophy of mind and phenomenology. While that chapter remained more or less within the field of philosophy itself, the current chapter focuses on the relationship between philosophy and science. The American philosopher, Willard Quine, was instrumental in linking the domains and methods of philosophy to the sciences. By questioning the validity of both the analytic–synthetic distinction and the a priori–a posteriori distinction, he called into question the principal methodological difference between philosophy and science. In doing so, Quine considered philosophy and science to be on a continuum within the natural world. This means that philosophy became naturalized in both its domains (metaphysical and epistemological naturalism), and its method (methodological naturalism). This is important to consider in the context of the mind and how it relates to the brain, especially when the mind and the brain have been associated with philosophy and neuroscience, respectively. Thus any neuroscientific investigation of the mind presupposes the possibility of both metaphysical and methodological naturalism for which Quine provided the most general foundation. However, both metaphysical and methodological naturalism have been considered in different varieties: they can be more radical, as is the case in replacement naturalism where the philosophical domains and method are eliminated completely in favour of science's domain and method – or they can be more moderate, as in cooperative naturalism, where philosophy and science methodologically and practically interact and support one another. Both forms of naturalism – replacement naturalism and cooperative naturalism – are important to consider because they entail different concepts of neurophilosophy, reductive and non-reductive, with different relationships between philosophy and neuroscience.

Revision notes

- Describe the developments at the beginning of the twentieth century and the changes in the relationship between philosophy and science.
- How can you distinguish between philosophy and science?
- What are domains in general and those of philosophy and science in particular? What are their respective methods?
- Who was Willard Quine and what did he do?
- Explain the distinction between analytic and synthetic sentences and provide some examples.
- How did Quine undermine the analytic–synthetic distinction?
- What do a priori and a posteriori mean?
- How did Quine undermine this distinction, and why is that relevant for philosophy?
- How did Quine view the relationship between philosophy and science?
- What is naturalized epistemology? What are observation sentences?
- How can you explain the distinction between methodological and metaphysical naturalism?
- Explain the different forms of metaphysical naturalism.
- What is the difference between replacement and cooperative naturalism?
- What is cooperative naturalism, and how can one link philosophy and science without eliminating either domain or method?
- Why is the distinction between replacement and cooperative naturalism relevant for neurophilosophy?

Suggested further reading

- de Caro, M. and Macarthur, D. (eds) (2004) *Naturalism in Question* (Cambridge, MA: Harvard University Press).
- de Caro, M. and Macarthur, D. (eds) (2010) *Naturalism and Normativity* (New York: Columbia University Press).
- Northoff, G. (2011) *Practice of Neuropsychoanalysis* (Oxford/New York: Oxford University Press).
- Quine, W. v.O. (1951) 'Two dogmas of empiricism', *The Philosophical Review*, 60, 20–43. *This is the seminal and classic paper in which Quine rejects the dichotomy between analytic and synthetic sentences as well as the distinction between a priori and a posteriori knowledge.*
- Quine, W. v.O. (1969) 'Epistemology naturalized', in W. Quine, *Ontological Relativity and Other Essays* (New York: Columbia University Press). *This is a collection of essays by Quine that contain his main views on the relationship between philosophy and science as well as how he considers epistemology and ontology including his classic paper 'Epistemology naturalized'.*

3
Mind, Brain and Science: Psychology and Neuroscience

Overview

The previous chapters highlighted developments in philosophy during the twentieth century. Chapter 1 focused on how the mind was thematized and discussed, especially in philosophy of mind and phenomenology. In Chapter 2, this was complemented by considering methodological developments and how philosophy and its method were reinterpreted as a specific subset of the scientific method. This discussion opened the door to naturalism, which claims that both philosophy and science operate on a continuum within the natural world, or the world as we can observe it. While philosophy opened itself to the sciences, psychology and neuroscience reached out toward the mind. The development of neuroscience in the nineteenth and twentieth centuries and its more recent extension from the brain toward the mind are the focus of this chapter.

Objectives

- Understand the difference between philosophy and psychology, and how the latter was emancipated from the former
- Explain cognitive psychology, what it is about, and why it is relevant for the mind
- Understand the origins of neuroscience
- Explain the developments of neuroscience at the beginning of the twentieth century
- Understand functional brain imaging
- Follow the developments of neuroscience in the twentieth century and its heterogeneous origin in different disciplines
- Number and recall the disciplines to which neuroscience recently extended

Key concepts

Neuroscience, psychology, mind, brain, functional brain imaging, neuron doctrine, clinical neuroscience, critical neuroscience.

Background

From philosophy of mind to science of mind

Chapter 1 focused on the development of philosophy during the twentieth century and in particular on the mind. Phenomenology was a branch of philosophy developed at the beginning of the twentieth century, mainly in continental Europe.

The main target of phenomenology is consciousness, and its aim is to investigate the structure of experience in a first-person-based way. As detailed in Chapter 1, the focus here is on the structure and organization of consciousness, how and in which ways we experience the various contents in first-person perspective. This revealed subjectivity (subjective experience in first-person perspective) and intentionality (directedness of mental states toward contents) as the phenomenal hallmark features of consciousness. This presupposes what can be described as the phenomenal domain.

In addition to phenomenology, philosophy of mind as it developed during the second half of the twentieth century was another branch in philosophy that saw the mind as its focal point. Based on the linguistic tradition of the Anglo-American world, philosophy of mind targets the nature and features of the mind. The nature of the mind concerns its existence and reality, and how it is related to the nature and reality of the body and the brain. How are mind and body, and their respective existences and realities, related to each other? This is the question for the mind–brain problem as the core issue in philosophy of mind. In addition to the mind–brain problem, philosophy of mind also focused on the various mental features such as consciousness, self and intersubjectivity.

Both phenomenology and philosophy of mind pay tribute to the fact that the mind was (and still is) a central focus in philosophy. This focus on the mind was accompanied by parallel developments during the twentieth century concerning the relationship between philosophy and science. This was discussed in Chapter 2. Here, relying on Quine, we showed that the traditional philosophical method – the **rational-argumentative (or conceptual-linguistic)** strategy – can be considered a subset of the scientific method which employs an observational-experimental strategy.

Following Quine's lead, philosophy would in this case no longer be considered as principally different from science and its focus on the natural world. In other words, the traditional domains of philosophy – the metaphysical and the epistemological domains – can be naturalized. To be naturalized means to become linked to the natural world as we can observe it. This in turn opens up the possibility for investigating metaphysical and epistemological (and ethical) domains using the scientific method.

In sum, developments in twentieth-century philosophy allowed for the opening of the metaphysical and epistemological domains and their strategies to the sciences, the empirical domain, and the observational-experimental method. This allowed researchers to tackle the question of the mind and its mental features in a scientific way and paved the way for the development of psychology as a science of the mind. The present chapter focuses therefore on two issues: (i) the shift of the mind and its mental features from the metaphysical and epistemological domains of philosophy to the empirical domain of science; and (ii) the application of the scientific method to the mind in psychology and later neuroscience.

Abbreviated history of psychology 1

Psychology as science of mind

Wilhelm Wundt (1832–1920) was a German medical doctor, physiologist and philosopher who was particularly interested in consciousness. One of his major works, *The Principles of Physiological Psychology* (1874), discussed various mental features of consciousness like feelings, emotions, volitions and ideas. Wundt sought to investigate these mental features using observation and experiments that he devised. He claimed that the investigation of the mind needed to be liberated from the metaphysical speculations of philosophy and replaced by an observational-experimental approach to the mind. Hence, Wundt is considered one of the founding fathers of psychology as science of the mind.

The introduction of the observational-experimental method in the investigation of the mind let Wundt show that the phenomena of consciousness are just the tip of the iceberg. Beneath the visible iceberg, many processes go on which can be described as unconscious. What we experience and introspect as consciousness must be a composite product of the unconscious psyche and its unconscious processes. The higher-order cognitive functions like consciousness so much emphasized by philosophy were thereby undermined and put into an unconscious context of mere physiological processes.

Wundt is generally considered one the main fathers of modern scientific psychology, which makes it clear that psychology is a rather young discipline, especially when compared to philosophy.

How can we describe the relationship between philosophy and psychology? First and foremost both focus on the mind; they thus share the thematic field. However, both psychology and philosophy approach the mind in different ways, and using different methodological strategies. While philosophy relies on conceptual-linguistic methods to explore and understand the mind, psychology investigates the mind in an observational-experimental and thus scientific way. Hence, while they overlap in their thematic domain (the mind), philosophy and psychology diverge in their respective methodological strategies.

Abbreviated history of psychology 2

Psychology and consciousness

Another pivotal figure in the history of psychology was the American philosopher and psychologist, William James (1842–1910). He was born in New York City into an independently wealthy family with strong connections to Europe. His brother was the famous novelist Henry James. Rather than devoting his life to writing novels as his brother did, William James wrote extensively on philosophy and psychology, which he taught at Harvard University in Cambridge, Massachusetts.

William James published a two-volume work, *Principles of Psychology* (1890). This work contributed greatly to laying the foundation for the scientific study of the mind as well as the linkage of psychology to the scientific method. In his book, James discusses several functions such as consciousness, emotions and attention, and describes how they can be approached and investigated experimentally.

James spoke of what he described as the 'stream of consciousness', comparing it to a continuous flow of water in a river. How can we characterize the 'stream of consciousness'? James distinguished between substantive and transitive parts in the stream of consciousness.

Substantive parts reflect the different contents of consciousness like the book, the laptop, the table, the chair, etc. you experience while reading this book. While acknowledging their differences, your nevertheless experience these different contents without transition. According to James, this is because of the transitive parts in consciousness. Together, substantive and transitive parts form the 'stream of consciousness'.

How can we further illustrate this? Imagine boats traveling up and down a river. The boats represent the substantive parts of consciousness: the contents. The water of the river, which makes it possible for them to travel smoothly up and down the river, corresponds to the transitive parts of consciousness.

How can we describe the 'transitive parts' in further detail? James points out several features, of which two – 'sensible continuity' and 'continuous changes' – are particularly relevant. Let us start with 'sensible continuity'. 'Sensible continuity' means that no phenomenal state vanishes or perishes instantaneously and immediately. Instead, there are continuous transitions between different states which slide into each other: there is a transition from moment-to-moment allowing for smooth transition.

Why are these different features of consciousness relevant in the present context of linking the mind and the brain? If we want to link consciousness and its phenomenal features to the brain, we need to account for how the stream of consciousness – including its substantive and transitive parts – is constituted on the basis of specific neuronal mechanisms. While this will be discussed in more detail in Part IV, it should be noted that James (1890, Vol. I, pp. 248, 82) assumed that the transitive parts of consciousness linked such transitions to the brain when he speaks of a 'summation of stimuli in the same nerve tract'.

Abbreviated history of psychology 3

From psychology over cognitive science to cognitive neuroscience

After its initial beginnings, psychology developed rapidly into different fields and sub-branches. With regard to the mind, behaviorism was the dominant position in the first half of the twentieth century. In essence, behaviorism argues that mental states are nothing but behavioral states; there is no more to the mind and its mental features than the behavior and its external states we can observe (see Parts II and III for more details). We do not need to assume special internal states – mental or cognitive states – to explain our mental states.

However, behaviorism, with its reduction of the mind to observable behavior, was challenged by the development of cognitive psychology and cognitive science during the second half of the twentieth century. More and more, psychology targeted cognitive functions like attention, memory and executive functions. Attention describes the focus on a particular content as, for instance, if I were to focus on the black color of my laptop while its red frame remained in the background of my consciousness.

Memory concerns the encoding, storage and retrieval of events, objects and people from the past. Subsequent psychological research led to the differentiation between different kinds of memory, like short-term memory (within the range of seconds as described by working memory) and long-term memory (within the range of years and decades as, for instance, autobiographical memory). Finally, executive functions concern the development and realization of goals and goal-directed behavior. Cognitive psychology focuses on exploring the psychological mechanisms underlying the various cognitive functions in an observational-experimental way.

While focusing on various cognitive functions, cognitive psychology converged with other disciplines like computer science and artificial intelligence that modeled the cognitive functions like memory on computer. Different kinds of computer models like the digital-symbolic model versus the connectionist model were discussed to simulate cognitive functions like memory. Another discipline linking to cognitive psychology is linguistics, that aimed more and more to reveal the psychological mechanisms underlying language acquisition.

The different disciplines like psychology, artificial intelligence, computer science and linguistics (and others like physics and biology) converged and were linked under the umbrella of cognitive science. *Cognitive science can be described as an umbrella discipline that includes different single disciplines like computer science, artificial intelligence, etc. all being unified in their aim to explain and understand cognitive functions.*

The most recent inclusion into cognitive science is neuroscience. While initially focusing mainly on sensorimotor functions (see below), neuroscience has now extended its reach and investigated the neuronal mechanisms underlying cognitive functions. Cognitive functions include, for instance, attention, working memory, episodic memory and executive functions; these are the subject of what is described as 'cognitive neuroscience' (Gazzaniga, 2008).

For example, many brain-imaging studies have recently been conducted to search for the regions of the brain that are recruited by the various cognitive functions. This led to the neuronal distinction between working memory and autobiographical memory. Working memory – as short-term memory – is mainly associated with the lateral prefrontal cortex. Autobiographical memory – as long-term memory – is more related to neural activity in the medial temporal lobe, also known as the hippocampus.

Abbreviated history of neuroscience 1

Neurons as basic functional unit of the brain

In the days of ancient Egypt and Greece (1000–300 BC), the heart was considered the seat of intelligence and consciousness. The Greek physician Hippocrates (460–377 BC) was convinced, however, that thoughts, feelings and perceptions were related to activity in the brain, rather than the heart. He observed that brain damage was associated with paralysis, the loss of movement, or dementia, the loss of memory. For Hippocrates, this implied that movements and speech must have their originating causes in the brain.

The famous Greek philosopher, Aristotle, a student of the philosopher Plato, discarded Hippocrates' lessons about the brain. Unlike Hippocrates, Aristotle shifted intelligence and consciousness from the brain back to the heart, which he considered the seat of consciousness. Aristotle believed that the brain serves only to cool the blood.

This view of the heart as the seat of intelligence and consciousness was the predominant view in ancient Egypt and Greece. That, however, changed with Galen (AD 130–200), a physician of the Roman Empire, who again made the case for the brain. He observed that some of his patients showed severe changes in their mental states and consciousness when they suffered brain damage. Galen, and much later Vesalius (1514–64), thought that the ventricles of the brain and their fluids were the seat of higher-order cognitive functions like perception, thought and consciousness. It was assumed that the surrounding tissue – the grey and white matter (see below) – served only a supportive role. Because of their association with the ventricles and the brain's fluids, intelligence and consciousness were also considered fluid and watery in a literal sense.

Real progress started later. At the end of the nineteenth century, many new technical inventions were made. This was the time when the science of the brain, that is neuroscience, underwent major developments. Among the many technical inventions was the microscope. This allowed the Italian scientist, Camillo Golgi, (1843–1926) to visualize cells in the body and to distinguish them from their surroundings.

Golgi's coloring technique – using silver chromate salt that tinted the cells, allowing one to visualize them – was employed by the Spanish scientist, Santiago Ramón y Cajal (1852–1934). Ramón y Cajal formulated the 'neuron doctrine', which hypothesizes that the neurons, as the cells of the brain, provide the relevant functional unit of the brain and its functions. This shifted the former focus on the fluid in the ventricles to the cells (neurons) of the brain and their location in the so-called grey matter.

Let's go back to the beginnings of neuroscience. Neurons were discovered to show electrical excitability called action potentials. More specifically, neurons displayed specific electrical states that changed and could be predictably impacted by the electrical activity of the adjacent neurons. It became clear that neurons exhibit electrical activity that can be induced by, for example, sensorimotor and cognitive stimuli or tasks.

This led to the development of cell recording, which investigates and records the electrical activity of neurons. The longstanding and predominant methodological approach in neuroscience, cell recording, was recently complemented by the development of brain imaging, which allows for the visualization of neural activity on a more macroscopic level (see below).

Neurons account for what is called the grey matter of the brain. As mentioned above, neurons reflect the single cells of the brain that show electrical activity with spikes during particular sensorimotor or cognitive stimuli or tasks. However, recent discoveries show that there seems to be more to the brain than just neurons. Neurons are surrounded by cells called glial cells, the functions of which were unclear until recently. Glial cells seem to be involved in the metabolic supply of neurons, as well as in cognitive functions. Further study is required to understand their function in the brain fully.

Finally, where there is grey matter there is also white matter. The white matter of the brain is the largest portion of the brain and consists of various fibers and tracts that connect the different regions and their respective white matter.

Abbreviated history of neuroscience 2

Clinical disorders and the brain

Another line of research at the turn of the nineteenth century was the investigation of brain-damaged patients. Paul Broca (1824–80) showed that lesions in a particular brain region, the left lateral frontal cortex, lead to impairment in speech and language. Other scientists, including the American neurologist, Hughlings Jackson (1835–1911), demonstrated the regional localization of specific epileptic seizures in particular regions of the brain such as the temporal lobe.

Other neurological disorders like Parkinson's disease, and psychiatric disorders like depression and schizophrenia, were also associated with the brain, as argued for by German psychiatrists like Carl Wernicke (1848–1905), Emil Kraepelin (1856–1926), Eugen Bleuer (1857–1939) and Karl Kleist (1879–1960). All this led to furthering support for the assumption that the brain is crucially involved in higher-order cognitive functions like consciousness, language, thinking, emotions, etc.

Another approach was taken by the American-born neurosurgeon Wilder Penfield (1891–1976) who mainly worked in Montreal at McGill University, where he established the Montreal Neurological Institute. He operated on patients' brains if they suffered from epilepsy. Before removing the supposedly pathological cells that caused the seizures, he stimulated the cells and the surrounding regions in order to observe the behavioral and subjective-experiential effects of the various regions (while the patients were fully awake, being only in local anesthesia). By observing his patients' behavior and their consciousness, he could infer which region's neural activity is in charge of what function, and how the respective regions are related to consciousness and its various contents.

In sum, neuroscience considered clinical disorders in neurology and psychiatry from early on in its history. By showing either lesions in the brain as neurological disorders or dysfunction in mental features like psychiatric disorders, both neurologic and psychiatric disorders provide useful information. This concerns the association of certain regions with particular functions like sensorimotor or cognitive functions. At the same time, it can also reveal some insight into how the phenomenal features of consciousness are related to the functions of particular regions in the brain.

More specifically, these clinical disorders provide useful starting points for generating hypotheses about the linkage between specific neuronal mechanisms and particular phenomenal and mental features. This provides an ideal starting point for subsequent experimental testing of the supposed neuronal mechanisms in healthy subjects. This is a strategy that is most prevalent in current neuroscientific research on consciousness. It relies strongly on insights and findings from patients with disorders of consciousness (i.e. anesthesia, vegetative state, sleep) (see Part IV for details).

Abbreviated history of neuroscience 3

From clinical disorders of the brain to consciousness

Another development occurred mainly in pre-Second World War Germany. Neuroscientists and neurologists like Erwin Strauss (1891–1975), Kurt Goldstein (1878–1968) and Hartmut Kuhlenbeck (1897–1984), all of whom later emigrated to the United States, were very interested in how the brain functions and operates in relation to consciousness. In order to study this, they investigated patients with lesions in the brain that they had acquired during the First World War. These early neuroscientists investigated these patients in different ways. First, they localized the lesion in their brain, including which regions were affected and which regions were not. Second, they observed the patients' behavioral deficits and the changes in their sensorimotor and cognitive functions. This allowed these scientists to link deficits to regional lesions in their patients' brains. This in turn provided some indirect information about the function of the region.

Finally, these neurologists also asked their patients about their experiences and thus how the brain lesion had changed the way they experienced themselves, their body and the world. Phenomenal analysis as the method of phenomenology, as explained in Chapter 1, was thus applied to the clinical-pathological realm of consciousness in brain-lesioned patients. This allowed these scientists to link the phenomenal features of consciousness indirectly with the neuronal features of the brain.

Let's be more specific. Based on phenomenological philosophy (the study of the structure of experience – consciousness – in a first-person-based, phenomenal way; see Chapter 1), these scientists explored the subjective experiences of these patients. This let them conclude that these patients did not suffer merely from deficits in their sensorimotor functions. The patients had to report how they subjectively experienced their sensorimotor deficits. The neurologists were interested in learning about how the patients experienced in first-person perspective what, in third-person perspective, is described as sensorimotor or cognitive deficits. They thus combined first- and third-person-based accounts and therefore phenomenal and psychological analysis.

What did the patients experience in first-person perspective? The different brain lesions led to different subjective experiences of the patients in relation to both their body and environment. What does this tell us about consciousness in general? First and foremost it tells us that consciousness is related to the brain. While being most evident now, it was a major insight in the first half of the twentieth century.

Critical reflection 1

Sensorimotor functions and consciousness

What can we learn from these pioneers? These early investigations tell us that consciousness is about the subject's relation to body and environment. Consciousness is always integrated into both the body and the environment; it is thus embodied and embedded, implying what is called 'embodiment' and 'embeddedness' today (see Chapter 1, as well as Chapters 12 and 14).

What does this tell us about the brain and its relationship to consciousness in general? Scientifically, the assumption of a basic deficit in relation to the body and environment gives us some clue that the brain might be central in constituting such a relationship, e.g. embodiment and embeddedness. Otherwise the brain lesion could not have led to deficits in these patients. This means that the brain itself must be characterized by some kinds of neuronal mechanisms that make its close relationship to the body and brain possible as we experience it in consciousness. What are these neuronal mechanisms that integrate the brain into the body and the environment? We currently do not know (see Parts III and IV for more detailed discussion).

We can ask the same question in a different way. One can suggest that the neuronal mechanisms underlying the sensorimotor functions themselves account for the subjective experience and thus consciousness. In this case, consciousness should come with the sensorimotor functions. This is what the recently developed field, neurophenomenology, suggests.

What is neurophenomenology? *Neurophenomenology describes a predominantly empirical research strategy that aims to link first-person-based accounts of subjective experience, including its phenomenal features, to the brain's neuronal mechanisms as observed in third-person perspective (see Chapter 4 for more details).* The focus here is especially on the neuronal mechanisms of sensorimotor function that are supposed to account for the subjective experience in first-person perspective; that is, consciousness.

Alternatively, one may assume that different neuronal mechanisms underlie the sensorimotor functions and the phenomenal features of consciousness. Both sensorimotor and phenomenal functions may then be subserved by different underlying neuronal mechanisms. Between these two options – phenomenal features either coming with sensorimotor function or being distinct from them – we currently do not know which is correct (see Chapters 12 and 14 for more details).

Abbreviated history of neuroscience 4

Association of brain science with different disciplines

We read that the brain did not acquire a central role in scientific investigation before the end of the nineteenth century. Though we take it for granted now, there was no neuroscientific discipline in those days. Instead, there were other disciplines. One central discipline was anatomy. Anatomy is concerned with the structure of the body, its parts, its various organs, and how these parts are structured and organized in their microscopic and macroscopic appearance. Anatomists who studied the brain and its different macroscopic structures (regions and their connecting tracts) were called neuroanatomists.

The neuroanatomists divided the brain into different regions and parts. Besides right and left hemispheres, they distinguished regions on the outer surface, the so-called cortex, from the ones in the interior lying beneath the cortex, the subcortical regions. The cortex was divided into different regions including the frontal cortex (at the front of the brain), the parietal cortex and the occipital cortex (at the back of the brain). Subcortical regions include different structures, parts and nuclei including the striatum, the globus pallidum, the thalamus, the locus coerulus (LC), the raphe nuclei (RN) and the ventral tegmental (VT) area. The latter three, LC, RN and VT, were later also associated with different neurochemicals like adrenaline/noradrenaline, serotonin and dopamine, respectively.

One notable neuroanatomist at the time was Sigmund Freud (1895–1939), the founder of psychoanalysis, who aimed to investigate the psychological mechanisms of the unconscious. It is often forgotten that Freud started his career in an anatomical laboratory and investigated nerve cells using a microscope. However, his interest in complex psychological phenomena was stronger than his fascination for single nerve cells. Freud abandoned the anatomy of the brain and replaced it by the structure of the psyche as dealt with in psychology, and more specifically, psychoanalysis.

Interestingly, there is a strong movement today to establish links between Freud's psychoanalysis and the neural mechanisms of the brain. This movement is called neuropsychoanalysis. More and more it is developing into a discipline of its own

78 *Mind and Brain*

(see Kaplan-Solms and Solms, 2000; Solms and Turnbull, 2003; Northoff, 2011). Let us go back to Freud's time, though, for now.

Abbreviated history of neuroscience 5

Institutionalization of neuroscience as discipline

At the turn of the nineteenth century, the discipline of neuroscience did not yet exist. The term neuroscience and the establishment of a separate discipline was coined later (see below). Researchers originating from different disciplines extended their search from other organs of the body to the brain. For example, the early anatomists investigated not only the anatomy of organs like the stomach and the pancreas, but also of the brain. Early physiologists like Helmholtz (1821–94), Mueller (1801–58) and Bruecke (1819–92) focused on the brain. Researchers from other disciplines like pharmacology also focused on the brain.

Unlike today, where the brain and its corresponding discipline, neuroscience, are a major focus at almost any university, the brain was a new and largely unexplored topic of research in the nineteenth century. Despite this lack of institutionalization, research on the brain was taking place in various disciplines ranging from anatomy, physiology, pharmacology to the more clinical fields of neurology and psychiatry.

What does this institutional overview tell us? It tells us that scientists investigating the brain were spread across different disciplines as heterogeneous as anatomy, pharmacology, physiology, neurology and psychiatry. Despite the fact that neuroscience had not yet been invented as a discipline, there were still various scientists with different backgrounds and methodological strategies who shared a common interest in the brain and how it functions.

This is important when considering current neuroscience. Based on its origins, neuroscience is a mixture of different fields (domains) and methods of investigation. One can say that neuroscience is metaphorically a 'mélange de tout' ('mixture of everything'): a mixture of different questions and methods unified by a common interest in the brain. This 'mélange de tout' implies that different concepts with different backgrounds, all stemming from the different disciplines, converge in current neuroscience. This is of utmost importance, especially when one wants to merge neuroscience and philosophy. Such a merger is confronted with a conceptual heterogeneity that has to be integrated in a systematic way to allow close ties between philosophical and neuroscientific concepts.

Neuroscience 1

Biochemistry of neural activity

The brain and its neural activity are not only determined by the neurons and their electrical activity; they are also subject to modulation by biochemical substances. These discoveries were initiated by observations in neurological and psychiatric disorders around the years 1950 to 1960.

Certain drugs, such as those that increase the biochemical dopamine, were found to be therapeutically effective in treating Parkinson's disease, a disorder characterized by tremors and mobility disability. Drugs that decrease the level of dopamine, such as dopamine antagonists, were therapeutically successful in treating psychiatric disorders including schizophrenia. Schizophrenia is characterized by abnormal hallucinations (mainly auditory), delusions and severe thought disorders.

Other drugs tested positively in managing the effects of depression. These drugs targeted specific substances in the brain, including mono-amines like adrenaline, noradrenaline and epinephrine/norepinephrine. Drugs such as 'Prozac', that were developed later, modulated the biochemical serotonin and were also found to be beneficial. Most recently, a drug that lowers the biochemical glutamate (ketamine), was observed to show immediate therapeutic effects in treating severe depression.

These clinical-therapeutic observations paved the way for much research into biochemicals such as dopamine, adrenaline, glutamate, serotonin (and others), and how they impact neural activity and behavioral and mental functions in animals and the healthy brain. Dopamine is a biochemical substance that seems to mediate in particular reward in subcortical regions like the ventral striatum. Adrenaline is another substance that is present strongly in a subcortical region, the locus coerulus, and mediates in particular somatic and vegetative functions.

Serotonin is a subcortical substance associated with another region, the raphe nucleus, and seems to be of particular importance in mediating emotions and their alterations, as in depression. Glutamate can be found throughout the whole brain and generates neural excitation in the brain. Its counterpart is *gamma-aminobutyric acid* (GABA) that, more or less equally ubiquitous in the brain, generates neural inhibition, the inhibition of neural activity. The investigation of these substances, including their psychological and behavioral effects, led to the development of psychopharmacology and behavioral pharmacology as major fields within neuroscience.

In sum, neural activity in the brain is modulated by various biochemical substances such as dopamine, adrenaline, serotonin, GABA and glutamate. The detection of these substances, and their abnormal changes in neuropsychiatric disorders, has led scientists to investigate their impact on neural activity, including their molecular and genetic mechanisms.

Neuroscience 2

Genes and neural activity

The study of biochemical mechanisms has been complemented by insightful discoveries about molecular mechanisms. For example, specific proteins and their genetic regulation have been investigated, and this has led to the development of neurogenetics and neuroproteonomics. These novel fields investigate how genes (reflecting DNA, deoxyribonucleic acid) and the resulting proteins control neuronal activity.

For example, the impact of certain genes that modulate the expression of serotonin has been investigated at both biochemical and behavioral levels, as well as in neuropsychiatric disorders, such as depression, where serotonin seems to play a central role. The expression of substances like serotonin is controlled by certain genes. Each of these genes may be present in different variants, which would entail slightly different expression of substances like serotonin.

These so-called polymorphisms may be risk factors for certain disorders like depression if certain variants predominate over others. Recent investigations have demonstrated that patients with depression have a higher incidence of a certain polymorphism coding for a specific serotonin transporter (promoter polymorphism (5-HTTLPR) of the serotonin transporter gene (SLC6A4)) when compared to non-depressed subjects.

This particular polymorphism also impacts neural activity. Imaging studies have shown that healthy subjects with the respective polymorphism – the S-allele – show increased neural activity in the amygdala, a region specifically involved in processing emotions. This is not the case in those subjects carrying the L-allele. This is even more interesting when considering that subjects with severe depression show increased neural activity in the amygdala during emotional tasks. It could be that the genes (and more specifically, particular polymorphisms) put subjects at risk for increased amygdala activity and consequently for depression.

There is, however, more to the genome than polymorphisms. Initially it was thought that the genes directly control the expression of proteins (that is, the mRNA – messenger ribonucleic acid) that regulate specific substances like serotonin. Today we know better. Besides the polymorphism, it is clear that there are multiple copies of one and the same gene, the so-called copy number variants (CNV). However, as in real life, the copying process may not always go smoothly, meaning that there may be some defects hidden in the various copies of one and the same gene; the different defects are called deletions, insertions and duplications. This may be particularly relevant in psychiatric disorders like schizophrenia and depression. What exactly happens in these cases we currently do not know. It is clear, however, that the genome is far from being understood at this point in time, and that it may hide some secrets and surprises.

Most interestingly, these copying processes may be altered and disrupted by severe life events. Particular stressful (like bombing during war) and early traumatic life events (like maltreatment or sexual abuse) may leave their traces in the genome in yet unclear ways. This may, for instance, be manifest in an abnormally high number of defects in the earlier described CNV. Exact mechanisms are, however, unclear.

What is clear, however, is that the genome is tightly interwoven with the environment such that the latter can exert a direct impact on the former. The experts speak therefore of gene x environment interaction. We will see later that such a close and tight relationship holds true also for the brain itself. Here, one may want to speak of brain–environment interaction (see Northoff, 2012a, 2012b, 2012c). This, as we will see, will be of particular importance in understanding how the brain can contribute to yielding consciousness (see Part IV).

This progress, while still ongoing, has resulted in the development of various subdisciplines in neuroscience including behavioral pharmacology, neuropharmacology, psychopharmacology, neurogenetics, neuroproteonomics and neuroimaging. These studies have vastly increased our understanding of the biochemical, molecular and genetic mechanisms of neural activity and associated behavioral and cognitive functions.

Biography 1

Donald Hebb and synapses

Let us now turn from the biochemical, molecular and genetic modulation of neural activity to the cognitive functions themselves and their underlying neural mechanisms.

For this I turn to the Canadian psychologist Donald Hebb (1904–85). Donald Hebb was born at the Atlantic coast in Nova Scotia, Canada, and later became a teacher at a high-school. After that he worked as a farmer and traveled around Canada before going to Chicago and later Harvard in the USA to study with the prominent neuroscientist and psychologist, Karl Lashley. After he completed his PhD thesis, he joined McGill University in Montreal, Canada, where he taught and researched in psychology.

Hebb asserted in his influential book, *The Organization of Behavior* (Hebb, 1949), that higher cognitive functions like learning and memory are associated with the brain. More specifically, higher cognitive functions are associated with the transmission of information between different neurons, across the gaps between their cell bodies called the synapses. *Synapses describe the links between the different neurons; these links are the locus of much electrical and chemical activity as we will see in the following.*

Hebb observed that the activity of one neuron may impact the activity of another via their common synaptic connection, called 'Hebb synapses'. In short, the more correlated the activity of two neurons, the higher their synaptic strength and the higher the degree of transmitted information, to quote Hebb himself, 'Neurons that fire together, wire together'. If, in contrast, the two neurons' activity is uncorrelated, the strength of their synaptic connection decreases, resulting in less information being transmitted.

Neuroscience 3

Neurons and cognitive functions

Most importantly, Hebb associated such modulation of the synaptic connection's strength with learning and memory. By modulating the synaptic strength, the brain can encode and store novel information. This sets the stage for both memory and learning. More generally, this allowed Hebb to relate higher-order cognitive functions like learning and memory to particular neuronal mechanisms and, more specifically, to changes in a single neuron's electrical (also biochemical; see below) activity. This led to numerous studies using cell recording to investigate the electrical activity of neurons and their connections during cognitive functions like learning and memory (and later others like attention).

Hebb drew a direct link between the microscopic level of single neurons and cognitive functions. This left open for discussion the impact of the level between single neurons and cognitive functions – the regional level of the brain's neural activity. The regional level concerns the neural activity of the different regions and how they interact with each other by, for instance, connecting and forming neural networks. Particularly the latter, where the connections between the different regions and the formation of neural networks has most recently been subsumed under the concept of 'connectome'.

At the time of Donald Hebb, the only way to investigate the association of particular regions with cognitive functions was through neurologic and psychiatric disorders. The cognitive dysfunctions in these patients were associated with lesions in particular regions (and networks) in their brains; this allowed researchers to infer what kinds of function that particular region generates in the healthy brain. This is called the 'lesion-based method'. Since here psychological functions are directly related to the brain, the corresponding disciplines were called neuropsychology, neuropsychiatry, cognitive neurology and behavioral neurology.

Biography 2

Hans Berger and the EEG

While providing much insight, the lesion-based method only allowed for indirect inference of the association of a particular region with a specific cognitive function. More direct evidence was, however, warranted that allowed researchers to directly measure and visualize how the neural activity in a particular region or network is related to the cognitive function in question. Technical innovation proved pivotal for neuroscience. In the 1980s the technical invention of brain imaging paved the way for accessing cognitive functions and their relation to the brain on the macroscopic level of the brain's different regions.

Hans Berger (1873–1941) was a German researcher who was the first to record electrical activity from the skull of human subjects. After recording the brain's electrical activity for the first time in 1924, it took him five years to write everything down and to finally publish a paper about it in 1929. Recognition took even longer. As is so often the case with new developments, Berger's electroencephalography (EEG) was met with widespread skepticism within the community of scientists. However in the 1930s, he and his method, EEG, were acknowledged. He also observed particular rhythms, fluctuations, in the electrical activity in a particular frequency range, 8–12Hz, which was called the 'Berger wave' or alpha wave. After an episode of clinical depression, Berger committed suicide by hanging himself in the psychiatric clinic where he had worked for many years.

After the discovery of EEG and an initially sluggish start, EEG gained widespread recognition and was also used in the clinical diagnosis of epilepsy and other disorders. EEG was, for a long time, the only imaging technique available for the brain. This changed in the 1980s when other techniques were introduced, included functional magnetic resonance imaging (fMRI) and positron emission tomography (PET).

Neuroscience 4

Regions and cognitive functions

Unlike EEG, the new techniques of PET and fMRI do not measure electrical activity. Instead they measure metabolic, biochemical and neurovascular activity in the brain. While the exact neurophysiological mechanisms underlying the signals measured especially with fMRI remain undiscovered (see Logothetis, 2008), these new imaging techniques, especially fMRI, found widespread application among researchers, especially those interested in higher-order cognitive functions.

What is the difference between PET and fMRI? PET uses radioactive substances to visualize glucose metabolism, cerebral blood flow, or specific receptors (entrance doors) of different biochemical substances (like the dopamine receptor, the serotonin receptor, etc.). This allows researchers to measure metabolic, vascular and biochemical activity in the brain. fMRI measures changes in the coupling between neural and vascular activity; how they are coupled and how that affects the neuronal activity remains unclear at this point in time. In addition to fMRI, researchers can also measure the concentration of biochemical substances like GABA and glutamate in specific regions of the brain using magnetoresonance spectroscopy (MRS).

What is the difference between PET/fMRI and EEG/MEG? The main difference between EEG (and the more recently developed MEG, the magnetoencephalogram) and fMRI and PET consists in their temporal and spatial resolution. EEG and MEG allow for an excellent temporal resolution of neural activity in the range of milliseconds, but its spatial resolution (the assignment of neural activity to particular regions in the brain) is rather low. Spatial resolution, in contrast, is much higher – within the range of mm – in techniques like fMRI and PET, which have low temporal resolution (within the range of seconds).

These techniques for the first time offered online visualization of metabolic, vascular, neural and biochemical activity changes during the performance of higher-order cognitive functions. Not only did these techniques introduce a new methodology, they also made it possible to study phenomena that were previously beyond the scope of experimental investigation. The neural effects of cognitive functions like attention, working memory, etc. could now be visualized. These imaging techniques allow researchers to observe online the changes in neural activity during particular tasks or stimuli. The introduction of new technological tools once more opened up novel ways to investigate neurons – this time with regard to the macroscopic (regional basis) of neural activity.

Neuroscience 5

Neural basis of cognitive functions

Before the introduction of imaging techniques of the brain, cognitive functions like memory, learning, attention, etc. were once considered the stronghold of psychology. Following Hebb's association of learning and memory with the synaptic strength between neurons, numerous studies using cell recording were conducted to investigate the cellular (microscopic) basis of learning, memory and other cognitive functions.

This was complemented in the 1980s and 1990s by using the newly invented imaging techniques like fMRI, PET and EEG/MEG to study the macroscopic effects of cognitive functions on the neural activity in different regions of the brain. Hence the subfield of psychology that deals specifically with cognitive functions – cognitive psychology – was increasingly aligned with neuroscience. This led to the development of the new field called cognitive neuroscience. Mainly using brain imaging, cognitive neuroscience investigates the neural basis of cognitive functions like memory, attention, learning and other similar executive functions.

As in the case of biochemical, molecular and genetic regulation of neural activity, the exact description of all results is beyond the scope of this book. We nevertheless want at least to highlight one development that is important for neurophilosophy. Concepts like memory, attention and learning were increasingly split into different subfunctions and subprocesses in orientation with different underlying neuronal mechanisms.

Critical reflection 2

Cognitive functions and consciousness

How are the various cognitive functions related to consciousness? For example, attention describes the focus on a particular content in consciousness. While originally describing one homogeneous process, the results from psychology and neuroscience led to the distinction between different attention systems as subserved by different neural systems (i.e. regions and networks) in the brain. Let us take memory as paradigmatic example. This will be described in more detail.

Based on psychological and neuroscientific results, different subtypes of memory are distinguished according to their contents, temporal dimensions (e.g. short- versus long-term) and underlying neural systems. Working memory describes short-term memory and is associated in particular with the lateral prefrontal cortex. Procedural memory concerns memory for actions and is associated with motor systems in the brain. Priming refers to memory effects in the sensory cortex. Working memory, procedural memory and priming operate on a short-term basis within seconds and even milliseconds.

This distinguishes them from long-term memory, which is based on hours, days, weeks and even years. Semantic memory concerns the long-term memory of personally unrelated semantic information, such as Paris is the capital of France. Neuronally, semantic memory has been associated with neural activity in the medial temporal cortex and the lateral prefrontal cortex.

This contrasts with episodic memory, and more specifically autobiographical memory. Episodic memory and autobiographical memory concern the storage and memory of episodes experienced by the person him/herself. One region in the temporal lobe, the hippocampus, has been shown to be central for episodic/autobiographical memory. In addition, regions in the middle of the cortex of the brain, the so-called cortical midline structures (CMS) (see Part V for details), have also been implicated in episodic/autobiographical memory.

In addition to their different contents, temporal dimensions and neural systems, different processes for memories can now be distinguished from one another. The encoding of memories describes the process of how a particular event, object, or person is translated into neural activity that can be maintained and persists. This is complemented by the storage of that event, object, or person within the neural activity. To do this, the brain seems to use neural processes distinct from those underlying the encoding.

Moreover, the event, person, or object may be remembered; this is possible by recalling the respective content from the memory. This is called retrieval. All three processes – encoding, storage and retrieval – are supposed to be mediated by different neural systems (regions and neural networks) which, at least in part, overlap with those associated with the different forms of memories as described above.

Finally, one may also distinguish different types of memories according to their association with consciousness. Memories associated with consciousness are described as declarative, for which the hippocampus and the medial temporal lobe seem to be essential. If, in contrast, memories are not associated with consciousness, thus remaining unconscious, they are characterized as non-declarative.

What can we learn from this short discussion? Cognitive functions seem to be somehow related to consciousness in an as yet unclear way. Consequently many neuroscientific theories of consciousness focus on cognitive functions these days (see Chapter 14 for details), which are complemented by sensorimotor theories of consciousness (see above and Chapter 14). This will be discussed in Chapter 14.

Critical reflection 3

Discrepancy between concepts and facts and the method of neurophilosophy)

What does the example of memory tell us? An originally homogeneous concept like memory is split into different subfunctions and subprocesses. There is no longer one homogeneous function called memory. Instead, the concept of memory must be considered an umbrella concept that covers heterogeneity of different functions and processes associated with different neural mechanisms and regions in the brain. Accordingly, what appears to be homogeneous on the conceptual level, turns out to be heterogeneous in neuronal, empirical terms.

Notably, the concept of memory is no exception. Other functions where the same occurred or is still ongoing are, for instance, emotions, reward, attention, aversion. All these were conceived initially as homogeneous entities. Now, based on the neuronal findings, they are split into a variety of different subfunctions and subprocesses.

Why is this important to consider in the context of linking the mind and the brain, and thus philosophy and neuroscience? The examples of memory tell us that there is no one-to-one correspondence between conceptual description and empirical findings. Several distinct empirical findings, though heterogeneous, may be lumped together under one concept that serves as an umbrella. This can give the illusion of homogeneity on the conceptual level but does not match up with the empirical level.

What does this imply for the linkage between philosophy and neuroscience? If we want to link philosophical concepts and neuroscientific findings, we will need to search for and consider possible disparities between conceptual and empirical levels. One may either reduce the conceptual to the empirical level, implying reductive neurophilosophy. Or one may link both conceptual and empirical levels in a non-reductive way, which leads to non-reductive neurophilosophy. As we will see in the next chapter, this distinction between reductive and non-reductive approaches will be a major factor for the development of a genuine neurophilosophical methodological strategy.

Let me preempt an example that will be discussed in full detail in Part IV. For instance, one may distinguish between different subtypes of consciousness, such as experiential consciousness (phenomenal consciousness), and more cognitive forms of consciousness like reflective consciousness. These different subtypes can then be related to different neural mechanisms and processes. One may also differentiate between different aspects of the self, like bodily, mental and social self. These, too, may be associated with neural activity in different regions and neural networks of the brain (see Part V).

Based on the experience of cognitive psychology and cognitive neuroscience, we can expect homogeneous mental concepts like consciousness, self, free will, etc. to be split and dissociated into different psychological and neuronal processes. Hence, conceptual homogeneity goes along with empirical heterogeneity. As such, there may be a discrepancy between conceptual and empirical levels of description as manifest in the difference between homogeneity and heterogeneity.

Since neurophilosophy aims to operate at the border and junction between the mind and the brain, it also is confronted with the differences between conceptual and empirical levels of description. How can neurophilosophy deal with this discrepancy between conceptual and empirical levels? To do so, neurophilosophy may need to develop a specific methodological strategy of linking and adapting conceptual and empirical levels of description. We will discuss neurophilosophy's methodology in Chapter 4. Such methodological strategy can either reduce concepts to facts, as in reductive neurophilosophy, or maintain and link both, as in non-reductive neurophilosophy.

Neuroscience 6

Introduction of neuroscience as a discipline

Recall the earlier sections of this chapter. At the turn of the twentieth century, the discipline of neuroscience had not yet been developed. It was during the 1960s that neuroscience began to emerge as a field of study. All scientists researching the brain, irrespective of the different departments with which they were associated, gathered under the common umbrella of the newly founded discipline of neuroscience.

Since the early 1960s neuroscience has been well institutionalized with rapidly developing departments at almost all universities worldwide. Despite being comparatively young when compared to the discipline of philosophy, neuroscience developed rapidly and is now one of the leading disciplines among the sciences.

Because of its origins in such varied departments as anatomy, physiology, pharmacology, neurology and psychiatry, neuroscience is built on an interdisciplinary foundation (Cowan et al., 2000). This means that neuroscience is far from being a homogeneous discipline with one specific topic and one particular, well-defined methodology. Instead, the opposite holds true. Neuroscience is highly heterogeneous, both thematically and methodologically. Let us explicate that heterogeneity in the following.

Thematically, neuroscience covers a wide range of different levels of investigation. Different levels of investigation include genetic, molecular and biochemical levels. This is complemented by the cellular level of neurons, the regional level and the level of networks. The latter especially are associated with higher-order cognitive functions. Methodologically, the scientific investigation of each level requires specific technological tools and approaches. There are consequently a wide variety of different methodologies, technologies and techniques in current neuroscience with each level developing its own methodological specifics.

Let's go back to neuroscience itself and start with an analogy. Every mother knows it very well. Her young infant develops rapidly and acquires new abilities almost every month as they extends their reach toward the world. First, the infant is intimately related to the mother and her breastfeeding. Then the father comes into the picture, followed by the siblings and finally by other people. The same holds for the young infant called neuroscience. Initially, it was restricted to anatomy and physiology, the founding pillars of the scientific investigation of the brain. Then it extended its scope by reaching towards pharmacology and psychology. This led to new subdisciplines like neuropharmacology and neuropsychology.

Further developments, including brain imaging, made the integration of cognitive psychology into neuroscience possible, which resulted in the field of cognitive neuroscience (Gazzaniga, 2008). Cognitive neuroscience is the subdiscipline of neuroscience that deals with the investigation of the higher-order cognitive functions in relation to the brain and its neuronal mechanisms. Hence, cognitive neuroscience may be regarded as the ideal mediator or bridge between philosophy and neuroscience. No wonder that Anglo-American neurophilosophy (see Chapter 4) considers cognitive neuroscience as its empirical-experimental sibling.

Neuroscience 7

Extension of neuroscience into other disciplines

We showed that despite being a relatively young discipline, neuroscience has developed enormously over the past 10–20 years. This has led to the extension of neuroscience into other thematic fields and disciplines with the development of several novel disciplines or subdisciplines. Venturing into the territory of molecular and genetic mechanisms led to molecular and cellular neuroscience; the focus on affect and emotions triggered the development of affective neuroscience. Relating social phenomena such as empathy to the brain yielded social neuroscience, and even cultural differences have been associated with neural differences.

The computer-based simulation of neural networks and their neural instantiation of higher-order cognitive functions like attention or memory has led to the development of computational neuroscience. The introduction of the new imaging methodologies like PET and fMRI has led to a subdiscipline of its own, neuroimaging. Focus on neural plasticity and development triggered the emergence of developmental neuroscience. Relating neural function to endocrine and immunological processes yielded neuroendocrinology and neuroimmunology – the most recent subdisciplines within neuroscience.

Neuroscience is expanding even further as it ventures into disciplines and territories associated with the humanities. For instance, the neuroscientific mechanisms underlying education are targeted in the field of neuropedagogics, while religious belief and its neural mechanisms are investigated in neurotheology. Even economics has been invaded by neuroscience with the development of neuroeconomics. Neuroeconomics focuses especially on the neural mechanisms of rational (and irrational) decision making as studied in economics as buying and selling.

Finally, as discussed above, neuroscience has increasingly expanded into the territory of the mind and its mental features, topics that originally belonged exclusively to philosophy. This has led to the emergence of neurophilosophy and will be discussed further in the next chapter.

What does this impressive list of new and recent neuroscientific subfields tell us? First and foremost, this list demonstrates that neuroscience is developing rapidly by continuously expanding its range. One can predict that this will continue into the future, giving rise to new disciplines with the prefix 'neuro'. It also tells us that neuroscience is not thematically or methodologically homogeneous. Instead, it is a highly heterogeneous discipline with different thematic fields and various methods. This means that, when talking about neuroscience, we need to specify the thematic field and methodological approach.

There is no homogeneous entity called neuroscience. Please specify what subfield and which method you mean when you talk about neuroscience. This shows clearly that neuroscience is historically not only based on an interdisciplinary foundation, but also that it builds and develops itself in a highly interdisciplinary way. Hence neuroscience presents us with the challenge to link these different levels and disciplines from which it emerges. This will be discussed in Chapters 9–12. That also has major implications for how we can link neuroscience to philosophy, for which, as already indicated in this and the previous chapter, different approaches may be pursued, both reductive and non-reductive. This will be the topic of the next chapter, on the concept of neurophilosophy.

Take-home message

The previous chapters focused on how the mind and its mental features were studied in philosophy and science. This chapter has complemented this survey by considering the scientific investigation of the brain, and how its associated discipline – neuroscience – extended its scope to include the mind. The first scientific discipline that explored the mind was psychology. Psychology investigates the mind and its mental features in a third-person-based observational-experimental way. As psychology explored cognitive functions like working memory, attention and executive functions, cognitive psychology, as its own subdiscipline, was born. Parallel to the development of psychology as the scientific investigation of the mind, the scientific investigation of the brain started at the turn of the twentieth century. Technical inventions allowed researchers to visualize neurons as the basic functional microscopic unit of neural activity. Later, in the 1980s and 1990s, brain imaging techniques like fMRI and PET were developed. These techniques allowed scientists to visualize neural activity on a more regional macroscopic level. This made it possible to investigate the regional macroscopic effects of cognitive functions, which soon led to the merger of cognitive psychology and neuroscience, resulting in cognitive neuroscience. In the past 20 years, neuroscience has reached out toward other disciplines like theology, economics, social sciences, anthropology and philosophy so that many novel fields with the prefix 'neuro' have developed. This shows that neuroscience is a highly heterogeneous discipline, under whose umbrella a variety of different thematic fields and methodological strategies are subsumed. All are unified through a common interest in the brain and how it functions.

Summary

The previous chapters focused on developments in philosophy during the twentieth century. Philosophy focused intensely on the mind, its nature and features, as manifested in phenomenology and philosophy of mind (see Chapter 1). At the same time, philosophy approached science by transforming and reinterpreting its methodological approach. Philosophy began to orient itself toward the scientific method of observational-experimental investigation (see Chapter 2). How about the developments from the other side, the side of science? How did the science of mind and ultimately the science of brain develop in the twentieth century? This is the focus of the present chapter. Psychology, as the study of mind, became aligned with the observational-experimental strategy of science at the end of the nineteenth century. This approach was introduced by a German psychologist named Wilhelm Wundt. This resulted in the emancipation of psychology from philosophy at the beginning of the twentieth century. Mental states and the mind were investigated on the basis of the third-person perspective, relying on an observational-experimental approach. This led to the development of cognitive psychology during the second half of the twentieth century. Cognitive psychology focuses on cognitive functions of the mind such as attention, working memory, etc. Cognitive psychology works closely together with related disciplines like computer science, biology, physics and ultimately neuroscience. Neuroscience itself developed at the end of the nineteenth century when, in the context of novel technological tools, the study of the brain relied more and more on observational-experimental investigation. Clinical alterations in psychiatry and neurology, as well as anatomical investigation of the brain, served as main starting points. The early researchers who studied the brain scientifically were associated with various disciplines like anatomy, physiology, neurology, psychiatry, etc., during the first half of the twentieth century. It was only during the mid-twentieth century that these different approaches were subsumed under the umbrella term neuroscience. New techniques, including functional imaging, were introduced and allowed researchers to scan the brain and its neuronal activity as research participants performed tasks. Moreover, neuroscience integrated concepts and assumptions from different disciplines such as philosophy, economics, psychology, anthropology, etc., thereby significantly extending its reach. This makes it clear that neuroscience is not a homogeneous discipline with one thematic field and a specific method. Instead, the concept and discipline of neuroscience is varied and broad, and includes a wide range of different thematic fields and methodologies.

Revision notes

- Why did psychology develop as a different discipline from philosophy?
- Who was Wilhelm Wundt, and what did he do?
- How did cognitive psychology and science develop? Why? Which other disciplines do they consider?
- What are the origins of neuroscience? Explain the main techniques and methods neuroscience relies on.
- When and how did neuroscience become a discipline on it is own?
- What is functional brain imaging? When was it introduced?
- Why and how, and into which disciplines, did neuroscience expand most recently?
- Explain the discrepancy between the conceptual and empirical levels of description.

Suggested further reading

- Gazzaniga, M. (2008) *Cognitive Neuroscience* (Cambridge/MA: MIT Press). *This is an extensive compilation of the different cognitive functions and their underlying neural mechanisms written by the respective experts in the field.*
- Goldstein, K. (1939) *The Organism: A Holistic Approach to Biology Derived from Pathological Data in Man* (New York: American Book Company). *This is an important book that links first-person description of brain-lesioned patients from the First World War with a third-person-based account of their brain lesions. One may thus speak of a neurophenomenological investigation, a forerunner to the later-developed concept of neurophenomenology (see Chapter 4).*
- James, W. (1890) *The Principles of Psychology*, 2 vols, Dover Publications 1950, Vol. 1: ISBN 0-486-20381-6, Vol. 2: ISBN 0-486-20382-4. *This is a pivotal work in the history of psychology, laying its ground and foundation as distinct from philosophy.*
- Kuhlenbeck, H. (1957) *Brain and Consciousness: Some Prolegomena to an Approach of the Problem* (Basel/Switzerland: Karger).
- Slaby, J. (2010) 'Steps towards a critical neuroscience', *Phenomenology and Cognitive Sciences*, 9, 397–416.

4
Brain and Philosophy: Neurophilosophy

Overview

During the twentieth century, philosophy focused greatly on the nature and the features of the mind. Philosophy of mind and phenomenology are two dominant branches of philosophy that exemplify this focus. Philosophy of mind focuses on the nature and features of the mind and how they relate to the brain. This is called the mind–brain problem (see Chapter 1). Phenomenology, in contrast, developed a first-person-based analysis of the phenomenal features of experience. This approach was intended to reveal the structure and organization of experience (see Chapter 1). During the twentieth century there was also much discussion about how philosophy is related to the sciences. Evidence of this can be found in what is called the 'naturalization of philosophy', where it is assumed that philosophy and science exist on a continuum (see Chapter 2). Finally, empirical (observational-experimental) investigation of the brain developed rapidly with the emergence of neuroscience as a major field that increasingly expanded its scope to include the mind and its mental features like consciousness, free will, self, etc. (see Chapter 3). Taken together, all three developments provided the background for introducing the brain and the various neuroscientific discoveries into the traditionally philosophical problem of the mind and how it is related to the brain, or the mind–brain problem. The brain and its empirical characterization were consequently placed at the center of metaphysical and epistemological discussions in philosophy. This led to a merger between philosophy and neuroscience and the development of neurophilosophy, which is the topic of the present chapter.

Objectives

- Understand how neurophilosophy developed and evolved
- Explain the role of Patricia Churchland in the development of neurophilosophy
- Understand the difference between reductive and non-reductive approaches to neurophilosophy
- Distinguish neurophenomenology from the Anglo-American approach to neurophilosophy
- Explain the aims and goals of neuroepistemology and neuroontology
- Understand what neuroethics is about
- Explain the required methodological strategy in neurophilosophy

Key concepts

Neurophilosophy, reductive neurophilosophy, non-reductive neurophilosophy, brain-reductive versus brain-based, domain linkages, neuralism, neuroethics, neuroepistemology, neuroontology, neurophenomenology, cross-disciplinary methodological strategies, concept–fact iterativity.

Background 1

Philosophy, mind and naturalism

Let us briefly recall the previous chapters. In the first chapter, we discussed the mind as the focus of twentieth-century philosophy, especially in philosophy of mind and phenomenology. Phenomenology aimed to investigate experience (consciousness) in a first-person-based way (phenomenal analysis) in order to reveal the structure of experience. This revealed the phenomenal features of experience like subjectivity and intentionality. Subjectivity describes the involvement of the subject and its first-person perspective. Intentionality refers to the directedness of our experience towards meaningful (semantic) contents.

The predominantly continental European phenomenology was complemented in the second half of the twentieth century by the mainly Anglo-American philosophy of mind (see Chapter 1). Rather than analyzing experience in a first-person perspective (phenomenal analysis), their method was based on the linguistic analysis of mental concepts and sentences. Thematically, such linguistic analysis focuses on the concept of mind in order to reveal its existence and reality and its relation to brain and body. The thematic focus is here thus on the mind–brain relationship as a metaphysical problem. In addition to the mind–brain problem, philosophy of mind also discusses the features of the mind like consciousness, self, intersubjectivity, free will, etc.; how they can be defined; and how they are related to the brain's neuronal features.

The strong focus on the mind in twentieth-century philosophy was paralleled by much discussion about the relationship between philosophy and science. Due to the rise and success of sciences like physics, chemistry and biology (and others) in the nineteenth and especially twentieth centuries, the role of philosophy and its relationship to the sciences was questioned. There was much discussion in the Anglo-American world especially about the domain and method of philosophy, and how it is related to science.

This is exemplified by the American philosopher Willard Quine, who rejected the traditional conceptual-linguistic method of philosophy (see Chapter 2). Most importantly, he reinterpreted philosophy as an abstract version of the scientific enterprise and its observational-experimental method. Thanks to Quine, philosophy became part of the natural world (the world as we can observe it). The naturalization of philosophy made it possible to link the traditional domains of philosophy – metaphysics and epistemology – with the empirical domain of science. One can consequently speak of metaphysical (and epistemological) naturalism.

In addition to domain, the conceptual-linguistic method of philosophy was also naturalized and thus linked to the observational-experimental method of science. This is called methodological naturalism. There are two types of methodological naturalism, replacement naturalism and cooperative naturalism. In the former, the conceptual-linguistic method of philosophy is completely replaced by the observational-experimental investigation of science. In the latter, both conceptual-linguistic and observational-experimental methods are allowed to coexist and cooperate with one another.

Background 2

Neuroscience and neurophilosophy

Parallel to these discussions about the mind and the sciences in continental European and Anglo-American philosophy, the discipline of neuroscience started to develop in the second half of the nineteenth century (see Chapter 3). New inventions allowed researchers to access and investigate the brain in a scientific way. As a relatively new discipline, neuroscience has grown rapidly.

While neuroscience initially focused solely on sensory and motor functions and their neural basis, it has continuously extended its reach. Neuroscience – in close collaboration with cognitive science – started to investigate cognitive functions like memory, attention, etc. This led to the establishment of cognitive neuroscience. In addition to cognitive functions, neuroscience also targets affective and social functions. Affective functions concern our emotions, while social functions are about social communication and interaction with others; the search for their neuronal mechanisms is the focus of what these days is called 'affective neuroscience' and 'social neuroscience' (see Chapter 3).

Extending its reach still further, neuroscience started investigating features of the mind like consciousness, free will, self and intersubjectivity. Thanks to the development of functional imaging techniques like functional magnetic resonance imaging (fMRI), researchers were able to visualize and access the neural activity of participants undergoing conscious experiences. Furthermore, observations in neurological and psychiatric patients provide paradigmatic examples for studying the relationship between mental and neuronal functions. Not only do these patients suffer from changes in self and/or consciousness, but they also demonstrate alterations in their brain and its neuronal states (see Chapter 3).

Where does the impressive progress in neuroscience leave us? As neuroscience extends its reach from the brain into concepts like the self, consciousness and free will, neuroscience has ventured more and more into the territory of philosophy. As we will see below, this inclined some philosophers to argue that philosophy can more or less be replaced by neuroscience.

Taken together, one can observe convergent developments in philosophy and neuroscience. Neuroscience, as the scientific study of the brain, is extending its reach more and more toward those neuronal mechanisms that underlie the mind and its mental features. Meanwhile, philosophy, coming from the opposite end – that of the mind – extends increasingly toward the brain and how it is investigated in neuroscience.

Such rapprochement between neuroscience and philosophy went along with the convergence of their respective methods, the observational-experimental method of science and the conceptual-linguistic one of philosophy. In short, the domains and methods of both neuroscience and philosophy converged despite their development from opposite starting points. This most recent development of 'neurophilosophy' is the focus in this chapter.

Biography

Arthur Schopenhauer

The Canadian-born American philosopher, Patricia Churchland, explicitly introduced the term 'neurophilosophy', in her book *Neurophilosophy* that was published in 1986. Despite this, earlier philosophers had already introduced the brain into philosophical discussion, without calling it 'neurophilosophy'. Reverting to French, one can speak of what could be called '*neurophilosophie avant la lettre*' (translated as 'neurophilosophy prior to the term'). This will be described briefly below (see Breidbach (1997) for an excellent summary).

This is the point at which to introduce Arthur Schopenhauer (1788–1860). He was born in 1788 to a rich merchant family in Danzig, Germany. After studying philosophy, he took a position at Humboldt University in Berlin, Germany. At the same time, however, the famous German philosopher, Georg Wilhelm Friedrich Hegel, was teaching there too. He was very famous and considered the most important philosopher of his time. Schopenhauer, in contrast, was not acknowledged at all. His seminars, unlike those by Hegel, were only attended by a few students. This left Schopenhauer embittered and with a life-long grudge against academic philosophy. Fortunately, he was independently wealthy and could afford to withdraw from the field.

Abbreviated history of neurophilosophy 1

Implicit neurophilosophy: Schopenhauer

How did Schopenhauer introduce in the brain into philosophy? In contrast to his archenemy Hegel, Schopenhauer seemed to be very attracted to the brain and to the idea of giving it a central role in philosophy. He claims that the various cognitive faculties like the categories the German philosopher Immanuel Kant attributed to the mind (see Chapters 8 and 12 for more details) must be related to the brain. Schopenhauer therefore believed that the brain and its neuronal processes account for our cognition, knowledge, and ultimately the first-person perspective.

This is a radical thesis that we need to explain further. Any cognition and knowledge requires a subject who cognizes/knows in first-person perspective. Traditionally, the subject of cognition and knowledge was associated with the mind, as was the case for the philosopher Descartes (see Chapter 1). Many other philosophers including Schopenhauer's German predecessor Immanuel Kant tacitly and silently presupposed such mind to account for the subject of cognition and knowledge. Such tacit and silent presupposition of a mind reaches deeply into even our current philosophy as for instance in philosophy of mind. Therefore philosophy in general and philosophy of mind in particular can be characterized as mind-based, as will be described in Chapters 7 and 8 in more detail.

Schopenhauer departed from that tradition. He argued that it is not some kind of mind that cognizes/knows. Instead of the mind, it is the brain. The brain itself provides the first-person perspective and is consequently the subject of cognition and knowledge. In short, the brain is the subject of cognition and knowledge. This, as we will see, opens the door to replace the traditional mind-based approach in philosophy by a truly brain-based approach as the hallmark of neurophilosophy.

However, things are not easy when it comes to the brain. The 'brain as the subject of cognition' must be distinguished from what Schopenhauer described as the 'brain as the object of cognition'. Unlike the brain as the subject of cognition, the brain as the object of cognition can be observed as such. This is the brain we perceive as a gray pulpy mass, as Schopenhauer called it. This is likewise the brain we visualize in the brain imaging techniques we use today.

Abbreviated history of neurophilosophy 2

Implicit neurophilosophy: brain-paradox

One may be puzzled now. Schopenhauer distinguishes between the brain as the subject of cognition and the brain as the object of cognition? Are there two different brains? Certainly this is implausible since one only has one brain. Is there only one brain? This is also not possible because the same brain cannot be both the subject and the object of cognition at the same time. As such, consideration of the brain and its double characterization as both the subject and the object of cognition lead to logical problems and contradictions. This has been described as the 'brain paradox' (see Chapter 8 for details on both Schopenhauer and the 'brain paradox').

While leaving us with the problem of the 'brain-paradox', Schopenhauer must nevertheless be given credit for being the first who attributed a central role to the brain in philosophy. As such, he can be regarded the first neurophilosopher in the history of philosophy, a *'neurophilosophe avant la lettre'* (neurophilosopher before the introduction of the term). Despite Schopenhauer's introduction of the brain into philosophy, the brain nevertheless still did not become really popular in philosophy. It took almost another 100 years until the mid-twentieth century when another philosopher, the French phenomenologist, Maurice Merleau-Ponty (see Chapter 1), shifted the brain into the center of philosophy.

In his main work, *Phenomenology of Perception* (1962[1945]), Merleau-Ponty strongly relied on the work of German neurologists like Kurt Goldstein and Erwin Strauss (see Chapter 3). These neurologists related the brain's sensorimotor functions to consciousness and thus to the mind. They related the brain and its sensorimotor functions directly to the subjective experience in patients with brain lesions from the First World War. The third-person-based observation of the brain and its lesions was here linked to the phenomena in consciousness as experienced in first-person perspective. This approach can be regarded a predecessor to what is called 'neurophenomenology' today, which emphasizes the need to consider the brain in relation to the body and thus as embodied and embedded (see below and Chapter 8).

Abbreviated history of neurophilosophy 3

Explicit neurophilosophy: Popper and Eccles

As detailed above, the phenomenological approach and its expansion into to neuroscience occurred predominantly in continental Europe during the first half of the twentieth century. At the same time, there was much discussion in the Anglo-American world about the metaphysical relationship between mind and brain (the mind–brain problem). Is the mind's existence and reality identical to that of the brain? Can the mind be reduced to the brain? This led to the emergence of different theories about the mind–brain relationship that strongly considered the empirical results (though not yet in full detail) from neuroscience (see Part II for details). These earlier authors and their lines of thoughts can be considered predecessors of neurophilosophy.

In the 1970s, Karl Popper (1902–94), a renowned philosopher teaching in Britain, and the Australian, John Eccles (1903–97), a Nobel Prize-winning neuroscientist, published a book, *The Self and its Brain* (1989[1977]). They assumed three different worlds – the physical world, the mental world and the socio-cultural world – to account for the mind–brain problem. They associated the brain with the physical world, while the mind was related to the mental and socio-cultural world. This position advocated for some kind of dualism.

As we will see in Part II of this book, the neuroscientist, John Eccles, surprisingly advocated this kind of mind–brain dualism, while the philosopher, Karl Popper, argued against it. Regardless of the concrete mind–brain positions they held, by publishing this book together, Popper and Eccles brought the brain itself – and various neuroscientific findings – directly and explicitly into the center of the mind–brain debate in philosophy.

Abbreviated history of neurophilosophy 4

Explicit neurophilosophy: Churchland

The book by Popper and Eccles was followed in 1986 by Patricia Churchland's publication of her book *Neurophilosophy* in the USA. Patricia Churchland (1943–) was born into a farming household in British Columbia, Canada. She became interested in philosophy early on in her childhood and studied it at university. There she met her future husband, Paul Churchland, who became a famous philosopher himself. Both now teach in California at the University of San Diego. Together, they strongly advocated the naturalization of the mind by bringing it together with the brain and neuroscience.

In her book *Neurophilosophy* (1986), Churchland explains the current state of neuroscience to philosophers and the philosophical mind–brain discussion to neuroscientists. Here, the term neurophilosophy stands loosely for the encounter and overlap between philosophical issues that deal with the mind, and neuroscientific discoveries that concern the brain. Churchland devotes one part of the book to the discussion of philosophical issues, like the current state of the mind–brain discussion. The other half of her book consists of a detailed account of the brain and how it functions.

Churchland's book and approach laid the foundation for neurophilosophy in especially the Anglo-American world. How can we now determine neurophilosophy? In the present Anglo-American world, neurophilosophy is defined as the 'application of neuroscientific concepts to traditional philosophical questions' (Bickle *et al.*, 2010). This means that metaphysical questions like the mind–brain problem that were originally discussed in a purely conceptual-logical way are now approached in a more scientific, observational-experimental way. In short, metaphysical questions are no longer discussed in merely conceptual terms, but rather are investigated experimentally.

Concept of neurophilosophy 1

Dissociation between domain and method

The application of the observational-experimental approach to the study of metaphysical questions implies convergence, if not reduction (see below), between different domains. We recall from the Introduction and Chapter 2 that the concept of domain describes particular fields, issues or questions and thus the subject or matter of investigation as distinct from the method of investigation, the discipline.

How is the distinction between domain and discipline related to neurophilosophy? Neurophilosophy suggests approaching metaphysical questions like the existence and reality of the mind and its relationship to the brain, the mind–brain problem, in observational-experimental terms rather than in the traditional rational-argumentative way as in philosophy. The metaphysical domain is here thus dissociated from its traditional method and instead associated with a different method; this implies a shift of the mind–brain problem from the discipline of philosophy to that of neuroscience. Neurophilosophy in this sense is consequently regarded as the application of the neuroscientific method (and concepts) to traditionally philosophical questions (see above).

However, nothing is simple, as we all know. The exchange of methodological strategy in investigating the metaphysical domain has major ramifications. The metaphysical domain of philosophy is now linked and integrated with the empirical domain of science and its observational-experimental methodology. Metaphysical issues are no longer dealt with in purely logical-conceptual terms, but rather in terms of what we can observe in the natural world. This leads to what we described in Chapter 2 as naturalization, implying metaphysical (and epistemological) naturalism. This is accompanied by the replacement of the conceptual-linguistic method of philosophy by the observational-experimental approach of neuroscience.

In sum, one can characterize neurophilosophy by dissociating the philosophical domains like the metaphysical (and epistemological) domains from their traditionally associated rational-argumentative and conceptual-logical methods. One can thus characterize neurophilosophy by dissociation between domain and method when compared to traditional philosophy. Such dissociation, in turn, opens the door to associate the metaphysical and epistemological domains with a different method of investigation, the observational-experimental approach, and to shift them from the discipline of philosophy to neuroscience.

Concept of neurophilosophy 2

Intertheoretic reduction and replacement naturalism

Neurophilosophy in this sense is quite a radical approach. What was formerly discussed in philosophy in logical-conceptual terms is now investigated scientifically by using the observational-experimental approach. This means that we have to reduce the theories of philosophy (and psychology) to those of neuroscience. This is called 'intertheoretic reduction' (see Part III).

Such 'intertheoretic reduction' has major implications for how we conceive the mind. 'Intertheoretic reduction' entails the description of the mind and its mental capacities in a purely scientific and thus objective way. This leaves no place for any phenomenal features of consciousness like subjectivity and intentionality. Rather than in phenomenal terms, these are now supposed to be better described by concepts indicating neuroscientific mechanisms. On the metaphysical side this goes along with the rejection of any form of dualism between mind and brain. Rather than presupposing different existences and realities, we have to opt for monism and more specifically for materialism. Following Churchland, the mind simply does not exist; there is nothing but the brain. This position is called 'eliminative materialism' (see Chapter 6).

In addition to existence and reality, our knowledge can also be reduced to the brain. This means that the subject of knowledge and knowledge itself can no longer be associated with the mind, but must be associated with the brain. The brain is the subject of knowledge and harbours all the knowledge we can possibly acquire. This raises the question of how the brain acquires knowledge, for which Churchland and many other philosophers assume representation to be central. Representation suggests that the world we know is encoded and somehow reconstituted in the neural activity of the brain. How the brain represents the world is thus a central question in this kind of neurophilosophy.

This short and rather abbreviated overview already indicates that Churchland stands very much on the shoulders of the naturalization of philosophy and its metaphysical and methodological replacement by neuroscience. Quine opened the doors to setting philosophy and science in direct contact with each other by making naturalization possible (see Chapter 2). Churchland followed up on this and specifies convergence with regard to the mind and the brain by eliminating the mind and subsuming philosophy of mind under neuroscience, entailing what she describes as 'neurophilosophy'.

How can we describe the naturalization in more detail? Churchland's positions about the mind presuppose what Quine and other philosophers described as 'metaphysical and epistemological naturalism' and 'methodological naturalism'. She seems to prefer the more radical variants of analytic naturalism and replacement (or incorporation) naturalism (see Chapter 2) with regard to the relationship between the mind and the brain. As we shall see, this has major implications for the definition of neurophilosophy as a discipline, its domains and method.

Concept of neurophilosophy 3

Reductive neurophilosophy

We have seen that the Churchlandian concept of neurophilosophy ultimately reduces philosophy and its traditional domains, metaphysical, epistemological, ethics, etc., to the empirical domain of neuroscience. This is well reflected in the emphasis on intertheoretic reduction and replacement naturalism. There is thus not only an exchange of method but also a reduction (and ultimately elimination) of the traditionally philosophical domains to the empirical domain of neuroscience; one can characterize such a concept of neurophilosophy as reductive. In what follows, I refer to '**reductive neurophilosophy**'.

How can we characterize such 'reductive neurophilosophy' in further detail? Taken altogether, one may characterize Churchland's concept of neurophilosophy, (which is especially prevalent in the Anglo-American world), by: (i) **domain monism** in that it exclusively associates the brain with the traditional domain, the empirical domain, which makes its approach rather conservative; (ii) **methodological monism** by claiming for the observational-experimental method as the sole viable methodological approach, thus being rather narrow; and (iii) as **brain-reductive** by postulating the complete elimination of the mind such that all mental states are reduced exclusively to the brain itself and its neural activity.

Let us explain these features of the concept of reductive neurophilosophy as it is prevalent these days, especially in the Anglo-American world:

(i) It is conservative in the sense that it presupposes one single domain – the empirical domain – as the main domain. Meanwhile all other domains, like the metaphysical or the epistemological domain, are either neglected or reduced to the empirical domain. There is thus what can be described as '***domain monism***' in

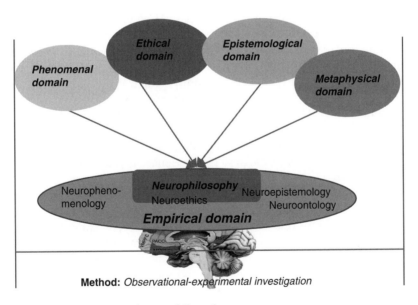

Figure 4.1 Different concepts of neurophilosophy
Figure 4.1a Reductive neurophilosophy
The figure illustrates different concepts of neurophilosophy.

(a) The figures illustrates the concept of reductive neurophilosophy. It presupposes domain monism with the reduction of all other domains on to the empirical domain as it is indicated by the downward arrows. There is methodological monism because of the exclusive focus on the observational-experimental method of neuroscience as indicated in the lower part of the figure. And it is brain-reduced because it reduces all mental capacities to the empirically observable neuronal processes in the brain itself.

current neurophilosophy, meaning the dominance and priority of one particular domain at the expense of others. Such domain monism of the empirical domain in neurophilosophy is supposed to replace the domain monism of traditional philosophy where, historically, either the metaphysical or epistemological domain was dominant. Such domain monism is closely associated with neuralism, the exclusive consideration of metaphysical, epistemological, phenomenal and ethical issues in solely neural terms. While the brain-reductive approach may consider such neuralism as a virtue, it may be regarded as a threat from the perspective of both traditional philosophy and non-reductive neurophilosophy.

(ii) The Anglo-American concept of neurophilosophy is narrow in that it relies methodologically on only one particular methodology: the observational-experimental approach of science. It excludes and replaces other methodologies like the conceptual-linguistic method of philosophy and the phenomenal analysis of phenomenology (see below). This is a form of replacement naturalism and can be described as *'methodological monism'*, the reliance and use of only one particular methodological strategy. Such methodological monism of neurophilosophy is supposed to replace the methodological monism of traditional philosophy where one particular method like the conceptual-linguistic or the phenomenal method were considered the main methodological strategy to investigate the mind and its mental capacities.

(iii) Finally, the Anglo-American concept of philosophy assumes that the mind and its mental features can be completely reduced to – and ultimately replaced by – the brain and its neuronal features. One can say that it is **brain-reductive** in that it no longer gives a role to some kind of mind (or mental property) in constituting the various mental capacities like consciousness, self, free will, etc. Any assumption of a mind is considered simply wrong, if not illusory, thus entailing a mind-aversive stance. The brain-reductive and mind-aversive stance of neurophilosophy is supposed to replace the mind-reductive and brain-aversive stance of traditional philosophy. Historically, philosophy neglected the brain and considered it to be merely empirical, which therefore stands square to the metaphysical and epistemological goals. Hence, to introduce the brain into philosophy was (and still often is) considered simply a category error where one confuses empirical and metaphysical/epistemological domains (see Chapter 8 for more details on such a category error). This led to a brain-averse stance in philosophy. At the same time, mental capacities were associated with some kinds of mind or mental features entailing a mind-reductive stance. In sum, the brain-reduced and mind-aversive stance may be considered the neurophilosophical reaction to the brain-averse and mind-reduced stance of traditional philosophy.

Concepts of neurophilosophy 4

Non-reductive neurophilosophy

Reductive neurophilosophy presupposes analytic and replacement naturalism. One may, though, imagine another concept of neurophilosophy, one based on different forms of metaphysical and methodological naturalism (see Chapter 2). Rather than opting for analytic naturalism and replacement naturalism in the relationship between philosophy and neuroscience, one can take a less radical stance. One can then presuppose synthetic non-reductive naturalism, rather than analytic naturalism. Similarly, one can adopt cooperative naturalism, rather than replacement naturalism.

This has major ramifications for the concept of neurophilosophy. The emphasis on reduction and incorporation would then be replaced by non-reduction and cooperation. Such neurophilosophy could then no longer be characterized as reductive but rather as non-reductive. I henceforth speak of '**non-reductive neurophilosophy**' as distinguished from reductive neurophilosophy. How can we characterize such non-reductive neurophilosophy? This concept of neurophilosophy would be characterized by (i) *domain pluralism* rather than domain monism, (ii) *methodological pluralism* rather than methodological monism, and (iii) as *brain-based* rather than brain-reductive.

Note that the arrows from neurophilosophy always target the borders of the empirical domain to the other domains. The border between the domains is exactly where neurophilosophy can be situated. This distinguishes neurophilosophy from both neuroscience and philosophy, which are situated within their respective domains rather than at their borders with other domains. This also distinguishes the non-reductive

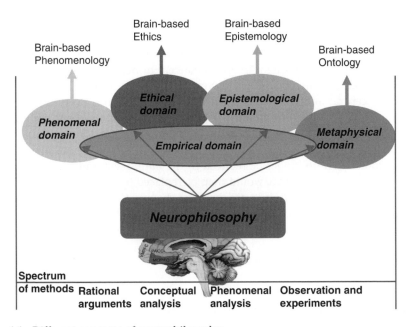

Figure 4.1 Different concepts of neurophilosophy
Figure 4.1b Non-reductive neurophilosophy

(b) This is different in non-reductive neurophilosophy. There is domain pluralism rather than domain monism because it includes different domains, e.g. empirical, epistemological, ethical and phenomenal (as well as metaphysical) (upper part). It presupposes methodological pluralism rather than methodological monism because it relies on a spectrum of different methods (lower part). And it is brain-based rather than brain-reduced, with the brain as the commonly underlying basis which serves only as a starting point but not as an end point.

neurophilosophy suggested here from its reductive sibling that can be situated within the empirical domain itself.

(i) Instead of one single domain, several domains may be considered. Instead of considering either the metaphysical/epistemological or the empirical domain as the ultimate domain, one may consider the various domains on an equal basis. No domain predominates. The domain monism of both traditional philosophy and current neurophilosophy would then be replaced by what can be described as '*domain pluralism*'. The conservative stance of both philosophy and reductive neurophilosophy, that still claim for one superior or predominant domain, would then be replaced by a more liberal stance that includes the various domains and attributes equal rights to them. Unlike in reductive neurophilosophy and its domain monism, domain pluralism does not entail neuralism, the neuralization of metaphysical, epistemological, phenomenal and ethical questions and their respective domains.

(ii) Replacement naturalism would be replaced by cooperative naturalism. This means that different methods – observational-experimental, conceptual-linguistic and phenomenal (see below) – can be used at the same time and are no longer mutually exclusive. For example, the conceptual-linguistic method could be applied to the brain and the observational-experimental approach could be applied to the mind. Hence, the spectrum of the methodological strategies that are applied is here much wider, so that the methodological monism of the narrow neurophilosophy is replaced by **methodological pluralism**.

(iii) Finally, unlike in the Anglo-American concept of neurophilosophy, the mind is not reduced completely and exclusively to the brain. Instead, the brain is considered central and the basis, but neither exhaustive nor everything by itself. For example, besides the brain, the body and the environment can also play essential roles. Put more technically, the brain is regarded a necessary, but not sufficient, condition of the various mental capacities. The various mental capacities could then no longer be reduced to the brain itself and its neural activity. The brain-reductive stance would then be replaced by a **brain-based** stance. The concept of the brain-based stance describes the way that the brain and its neural activity are no longer considered sufficient by themselves but only necessary for constituting the various mental capacities like self, consciousness, free will, etc.

Concepts of neurophilosophy 5

Neurophenomenology

Can we give a more specific example for a non-reductive neurophilosophy? One step in this direction is the recent linkage between phenomenology and neuroscience. We have already discussed phenomenology, the first-person-based analysis of experience (consciousness) and its structure and phenomenal features (subjectivity and intentionality) in detail (see Chapter 1). We referred to phenomenology again when discussing neuroscientists such as Strauss, who related the structure of experience and its 'being in the world' to neurological processes (sensorimotor functions) in the brain (see Chapter 3).

This served as the basis for the French philosopher and phenomenologist, Merleau-Ponty, who considered the body and its brain-based sensorimotor function to be central for consciousness. In this model, consciousness was considered necessarily dependent on the body. This is called 'embodiment' (see Chapter 1 in this part). More recent neuroscientists like Francisco Varela (1946–2001) and philosophers like Evan Thompson (1962–) consider embodiment central for the constitution of consciousness. Moreover, Varela and Thompson suggested the direct linkage of first-person accounts of experience (and its phenomenal features like subjectivity and intentionality) to third-person-based observation of neuronal states in the brain. This led them to coin the term 'neurophenomenology', which mainly describes a research strategy that links first- and third-person perspectives and thus the experience of content in consciousness and observation of the brain's neuronal mechanisms.

What is the difference between the neurophenomenological approach and the reductive approach to neurophilosophy? Unlike in reductive neurophilosophy, neurophenomenology takes the phenomenal features of experience, like subjectivity, intentionality and the sense of self, seriously, and does not eliminate them by declaring them to be illusory. Instead, the phenomenal features are taken as a starting point and thus as a template for the subsequent observational-experimental investigation of the brain.

How does this stand in relation to reductive neurophilosophy? Reductive neurophilosophy ultimately eliminates the first-person perspective and its phenomenal features in favor of the third-person perspective and a purely neuronal account. This is different from neurophenomenology. Neurophenomenology aims to link both first- and third-person perspectives as well as phenomenal and neuronal features. As such, neurophenomenology can be considered a first step toward the future development of a non-reductive neurophilosophy that includes phenomenal features in a non-reductive way. Moreover, neurophenomenology presupposes a brain-based rather than brain-reductive stance in that it includes the body in its account of the mind.

Critical reflection 1

Reductive versus non-reductive neurophilosophy

In order to contrast the thematic and methodological differences between reductive and non-reductive neurophilosophy, I set up an imaginary dialogue between their respective proponents. RN stands for a reductive neurophilosopher, while NN refers to a non-reductive neurophilosopher.

RN: You argue for a wide version of neurophilosophy that does not reduce mental states to the brain but suggests that mental states be based on the brain. Brain-based rather than brain-reductive, that is the question here. What is your argument for that?
NN: When you reduce the mental features of the mind to the brain and its neuronal features, you refer to the brain as observable in third-person perspective. Correct?
RN: Yes, off course, that is what the brain is all about.
NN: How do you know that? Maybe there is more to the brain than just what we can currently observe. There may be additional features that we don't know about yet. And most importantly, these additional features may not need to be

beyond the neuronal features. There is thus nothing metaphysical about my assumption.

RN: But all those supposedly unknown neuronal features are, in principle, observable in third-person perspective. If so, your concept of the brain does not differ essentially from mine. Are you saying that my concept of 'brain-reductive' is the same as your concept of 'brain-based'?

NN: Wait, wait, wait. Not so quick. When I talk of brain-based, I leave open certain properties and features of the brain. These may indeed be neuronal but their realization and manifestation may exceed the boundaries of the brain itself. They may, for instance, link the brain intrinsically with the body and environment. Consider, for instance, the heart. Because it pumps blood, the heart cannot avoid being linked to the whole body. In the same way, it is possible that some yet unknown neuronal mechanisms might make it necessary that the brain be intrinsically linked to body and environment by default.

Critical reflection 2

Brain-reductive versus brain-based neurophilosophy

RN: How does that work? Sounds rather mysterious to me!

NN: Let me give you an empirical example. The brain shows intrinsic activity. And that intrinsic activity, generated in the brain itself, fluctuates in different frequencies. For example, between 1 and 4 Hz (delta frequency). Such fluctuations show positive and negative phases as well as certain onsets when the phases begin and are at 0. Recent findings demonstrated that the phase onset of these intrinsic fluctuations can be shifted to the onset of the extrinsic stimuli in the environment. If, for instance, extrinsic stimuli occur in a rhythmic way like in a melody, it is easy for the intrinsic fluctuations to set their phase onset in the delta frequency range accordingly, while this is much more difficult in the case of non-rhythmically presented extrinsic stimuli, as for instance when there is no melody.

RN: Everything is purely neuronal and thus within the brain. Nothing else is needed.

NN: That is not true. The degree of phase shifting does not only depend on the brain itself and its intrinsic activity, but also on the occurrence of the extrinsic stimuli and their respective timing and rhythm. Hence, the brain also depends on the environment. If the brain's dependence on the environment can be shown to predict the degree of consciousness, as does some initial tentative support, you could no longer say that consciousness can be reduced to the brain and be 'located' within the brain.

RN: You mean that the findings provide an argument in favor of a brain-based rather than brain-reductive approach to neurophilosophy?

NN: Yes, I would claim exactly that. If this intrinsic linkage between environment and brain is central in constituting mental features like consciousness, you can no longer reduce consciousness to the brain itself. Instead, it is based on the brain and thus brain-based rather than brain-reduced. The same may hold true in the case of the self, the subject of experience. It may be that the self cannot be found within, and thus reduced to, the brain and some specific neuronal states. Instead, it may rather be localized in a specific form of the environment–brain relation and thus be brain-based rather than brain-reductive.

RN: I am still not sure about this point. Isn't the reference to the environment–brain relationship just a dummy argument, an argument on behalf of a straw man, behind whose outer façade you will find nothing but the brain itself and its neuronal states?

NN: Yes, you are right with your argument when you consider exclusively the empirical domain and its third-person-based approach. From that point of view, the brain is indeed clearly segregated from the environment, implying that everything can be accounted for by the brain alone and its neuronal state. But the price you pay for that is rather high; you exclude consciousness and basically everything that distinguishes our mind and its mental features.

RN: You mean that I cannot explain consciousness and the mind's mental features because I eliminate them from the very beginning?

NN: Exactly that. Consciousness and the mind's mental features must ultimately remain mysterious to you because you cannot close the gap between mind and brain. Isn't that almost ironic? You started out with neurophilosophy to abandon finally the rather mysterious assumptions of metaphysics and its mind-reductive accounts. And then you replace them with the more transparent findings of neuroscience. Now, though, by shifting to the opposite extreme of a brain-reductive account, you, too, cannot avoid mysterious assumptions.

RN: You say that I end up exactly where I wanted to escape from? I never thought that my approach that aims to solve the problems of mind and consciousness actually causes them. I always thought that by bringing in the brain as much as possible and thus to opt for a brain-reductive strategy I would be able to finally get rid of the old metaphysical problems. A trap, apparently…

Domains of neurophilosophy 1

Neuroepistemology: neural functions and knowledge

After having sketched different concepts of neurophilosophy, reductive and non-reductive, we now want to describe neurophilosophy in further detail with regard to both its subject or matter of investigation and its methodological strategy. What are matter and method of neurophilosophy? This question may be answered slightly differently in reductive and non-reductive approaches, which we will hint about in the following.

Let us start with the matter or subject of neurophilosophy. This leads us to the fields and issues of neurophilosophy and thus to what I characterized as domains. What exactly is (are) the domain(s) of neurophilosophy? Recall from Chapter 2 that the concept of domain describes problems, issues, topics and questions that are dealt with in a discipline. For instance, philosophy can be characterized by the metaphysical domain that asks questions about the existence and reality of beings and the world. In addition, philosophy also focuses on the epistemological domain that deals with questions about knowledge, what can we know and how can we know it? Finally, philosophy also discusses questions about norms and moral behavior. These problems belong to the ethical domain.

In contrast to philosophy, neuroscience is more concerned with the empirical domain and what we can observe in third-person perspective about how the brain works and functions. Metaphysical, epistemological and ethical questions were, until recently, not part of the empirical domain of neuroscience. That, however, as we have seen in Chapter 3, has changed in recent neuroscience. Current neuroscience extends and reaches out to neuronal functions relevant to epistemological and ethical issues. This encounter between neuroscience and philosophy raises the question: how can different domains be linked to each other (domain linkage)?

One such domain linkage is that between empirical and epistemological domains. The American philosopher Willard Quine – as one of the main proponents of the naturalization of philosophy – also suggested the naturalization of epistemology in his famous paper 'Epistemology naturalized' (see Chapter 2). In a nutshell, Quine argued that we need to investigate the neuronal processes underlying our acquisition of knowledge and that this should be distinguished from knowledge justification (the stronghold of traditional epistemology). Quine considered observation sentences – basic sentences about our observations of the world – to be the most basic building blocks of knowledge acquisition.

Following Quine and his focus on knowledge acquisition, Churchland and other Anglo-American neurophilosophers emphasized higher-order cognitive functions like attention, working memory, etc. as central for knowledge acquisition. Why? Because they allow the content that is primarily represented in the sensory functions to be re-represented, which in turn may yield knowledge. Rather than on observation sentences as being closely linked to basic lower-order representation via sensory functions, the focus here is on higher-order cognitive functions and higher-order representation (see Part III for a more extensive discussion of the issue of representation).

The focus of neuroepistemology is: how do brains represent the world (reality) with neuronal activity, and how do brains learn to acquire knowledge? This is well expressed in the following quote by Churchland: 'As a bridge discipline, neuroepistemology is the study of how brains represent the world, how a brain's representational scheme can learn, and what representations and information in nervous systems amount to anyhow. This characterization must be seen as provisional, however, for it is too early in the game to be very confident that "representing reality" is the right way to describe the central function of the mind-brain' (Churchland, 2002, p. 270).

How does that relate to the distinction between the brain as the subject of cognition and brain as the object of cognition as advanced by the philosopher Schopenhauer (see above)? We recall: the brain as the object of cognition is the brain that we can observe in third-person perspective; that brain appears to us as gray matter and can today, with our technology, be visualized using functional brain imaging. The brain as the subject of cognition is the brain that perceives, that cognizes and acquires knowledge.

Domains of neurophilosophy 2

Neuroepistemology: neuroepistemic limitations

Churchland and other Anglo-American neurophilosophers assume that the brain as the object of cognition (the brain as investigated and observed in third-person perspective) can account for the brain that is associated with cognition and knowledge (the brain as subject of cognition). However, this position relies on two assumptions: the first assumption is that the brain as the subject of cognition can be fully apprehended and known by mere observation in third-person perspective.

The second assumption is that there is no principal difference between the brain as the subject of cognition and the brain as the object of cognition. Taking both assumptions together implies that the difference between the epistemic domain, as presupposed by the subject of cognition, and the empirical domain, as associated with the object of cognition, is resolved and replaced by the empirical domain.

How, though, can Churchland and others be sure that there is no principal limitation in our access to and knowledge of the brain as the subject of cognition? One cannot exclude that we, as a knowledge apparatus based on our brain, suffer from a principal limitation in accessing the brain as the subject of cognition. This epistemic limitation has indeed been assumed by current philosophers like Colin McGinn, who claim that consciousness is based on a specific property of the brain which we cannot access and thus know (see Chapter 8).

Furthermore, we may suffer from certain epistemic limitations with regard to our own brain. First, we remain unable to experience our brain as such (and thus its neuronal activity) in first-person perspective in consciousness. Hence, we have no direct (or online) access to our brain as content in our consciousness; one's own brain can never be a content of one's consciousness. Furthermore, we remain unable to observe the brain's neuronal states directly in third-person perspective without any technical means that allow for indirect observation.

One may go even one step further. One could assume that epistemic limitations are based on the brain itself – on its specific design and intrinsic features. This epistemic limitation is then the default position, meaning that the brain, by its very nature, cannot avoid it. In this case, the epistemic limitation is a truly brain-based and thus neuroepistemic feature. Accordingly, a future neuroepistemology may also want to investigate the neuronal mechanisms of our epistemic limitations in addition to those underlying our epistemic abilities or capacities.

Neuroepistemology could widen its scope by searching for the neural basis of both epistemic abilities and limitations. More specifically, we may want to search for the neuronal mechanisms that prevent us from observing the brain and its neuronal states directly in third-person perspective. We may also want to investigate why the brain and its neuronal states cannot become the conscious content of our experience in first-person perspective in consciousness. Hence, neuroepistemology will not only be about neuroepistemic abilities, but also neuroepistemic inabilities, the principal limitations and borders in our possible knowledge of ourselves and the world as they may be traced back to the brain and its particular features.

Domains of neurophilosophy 3

Neuroontology: physical characterization of the brain

What about the metaphysical domain? The metaphysical domain concerns the existence and reality of being and the world. Ontology is a branch within the metaphysical domain that focuses on the different categories of being and how they relate to each other. How can we link metaphysics and ontology to the brain? One way is to characterize the existence and reality of the brain by itself and to describe how it is related to what philosophy describes as the mind. This issue is also known as the mind–brain problem of philosophy of mind (see Chapter 1 and Part II for details).

Within this context, the existence and reality of the brain is usually presupposed as merely physical. The brain is part of the body, and the body is supposed to be no more than a physical machine. How are the brain's physical features related to the mind's features, including consciousness, self and intersubjectivity? The proponents of a brain-reductive neurophilosophy assume different neuronal mechanisms underlie the different mental features. For instance, neuroscience has discovered certain neurons in the brain that become active during observation of other people's actions and movements. These neurons are called 'mirror neurons' and may be central in establishing intersubjectivity.

These findings do not mean that we can avoid talking about the definition of the concept of intersubjectivity. Intersubjectivity may be determined by different features, including the mirroring of other people's actions as related to the mirror neurons. Most importantly, these findings are merely empirical and rely on observation. In contrast, they do not imply anything about the existence and reality of intersubjectivity itself. Put more technically, we have to be careful when inferring from the empirical domain of observation to the metaphysical domain of existence and reality.

This also applies to the brain. What we observe are empirical findings that we frame in physical terms. This, however, does not imply anything about the existence and reality of the brain itself and whether it is physical or not. The brain-reductive approach to neurophilosophy claims that the empirical findings of particular neuronal mechanisms (like the mirror neurons and their relation to specific mental features like intersubjectivity) make the assumption of a purely physical characterization of the brain's existence and reality likely. They therefore argue in favor of eliminative materialism.

In sum, the reductive approach in neurophilosophy argues for the ontological determination of the brain as purely physical. This is based on the empirical findings which, the proponents say, do not lend any evidence in favor of a mind as being distinct from the brain in its existence and reality. There is thus a more or less direct inference from the empirical to the metaphysical domain. This can be described as 'empirical-metaphysical inference'.

Domains of neurophilosophy 4

Neuroontology: mental characterization of the brain

Current brain-reductive neurophilosophy characterizes the brain's existence and reality in merely physical terms. This is based on the various empirical findings in neuroscience. Besides an empirical starting point, one may also take an epistemological starting point when determining the existence and reality of the brain. This approach is taken by philosophers like Colin McGinn and Thomas Nagel. They suggest that our purely observational access to the brain in third-person perspective leaves open the neuronal mechanisms that underlie consciousness and other mental features of our mind. Apparently, according to Nagel and McGinn, we have some epistemic limitation which makes it impossible to access directly those properties in our brain that underlie our mental features.

This position concerns our epistemological access to the brain and our epistemological capacities. What does this imply for the metaphysical characterization of the brain? This is where McGinn and Nagel depart from each other. McGinn argues that this shows that our brain must possess some additional properties besides the physical ones. He thus assumes what he describes as property P in the brain. Property P, according to McGinn, underlies consciousness and other mental features; this property P cannot be purely physical, since then we could access it. However, it is not yet fully mental either (see Chapter 8 for details).

Let's leave the exact description of McGinn's position for Part II. What is important here is that McGinn, considering the presence of an epistemic limitation in our ability to access the brain, infers the metaphysical domain of existence and reality of the brain. He thus commits what can be described as epistemic-metaphysical inference. Nagel, in contrast, rejects any epistemic-metaphysical inference. He only assumes that there may be some features in the 'deep interior of the brain' that we haven't discovered yet. Nagel considers these unknown features physical, and thinks that they may turn out to be central for consciousness and other mental features. Nagel's characterization of the brain may open a new and different way of describing and conceptualizing the brain and its various features. This in turn may be important for addressing the mind–brain problem in a different way. Since the detailed description of McGinn's and Nagel's position extends far beyond the characterization of neurophilosophy, I devote Chapter 8 to the discussion of their positions including the implications for the mind–brain problem.

Domains of neurophilosophy 5

Neuroontology: brain-based neuroontology

How does the brain-reductive neurophilosopher view McGinn's and Nagel's position? They consider it to be mysterious, since they assume certain features beyond the physical ones that we can detect and access. From the brain-reductive neurophilosophical perspective, there is no essential difference between McGinn and Nagel because both refer to properties in the brain that we currently (or even in principle) remain unable to observe. This, however, looks rather mysterious to the empirically-minded brain-reductive neurophilosopher who might therefore declare Nagel and McGinn to be 'mysterians'.

Despite their differences, the brain-reductive neurophilosopher and the mysterians share the way they come to their metaphysical characterization of the brain. Both infer from one domain to another: the neurophilosopher infers from the empirical to the metaphysical domain, while the mysterian infers from the epistemic to the metaphysical domain. One may thus want to speak of empirical-metaphysical and epistemic-metaphysical inferences (see Chapters 7 and 8 for details on these and other kinds of inferences).

Are such inferences from one domain to another valid? Are these types of inferences good methodological strategies? One may, for instance, claim that the empirical domain and its empirical observation do not imply anything about the metaphysical domain. And the same can hold true for the epistemological domain, which only provides knowledge but no information about existence and reality (see Chapters 7 and 8 for details).

What does this tell us about neuroontology in general? Neuroontology focuses on the metaphysical characterization of the brain, its existence and reality. We will see later that a proper neuroontological characterization of the brain may need to go beyond the traditional physical and non-physical dichotomy. Furthermore, any neuroontological investigation of the brain must be accompanied by reflection about the kind of methodological strategy that works, and especially the kinds of inferences like empirical-metaphysical or epistemological-metaphysical.

Finally, one may also want to search for the conditions underlying the various inferences we draw in order to come to neuroontological assumptions about the brain. If these conditions themselves can be related to particular neuronal mechanisms and thus the brain itself as their necessary (but non-sufficient) basis, we will be able to develop a truly brain-based neuroontology of both the brain and the world.

Domains of neurophilosophy 6

Neuroethics

In addition to epistemological and metaphysical domains, the ethical domain is another classical domain of traditional philosophy. The peculiarity of the ethical domain is the introduction of the normative dimension – the inclusion of norms and prescriptions as distinguished from the descriptive dimension. Rather than asking for a mere description about how something like the brain works and functions, the normative dimension focuses on how a person *should* behave.

How can the purely descriptive ('what is') empirical domain of the brain and neuroscience be related to the normative dimension ('what ought to be') and thus the ethical domain? Recent progress in neuroscience has led to various questions that touch upon ethical issues and to the emergence of a novel field, neuroethics. For the sake of space, neuroethics as a field of study will only be discussed briefly. For further details, refer to the suggested reading given below.

Neuroethics can broadly and preliminarily be defined by the drawing of relationships between neuroscientific observations and ethical concepts. Adina Roskies (2002) distinguished between the ethics of neuroscience and the neuroscience of ethics. The ethics of neuroscience deals with ethical problems in neuroscience and thus with issues like validity of informed consent in psychiatric patients, enhancement of cognitive functions by neuroscientific interventions, and coincidental findings in neuroimaging, etc. This may be subsumed under the concept of 'practical neuroethics'.

Meanwhile, the 'neuroscience of ethics' investigates the psychological and neural mechanisms that may possibly underlie ethical concepts like informed consent, moral judgment, free will, etc. There is, however, no sharp distinction between the neuroscience of ethics and the ethics of neuroscience. Consider the example of informed consent. How valid is the informed consent of patients whose cognitive and emotional capacities are altered or deteriorated by the nature of their very disorder as in, for instance, dementia or schizophrenia? This is an issue in the domain of the ethics of neuroscience. At the same time, it requires empirical investigations of those cognitive and neural functions that are implicated in giving valid informed consent which, in turn, falls more into the domain of the neuroscience of ethics (Northoff *et al.*, 2010a, 2010b).

Another example where the neuroscience of ethics and the ethics of neuroscience converge is in the recent discussion about moral judgment. What is moral judgment and how does it affect our ethical decisions in current neuroscience? This question falls into the ethics of neuroscience. At the same time, many imaging studies have been conducted to investigate the neural basis of moral judgment, and how they differ from non-moral but emotional judgments. This falls within the domain of the neuroscience of ethics.

The topic of brain function enhancement is a highly discussed subject. Because of its rapid development, neuroscience develops novel tools and drugs to modulate brain activity and its various cognitive functions, not only in the diseased brain, but also in the healthy brain. Does this lead to changes in our self and personal identity? And what are the ethical boundaries of such enhancement? We may, for instance, use pharmacological drugs to enhance our memory and attention skills and to improve our job performance.

Finally, one may also raise a more methodological issue. How can we make the translation between ethical concepts, including their normative dimension, and the descriptive dimension of neuroscientific concepts and findings? We need predefined rules and valid methodological strategies for linking ethical concepts and neuroscientific findings. These conceptual and methodological issues may be subsumed under the umbrella of theoretical neuroethics, as distinguished from a more empirical neuroethics.

Method of neurophilosophy 1

Is there a specific method in neurophilosophy?

Disciplines like philosophy and neuroscience can be characterized by specific domains and methods. Domains describe the kind of problems, issues and questions a discipline deals with. For instance, philosophy focuses on the metaphysical domain, which deals with questions of existence and reality. Philosophy also discusses questions of knowledge. These are part of the epistemological domain. Questions about moral behavior are part of the ethical domain. In contrast, neuroscience operates within the empirical domain, which focuses on questions about the world as we can observe it in third-person perspective.

Neurophilosophy, as we have seen, also presupposes a specific domain. Reductive neurophilosophy replaces the metaphysical and epistemological domains of philosophy with the empirical domain of neuroscience. This results in domain monism. Meanwhile, non-reductive neurophilosophy prefers domain pluralism. It includes all domains – empirical, metaphysical and epistemological (and ethical) – and focuses specifically on their overlap and linkages, also known as domain linkages. As such, the domain of non-reductive neurophilosophy consists in domain linkages, rather than a specific domain by itself.

How about the method of neurophilosophy? Besides the domain(s) it covers, a discipline can be characterized by the specific methodological strategy it applies to its domain(s). Philosophy applies conceptual-linguistic (and/or phenomenal) analysis, while science/neuroscience is characterized by the observational-experimental approach.

What is the method specific to and characteristic of neurophilosophy? Neurophilosophy seems to stand in between both conceptual-linguistic and observational-experimental methodological strategies. One may then put the emphasis on the linkage between the different methods, implying methodological pluralism.

In that case neurophilosophy would be a distinct discipline distinguished from both neuroscience and philosophy on methodological grounds: the methodological monism of both philosophy and neuroscience would be replaced by methodological pluralism and linkage in neurophilosophy. That would amount to non-reductive neurophilosophy that then could be characterized by a specifically neurophilosophical methodology.

However, one may also shift the balance completely toward the observational-experimental pole of neuroscience, with the conceptual-linguistic method subsumed by the primacy of the scientific method. In that case there would be methodological monism rather than methodological pluralism. And, most important, neurophilosophy could then no longer be considered a separate discipline as distinct from both philosophy and neuroscience. Instead, neurophilosophy would be subsumed under neuroscience where it would be basically the theoretical branch of neuroscience. This would thus amount to reductive neurophilosophy that then would no longer be characterized by a specific neurophilosophical method as distinguished from neuroscience.

Method of neurophilosophy 2

Can we unilaterally infer concepts *from* facts?

Let us sketch the methodological strategy of reductive neurophilosophy in further detail. The unilateral methodological focus leads to what I described above as reductive neurophilosophy characterized by methodological monism. There is methodological monism because it ultimately restricts neurophilosophy to one specific methodological strategy. The advocates of the reductive concept of neurophilosophy claim Quine to be their methodological godfather. According to them, Quine opened the door for their methodology when he eliminated the principal distinction between conceptual-logical and observational-experimental methodological strategies.

The current neurophilosophers follow up on this idea and apply it to the relationship between mind and brain, and between philosophy of mind and neuroscience. Methodologically, reductive neurophilosophy can be seen as a theoretical abstraction of the empirical data provided by the observational-experimental method of neuroscience. Such theoretical abstraction is described in term of concepts.

How are concepts generated? Recall from Chapter 2: traditionally, philosophy presupposed the meaning and definition of specific concepts in order to make logical inferences while relying exclusively on analytic sentences ('All bachelors are unmarried') and a priori knowledge (knowledge based on logical inference only without any additional observation). This contrasts with science, whose data and facts yield synthetic sentences ('The tree and its leaves are green rather than yellow') and a posteriori knowledge (knowledge-based observation rather than pure logical inference).

Based on Quine and his rejection of this dichotomy between philosophical concepts and scientific data/facts, reductive neurophilosophy considers philosophical concepts as mere abstractions of scientific data/facts. Philosophical concepts do not refer to something the scientific data/facts describe; instead, philosophical concepts present the same material and content but in a more abstract way.

According to this model, philosophical concepts are part of the empirical domain of scientific data/facts. While this resolves the difference between logical and empirical domains, it leaves open how the differences between concepts and facts within the empirical domain itself can be bridged. This is a central question for any neurophilosophy that aims to avoid reducing philosophical concepts to scientific data/facts. Only by relying on a methodological strategy that allows for coexistence between philosophical concepts and scientific data/facts will we be able to develop a truly brain-based and thus non-reductive neurophilosophy as distinguished from its more brain-reduced and thus reductive Anglo-American sibling.

Method of neurophilosophy 3

'Bootstrapping' with concepts as *outputs* (Churchland): definition

How can we link data/facts and concepts in a systematic way? Patricia Churchland, the originator of the term 'neurophilosophy' (see above) suggests a pragmatic strategy with a specific method called 'bootstrapping'. In order to understand 'bootstrapping' as a distinct methodological strategy, we need to briefly return to philosophy in general and philosophy of mind in particular.

Traditionally, philosophical investigation starts with the definition of a particular concept; for example, the concept of the mind. The mind may be defined as a substance and by specific phenomenal features like subjectivity (experience in first-person perspective) and intentionality (directedness of mental states towards contents) (see Chapter 1). These definitions can then serve as a starting point to discuss whether or not the mind is a purely physical substance and can thus be traced back to brain and body. One could equally ask whether or not the mind is a mental substance, distinct from the body and the brain.

Churchland changes the starting point in her 'bootstrapping' method. Rather than starting with a clear-cut definition of the phenomenon and its concept in question, Churchland suggests that one starts with typical empirical cases where everybody agrees that these reflect the phenomenon in question. For example, one might start with patients who have brain lesions. One may consider neurological patients with brain lesions in selective regions and concurrent changes in the contents of consciousness, like in visual content (see Part IV). Or, even more extreme, one may start with patients suffering from loss of consciousness, as in a vegetative state (VS) (see Part IV).

One may now investigate these patients in a purely empirical way by using the scientific method (observational-experimental investigation). This will ultimately increase our knowledge about the neuronal mechanisms underlying the changes in or loss of consciousness in brain lesion or VS patients. That in turn may then also provide some insight into how to define the concept of consciousness and ultimately the mind and how it stands in relation to the brain. The definition of the concept of consciousness is, here, not presupposed. Instead it is taken as a starting point, as in traditional philosophy. The definition of the concept is inferred from the empirical data and facts. In short, the (definition of the) concept is no longer regarded as input but rather as output.

Method of neurophilosophy 4

'Bootstrapping' with concepts as *outputs* (Churchland): *example*

Let us discuss Churchland's method of bootstrapping by using a specific example. Besides subjectivity (subjective experience in first-person perspective) and intentionality (directedness towards contents), unity is often regarded as another phenomenal feature of consciousness. Unity describes that we experience the different contents of consciousness in a unified form. For instance, despite there being many objects lying in front of me on the table – a laptop, pens, various papers, several books, etc. – I nevertheless experience all these distinct contents as contents of one and the same, unified, consciousness. Where does this unity come from? If the mind is to be replaced by the brain, the brain itself must provide such unity.

The brain consists of two halves – right and left hemispheres – that are closely connected to each other via a structure called the corpus callosum. This structure is assumed to provide the unity of the brain. In some cases, however, the corpus callosum is disrupted so that the two hemispheres are no longer connected and the unity of the brain is disrupted. This is described as 'split brain'. Do split-brain patients no longer have unity in their consciousness? Rather than discussing the concept of unity, as a traditional philosopher of mind will do, Churchland suggests that we start with the empirical observation and experimental investigation of these patients. This will then also impact and reveal the kind of concept of unity that is compatible with and suits the facts.

Let's listen to Churchland herself: 'A standard principle, illustrated by the split-brain results, is that the definition of the phenomenon to be explained coevolves with experimental discoveries. In the early stages of the scientific attack on any problem, accurate definition of the phenomenon is hampered precisely because not enough is known to permit an accurate definition. A pragmatic strategy is to begin by studying those cases agreed to be obvious examples of the phenomenon. Powered by this agreement, provisional, rough characterization can leverage the science's first stages, with refinements in the phenomenon's definition emerging as the surrounding facts become clear' (Churchland, 2002, p. 309).

On the whole, 'bootstrapping' provides a methodological strategy to infer the definition of concepts from neuroscientific data and facts. This, according to Churchland and many other Anglo-American colleagues, distinguishes neurophilosophy from philosophy. Rather than starting with the concept in question and its definition, neurophilosophy (in a reductive sense) reverses the methodological starting point by beginning with the observational-experimental investigation of an apparently related empirical phenomenon. This in turn allows one to unilaterally infer the definition of the concept in question from the (empirical observation of the) respective neuroscientific data/facts.

Method of neurophilosophy 5

Unilateral *adaptation* of concepts *to* facts (Searle)

Let us introduce John Searle (1932–), a well-known philosopher who teaches at University of California, Berkeley in the USA. What does Searle think about the relationship between concepts and facts? We recall from above: Churchland suggests that one infers the definition of philosophical concepts from neuroscientific data and facts.

However, data and facts by themselves already presuppose concepts and a specific definition. For instance, the investigation of vegetative patients with loss of consciousness already presupposes a certain meaning of the concept of consciousness. Is Churchland's methodological strategy of 'bootstrapping' thus circular? The charge of circularity may come to mind since any data and facts presuppose some kind of concept. At the same time, though, the method of 'bootstrapping' aims to infer concepts from the data/facts. This is a circular position since concepts cannot be inferred from data/facts which by themselves presuppose concepts. In short, concepts may be presupposed by data and facts, rather than being inferred from them.

This is the point where Searle steps in. He suggests that we adapt the concepts and their definition to the data and facts rather than inferring the former from the latter. Let us exemplify his approach using the concept of the mind and the metaphysical dichotomy between the mental and the physical. Recall that traditional philosophy starts with the definition of a particular concept. For instance, philosophy of mind starts with and (more or less) presupposes the metaphysical dichotomy between the mental and the physical. As such, the metaphysical concept of the mental is determined by experience (consciousness) in the first-person perspective while the metaphysical concept of the physical is associated with observation in the third-person perspective.

How is the first-person perspective related to the notion of the physical? If we believe in the physical–mental dichotomy, first-person perspective has no place in the physical. Any physical account of the mind in terms of the brain must leave out the first-person perspective and ultimately experience and consciousness. Searle now disputes that claim. He argues that there are differences between the physical realm itself and the way we conceive and conceptualize it. The first-person perspective can have a place in the physical realm itself, whereas this seems to be impossible in our conception and understanding of the physical reality. We thus, according to Searle, need to change our conception of physical reality in order to accommodate the first-person perspective.

How exactly do we conceptualize and understand physical reality? Searle argues that we usually take for granted the conceptual dichotomy between the mental and the physical. This, however, makes any account of the mind in terms of the brain and its neuronal (physical) functions impossible from the very beginning. By definition, the concept of the physical excludes the first-person perspective and thus consciousness is excluded. This makes it impossible for neuroscience to account for consciousness.

This, however, contradicts the facts. In recent research on consciousness, specific neuronal mechanisms may be related to the first-person perspective itself and its association with experience (see Part IV). The empirical observations and thus the neuroscientific facts contradict our concept of the physical as being associated only with the third-person perspective. What can we do? We cannot change the facts. But we can modify our concepts to accommodate the facts. This means that the neuronal data/facts are compatible with consciousness and the first-person perspective. Thus, the concept of the physical used to describe the brain must be extended to include the first-person perspective.

Method of neurophilosophy 6

Conceptual propaedeutic as *input* (Searle)

What does this unilateral adaptation look like? In the case of the mental–physical dichotomy, we need to redefine the concept of the physical by associating it with both first- and third-person perspective, resulting in what Searle calls 'first- and third-person ontology' (see Chapter 7 for a more detailed account of Searle's position on this issue): 'Once you revise the traditional categories to fit the facts, there is no problem in recognizing that the mental qua mental is physical qua physical. You have to revise the traditional Cartesian definitions of both "mental" and "physical", but those definitions were inadequate to the facts in any case' (Searle, 2004, pp. 82–3).

By adapting the definitions of originally philosophical concepts to the empirical findings of neuroscience, the revised concepts open the door for linking philosophical concepts to neuroscientific facts in novel ways. This empirical adaptation of the originally metaphysical (or epistemological or ethical) concepts makes it possible to link their respective domains (metaphysical and epistemological) to the empirical domain of neuroscience. Hence, by shifting the initial conceptual definitions towards the empirical domain, different domains, like the metaphysical and the empirical, can be linked to each other.

Searle argues that by removing such conceptual confusions about the mind, philosophy of mind opens the door for science and more specifically for neuroscience. This results in questions about the nature of the mind that 'cease to be philosophical and become scientific' (Searle, 1999, p. 2070). For example, Searle argues that the problem of life was once considered to be a philosophical problem. Now, after scientific progress, the problem of life is no longer considered as such, but, instead, as a biological one (Searle, 1999, p. 2070).

How does such unilateral adaptation of concepts to facts impact the relationship between philosophy and neuroscience? Following Searle, philosophy will then become a conceptual propaedeutic to neuroscience. Proper definition of the concepts is necessary to make them amenable and accessible to scientific investigation. As such, rather than being the output – as in Churchland's bootstrapping – the concepts and their definitions provide the input to subsequent observational-experimental investigations and the generation of data and facts. This is, as we will see in the following, the first step toward establishing a less reductive and thus non-reductive methodological strategy that is specific to neurophilosophy as distinct from both neuroscience and philosophy.

Ideas for future research 1

Churchland versus Searle

What does Searle do, and can his approach be distinguished from traditional philosophy? Unlike traditional philosophy, he no longer determines the concepts of the physical and the mental exclusively using conceptual and logical requirements. This signifies what may be defined as conceptual plausibility. Conceptual plausibility describes the fulfillment of conceptual and logical criteria by the definition of the concept in question.

In addition to conceptual plausibility, Searle also requires the concept to be in accordance with the empirical data and facts. They need to be not only conceptually plausible, but also empirically plausible. What do I mean by the terms 'conceptual and empirical plausibility'? *The concepts of empirical and conceptual plausibility describe the accordance and correspondence of conceptual determination with the currently available empirical data.*

How does such conceptual and empirical plausibility stand in relation to Churchland's bootstrapping approach? She considers concepts and their definition as mere output, rather than as input, as Searle does. This means that the plausibility of concepts is completely and exclusively determined by their degree of accordance with the data and facts from which they are inferred. Hence, the only criterion of plausibility Churchland seems to presuppose is empirical plausibility. In her model, conceptual plausibility is neglected and discarded. In this light, Searle's and Churchland's methodological strategies seem to be incompatible with each other.

Ideas for future research 2

'Conceptual feedforward input' and 'feedback output'

Are Searle's and Churchland's methodological strategies really incompatible? Let us have a look at how neuroscience proceeds. Take the example of consciousness. A neuroscientist wants to do a functional imaging study of consciousness and aims to design a corresponding experiment. For that he must have some kind of definition of consciousness, even if only an operational definition. This operational definition provides some criteria according to which the neuroscientist can, for instance, distinguish between the presence and absence of consciousness (and thus between consciousness and unconscious). Without this operational definition and its criteria, the neuroscientist remains unable to make sense and interpret his data and to associate specific neuronal mechanisms with particular phenomenal features of consciousness.

Based on the definition of consciousness, the neuroscientist will select the criteria. If, for instance, consciousness is defined in a phenomenological way (as phenomenal consciousness), he will need criteria for its phenomenal features, like subjectivity and intentionality. If, in contrast, the researcher defines consciousness in a cognitive way, as a higher-order meta-representational function, he needs cognitive criteria (see Part IV).

What does this example imply for the methodological strategy? This example makes it clear that the definition of the concept in question biases and strongly impacts the kind of criteria around which the scientist builds his experimental design. As such, we cannot discard (and eliminate) conceptual input in the observational-experimental investigation. There is a need to consider what Searle calls 'conceptual propaedeutic' which may be described as *'conceptual feedforward input'* to observational-experimental investigation.

In addition to such 'conceptual feedforward input', the data and facts yielded by the observational-experimental investigation may also be compared and matched with the initial concepts and its definition. This is the step Churchland described in her method of bootstrapping. While she claims that the definition of the concept can be inferred from the data and facts, it may consist in comparing and matching data/facts and concepts. Such matching may then lead to the redefinition of the concept in question in order to increase empirical plausibility in the case of a mismatch between the initial conceptual definition and the acquired data/facts. The redefined and empirically plausible concept may then be regarded as the *'conceptual feedback output'* of the observational-experimental investigation.

One can go even one step further. The redefined empirically plausible concept may now be put back into the originally philosophical context and its respective domains like the metaphysical and epistemological domains. For instance, the redefined empirically plausible concept of consciousness may now be investigated with regard to its implications for the mind–brain problem as a metaphysical problem. The redefined empirically plausible concept thus re-enters the philosophical context – the metaphysical domain in this case – from where it originates. In this case, one may speak of, put more technically, a *'conceptual re-entrant loop'*.

Ideas for future research 3

'Conceptual re-entrant loop' and 'concept–fact iterativity'

Here, we distinguished between different stages in neurophilosophical investigation. The first stage consists of 'conceptual feedforward input', where an originally philosophical concept is taken as input for developing operational criteria for subsequent observational-experimental investigation. This is followed by the observational-experimental investigation itself, whose data and facts can then be compared and matched with the original definition of the concept. That leads to 'conceptual feedback output' in which the concept is redefined in an empirically plausible way.

Finally, the redefined concept may be put back into its original philosophical space (the metaphysical or epistemological (or ethical) domain) in order to investigate its implications for the respective philosophical questions. More specifically, so far the concept is redefined within the empirical context of neuroscience; the redefined concept is thus an empirical concept that as such can be situated within the empirical domain. However, the neurophilosopher does not only aim for empirical concepts, as the neuroscientist does, but also for concepts that are applicable in other domains, like the metaphysical and the epistemological domains. He/she thus has to put his/her redefined concept from the empirical domain back into the context of the metaphysical or epistemological domain from where it was originally derived.

How can the neurophilosopher do that? The neurophilosopher must now investigate what his/her redefined concept in the empirical domain implies for the other domains, like the metaphysical and epistemological domains. More specifically, he/she now has to directly compare and match his/her redefined concept in the empirical domain with its original definition in the metaphysical or epistemological domain. The original metaphysical or epistemological concept that has been redefined in the empirical domain is thus put back or re-entered into its original domain, the metaphysical or epistemological domain. In Chapters 13–20 we will see concrete applications of such a methodological approach.

This methodological step can be described as conceptual re-entrant loop that yields truly neurophilosophical concepts distinguished from merely neuroscientific or philosophical concepts. These can then be compared and matched with the initial philosophical concepts. From there, one can start again and feed the concept as input into further observational-experimental investigations. *Hence, there will be continuous exchange with an iterative movement between concepts and facts where both continuously and reciprocally modify and redefine each other. One may thus want to speak of what can be described as 'concept–fact iterativity'.*

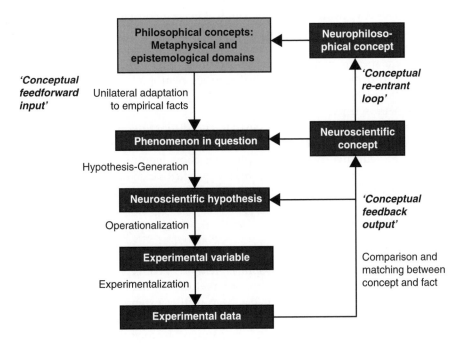

Figure 4.2 Concept–fact iterativity
The figure illustrates concept–fact iterativity as a bilateral transdisciplinary methodological strategy in non-reductive neurophilosophy. Unlike in Churchland, philosophical concepts are still considered, and unlike in Searle they are bilaterally connected with the data/facts. The philosophical concepts impact data/facts and are presupposed by the latter as indicated by the arrow to the phenomenon in question. At the same time the philosophical concepts receive some feedback from the data/facts themselves as indicated by the arrow from the neurophilosophical concepts in the right upper part and the gray shading in the philosophical concepts box. There is thus double feedback, from the neuroscientific concept to the phenomenon in question as well as from the neurophilosophical concept to the philosophical concept.

Ideas for future research 4

'Concept–fact iterativity' as transdisciplinary methodological strategy

Including different steps like 'conceptual feedforward input', 'conceptual feedback output' and 'conceptual re-entrant loop', concept–fact iterativity crosses the boundaries between different methods and different domains. By conducting the different steps, concept–fact iterativity crosses and links originally philosophical domains, like metaphysical and epistemological domains, to the empirical domain of neuroscience. This is called 'domain linkage' (as distinguished from 'domain separation') and 'domain pluralism' (as distinguished from 'domain monism').

Furthermore, concept–fact iterativity crosses the boundaries between different methodological strategies, without forcing any given strategy to be subservient to any other. There is thus methodological pluralism rather than methodological monism. This makes concept–fact iterativity a truly cross-disciplinary methodological strategy. As such, concept–fact iterativity is the methodological ally of the non-reduced and thus brain-based neurophilosophy as distinguished from the reduced and brain-reductive neurophilosophy.

In sum, concept–fact iterativity characterizes the methodological strategy that is specific to non-reductive neurophilosophy. This will make it possible for neurophilosophy to establish itself as a separate discipline distinct from both neuroscience and philosophy. Neurophilosophy as a separate and distinct discipline can then be characterized by a specific method, concept–fact iterativity. Concept–fact iterativity as a specific methodological strategy distinguishes neurophilosophy from both the conceptual-linguistic approach of philosophy and the observational-experimental method of neuroscience.

In contrast, such distinction as a separate and distinct discipline remains impossible in reductive neurophilosophy that aligns both its domains and methods with neuroscience. Reductive neurophilosophy can consequently not be regarded as a separate and distinct discipline but, at best, a specific branch of neuroscience and its exclusive focus on the empirical domain and the observational-experimental strategy.

Ideas for future research 5

'Concept–fact iterativity' versus 'concept–argument iterativity'

How does the proposed neurophilosophical method of concept–fact iterativity stand in relation to the method of philosophy? Philosophy traditionally relies on the rational-argumentative method that has been further extended into a conceptual-logical approach, especially in the twentieth century. The hallmark of such a method is the use of definition of concepts and their support by logical plausibility as set up in arguments. Roughly, different definitions of the concept in question are tried out and compared with each other by using arguments. The definition for which the arguments are logically consistent is then the definition that is accepted.

Most important, the emphasis here is not so much on the definition itself but on the arguments and their logical plausibility. How can we provide a logically sound argument that fits the definition of the concept in question? Usually, the arguments and their logical plausibility are here adapted to the definition of the concept: the arguments and their logical consistency and plausibility are adapted and worked on to fit and match the definition in an iterative way. One may thus want to speak, analogously to concept–fact iterativity, of 'concept–argument iterativity' that describes the iterative adaptation of the logical arguments to the definition of the concept.

How does this concept-argument iterativity in philosophy stand in relation to the proposed concept–fact iterativity in neurophilosophy? Taking the perspective of the philosopher, concept–fact iterativity must simply be regarded a category error where logical and natural world, arguments and facts, are simply confused. 'Wrong category, Mr Neurophilosopher,' the philosopher will proudly throw into the room, 'what you do has nothing to do with philosophy at all.'

Given his perspective and presupposition of concept–argument iterativity, the philosopher is right. However, that changes as soon as one changes the presuppositions. Instead of logical consistency and plausibility, empirical consistency and plausibility may be more important. This entails the need to shift the focus from arguments to data and facts: the focus is then no longer on the logical consistency and plausibility of arguments but on the empirical consistency and plausibility of the concepts and their definitions with the empirical data and facts. In short, we need to shift our emphasis from arguments to facts.

Ideas for future research 6

Philosophy versus neurophilosophy

Such a shift from arguments to facts implies two major changes in methodological strategy. First, we need to replace arguments with data and facts. We thus need to replace the traditional philosophical method of concept–argument iterativity by the non-reductive neurophilosophical method of concept–fact iterativity. Though wrong and considered a category error from the perspective of the philosopher and his/her concept–argument iterativity, the method of concept–fact iterativity must be considered valid once one puts the priority on empirical rather than logical plausibility.

Second, we need to make it possible for the definition of the concepts to be changed by themselves and adapted to the empirical data and facts. For that we include the re-entrant loop from the neuroscientific concepts back to the originally philosophical concepts. This sounds strange to the philosopher. He/she does not change the definition of his/her concepts but only his/her arguments to make the definition logically consistent and plausible.

However, unlike the philosopher who can change his/her arguments to account for logical consistency and plausibility, the neurophilosopher cannot change his/her data and facts. He/she must thus change the definition of the concepts themselves in order to achieve empirical consistency and plausibility; this is achieved by the final re-entrant loop from neuroscientific to philosophical concepts in our concept–fact iterativity. While such a final re-entrant loop and the change in the definition of concepts may be considered flawed from a philosophical perspective, it must be considered necessary in a truly non-reductive neurophilosophical framework.

The shift from concept–argument iterativity of philosophy to concept–fact iterativity in neurophilosophy has major implications for how to discuss philosophical issues. The philosopher puts the focus on arguments and their logical consistency;

the definition of the concepts are secondary here. The philosopher tries out different arguments and sees whether they fit the definition. Logical plausibility is the priority here; empirical plausibility comes, at best, second or, at worst, is neglected completely.

This is different in neurophilosophy. Here the focus is on the definition of the concepts, since these provide the starting point for their subsequent comparison with the data and facts. Different definitions may thus be directly compared with each other and ultimately with the empirical data and facts. In contrast, the arguments on which the philosopher puts such strong emphasis are not of major importance here. Instead of arguments and their logical plausibility, the neurophilosopher focuses more on the definition of the concepts and their empirical plausibility. Following such a methodological strategy, we will, for instance, in the following part about the mind–brain problem put the emphasis on the definition of the different mind–brain theories rather than on the arguments for the different positions. While this may confuse the traditional philosopher, who is used to arguments rather than definitions, it may open the door for a truly neurophilosophical approach to the mind–brain problem.

Take-home message

In the previous chapters we discussed the twentieth-century developments in both philosophy and neuroscience. Philosophy was very concerned with the mind (Chapter 1) and its rapprochement to the sciences (Chapter 2). Meanwhile neuroscience developed as separate discipline, blossoming more and more by providing novel insights about the brain and how its neuronal states are related to mental features like consciousness, self, free will, etc. which were originally discussed in philosophy (Chapter 3). These convergences from both sides – philosophy on the one hand, and neuroscience on the other – provided the ground for their interdisciplinary collaboration which resulted in the development of neurophilosophy. Depending on the balance between philosophy and neuroscience (their domains and methods), one may define different concepts of neurophilosophy. Reductive neurophilosophy subsumes the philosophical domains (metaphysical, epistemological and ethical) and methods (conceptual-linguistic) to the empirical domain and the observational-experimental method of neuroscience, while non-reductive neurophilosophy takes domains and methods from both philosophy and neuroscience and links them, rather than reducing one to the other. Non-reductive neurophilosophy in this sense opens up novel fields or domains of investigation, like neuroepistemology, neuroontology and neuroethics. At the same time it makes necessary the development of a methodological strategy like concept–fact iterativity that is truly cross-disciplinary and thus neurophilosophical as distinguished from both neuroscientific and philosophical methods.

Summary

The previous chapters in this part focused on the philosophical investigation of the mind (Chapter 1), how philosophy stands in relation to science (Chapter 2), and the development of neuroscience (Chapter 3). This provided the background against which the brain and its investigation in neuroscience entered and permeated traditional philosophical issues like the mind, knowledge and ethical norms. The developments on both sides, philosophy and neuroscience, increased the potential for interdisciplinary collaboration, which eventually led to the formation of what is described as neurophilosophy. The concept of neurophilosophy, its domains and methods, is the focus of the present chapter. Non-reductive neurophilosophy describes the linkage between philosophical issues and neuroscientific findings. More reductively defined, as in the Anglo-American context, reductive neurophilosophy is described as the application of the scientific method (the observational-experimental investigation of the brain) to problems and questions originally discussed in philosophy. This implies that ultimately all philosophical concepts, irrespective of their original domain, are reduced to the empirical domain of neuroscience and the brain. I describe this approach as reductive neurophilosophy because it ultimately amounts to what has been described as replacement naturalism or neuralism, implying a brain-reduced stance. This model may be contrasted with a non-reductive neurophilosophy that is brain-based rather than brain-reduced. This alternative approach considers, in addition to the empirical domain, other domains such as the phenomenal, epistemological, metaphysical and ethical, implying domain pluralism rather than domain monism. This led to the developments of novel approaches like neurophenomenology, neuroethics, neuroepistemology and neuroontology, which are briefly described and characterized. How can we link the different domains and thus ultimately neuroscience and philosophy? To do so, a cooperative strategy, as suggested in 'cooperative naturalism' (see Chapter 2) is needed. Such a methodological strategy must allow for dissociation between the traditional linkages of particular domains with specific methodological strategies. At the same time, it must allow for the different domains to be linked to each other in a non-reductive and thus transdisciplinary way. This is possible by linking and integrating concepts and facts by moving continuously back and forth between them. This leads to concept–fact iterativity as a truly transdisciplinary methodological strategy. This allows for the establishment of non-reductive neurophilosophy that, unlike reductive neurophilosophy, is a separate discipline and distinguished from both neuroscience and philosophy.

Revision notes

- What is neurophilosophy?
- How did it evolve? Name some of the predecessors and their claims.
- What does the concept of neurophilosophy stand for in the Anglo-American world?
- Why is that concept of neurophilosophy reductive?
- What is neurophenomenology? How can it be distinguished from the Anglo-American approach to neurophilosophy?
- What is non-reductive neurophilosophy? How can it be characterized and distinguished from reductive neurophilosophy?

- What is neuroepistemology?
- Define some approaches to neuroontology.
- What are the difficulties of linking neuroscience and ethics?
- Why is the difference between replacement and cooperative naturalism relevant for neurophilosophy?
- What is bootstrapping?
- How can bootstrapping and concept–fact iterativity be compared?
- Why is concept–fact iterativity so important for defining neurophilosophy?

Suggested further reading

- Bickle, J. (2009) *The Oxford Handbook of Philosophy and Neuroscience* (Oxford/New York: Oxford University Press). *This provides a good overview of the current themes and trends in reductive neurophilosophy.*
- Churchland, P. (2002) *Brain Wise. An Introduction into Neurophilosophy* (Cambridge, MA: MIT Press). *This is a later version of Churchland's approach to neurophilosophy with many neuroscientific details and how they are relevant for metaphysical and epistemological issues.*
- Northoff, G. (2013a, 2013b) *Unlocking the Brain. Vol. I: Coding, Vol. II: Consciousness* (Oxford/New York: Oxford University Press). *Though not outlining non-reductive neurophilosophy, these two volumes provide an excellent application of such methodological approach to the question of consciousness.*
- Petitot, J. and Varela, F. (1999) *Naturalizing Phenomenology* Stanford, CA: Stanford University Press).
- Roskies, A. (2002) 'Neuroethics for the new millennium', *Neuron*, 35, 21–3. *This is a seminal paper which provides an excellent overview and definition of neuroethics.*

Part II
The Mind–Brain Problem: From Philosophy of Mind to Philosophy of Brain

The first part of this textbook focused on developments in philosophy and neuroscience that led to the establishment of neurophilosophy as a separate discipline. One central topic in neurophilosophy in this sense is the relationship between mental capacities and the neural functions of the brain. This relationship has traditionally been discussed in philosophy, and especially in current philosophy of mind, as the mind–brain problem: what is the existence and reality of the mind and how is it related to that of the brain? This is described as the mind–brain problem – a metaphysical problem that explores the existence and reality of the mind and its relationship to the brain.

Different theories about the mind–brain relationship have been suggested in philosophy of mind. These different theories and how they relate to a brain-based (rather than brain-reductive; see Chapter 4) neurophilosophy are the focus of this part. More specifically, based on the metaphysical characterization of the mental capacities as mental, physical, neutral, or neural, I will distinguish different chapters: Chapter 5 is about mental approaches like mind–brain dualism, Chapter 6 concerns physical characterization like materialism, and Chapter 7 describes approaches that stay neutral to the mental–physical dichotomy. Finally, Chapter 8 focuses on the brain itself and its metaphysical (and epistemological) characterization with regard to the mind–brain problem.

Before going ahead, we need to make it clear that the presentation of the different mind–brain theories relies on a succinctly neurophilosophical rather than either philosophical or neuroscientific methodological strategy. The philosopher focuses on the various arguments pro and contra a particular mind–brain position like dualism or materialism. The main goal of these arguments is to show the logical-conceptual consistency or inconsistency of a particular mind–brain theory. Hence, the central arguments take on a central role in a philosophical approach to the mind–brain problem.

This is different in the neuroscientific approach to the mind–brain problem. Here the main goal is not so much logical-conceptual plausibility and thus different arguments, but rather the observation of certain neuronal processes and mechanisms as they may be related to mental features like consciousness and self. There is thus a shift from mere logical-conceptual plausibility to empirical plausibility in the neuroscientific approach to the mind–brain problem and mental features. This will be discussed in detail in Parts IV and V.

How about a genuinely neurophilosophical approach to the mind–brain problem? In this case the focus is not so much on mere logical-conceptual consistency and thus on the related arguments. Instead, the focus shifts here to the question of whether the proclaimed metaphysical features that are supposed to characterize the mind and its mental features are compatible and thus plausible with the conceptual, epistemological, metaphysical, and empirical features of the brain. For that we need to characterize the mental features themselves in metaphysical terms and investigate how that is compatible with the characterization of the brain.

In order to make such a neurophilosophical approach possible, here we shift the focus from the purely philosophical arguments pro and contra the different mind–brain positions to the characterization of the mental features themselves and what they imply for the characterization of the brain. Accordingly, unlike purely philosophical approaches, we focus more on characterizing the mental features rather than on describing in full detail all logical-conceptual arguments. In short, our focus is more on definitions than arguments (see also Chapter 4 for this point).

Therefore, in structuring and organizing the chapters in this part, we replace the traditionally philosophical distinction between dualism and materialism with the more neurophilosophical one between mental and physical approaches. Dualism is then subsumed under mental approaches (see Chapter 5), while materialism will resurface in Chapter 6, about physical approaches. This begs the question for non-mental and non-physical approaches to the mind–brain problem, as will be discussed in Chapter 7. Finally, a neurophilosophical approach to the mind–brain problem must include a discussion of the metaphysical, epistemological, and conceptual characterization of the brain itself; this will be the focus in Chapter 8, which will also be strongly impacted by the distinction between reductive and non-reductive approaches to the brain in particular and neurophilosophy in general.

5
Mental Approaches to the Mind–Brain Problem

Overview

Part I of this textbook focused on the mind as a concept in philosophy and how empirical research on the brain in neuroscience has extended increasingly toward investigating this concept of the mind and its mental features. We also observed a convergent movement in philosophy and neuroscience: philosophy opened itself up to science and the scientific method, and neuroscience linked its empirical investigation of the brain with mental features. As discussed in the previous chapters, these developments led to the birth of neurophilosophy. Of primary importance in neurophilosophy is the metaphysical problem of the relationship between the mind and the brain. Various theories have been put forward as solutions to the mind–brain problem: mental solutions consider the mind and its existence to be different from the brain and its physical reality. In contrast, physical solutions assume the mind to be physical like the brain. This chapter focuses on the mental solutions, which include options described as dualism, epiphenomenalism and panpsychism. Physical solutions to the mind–brain problem will be discussed in the next chapter.

Objectives

- Understand the nature of the mind–brain problem
- Understand the historical origins of the mind–brain problem
- Understand the dualism of Descartes and its problem
- Understand the different forms of dualism
- Understand the problem of mental causation
- Differentiate dualism from epiphenomenalism and panpsychism
- Understand the difference between empirical and metaphysical approaches

Key concepts

Mind–brain problem, mental causation, metaphysical problem, Descartes, interactive dualism, epiphenomenalism, panpsychism.

Background 1

Differences between mental and physical states

We are conscious and have experiences. For instance, I – as the author of this book – experience writing these lines on my laptop that sits in front of me on my desk.

I am conscious of myself and my mental state as I write these lines. Most importantly, my phenomenological experience and my consciousness of this experience does not include anything physical like size, weight, motion, shape or color, and their manifestation in space and time. There is no size, weight, shape, or otherwise in my experience so that consciousness can seemingly not be measured by physical variables. My mental states do not seem to be constrained by physical time and space: they are not spatial in that they are not extended in physical space and time like physical objects are (see Part IV for more details on the experiential features of consciousness).

This example highlights the contrast between experience and consciousness and our observation of the physical world. Unlike mental states, physical states are spatially extended and located in distinct, discrete points in physical time and space. For instance, atoms as parts of molecules occupy particular discrete points in physical time and space; they can be located and 'pinned down'. This is different in consciousness. We do not experience single discrete points in time and space. Instead, we experience a spatial and temporal continuum, a 'stream of consciousness', as William James, one of the founders of psychology (see Chapter 3), said.

These are my mental states. How about the brain and its neural states? Unlike the contents in my mental states like me writing this book, we remain unable to experience our own brains and their neuronal processes (as well as our own body and its physiological processes) as such by themselves, e.g. as the content of our mental states. Nobody has ever experienced his/her own brain and its neuronal states as brain and neuronal states in consciousness. In other words, we cannot become conscious of our brain and its neuronal states as such and experience neuronal contents and states. Instead, we experience what we call mental contents and thus mental states (rather than neuronal contents and neuronal states).

How would it be imaginable, even in principle, to experience our own brains as brains? We simply cannot imagine or conceive experiencing our own brain as content of our consciousness. Consciousness is therefore by default and thus by definition characterized by mental rather than neuronal contents. Taken together, there seem to be major differences between mental and neuronal states. Mental states do not seem to operate in physical time and space, while neuronal states (and physical states in general) do.

Background 2

Mind–brain problem

Due to these differences, which will be further explained and accounted for throughout this book, mental states have been attributed to a mind, while physical states have been related to the body (including the brain). *What is the existence and reality of the mind and how is it related to the one of body and brain? This is what is described as the mind–brain problem, a metaphysical problem that as such concerns the existence and reality of mind and brain.*

The mind–brain problem is a metaphysical problem that seeks to understand the existence and reality of the mind and its relation to the brain, a problem that asks about the possibility and extent of existence and reality. As such, the mind–brain problem includes two main facets: (i) What are mental states and what are physical states? How we can characterize the existence and reality of mental and physical states? These questions illustrate how the mind–brain problem is a metaphysical problem: they ask about the existence and reality of the mind and the brain; (ii) The metaphysical characterization of the mind and the brain also raises questions about their interaction. How can physical states cause mental states? This is especially puzzling because, as discussed above, the brain's physical states seem to show completely different features than the mental states of the mind. There seems to be a categorical difference between mental and physical states, which suggests their independence. At the same time, neuroscience demonstrates that mental states are dependent upon the brain and its neuronal states. Neuroscience demonstrates that neural activity in specific regions of the brain often accompany specific mental states (see Chapter 3). In addition, as observed in neurological and psychiatric patients, lesions in specific regions of the brain lead to changes in mental states. This suggests that mental states might be dependent on the brain's neuronal states, and that their assumed independence as based on their differences is incorrect. How are mental states dependent on the brain and its neuronal states? *In other words, how does the brain and its neuronal states generate and cause mental states? This is called 'physical (or neuronal) causation'.*

However, we also have to consider the reverse position: how can mental states cause physical states? For example, we might experience a mental state like the wish to move, and then we do indeed perform the respective physical movement. This suggests that *mental states can cause my physical states and thus my behavior. This is called 'mental causation' in current philosophy of mind.*

Background 3

'Causal closure' of physical causes

These descriptions of physical and mental causation seem to suggest that there is a double causation: causation from both the physical brain and the mental mind. The double occurrence of mental and physical causation makes the mind–brain relationship even more puzzling, since a proper solution should ideally be able to accommodate both mental and physical causation. More specifically, it seems that physical causation is closed in itself, physical causes cause physical effects, whereas physical causes apparently do not cause mental effects. If, however, mental effects are not caused by physical causes, they apparently cannot be characterized in physical terms. Does this mean that the existence and reality of mental states is not physical and thus mental? The 'causal closure' of physical causes thus poses a serious challenge to the metaphysical characterization of the mind and its relationship to the brain. In this and the following chapters, we will see how the different solutions to the mind–brain problem address both mental and physical causation between the mind and the brain. These range from the mental and physical characterization of the mind to alternative specifications.

Finally, this and the subsequent chapters in Part II are about the existence and reality of the mind and its mental states as distinguished from the brain and body. As such these chapters concern the metaphysical domain of existence and reality as discussed in philosophy of mind. This is accompanied by the application of a more traditional philosophical method – conceptual-logical reflection and argumentation.

Such metaphysics of the mind and the body must be distinguished from the discussion of mental features like consciousness, self and intersubjectivity. Here, the metaphysical question of existence and reality recedes into the background, while the empirical domain – the domain of observation in third-person perspective – shifts into the center. The shift from the metaphysical to the empirical domain opens the door to including concrete empirical findings and data from recent neuroscience about mental features like consciousness, self and intersubjectivity. This will be discussed in Parts IV and V.

History of dualism 1

Allegory of the cave in ancient Greece (Plato)

Below I will give a very brief history of the mind–brain problem in philosophy. As in all Western philosophy, the discussion about the existence and reality of the mind starts in Ancient Greece. At that time – 600–500 years before the birth of Christ – the brain was not the subject of any research. Hence what we describe as the mind–brain problem in our time was not yet considered as such. Instead, the mind was contrasted with the body. As such, what we now call the mind–brain problem was called the mind–body problem.

Western philosophy originated in Ancient Greece around 600–300 BC. Towering philosophical figures of the time were Socrates (469–399 BC), his student, Plato (428/7–348/7 BC), and Aristotle (384–322 BC) who was a student of Plato. Socrates was a great talker, known for going to the market places in search of discussions with all types of people. While he apparently never wrote anything down, his philosophical thoughts are documented in the dialogues written by Plato. Plato's dialogues portray Socrates as engaging in discussions with different people, where he raises various questions about man, the world and the universe in general: what are the fundamental causes and principles of the universe as a whole? How does the universe function? What is the role of man in such a universe and how can he access and know it?

Rather than focusing on the mind, philosophy in Ancient Greece focused mainly on the world and man's relationship to it. One of the most famous examples is Plato's allegory of the cave. In this allegory, Plato assumed that humans lived (metaphorically speaking) in a dark cave devoid of access to reality. Plato argued that what we can observe in the world and ourselves are nothing but mere shadows of its real objects and events. This means that we remain unable to perceive and cognize the real world as it is independent of our cognition and the shadows it throws upon the objects and events in the world. All we can see are the shadows of objects and events, while the objects and events themselves – independent of their shadows – remain inaccessible to us.

Why and how these shadows are generated is hidden from us. What reality hides behind them is also hidden from us. This means that we have neither direct access to the objects and events themselves, nor any knowledge of their reality and existence. There is thus no escape from the cave and the shadows of the objects and events we are confronted with in the cave.

Metaphorical comparison

Skyscraper as modern cave

How can we further illustrate the allegory of Plato's cave? For that, we can compare Plato's cave with a modern skyscraper. Imagine you are inside a skyscraper. You are looking out of the windows. The windows are tinted black and are tilted at odd angles so that everything you perceive is – to some degree – distorted. You see the sky, you see people, you see cars moving, and you see other skyscrapers through the windows of your own skyscraper. But because of the windows you are looking through, everything you see outside the skyscraper is not exactly as it is in reality.

Now imagine that you grew up in this skyscraper. You have never left and experienced or seen the world from outside the skyscraper. What you see through the skyscraper's windows you take for granted. But what you see are what Plato called 'shadows' – the reflection of the objects and events – rather than the objects and events themselves. In your case it means that you can only perceive (and have never perceived anything else) what the windows of your skyscraper reflect to you: you perceive and experience the objects and events outside in the world as removed from reality because of your skyscraper's windows. You can never be sure whether the cars, people, houses, etc. you perceive through the windows are that way in reality, and because you cannot leave, you have no direct access to reality.

You are trapped. Even if your skyscraper is light, bright and airy with its abundant windows, you nevertheless remain in the dark when it comes to knowledge of the objects and events themselves, independent of the skyscraper's windows. This was Plato's view of human experience. Plato argued that we, simply because we are humans, experience life as if trapped in a dark cave; we have no direct knowledge of the reality and existence of objects and events themselves.

What can we learn from Plato and his allegory of the cave? We can see that early on, in Ancient Greece, our ability to acquire knowledge was characterized by constraints or limitations. These epistemic limitations were associated with the senses – the sensory system blurs and confounds our access to reality and the existence of the objects and events in the world. We will see later that the issue of epistemic limitations will resurface in the characterization of both mind (see Chapter 7) and brain (see Chapter 8).

History of dualism 2

Concept of mind in ancient Greece: Plato

Why are we trapped in the cave? Why can we not access and know directly the objects and events themselves? Plato argued that it was the senses that distort the reality and existence of objects and events in reality. Our senses (tactile, auditory, visual, olfactory, gustatory) put us in chains – 'sensory chains' – that limit and constrain our possible knowledge of the world. Since our knowledge is based on our senses and cannot go beyond them, our senses are the main obstacle.

In the case of the skyscraper, the senses can only detect what the skyscraper's windows allow. The senses cannot experience and perceive beyond the windows. Here, what I designate as the window and its reflections corresponds to Plato's understanding of our senses. In the same way that we cannot go beyond the window in the skyscraper, we cannot go beyond our senses in our access and knowledge of the world.

What is behind the shadows of the objects and events? The objects and events themselves as they are, independent of our cognition of them. What do they look like? Plato assumed that the objects and events themselves consist of supernatural entities like mathematical objects and logical truths. This is what is described as Plato's 'theory of forms'. These immaterial forms structure and organize the material objects of the physical world.

How are these supernatural and immaterial entities related to the very material body in humans? Plato characterized humans by a soul with distinct parts: the sensible part of the soul determines our perceptions. The other part of the soul is the emotional part that allows us to feel honor, fear and courage. Finally, the soul has a third part: the rational part. The rational part is in charge of logical reasoning and making inferences. Most importantly, in the allegory of the cave it is this part that allows us to catch occasional glimpses of the light source outside the cave. This means that the rational part must be the most superior part of the soul since it accounts for thinking and rationality, as distinguished from affects and emotions.

This example illustrates that a theory of dualism was present as long ago as Plato. The rational part of the soul – which we today associate with the cognitive functions of the mind – stood side by side with the soul's sensible part, which we now associate with sensorimotor functions and the body. Meanwhile, in Plato's thought, the emotional part of the soul was regarded a mediator between both, in much the same way that we now regard affect/emotion as intermediating between sensorimotor and cognitive functions.

History of dualism 3

Concept of mind in ancient Greece: Aristotle

According to Plato, the soul can be characterized by both bodily and mental aspects. Bodily aspects concern sensorimotor functions of the body, while its mental aspects refer to cognitive functions that Plato considered necessary for accessing the form as supernatural entity. However, Plato's student Aristotle (384–322 BC) disagreed with his teacher. While he learned much from Plato, Aristotle did not always agree with him. Aristotle did not believe in characterizing forms as supernatural entities. While Aristotle believed in form, he did not 'locate' it as supernatural, as outside of the physical, natural world. According to Aristotle, form is part of our natural world.

How does Aristotle determine the form as part of our natural world? Is the form as material as the body? There has been much debate about Aristotle's concept of form and whether it can be interpreted either mentally or physically (see, for instance, Nussbaum and Rorty, 1992). While we cannot discuss this debate here, what is clear is that the form cannot be identified with the material, physical substrate of the body. This implies that the existence and reality of the form itself must be characterized metaphysically in a different way when compared to the body: since the body is material and physical, the form must be described as immaterial and nonphysical (i.e. mental) in order for it to be properly distinguished from the material and physical body.

Mental Approaches 133

How do Plato's and Aristotle's accounts relate to each other with regard to their mind–body theories? Plato suggested dualism between natural and supernatural parts of the soul. The natural part consists of the body, whereas the supernatural part must be 'located' outside the body. Plato's dualism resurfaces in Aristotle, though in a different way. There is no longer dualism between natural and supernatural, between body and universe. Instead, the dualism is shifted into the body itself that is considered dualistic and thus a composite by itself consisting of content/substrate and form/structure. The dualism is here no longer between the natural body and a supernatural non-body, but rather between form and content within the body itself.

Taken together, we can already see two kinds of dualism in Ancient Greece. One kind describes the dualism between supernatural and natural entities which extends beyond the natural world including our bodies. In contrast, the other kind refers to the dualism of form and content within the body itself as a natural entity. We will see below that both positions resurface in some way or other in later philosophers' assumptions about the mind–brain problem. We will now move on to shed some light on the concept of the mind in medieval philosophy.

Biography

Thomas Aquinas

The concept of form as developed by Aristotle served as a stepping stone for the medieval philosopher Thomas Aquinas (AD 1225–75). Thomas Aquinas was born and raised in Aquino, a village in Italy. Aquinas had a strong belief in God, which led him to enter the Dominican monastery in Naples at the age of nineteen. His relatives did not agree with this decision and decided to lock him in a tower, where they further tried to entice him away from religion with the beauty of pretty girl. Even the pretty girl could not change Aquinas' mind and his attachment to God, though. After a year he was finally able to escape from the tower and pursue his spiritual vocation. He entered the Dominican monastery, where he wrote several books about man and his relationship to God. These texts would make him one of the most famous philosophers of the medieval age.

History of dualism 4

Mind and God in medieval philosophy (Thomas Aquinas)

Like Aristotle, Thomas Aquinas rejected Plato's idea of supernatural mathematical objects and logical truths that are beyond our knowledge. Aquinas argued that we can access and know those objects and truths by means of reason. Unlike Plato, he assumed that we are not trapped in the shadows of our own cognition. We can go beyond them by continuously detaching our perception and cognition from the contents in the natural world; what will remain are the forms themselves independent of their association with the contents of the natural world. This detachment can thus be described as abstraction from the contents, which allows us to access and know the things in themselves as they are. This means, if we think back to the allegory of the cave, that we are able to leave Plato's cave and get a glimpse of the objects and events outside the cave, meaning reality itself.

Aquinas goes one step further and his reason for doing so is closely related to his intimate relationship with God. We continuously detach our perceptions and cognitions from their external contents. We are, for instance, able to hear voices and see objects during our dreams even though neither those voices nor the objects are present by themselves in the external world. Perceptions and cognitions must thus have an origin that remains independent of the external contents in our natural world. Where do perceptions and cognitions originate? This is the question for the ultimate and final form. That final origin or cause is nothing but God. According to Aquinas, continuous abstraction will ultimately lead our cognition to God.

How can we illustrate the role of God in a more metaphorical way? God himself is the source of light. As all light, God's light also casts shadows upon the events and objects in the natural world. According to Plato we can only perceive and cognize the shadows, but not the events and objects themselves independent of the shadows. This is why we are in a dark cave with shadows of light. Aquinas rejects Plato's approach. According to Aquinas, we can perceive and cognize the source of the light itself, namely God. This is possible by continuously detaching and thus abstracting the contents from our perception and cognition. This means that we can approach the light source itself, and in doing so approach God.

Put this into the context of our skyscraper analogy. We grew up in the skyscraper. However, we can detach and thus abstract from our existence within the skyscraper and figure out by logical reasoning that God himself built the skyscraper, including its windows and the reflections that account for the shadows in our perception. Once we have understood that God built the skyscraper, we are able to grasp the features of the windows, their dark tint and their specific angles, which, until now, blurred and constrained our perception. This makes it possible to go beyond the windows' reflections and to access direct perception of the world.

History of dualism 5

Mind and body in medieval philosophy (Thomas Aquinas)

What makes it possible for us to use abstraction and logical reasoning? If reason allows us to access and know God, reason itself must be divine. Reason is a function of the mind, and as such the mind must be immortal and immaterial since otherwise it could not provide access to and knowledge of God. This immaterial nature distinguishes the mind from its body, which is both mortal and material.

How are the material body and the immaterial mind linked to each other? Aquinas assumed that there must be a certain form that links both together. This form is what Aquinas described as 'soul': the soul, as different from the mind and the body, according to Aquinas, acted to link together these disparate parts. As soon as the material body dies, the mind is freed from the chains of the soul and the body.

This triple characterization by mind, soul and body situates man between animals and angels. Animals, like humans, possess bodies, but unlike humans, are mind-less. Angels, on the other hand, do not have bodies but do have minds. In this model, humans stand between animals and angels and must, therefore, be considered a composite of mind and body. This, as we will see below, sets the stage for raising the question about the relationship between the existence and reality of the mind and how it relates to that of the body.

History of dualism 6

Mind and existence in modern times: Descartes

Plato located man in the cave and regarded his knowledge of himself and the world as rather shadowy. His student Aristotle, and especially his theological follower, Thomas Aquinas, freed man from Plato's cave by refuting Plato's vision of the world. In doing so, they tried to gave man direct access to and knowledge of the objects and events themselves. Aquinas' position is possible only on the basis of his intimate relationship to and dependence on God, however. Man escaped Plato's cave but in exchange he is now chained to God. But one may want to go one step further and make man independent, able to stand on his own feet, independent of God. For this to happen, the chains of religion had to be cut. The famous philosopher and mathematician René Descartes (1596–1650) made some highly influential and far-reaching suggestions about the mind and how its existence and reality are related to the body.

Like his predecessors, Descartes was interested in understanding what we can know about the world. Unlike his predecessors, however, Descartes starts with doubt. Descartes argued that everything can be doubted; all our perceptions, all our reasoning, and all our knowledge can be called into question. This is called 'Cartesian or methodological doubt'. But there is something that cannot be doubted. When I am thinking, I become aware of myself. It is only me, myself, and no other self that thinks. If, therefore, it is only me and nobody else who thinks when I am thinking, I must exist. This led Descartes to the famous words 'Cogito ergo sum' ('I think, therefore I am'). Here, the existence and reality of the I – the self – is inferred from the activity of thinking; this will be specified below.

What exactly does 'cogito ergo sum' mean? There has been much interpretation and discussion about the exact meaning of Descartes' famous phrase. Due to space constraints, we will only provide a general and most basic explanation. Thinking is a process which must be attributed to somebody, to a specific person, and even more specifically, to me, to myself or I. Any kind of thinking activity must presuppose such I or self and since the thinking is real, the existence of the I must be real too. In other words, the activity of thinking implies the existence of the I, or self.

Why is all this important for our current discussion? Descartes introduces an element that will prove crucial for later developments, especially in neurophilosophy and philosophy of mind. Descartes infers the existence and reality of his self from his thinking activity. While thinking, I become aware or conscious of myself.

The self is thus identified and equated with consciousness of self, or self-consciousness. There is no difference between self and self-consciousness; both terms are used more or less synonymously. Descartes (and many other philosophers) presupposed, most often implicitly, this definition of the self in terms of self-consciousness. This position, however, is put into doubt by current empirical findings. Neuroscientific discoveries seem to support dissociation between self and self-consciousness instead (see Part V for details).

For instance, neural activity can be observed in the brain during the presentation of a person's own name or some autobiographical events even when consciousness remains absent, as in neurological patients in a vegetative state. The neuronal activity underlying the self, e.g. one's own name or autobiographical events, can thus apparently dissociate from the presence of the phenomenal features of consciousness. Does this mean that self and consciousness can dissociate from each other? We will discuss the empirical and conceptual details in Chapter 18.

History of dualism 7

Mind and body in modern times: Descartes

How can we think? Like Aquinas, Descartes assumed that the mind is characterized by reason and its specific cognitive functions, like thinking, logical reasoning and abstraction. He also, like Aquinas, believed that the mind must be distinguished from the body because the body is associated with sensory functions and perception. According to Descartes, the body is a machine and functions according to the causal laws of physics. In other words, the body is physical and material. As such, the body is extended in physical space, contains different parts, is divisible, and can be destroyed. Finally, the body obeys physical laws and is therefore deterministic. There is thus no room for spontaneous and therefore non-determined action within the body. This distinguishes the body from the mind in that the mind is non-physical. This means that the mental is thus non-material. Unlike the body, the mind is neither spatially extended, nor divisible. Instead, the mind is indivisible and contains no parts; it is a separately existing entity that as such has to be distinguished from the body. Finally, unlike the body, the mind does not obey the deterministic laws of physics and therefore remains non-determined. This is what we now describe as free will.

Descartes' theory about the mind and the body is often described as dualism. Dualism describes the assumption of two different entities or substances, in this case, the mind and the body. In this model, the mind can be traced back to an underlying mental substance, while the body is based on a physical substance. Both mental and physical substances are principally different and mutually exclusive. They are therefore thought to operate independently of each other.

What does this imply for the characterization of mental capacities? For example, mental functions like consciousness, reflective reasoning, logical reasoning, introspection and language are assumed to proceed independently of what happens in the body. The nature of mental functions means that they are not thought of as finding their origin in the body and its merely physical processes. This implies dualism between the mental and the physical: mental functions are attributed to the mind, whose existence and reality is different from the body and its physical processes.

History of dualism 8

Interactive dualism in modern times: Descartes

How are the mind and the body related to each other? Descartes assumed that the mind and the body interact with each other. How might they interact? Unlike Aquinas, Descartes no longer assumed a soul as an intermediate. Instead, he introduced the brain to philosophy. Why, though, does he single out the brain as the locus for such interaction? Descartes believed that the brain was the point where the body and the mind come closest to each other. More specifically, he assumed that the body and the mind could interact via a little structure in the brain located beneath the brain's ventricles – the so-called epiphysis or pineal gland.

Why the pineal gland? Descartes was well aware that the brain has two hemispheres and it could therefore not be considered a complete and comprehensive unity. This non-unity of the brain is important for our purposes because it contrasts with the mind and more specifically with our mental states. More often than not, we consider our mental states or consciousness to be unified and not split into two halves like the brain.

How can we reconcile the disunity of the brain with the unity of the mind? Descartes thought that we needed to look for structures in the brain that are not present in both hemispheres and are thus not replicated. Because the pineal gland is not replicated in each half of the brain, Descartes thought that it must be the perfect structure to allow for direct interaction between the brain and the supposedly unitary character of mind and consciousness. In Descartes' thought, the pineal gland becomes the meeting place between the mind and the body. This, however, shifts the problem into the brain: how can two different entities as disparate as the mind and the body interact with each other within the brain, or more specifically, its pineal gland? How can a non-physical substance that defies physical laws and causality interact with something completely physical?

Accordingly, Descartes was confronted with the problem of how to link the mind and the brain together despite their seemingly principally different natures. His characterization of the mind and the body by principally different substances – mental and physical – explicates what is called substance dualism, while his assumption of direct interaction between the mind and the body in the pineal gland of the brain specifies his position as interactive dualism.

Why is Descartes and his interactive substance dualism of the mind and the brain so important even for the mind–brain discussion in our time? Descartes' characterization of the mind and the body/brain set up the framework for the contemporary discussion of the mind–brain relationship. Even though our contemporary knowledge about the brain has increased considerably, the metaphysical issues remain essentially the same. How do we characterize the existence and reality of mental functions like consciousness and their relation to the brain and its neuronal states? Any current solution to the mind–brain problem can therefore only be understood in the context of Descartes. This will become clear when we look at the options for the mind–brain theories in the current philosophical discussion.

Dualism 1

Substance dualism

There are different forms or varieties of dualism. Descartes, for example, postulated two different substances, the mind and the body, each of which possess distinct properties, mental and physical. What is a substance? Properties – whether mental or physical – must be attributed to some kind of owner, an object, implying that properties are properties of objects. Descartes not only distinguished mental and physical properties, but also attributed them to different owners or objects, mental and physical substances as manifested in the mind and the body. Let's make this more concrete. The brain in particular and the body in general may be considered a substance to which certain properties can be attributed. One may even assume that the brain may possess two different properties (mental and physical) that are irreducible.

Descartes went one step further. He did not consider the assumption of mental and physical properties to be sufficient to account for mental states. The only way he could imagine the mind to exist and be real was to assume an owner or substance distinct from brain and body. He assumed a second substance besides the physical substance of body and brain, a mental substance. This mental substance can be characterized by a separate existence and reality that accounts for the mental properties and their irreducibility to the brain's and the body's physical properties.

Today, such a position sounds rather strange. We no longer believe in the existence and reality of mental substances. Why? We simply cannot observe any such mental substance in our natural world. No one has ever observed such a mental substance in the brain or the body, or the world – let alone provided any scientific proof that might validate its existence. The only way for such a mental substance to exist is outside our natural world and thus in a supernatural world. That, however, is not only beyond our scope, but also rather unlikely. Therefore the assumption of a mental substance is discarded.

We have to be careful at this point, however. The fact that we cannot observe such mental substances in either brain or body does not mean that they do not exist. Something that we cannot observe can nevertheless exist and be real. This is the point the philosophers make by applying their purely logical stance. Logically, it is possible that something like the mind can exist which, at the same time, we cannot observe. Since we cannot exclude such a discrepancy between the observational-experimental and conceptual-logical approaches, we remain unable to discard the existence and reality of the mind, on philosophical grounds at least.

The more empirically-minded neurophilosopher may now want to remark that such presupposition of mental substance is more a matter of belief than proof. Accordingly, dualist inclinations are strongly prevalent among religious people who, like Aquinas, believe in a direct connection to God that can be conceptualized only as mental and non-physical. That sounds implausible to any scientifically-minded person who is prone to reject substance dualism. Substance dualism is thus not a position that is favored among neurophilosophers, because of their strong reliance on scientific findings.

Dualism 2

Property dualism

Another way to escape the implausibility of substance dualism is to weaken the thesis. One may still uphold the assumption of mental and physical properties, while one may no longer attribute them to different owners or things (mental and physical substances) like mind and brain/body. Instead, one may associate different properties, e.g. mental and physical, with one and the same owner, the brain as physical substance. Any assumption of a mental and immaterial substance is eliminated here. There is only one substance, the physical body, which may have two properties, mental and physical. This is called 'property dualism'.

However, eliminating the mental substance does not imply that we've solved all the problems associated with dualism. Property dualism claims that mental and physical properties are attributed to one and the same substance; for instance, the brain as a physical substance. This raises the question: how can such mental properties be generated by a physical substance like the brain? How is it possible that a seemingly purely physical brain can generate something non-physical, namely mental properties that are irreducible to physical properties?

The assumption of such mental properties that arise within the physical brain is even more implausible, given that mental states and their respective contents cannot be detected in the brain at all. All we can observe in the brain are neuronal states, more specifically firing neurons, action potentials as their electrical discharges, and fluctuations of neuronal activity in different frequency dimensions. However, no mental states distinct from neuronal states (let alone mental properties) can be detected in the physical brain.

Even the images of the brain produced by functional imaging techniques like positron emission tomography (PET) and functional magnetic resonance tomography (fMRT) do not provide any evidence of mental states. All we can see in the colorful images are visualizations of neural activity changes. No mental states can be observed either in the images of the brain, or the brain's activity itself. There is thus no empirical evidence for such mental properties.

How are the supposed mental properties generated by the physical properties of the brain? This is the central question that needs to be answered in the model of property dualism. The property dualist may revert to the assumption that mental properties emerge out of the interplay between the brain's different physical properties as novel non-physical properties. *These are called emergent properties. Emergent properties are novel properties that supposedly arise out of the interactions among the physical properties. How exactly such non-physical properties can arise from the interactions among the physical properties remains unclear at this time. This position is referred to as 'emergence' or 'emergentism' and was supported by the nineteenth/twentieth-century philosopher C. D. Broad (1887–1971) and more recently by the Canadian philosopher Mario Bunge (1919–).*

Dualism 3

Problem of mental causation

Let's recall what Descartes argued for dualism and for the interaction between mental and physical substances. He opted for what is called 'interactive dualism'. Such interactive dualism is confronted with the question: how can the mind and the brain as different substances – mental and physical – interact with each other?

There is causal interaction within the realm of the physical. One physical particle causally impacts another particle and so forth. Most importantly, nothing non-physical is necessary to sufficiently explain the movement of the other particle, which can be traced back to the former physical particle and its causal effects. The realm of the physical is thus closed. This is what philosophers describe as 'causal

140 *The Mind–Brain Problem*

closure', meaning the exclusion of any other non-physical causes. It is also known as the 'causal exclusion principle'.

How does this relate to the interaction between mental and physical states? Causal closure is not compatible with the assumption of an interaction between mental and physical states. In causal closure, the domain of the physical is closed, which means that only physical states can interact causally with other physical states. Mental states, as non-physical causes, are thus excluded from causal interaction.

We may, however, leave both 'causal closure' and the 'causal exclusion principle' behind. In doing so, this opens the door for interaction between mental and physical states. There can be two kinds of interaction. Physical states may cause mental states, called 'physical causation'. Physical causation suggests that mental states like consciousness are caused by, and can thus be traced back to, the neuronal, e.g. physical, states of brain and body.

Conversely, mental states might also cause physical states. This is, for example, the case in experience. I have the idea of running down to the lake and taking a swim because it is so hot. I will consecutively pick up my bathing suit and running shoes and start running the 10k to the lake. In this case, a purely mental state – my idea – initiated a physical state, my behavior and actions. In this model, mental causation seems plausible when considering consciousness.

However, this contrasts with causation in the physical realm. First, both mental and physical states show different features (physical and non-physical), which makes causal interaction between them difficult. Even more tricky is the fact that the physical realm does not really need any additional causes that stem from the mental realm. All causality in the physical realm can be fully and thus sufficiently explained by merely physical causes. What this implies is causal closure and causal exclusion as described above. Here it seems that any mental causation is rendered superfluous.

Let's be more specific. My action and behavior of running the 10k to the lake can be explained in merely physical terms. For example, my prefrontal cortex – where thoughts are generated – was active. This activity caused the activation of the lateral prefrontal cortex, which is where goal-orientations are formulated. That in turn led to the activation of my motor cortex, which sent the signals to the individual muscles required to initiate the actual behavior and motor action. In this account, no mental cause is needed to explain my behavior and actions. Hence, the assumption of mental causes is superfluous.

How can we solve this apparent contradiction between mental causation in our experience and the superfluousness of mental causes in our brain and its neuronal states? For this we will want to turn to another variant of dualism, epiphenomenalism.

Dualism 4

Epiphenomenalism and parallelism

How can we resolve the contradiction between the apparent presence of mental causation in our experience and its absence in the brain's neuronal states? One way is to simply abandon the assumption of any kind of mental causation. In this model, dualism between mental and physical is assumed. Even causality from the physical to the mental – physical causation – can still be supposed. However, mental causation, as causality from the mental to the physical, is abandoned in order to accommodate for causal closure and the causal exclusion principle of the physical realm (see above).

In other words, the mental realm remains causally impotent in the physical world. Why is the mental realm causally impotent? Because its existence and reality is completely different from that of the physical world. There is thus a dualism between the mental and the physical. Such dualism goes along with what philosophers describe as epiphenomenalism, the absence of mental causation as causal impotence of the mental on the physical.

What is epiphenomenalism? Epiphenomenalism assumes that mental states are caused by physical states in the brain and body. This means that epiphenomenalism argues for the presence of physical causation. In contrast, mental events have no causal impact on physical states. This implies the absence of mental causation. This position within epiphenomenalism is logical in that it avoids the problems associated with mental causation by simply denying its plausibility and existence.

Despite its logical plausibility, epiphenomenalism seems to contradict the subjective experience of our mind as impacting our bodies. Our experiences, as discussed above, suggest that our mental states can cause changes in the physical realm – in our behavior and actions. This is denied and rejected in epiphenomenalism. In this model, mental states are regarded as mere by-products – as 'nomological danglers' or brute facts that are added as causally impotent features to the physical features of the body, as the American philosopher Herbert Feigl (1902–88) said.

How can we compare epiphenomenalism and interactive dualism? Interactive dualism assumes mental causation and is therefore in accordance with our experience and thus phenomenally plausible. Meanwhile, it is logically implausible because of the contradiction between mental and physical causation. The situation is reversed in epiphenomenalism. By denying mental causation, epiphenomenalism avoids the logical problems associated with the concurrence of mental and physical causation. Epiphenomenalism does not account for our daily experience of our mental states as often directing our physical states. As such, it remains phenomenally implausible.

How can we escape these problems? One way is to deny any kind of interaction between mental and physical processes. Instead, we can consider mental and physical processes as operating largely in parallel without any interaction. This is called parallelism between the brain and the mind. Parallelism was suggested by a German philosopher of the early eighteenth century, Gottfried Wilhelm Leibniz (1646–1716). Leibniz suggests that mental and physical realms are like two different clocks that operate in parallel and independently of each other.

Parallelism raises the question: how are the mind and body coordinated with each other if they cannot interact? Leibniz believed in what he described as a 'pre-established harmony' that ultimately could be traced back to God. Imagine two different clocks, one displaying the time in Tokyo, and the other the time in New York. Though both clocks indicate different times, they are nevertheless coordinated with each other. As the second hand of one moves, the second hand of the second clock also moves. There must thus be some coordination of the two clocks, which Leibniz described as 'pre-established harmony'.

Who and what coordinates the mind and the body? Leibniz refers to God as the ultimate source that coordinates mind and body with each other. Going back to our earlier mentioned example of the two clocks, one might think of it like this: despite the fact that the two clocks indicate the time in different time zones, they nevertheless must have been coordinated with each other at some point. Otherwise they would not be synchronized with one another.

Naturalistic escapes 1

Appeal of dualism (Searle)

Given the conceptual-logical problems and empirical implausibility of both substance and property dualism, as well as of epiphenomenalism, one may want to reject dualism as a metaphysical answer to the mind–brain problem altogether. However, this is not what has happened. Dualism is still considered a viable option within the current mind–brain debate.

What makes dualism so appealing? The American philosopher John Searle (1932–) (2004, p. 33) assumes that dualism appeals to what he calls a 'basic or primitive insight' – the fact that our experience and thus consciousness are real and existent. Nobody can deny that we experience and are conscious. Most importantly, we experience our mental states as principally different from physical objects and their physical states. In other words, dualism appeals to our intuition, and more specifically to our intuition of the distinction between the physical and the mental. Let's explain these intuitions in further detail. There is the intuition that the world as we observe it is purely physical, consisting of objects with physical measures like size, extent, weight, etc.

However, these intuitions about the physical existence and reality of the world contrast with our experience and consciousness of that very same world. There is no size, no weight, etc. to our experiences. As Descartes already pointed out (see above), mental states do not seem to operate in physical space and time. Nor are they characterized by physical measures like size, motion, weight, etc. As such, since they do not seem to be constrained by physical time and space, experience and mental states cannot be physical. Instead, their underlying existence and reality must be non-physical and genuinely mental, since otherwise we cannot account for them.

This non-physical and mental reality is usually characterized as lying outside the realm of the natural world. Mental existence and reality cannot be observed in the natural world and cannot, therefore, be considered part of it. Instead, mental existence must be something beyond the natural world. Dualism – as either substance or property dualism – can therefore not be regarded a naturalistic mind–brain theory subject to science and empirical evidence. This makes dualism implausible and untenable for any scientifically-grounded approach to the mind–brain problem.

Naturalistic escapes 2

Zombies and informational dualism (Chalmers)

We must now introduce the Australian philosopher David Chalmers (1966–). Chalmers argues in favor of dualism between the mind and the brain. What is different about his concept of dualism is that he considers dualism compatible with the natural world. What this means is that Chalmers thinks that there is dualism between the mind and the brain within the natural world itself. This distinguishes his version of dualism from his predecessors' who usually think of the mental realm as non-natural, or supernatural. Let us further explicate Chalmers' position below.

Chalmers made seminal contributions to the current philosophy of mind by formulating the so-called 'hard problem', as already raised in the Introduction. In a nutshell, the hard problem raises these questions: Why is there consciousness at all rather than nonconsciousness? Why there are mental states at all, rather than only physical states? (see Part IV for detailed discussion of the hard problem). Since Chalmers' introduction of the 'hard problem' in the mid-1990s, the 'hard problem' has been conceived of as a standard or measure for any solution to consciousness. Only if the respective proposal can address the hard problem can it be considered a viable answer or solution to the problem of consciousness (see, therefore, Chapters 17 and 20 for a more extensive discussion of the hard problem).

How does Chalmers himself address the hard problem? This leads us to another invention of his, the introduction of the concept of zombies into the debate about mind and brain. He argues that we can very well imagine a world without consciousness where the physical features remain the same, as it is the case in zombies. Let us describe zombies in further detail.

Zombies show exactly the same physical design and features as us. However, unlike humans, they do not show consciousness and thus have no mental states. If consciousness and mental states were purely physical, zombies should be impossible to even imagine. Since we can very well imagine and thus conceive of zombies as a logical possibility, consciousness and mental states cannot be physical. The fact that we can conceive of zombies, who have an absence of mental states in the presence of the same physical states as humans, thus provides evidence in favor of dualism.

What does the example of zombies imply for the relationship between mind and brain? Chalmers rejects any form of substance or property dualism where mental features are outside the natural world. This is called non-naturalistic dualism. He instead argues that there may be two kinds of features or properties within the natural world entailing a form of property dualism within the natural world itself. First, there are physical features as described by physics. And second, there are informational features that are accounted for by information theory. Both physical and informational features lie within the realm of the natural world and are thus subject to scientific examination. One can thus speak of naturalistic property dualism.

The central point in Chalmers' property dualism is the assumption of informational features. How can we characterize informational features in further detail? And what is their relationship to consciousness? Chalmers argues that informational features account for consciousness in humans (and other animals) while they are not linked to physical features. This is supported by his example of the possibility of zombies, who show similar physical features but lack the informational features. Consciousness consequently remains absent in zombies. These informational features complement the physical features. The dualism between the mental and the physical thus resurfaces here as property dualism between physical and informational features. By replacing mental features with informational features as the underlying existence and reality of consciousness, Chalmers shifts dualism into the natural world. One can thus speak of naturalistic property dualism as distinguished from the above-described non-naturalistic forms of both substance and property dualism that only apply to the logical but not the natural world.

Naturalistic escapes 3

Predicative or conceptual dualism (M. R. Bennett and P. M. S. Hacker)

How can we preserve the dualism between the mind and the body without making over strong metaphysical assumptions? The earlier versions of dualism associated mental and physical states with different substances, mental and physical, which implied the assumptions of two different worlds, mental and physical. That, however, is a rather strong assumption especially in the age of science that presupposes only one world, the physical world as we can observe it.

However, can we preserve the dualistic intuition with the difference between mind and brain within the physical world? Chalmers suggested one way when he proposed dualism between physical and informational features within the natural world. Another possibility may be to consider the different kinds of concepts we use to describe mental and physical features, which entails a dualism of concepts.

Mental concepts like belief, desire, hopes, etc. can simply not be reduced to physical concepts like weight, motion, etc. The difference between mental and physical states can, in this case, no longer be traced back to either different substances or properties, e.g. the metaphysical domain, but only to their description in terms of different concepts, e.g. the conceptual domain. *This is called 'predicative dualism'. Predicative dualism is a dualism of concepts for mental and physical features. Such predicative or conceptual dualism seems to be advocated by, for instance, Bennett and Hacker in their book Philosophical Foundations of Neuroscience that appeared in 2003.*

How can we further specify such dualism of concepts in predicative dualism? The classical example here is the analogy to water and H_2O. We use terms like transparent etc. to describe water and our perception and experience of it, but we do not use these terms to describe the chemical formula H_2O. Instead, H_2O is described by chemical and physical concepts. There is thus a dualism between the concepts that describe H_2O and those that describe water. In the same way, we also use different concepts – mental and physical – to describe mental and physical states.

The focus on the dualism of concepts implies a shift from the metaphysical to the conceptual domain. More specifically, predicative dualism shifts the mind–brain problem from the metaphysical domain of existence and reality to the conceptual domain of language and concepts. The mind–brain problem in this model is no longer regarded a metaphysical problem, but a conceptual one. The central question is thus no longer focused on the relationship between mind and brain themselves but rather on the relationship between mental and physical concepts. How are mental and physical concepts related to one another? One may be inclined to link both mental and physical concepts and to ultimately explain the former, mental concepts, in terms of the latter, physical concepts. This raises the problem of what is described as explanatory reduction, which will be discussed in Part III.

Neuronal escapes 1

Dualism and the brain (Sherrington and Penfield)

David Chalmers shifted the dualism between mental and physical states into the realm of the natural world by assuming informational features as distinct from physical ones. In doing so he shifted property dualism into the realm of the natural world – the observable world that we investigate in science. How does this naturalistic property dualism stand in relationship to neuroscience and the empirical findings of the brain? If dualism is compatible with the natural world, one could imagine that neuroscientists would also opt for this particular form of dualism; that is, naturalistic property dualism. The brain itself may then provide a neuronal escape from the problems inherent to the non-naturalistic forms of property and substance dualism.

At first glance one may be puzzled. All we can observe in the brain are neuronal states and these are purely physical. There is nothing mental about the brain which might suggest that neuroscientists should be prone to reject dualism, rather than supporting it. This is indeed the stance most neuroscientists and neurophilosophers take in the current debate. They opt instead for physicalism, rather than dualism (see Chapter 6). This was not the case around the turn of the nineteenth century, however. At that time, neuroscience focused on sensorimotor functions of the brain while cognitive functions and mental features like consciousness were way beyond experimental reach. This may have inclined some of the most prominent earlier neuroscientists to assume some version of mind–brain dualism. One early neuroscientist was the British neurologist and Nobel Prize winner Charles Sherrington (1857–1952) (see also Chapter 3). Besides his experimental discovery of the complex action of the reflex arc as an integrated action across brain and body, he also thought about the relationship between the brain and the mind. Sherrington suggests two different syntheses. There is a physico-chemical synthesis that allows integration between the various processes in motor action and behavior, the sensorimotor functions, as they are associated with the body. In addition, there is also a psychological synthesis. Rather than synthesizing action and behavior, the psychological synthesis integrates and unifies different perceptions into one consciousness. This is manifested in psychological functions like emotions, volitions and memories; these affective and cognitive functions are usually associated with the mind.

How are both physico-chemical and psychological syntheses related to each other? Sherrington suggests that both operate in parallel. The physico-chemical synthesis does not interfere with the psychological synthesis and vice versa. Accordingly, the dualism between the mental and the physical is here rephrased as dualism between the sensorimotor functions and the mind's affective/cognitive functions.

Another famous neuroscientist advocating mind–brain dualism was Wilder Penfield (1891–1976) who worked in Montreal, Canada. He performed many experiments with epileptic patients by stimulating different regions of the brain with electrodes. He was able to observe the reactions of his patients, some of whom experienced the recollections of vivid memories when a particular region of the brain, the hippocampus, was stimulated electrically. Memories are thus associated with the brain and stored there in the patterns of its neural activity.

Based on his experimental data, Penfield assumed dualism between the mind and the brain. Why? Because he supposed the mind to be continuously present, even during the temporary silencing of the brain by electrode stimulation or brain injury. Even when the neural activity of the brain is suppressed and silenced by the electrical stimulation, we are still conscious and have experiences. This, if one wants to say so, indicates the 'presence of the mind' during the 'absence of the brain'. This thesis, however, is only possible if the mind and the brain are distinct and do therefore remain independent of each other in their existence and reality.

Neuronal escapes 2

Psychons and the brain (Eccles)

More recent neuroscientists have also argued for dualism. The famous neuroscientist John C. Eccles (1903–97), who as a student of Sherrington earned a Nobel Prize for his neuroscientific discoveries, supported dualism between the mind and the brain. Together with the philosopher Karl Popper, Eccles wrote a landmark book, *The Self and its Brain* (1989[1977]), in which they assumed interactive dualism between the mind and the brain (see also Chapter 4). More specifically, Popper and Eccles assume three different worlds, the world of physical objects, the world of the states of consciousness, and the world of knowledge in an objective sense.

The primary reality in which we live is the second world – the world of the states of consciousness and the various mental features like free will, etc. This world of mental states is the basis of our access to the first world – the world of physical states – which we access via our thoughts and conscious experiences. These enable us to observe and discover the neurons of the brain and the various physical processes in our body as a whole. We are continuously going forward and back between the two worlds, between the world of mental states and the world of physical states.

How about Eccles' and Popper's third world? The third world is the world of culture, language and knowledge. Eccles and Popper consider this world to be specific for humans as distinct from animals. In the same way that there is continuous interaction between the physical and the conscious worlds, both physical and conscious worlds also interact with the cultural world of language and knowledge. There is thus trilateral interaction between the three worlds, physical, mental and cultural.

How does Eccles imagine the concrete interaction between the mind and the brain? He assumes so-called 'psychons' that mediate the interaction between the two. These psychons are supposed to work at certain synapses (gaps) or transfer points between different nerve cells or neurons in one particular region of the brain – the premotor cortex. Why does he suggest a central role for specifically the premotor cortex? The premotor cortex is the region where mental states formulate goals, and purposes are transformed into motor action and behavior. This functions as an interface between the mental and the physical.

However, Eccles' assumption of such interactive dualism is scientifically implausible. Nobody has ever provided any scientific evidence for such psychons, in much the same way that Descartes' assumption of the pineal gland as an interacting region has never been verified empirically. The location of the mind–brain interaction in one specific region or part of the brain is problematic because of the lack of scientific evidence.

Critical reflection 1

Dualism and science

What inclines neuroscientists to opt for dualism? We have already indicated that earlier neuroscientists like Sherrington and Penfield lacked empirical knowledge of the cognitive functions of the brain. This lack of knowledge may have prompted them to associate mental features like consciousness with non-neuronal, non-physical features. The same holds true, to a certain degree, for John Eccles who, in his empirical-experimental work, still focused on the sensorimotor functions of the brain. This left him unclear about the cognitive functions of the brain, which were explored in neuroscience only since the 1980s and thus after his death.

One may now want to raise several questions. First, can we provide scientific evidence for the existence and reality of mental features as distinct from physical features? Scientific evidence requires measurement, which is possible only if we can quantify what we want to measure. To provide scientific evidence for mental features thus means to measure and quantify them in the same way as when we quantify and measure physical features like velocity, gravity, etc. Some have even attempted to quantify the mind – for example, I cited the example of the Boston physician Dr Duncan MacDougall (1866–1920) in the Introduction. To determine the soul he weighed six of his patients shortly before and immediately after their deaths assuming that, with death, the soul leaves the body. If so the body must be lighter after death, which he indeed found in at least two patients. While failing to confirm this in a third patient, he sought further confirmation in dogs, who are presumed to have no soul. Based on his experiment, MacDougall assumed the weight of the soul to be 21 grams. For him, this proved that the soul/mind is a mental substance that is different from the physical substance of the body. He also undertook the same kind of experiment in 15 dogs, where he could not find any difference in weight. From that McDougall concluded that only humans and not animals have a soul.

One may also want to draw the analogy to another organ, the heart, which in ancient Egypt has also been associated with the mind (see Chapter 3). The British physiologist William Harvey (1578–1657) sought to discover the 'vital spirits' in the heart and aimed to locate them. He assumed the heart to produce the 'vital spirits' that keep the organism vital, existent and thus alive. However, as much as he searched, he did not discover any 'vital spirits' or other spirits like 'natural spirits' or 'animal spirits' in the heart or elsewhere. Instead, Harvey observed that the heart is a pump. The heart pumps blood throughout the whole body, which keeps the organism alive and existent. Hence, the central biological purpose of the heart, its pumping function, accounts for what was described as 'vital spirits' that turned out to be physiological in their existence and reality.

The same may now apply to the brain. We may not yet have discovered the purpose of the brain, an intrinsic feature that defines the brain as brain, in the same way that the pumping of blood defines the heart as heart. Once we are clear about the brain's intrinsic features and its overall biological purpose, we may no longer need to assume distinct mental existences and realities as the 'vital spirits' of the brain that produce the mind.

Critical reflection 2

Linkage between empirical and metaphysical domains

We can see that there is some interaction between the empirical discoveries of the brain and the metaphysical position on the mind–brain problem. Lack of empirical evidence may make one prone to believe in dualism, as was the case for earlier neuroscientists. Meanwhile, more detailed empirical insights may make non-physical assumptions, and thus dualism, superfluous. The metaphysical domain, and more specifically the assumption of a separate mental reality, is then curtailed by the empirical domain and its increasing insights. This suggests that there is some kind of relationship between the empirical and the metaphysical domains.

Are there direct linkages between empirical and metaphysical domains? We recall that the empirical domain concerns our observation; that is, what we can observe in the world. In contrast, the metaphysical domain is about the existence and reality of the world as it remains independent of our observation. In the case of direct linkages between empirical and metaphysical domains, we would be able to infer directly from, for instance, our observation to their underlying existence and reality, and thus from the empirical to the metaphysical domain. If so, one could assume that neuroscience will provide the 'correct' solution to the metaphysics of the mind–brain problem as discussed in philosophy. This position is assumed in what I described as reductive neurophilosophy, where the metaphysical domain is ultimately reduced to (or at best inferred from) the empirical domain (see Chapter 4).

However, if one considers neurophilosophy in a non-reductive way (see Chapter 4), it is not necessary to assume this type of reduction and inference. The assumptions of the metaphysical domain are here neither inferred from, nor reduced to, those of the empirical domain. Instead, metaphysical and empirical domains are considered as distinct, which prevents direct inference or reduction from one to the other. Such a version of neurophilosophy that entails domain pluralism can therefore be described as non-reductive (see Chapter 4 for details). This must be distinguished from the domain monism of reductive neurophilosophy that is reductive when it comes to the relationship between metaphysical and empirical domains.

Critical reflection 3

Differences between empirical and metaphysical domains

How can empirical and metaphysical domains be distinguished from each other? First, they differ in their subject matter. The empirical domain deals with neuronal (or other scientific) processes that can be observed in the natural world. The metaphysical domain, in contrast, is about existence and reality in the world, including both natural and logical worlds. In other words, the empirical domain focuses on how something works and functions, whereas the metaphysical domain is more about 'the what' that underlies those processes and functions.

One may now be puzzled. Isn't what we observe in science in general, and neuroscience in particular, like the brain's various neuronal processes during, for instance, consciousness real and existent? Without going into detail, we cannot make a direct inference from our observations to their underlying existence and reality.

We may, for instance, err in our observations and observe neuronal processes that are more related to our observations themselves and their technical instruments than to some kind of supposed underlying existence and reality. The possibility of such intrusions of the observer him/herself into his/her own observations (see Northoff, 2013a) makes direct inferences from observations in the empirical domain to existence and reality in the metaphysical domain impossible.

This, however, implies that neuroscience and its empirical domain will remain unable to provide a solution to the mind–brain problem by itself, independent of philosophy and its metaphysical domain. Rather than cashing out philosophy for neuroscience when replacing the metaphysical for the empirical domain (as in reductive neurophilosophy), we need to search for the link between empirical and metaphysical domains. This shifts the focus from either domain itself to the borders between metaphysical and empirical domains, where both interface with each other. Such border territory is the territory of neurophilosophy as understood in a non-reductive sense (see Chapter 4).

Critical reflection 4

Can we infer from the empirical to the metaphysical domain?

In addition to their subject matter, the empirical and metaphysical domains differ also in their methods. Investigation of the empirical domain is possible on the basis of experiments and observation in third-person perspective. This is different in the metaphysical domain that relies on conceptual and logical considerations rather than observation.

How are these differences between metaphysical and empirical domains related to the mind–brain problem? Given their differences, direct inference (or even reduction) to the empirical findings in neuroscience from the metaphysical characterization of the mind and brain seems to be rather difficult. Only if one disregards, and thus neglects, the differences between metaphysical and empirical domains, can one infer directly from the empirical to the metaphysical domains and replace the existence and reality of the mind by the brain.

This, in contrast, remains impossible if one considers the differences between the metaphysical and empirical domains. At the same time it opens the door for dualism even in the absence of scientific evidence. Such mind–brain dualism may be conceptually and logically plausible as a metaphysical assumption, even if it remains empirically implausible given the lack of scientific evidence. Hence, conceptual plausibility dissociates here from empirical plausibility.

Such possible dissociation between conceptual and empirical plausibility implies that there may thus be different layers of evidence or plausibility related to the different domains, metaphysical and empirical, and their associated methods. What is logically and thus philosophically plausible may not be naturally and thus empirically and neuroscientifically plausible. We thus need to be aware of these different layers of investigation, natural and logical, including their different methodological strategies, observational-experimental and conceptual-logical.

Critical reflection 5

Mind–brain problem as a 'border problem'

What can we learn from that for our investigation of the mind–brain problem in general? Our presuppositions about metaphysical and empirical domains strongly influence the kinds of inferences we can or cannot draw from the empirical findings of neuroscience. This means that we cannot consider the mind–brain problem as an exclusively metaphysical problem or just as a mere empirical problem. Instead, we may need to consider both domains and be clear about our presuppositions of their characterization when discussing the mind–brain problem.

More specifically, the mind–brain problem may be considered a problem that lies right at the border between the empirical and metaphysical domains rather than being situated exclusively within one domain. More specifically, the mind–brain problem may neither be considered a purely empirical problem, as is often suggested by neuroscientists and neurophilosophers, nor an exclusively metaphysical problem, as traditionally supposed by philosophers.

Instead, the mind–brain problem may be situated right at the border between the metaphysical and empirical domains. This, as noted above, is the specific territory of non-reductive neurophilosophy. The mind–brain problem may thus turn out to be a genuinely neurophilosophical problem rather than either a neuroscientific or philosophical one. The mind–brain problem may consequently neither be a metaphysical, that is, philosophical, nor an empirical, that is, neuroscientific, problem but a border problem and thus a neurophilosophical problem.

Mental escapes 1

Panpsychism: history of philosophy

We demonstrated that dualism between the mind and the body has strong historical predecessors in the history of philosophy. However, the current debate in philosophy of mind reveals both substance and property dualism as problematic. In response, alternative positions have been developed that aim to keep the original dualist intuition of the special nature of the mind and its mental states. These alternative approaches were described above when discussing epiphenomenalism, parallelism, informational dualism, and neuronally-grounded dualist approaches.

All these approaches take physical reality and existence as given. By taking physical existence and reality for granted, these approaches must consider how to account for the special nature of the mind and its mental features. This leads them to make assumptions about the specific nature of the mental as distinct from the physical. Different strategies, however, are also possible.

The usual strategy is to start by assuming physical reality and existence and then to raise the question: how can physical reality and existence give rise to the mental realm? One may, however, also reverse the starting point. Rather than starting with physical existence and reality, one could depart from the mental and take its existence and reality for granted. This leads one to ask the opposite question: how can mental reality and existence give rise to physical reality and existence? Panpsychism adopts this strategy.

Let us formulate the strategy of starting with the mental in a slightly different way. One way to escape the perils of dualism is to abandon it altogether and to replace it by monism. Monism claims that there is only one existence and reality. This singular reality consists in either the physical realm or, alternatively, in the mental realm. The first case is called physicalism or materialism, a theory which denies the mental realm altogether (this will be discussed in the next chapter in full detail). In the second case, the assumption is that existence and reality is mental. This is called panpsychism.

Panpsychism assumes that the mind, and thus the mental realm, is the fundamental feature of the world that exists throughout the universe. Our universe is not physical, but mental. This implies that there is no purely physical realm devoid of any mental features. The mind suffuses the physical world with its mental features. Historically, panpsychism and its assumption of the universe as intrinsically mental had important proponents in earlier centuries. The Dutch philosopher Baruch Spinoza (1632–77) considered mind and matter as distinct aspects of one mental substance which provides us with unity. Spinoza believed that this mental substance ultimately could be traced back to God. Since God is the ultimate origin, both nature and the world are suffused by his mental features. Nature and the world are therefore characterized as ultimately mental, which explicates panpsychism.

An even more radical version of panpsychism was proposed by the Irish philosopher George Berkeley (1685–1753). He and his idealistic successors assumed the world and its reality to exist completely in the mind and the mental states. This means that Berkeley denied any physical reality independent of the mind. This complete dependence on the mind leads to what is described as idealism. Idealistic philosophers are by default panpsychists in that they assume the world and reality as we perceive it to be intrinsically mental, rather than physical.

Mental escapes 2

Panpsychism: twentieth-century philosophy

What about panpsychism in the twentieth century? A philosopher in the twentieth century who argued in favor of panpsychism was the mathematician and philosopher Alfred North Whitehead (1861–1947). Rather than space, time and matter, he assumed processes and events to be the fundamental features of the world. He perceived of both processes and events as making up the world and accounting for mental features like creativity, spontaneity and perception. If this were the case, life and the world would be suffused by mental properties, implying panpsychism not only of the mind, but of the whole universe. This panpsychistic view of the universe has recently been linked with quantum physics, where the quantum is characterized in somewhat mental terms (see Part IV for details).

Whitehead's variation of panpsychism was quite radical. He has been interpreted by many as even assigning consciousness (and thus mental states) to stones. Why? Because stones are made of physical material. If the mind is assumed to pervade the physical realm with its mental features, any physical material including the stones must have mental features. Stones may consequently be assumed to show experiences and have consciousness. This sounds absurd and completely counterintuitive.

These counterintuitive assumptions contributed an argument for many to discard panpsychism as absurd and implausible.

One of the main arguments against such strong panpsychism is its apparent empirical implausibility. Physics, thus far, has been unable to provide any evidence that would suggest mental features suffuse the physical world. As such, it is rather difficult to imagine that reality and our world exist as completely mental and not at all physical. While panpsychism avoids the conceptual-logical problems of dualism, it creates a whole new set of problems including empirical implausibility.

Mental escapes 3

Panpsychism: strong and weak forms

One may distinguish between radical and moderate forms of panpsychism. Radical forms of panpsychism assume that mental features permeate any instance of reality and existence. In this case, even the table in front of you may show mental features and thus be conscious or experience something. There is no reality and existence that is not mental. This position leads to the seemingly absurd assumption that even stones experience reality and are conscious.

Such a radical form of panpsychism must be distinguished from its more moderate variations. In conservative models of panpsychism, the assumption of consciousness as omnipresent is rejected. Instead, one may assume certain features to be present that make possible and thus predispose mental features like consciousness. These predispositional features, so-called 'proto-mental features', are necessary but not sufficient by themselves to achieve full-blown mental features. For that, additional capacities are necessary.

What do the 'proto-mental features', the predispositional features, look like? In this case, neither the table nor the stone are conscious by themselves. Why? Because they either lack the features that predispose consciousness, or they do show the predispositional features, but lack the additional capacities that are necessary to turn them into full-blown mental features, including consciousness.

How does this more moderate version of panpsychism relate to the brain and its neuronal features? One could, for instance, imagine that there are certain unknown neuronal features in the brain that predispose it to generate mental states in specific contexts. One would then speak of neural predispositions, as has recently been introduced (see Northoff, 2013a, b, and Chapters 14 and 15 for details) that create a certain tendency in the brain and enable it to constitute mental states. This neural predisposition would then need to be characterized as proto-mental. In Parts IV and V, we will discuss some of the first empirical evidence in favor of proto-mental properties.

Take home message

Twentieth-century philosophy, including philosophy of mind and phenomenology, focused on the mind as a starting point for metaphysical and epistemological inquiry. At the same time, neuroscience developed and extended its empirical reach from the brain to the mind's various mental features like consciousness, self, free will, etc. Both disciplines – philosophy and neuroscience – asked questions about the relationship between the mind and the brain, known as the mind–brain problem. The mind–brain problem is a metaphysical problem that questions the existence and reality of the mind and how it is related to that of the brain. Since we do not experience the physical features of our brain as such in consciousness, mental features are often assumed to be special when compared to the physical features of body and brain. This is the basic premise that underlies the approaches that favor dualism between mind and body/brain, and sees them as distinct and separate existences and realities. This dual mental and physical existence and reality is supposed to be subserved by substances, properties, or information. However, any dualist solution runs into serious problems when analyzed for conceptual and/or empirical plausibility. How can we escape these problems, while at the same time preserving the special nature of the mental? This often leads one to panpsychism, the assumption that what we think of as physical reality and existence is actually mental, rather than physical.

Summary

The first part of this book focused on discussions in philosophy and neuroscience during the twentieth century and the development of neurophilosophy. We then turned to the core problem at the nexus between philosophy and neuroscience, the mind–brain problem. The mind–brain problem described is a metaphysical problem that asks about the existence and reality of the mind and how it is related to that of the brain. The metaphysical definition of the mind and its relationship to the body and brain has preoccupied philosophers since ancient Greece. Plato and Aristotle developed theories of an immaterial mind that resides either outside body and world (Plato) or inside the human body as its form (Aristotle). The framework for the current debate about the mind and the brain can be traced back to the French philosopher René Descartes. Descartes assumed two substances, one mental and the other physical. He postulated that they interacted with one another in the brain's pineal gland. Recent philosophy of mind has discussed various options for casting light on to the true nature of the mind–brain relationship. Descartes' substance dualism has been heavily criticized for various reasons, including its apparent empirical implausibility. Alternative forms of dualism that still preserve the special existence and reality of the mind and its mental features have developed. These include property dualism, informational dualism, and predicative dualism. Even earlier, neuroscientists working at the end of the nineteenth century and in the first half of the twentieth century advocated some form of mind–brain dualism. Besides dualism, panpsychism was discussed as another option that emphasizes the existence and reality of the mental realm. Panpsychism assumes that our reality and existence is inherently mental, and that its physical features contain some kind of mental or proto-mental property. Panpsychism has been attractive over the centuries and is still prevalent in the current mind–brain debate, albeit in different versions.

Revision notes

- Define the mind–brain problem.
- Sketch the main historical figures who argued in favor of different solutions to the mind–brain problem.
- Why did Descartes assume two different substances?
- Explain and define different forms of dualism and why they were developed.
- How can epiphenomenalism be distinguished from interactive dualism and emergentism?
- Why have philosophers come up with the idea of panpsychism as a monistic alternative to dualism?
- What is the difference between the metaphysical and the empirical domains and why is it important for the mind–brain problem?
- Is neuroscience compatible or incompatible with dualism? Why?
- What strategy does panpsychism deploy to counteract the problems of dualism?
- Why is panpsychism empirically implausible? Can there be forms of panpsychism that are empirically plausible?

Suggested further reading

- Chalmers, D. (1996) *The Conscious Mind* (Oxford/New York: Oxford University Press).
- Descartes, R. (1996 [1641]) *Meditations on First Philosophy*, translated by John Cottingham (Cambridge: Cambridge University Press).
- Searle, J. (2004) *Mind: A Brief Introduction* (Oxford/New York: Oxford University Press).
- Spinoza, B. (1985) *The Collected Works of Spinoza* (Princeton, NJ: Princeton University Press).
- Whitehead, A. (1929) *Process and Reality: An Essay in Cosmology* (New York: Macmillan).
- Whitehead, A. (1933) *Adventures of Ideas* (New York: Macmillan). (*My page references are to the 1961 Free Press (New York) edition.*)

6
Physical and Functional Approaches to the Mind–Brain Problem

Overview

The previous chapter discussed mental approaches to the mind–brain problem in philosophy. At the same time, the sciences, like physics and chemistry, emerged and made more and more important discoveries throughout the eighteenth and nineteenth centuries. By the end of the nineteenth century, the scientific investigation of the brain was developing rapidly and provided major insights into its sensorimotor and cognitive functions. How did the scientific investigation of the brain impact the metaphysical stance on the mind–brain problem? Based on the rapid development of neuroscience, the twentieth century witnessed a shift from dualistic and mental approaches to more monistic and physicalistic approaches to the mind–brain problem. These novel approaches to the mind–brain problem as they are discussed in current philosophy of mind are the focus of this chapter.

Objectives

- Understand the historical background behind the shift from mental to physical approaches
- Understand and explain behaviorism, its main thesis and its deficits
- Determine what the identity theory of mind and brain claims
- Define and understand functionalism
- Determine the different forms of physicalism
- Understand the limits of physicalism
- Understand the problems of functionalism

Key concepts

Behaviorism, identity theory, physicalism, supervenience, reductive and non-reductive physicalism, functionalism, hard- and software, multiple realizability, brain-based approach, intrinsic features of the brain.

Background

Mind–brain problem during the twentieth century

Mental approaches to the mind–brain relationship dominated until the beginning of the twentieth century. This resulted in various forms of dualism and panpsychism as discussed in the previous chapter. In contrast to mental approaches, purely physical approaches to the mind and the world were rare before the twentieth century. While the concept of materialism – which accounts for the world in terms of matter – can be traced back to ancient Greece, a purely materialistic account of the mind was almost unheard of. *The term physicalism, which describes everything in the world (including the mind) as purely physical was not introduced into philosophy until the twentieth century, when the philosophers Otto Neurath and Rudolf Carnap made such a claim.*

Today the terms materialism and physicalism are often used interchangeably. Following this precedent, I will do the same. *Both materialism and physicalism describe the existence and reality of the world and the mind as consisting of nothing but pure physical matter. There is no room for non-physical and mental features in either the mind or the world.* The world and our mind are understood as mere physical and materialistic entities. There is nothing else. This model denies the possibility of any kind of non-physical features, including those that were traditionally associated with religious belief and a mind – like God.

A materialistic-physicalistic view of world and mind reflects the increasing dominance and relevance of the sciences in our understanding of the world and ourselves. This dominance began with the scientific revolution and has continued throughout the centuries until the present day. In the twentieth century, for example, physics as a discipline was revolutionized, as was psychology (see Chapter 3).

Another important development mentioned in the first part of this textbook was the development of research on the brain and its functions. The turn of the nineteenth century saw the rapid development of a novel discipline – neuroscience – which focuses on the empirical investigation of the brain. Because of the introduction of novel techniques in the past 20–30 years like functional brain imaging, neuroscience is now able to investigate neural activity changes during various kinds of mental functions including different affective, cognitive and social functions. And even mental functions as complex as consciousness and self have become the subject of intense neuroscientific investigation. This will be discussed in Parts IV and V.

What does this imply for the mind–brain problem? First, the need for a materialistic theory of the world and the mind became even more urgent as the explanatory power of the sciences rapidly progressed. Second, the rise of neuroscience – and the increasing empirical research of the brain – left less and less room for the assumption of the existence and reality of a non-physical mind with some kind of mind–brain dualism. Instead, the empirical discovery of neuronal mechanisms in the brain prompted neuroscientists and many philosophers from the mid-twentieth century onward to argue in favor of a materialistic (rather than dualistic) approach to the mind–brain problem.

Current philosophy of mind therefore focuses on physical approaches to the mind–brain problem. Meanwhile, mentalist approaches like dualism and panpsychism are on the defensive; they are often considered strange and old-fashioned, given the seemingly overwhelming empirical evidence in support of a purely physical mind and brain. In this chapter, I will review the main physicalist approaches to the mind–brain problem.

History of physicalism 1

Methodological behaviorism

Behaviorism describes the assumption that the mind is nothing but the outer behavior of the body. There are no inner mental states inside our mind or brain. Instead they are behavioral states or behavioral dispositions of the body that emerge in relation to the environment. Hence, what is erroneously described as an inner mental state turns out to be nothing but a mere outer behavioral state.

Behaviorism denies the existence and reality of anything mental as distinct from the physical features that underlie the behavioral states. There are no inner mental states and there is no mental substance either. But how could Descartes come up with the idea of inner mental states and dualism between mental and physical substances? Behaviorism argues that Descartes committed a logical error by confusing outer behavioral states with inner mental states. He falsely associated inner mental states with outer behavioral states.

How does this behaviorism stand in relation to psychology? Recall Part I. Psychology is the science of the mind that aims to investigate the mind and its mental features experimentally. If, according to behaviorism, there is no mind and no possibility of any inner mental state, psychology must focus on outer behavioral states. Psychology, in this light, is characterized as a 'science of behavior', rather than a 'science of mind'. Behaviorism in this sense can be understood as a methodological strategy that details how to investigate the mind in a scientific way – by means of outer behavior. This is described as methodological behaviorism, or the empirical version of behaviorism.

Let us specify the strategy suggested by methodological behaviorism. Rather than studying inner mental states, psychology should focus on behavior as the output in response to input stimuli. The focus is then on how the input stimuli are related to the output, also known as the observed behavior responding to the input. Most importantly, these input–output relations or stimulus–response relations are supposed to be direct relations. In short, they are without any intervention or inference by a third variable, like some inner mental state. In this case, inner mental states are no longer needed in explaining our behavior – all that is needed can be found in the relation between input and output (or stimulus and response). The investigation of input–output relations would become the primary focus of psychology.

How can we study stimulus–response relations as input–output relations? Rather than relying on the introspection of inner mental states in first-person perspective (FPP), input–output relations can be investigated by mere observation in third-person perspective (TPP). Experience and consciousness in FPP are thus eliminated completely from psychology. By relying completely and exclusively on TPP, psychology can become a true science like physics and chemistry, both of which are also based on TPP.

History of physicalism 2

Rejection of inner mental states in behaviorism

The main advocates of this empirical or methodological version of behaviorism were John Watson (1878–1958) and B. F. Skinner (1904–90). Watson's and Skinner's assumptions were strongly fueled by the discoveries of the Russian neurophysiologist Ivan Pavlov (1849–1936). He showed that animals, such as dogs, show strong salivating, an unconditioned response, when exposed to stimuli like the sight of food. Pavlov trained animals to associate an initially neutral stimulus (the ringing of a bell) with the unconditioned stimulus, the sight of food. This eventually led to the production of a conditioned response (salivating) in response to the conditioned stimulus (the ringing bell). Even though there is no food in sight yet, the dog starts salivating when it hears the bell ringing. The input stimulus 'bell ringing' is thus associated with a behavioral output, the salivating as response.

How can we further illustrate this type of conditioning? Imagine the following scenario. You pass by a Starbucks sign, an announcement that 100m further down the road there will be a Starbucks café waiting for you. Since you love their latte macchiato so much, just by seeing the sign you start to feel like you are in need of caffeine. How is this possible? As in the case of the Pavlovian dog, a particular stimulus, the Starbucks sign, has been associated with a specific reward in your mind that leads to a specific behavioral response. According to the advocates of empirical or methodological behaviorism, this response can be explained without reference to any inner mental states.

Behaviorism rejects the assumption of any kind of inner mental state and the possibility of a corresponding mental substance. Instead of referring to inner mental states, behaviorism dictates that we should refer to the relations between stimulus and response. If psychology wants to be a serious science, it should focus exclusively on input–output as stimulus–response relations while discarding any mental features including consciousness. This type of methodological behaviorism was extremely popular in the first half of the twentieth century, especially in the United States, where two of its main proponents – Watson and Skinner – resided. Behaviorism as a theory expanded beyond the boundaries of psychology to the philosophical. How behaviorism entered the mind–brain debate will be discussed below.

History of physicalism 3

Logical behaviorism

Methodological behaviorism focuses on the empirical domain – on what is observable in third-person perspective. Behaviorism that focuses on the metaphysical domain is known as 'logical behaviorism'. *Logical behaviorism claims that it is a logical error to assume inner mental states. Instead, inner mental states are nothing but behavioral dispositions. According to this account, Descartes and other dualists committed a logical error when they confused behavior with inner mental states.*

How can we explain mental states in logical behaviorism? Mental states are nothing but behavioral dispositions. For example, I believe that it might rain today and this incites me to buy an umbrella, to wear my coat, and to walk beneath the shelter of the supermarket's roof. Buying an umbrella, taking my coat, and walking beneath the roof are behaviors that result from the association between input stimuli and outputs as the behavioral response. The input stimulus 'rain' is associated with 'umbrella' and that in turn triggers the indicated line of subsequent associations. Mental states like the belief that I might get wet are not really needed here; everything can be accounted for by the linkage between rain as input and umbrella as output. Mental states are thus to be discarded and can be replaced by behavioral dispositions.

The British philosopher Gilbert Ryle (1900–76) compared the assumption of a mind and its mental features to a 'ghost in the machine' of the body: the body as mere machine mediates the input–output relations and thus the association between stimulus and response. For that, no inner mental states are needed. The 'ghost' does not have any proper function.

Logical behaviorism is a radical thesis because it eliminates all inner mental states. This might seem implausible, given our common-sense intuition that mental states like free will and consciousness exist and impact our behavior. In this light, behaviorism seems counterintuitive to our experience.

Biography

Noam Chomsky and behaviorism

Let us discuss some of the objections to behaviorism. Noam Chomsky, the American linguist, proposed one of the main objections to behaviorism. Chomsky was born in 1928 and raised in Philadelphia, USA. His research focus was on linguistics. He made a seminal contribution when he suggested that our mind can be characterized by an innate structure that is organized along the lines of grammar and language. This will be discussed later, in Part III. Chomsky became well known to the broader public for his outspoken political activism and for his opposition to American foreign policy during the Vietnam War and after the 9/11 attack in 2001. Besides his strong political engagement he was, and still is, also philosophically active where he became known especially for his linguistics (see Chapter 14).

Philosophically and scientifically, Chomsky was vital in his rejection of behaviorism. Chomsky argued that behaviorism confuses the evidence about mental states – that is, behavior – with the existence and reality of the mental states themselves, the mental features. Let me explain this further. According to Chomsky, behavioral states indicate the presence of mental states; behavior thus provides evidence about mental states. Behavior only indicates mental states, but it is not the mental state by itself. This is analogous to, for instance, the smell of skunk that should not be confused with the skunk itself: in the same way, the rather interesting smell only indicates the (possible presence of a) skunk but cannot be identified with the (actual presence of the) skunk itself, the behavior indicates mental states but cannot be identified with them.

History of physicalism 4

Objections to logical behaviorism

Another objection that has been put forward against behaviorism is that it cannot properly account for behavior. By explaining inner mental states like beliefs in purely behavioral terms, one cannot avoid either implicitly or explicitly using some kind of mental terminology. My action of buying an umbrella is propelled by my desire and urge not to become wet. That in turn is to be traced back to my belief that it will rain. My belief that it will rain thus needs to be included in order to account for my behavior. This means that a mental state like belief and my urge to not get wet needs to be included in the explanation of my behavior. A purely behavioristic explanation of behavior thus remains insufficient, which puts the validity of behaviorism into doubt.

Why is the insufficiency of the explanation a problem for behaviorism? It is a problem because the concepts of urge and desire are mental terms rather than purely behavioral ones. This means that the original mental statement – my belief that it will rain – cannot be translated into a purely behavioral statement. Instead, my belief that it will rain and my desire not to get wet need to be linked to my behavior of going to buy an umbrella. This, however, undermines the main assumption of logical behaviorism, namely the complete elimination of mental statements in favor of behavioral statements. Because of the need for including beliefs and desires, behaviorism must include some inner mental features in order to explain behavior.

Most important is the fact that in logical behaviorism there seems to be no place for consciousness and its phenomenal features. Phenomenal features describe those features that characterize our subjective experience and thus consciousness. One such phenomenal feature is that experience is always tied to a first-person perspective and a point of view from which we experience the contents in our consciousness. We will come back to these and other phenomenal features in Part IV.

The same holds true for meaning (semantic content – see Chapter 10 for details), which has no place in the input–output behavioral account of logical behaviorism. Behaviorism goes against the intuition that our mental states, like my belief it will rain, cause our subsequent behavior. This is also called mental causation (see Chapter 5). Because of these various problems, behaviorism in both of its versions – empirical behaviorism and logical behaviorism – is no longer as popular in the current discussion of the mind–brain. Only a few philosophers today argue in favor of behaviorism. Max Bennett and Peter Hacker are two such thinkers (see Bennett and Hacker, 2004).

History of physicalism 5

Identity theory in philosophy

While behaviorism was revealed as problematic in the mid-twentieth century, other theories of the mind–brain relationship were developed. The aim was to take advantage of the 'good sides' of behaviorism, its rejection of dualism and a mentalistic approach to the mind, for example, while at the same time avoiding its shortcomings, like the complete elimination of the mind. American philosophers like H. Feigl, U. T. Place and J. C. Smart developed an alternative approach to the mind–brain problem called identity theory. This will be explained below.

Feigl and Smart claimed that mental states and brain states are identical (see below for further detail defining and explaining the term 'identity'). This is not a matter of logic, as is the assumption in logical behaviorism. Instead, it is a matter of empirical evidence. In fact, it is supposed that mental states are neuronal states and that future empirical discoveries in neuroscience will prove this fact. As such, this identity theory supports an empirically-based (rather than logically-based) identity between the mind and the brain.

How can we support this claim? In order to shed some light on the concept of identity, the proponents of identity theory often use analogous examples from the history of science where two seemingly distinct entities turned out to be identical. For example, chemistry found out that water is identical to the chemical formula H_2O. Water is H_2O and H_2O is water; both are thus identical. Another example stems from discoveries in thermodynamics that showed heat to be nothing but molecular motion.

How are these examples related to the question of the mind–brain relationship or the mind–brain identity? These cases illustrate that two different entities – like H_2O and water – are in fact identical. In the same way that water and heat were found to be identical to H_2O and molecular motion, future discoveries in neuroscience will show that mental states are identical to the neuronal states in the brain. If this were true, one could assume an empirically-based identity between mind and brain.

One example the proponents themselves like to discuss is the example of pain. There are C-fibers in the brain that mediate pain. What is the difference between C-fibers and pain? According to the identity theorists, there is no difference. Why? Because both are identical. If the C-fibers are stimulated, you will suffer pain. In short, pain is C-fiber stimulation and C-fiber stimulation is pain. Identity theorists take this as a paradigmatic example of the identity between mind and brain, the mind–brain identity theory.

History of physicalism 6

Different forms of identity in philosophy

How can we further determine the concept of identity? One can distinguish two different versions of identity, token and type identity. What do the concepts of token and type mean? Imagine you say the same word five times: pain, pain, pain, pain, pain. All five instances are different tokens of one and the same type of word: pain. The term token thus describes concrete particular objects or events, while types refer to more abstract general entities. There are thus five tokens and one type in our example.

How does the distinction between token and type apply to the identity theory? One could claim that every instance of an individual mental event may be identical to some neuronal event. For example, specific pain could be linked to a specific type of C-fiber stimulation. Another instance of pain may then be identical to a subtype of C-fibers, a different kind of C-fiber stimulation, or some completely different neuronal mechanisms outside the C-fibers. There would thus be identity between particular kinds of pain and specific neuronal events, with the identity varying from case to case. This describes token–token identity.

This is different in the case of type–type identity. Here any type of pain is supposed to be identical with the C-fibers. There is not only identity between single neuronal and specific mental events, as in token–token identity, but also the identity is specified on both sides, neuronal and mental. Pain of any type is identical with C-fiber stimulation. Unlike in the case of token–token identity, pain remains impossible outside of and apart from C-fiber stimulation.

One may now smile about the simplicity and naïveté of this example. As we know now, pain is far more complex than simply the firing of C-fibers. Pain includes distinct components – somatic, affective, cognitive and sensory – all of which are mediated by different fibers and regions. That, however, was not yet known at the time that identity theory was discussed, in the 1950s and 1960s. Today functional imaging techniques allow online visualization of neuronal activity, including while a test subject is experiencing pain. Functional imaging also shows that the regions involved in mediating pain may not only differ between different subjects, but within one and the same subject on different occasions, where different regions and networks may be recruited for one and the same kind of pain.

Such inter-individual variation seems incompatible with the assumption of type–type identity. Instead it seems to lend some empirical support to token–token identity that provides more variety in the relationship between mental and neuronal states. This makes it clear that modern neuroscience can inform philosophical assumptions and test for empirical (neuronal) plausibility. It also shows that neuroscience can provide a contribution to address the question: which of the two versions of identity theory is more empirically plausible?

Critical reflection 1

Different forms of identity in neurophilosophy

How can neurophilosophy contribute to addressing the question of the type of identity? Neurophilosophy may contribute to revealing what is consistent across individuals, thus being type-specific (as one may want to say) and what varies between individuals and is thus rather token-specific (as one may want to say). Let us start with type-specific neuronal features.

The central question here is the following: what neuronal features are shared across different individuals and their inter-individual differences? And how are these type-specific neuronal features necessary for the generation of mental states? One could, for instance, assume that type-specific neuronal features predispose the generation of mental states as the necessary conditions of their possibility. The type-specific features may then reflect what I later (see Chapters 15 and 16) describe as neural predisposition. In contrast, token-specific neuronal features may vary between different individuals. As such they may account for the actual (rather than possible) realization of mental states as the sufficient condition of their manifestation. One may then speak of neural correlates as distinguished from neural predispositions (see Chapters 14 and 15).

What does this imply for identity theory? Though logically contradictory, both token- and type-specific features may well be compatible with each other on empirical grounds. They may concern different neuronal features as related to neural predispositions and neural correlates. Token–token identity theory may apply to individual persons and their particular neuronal states (or neural correlates) that realize and manifest their actual mental states. Different individuals may use different neuronal mechanisms to actually realize and implement their mental states.

In contrast, type–type identity theory may rather apply to all individuals and concern the predisposition of all people to potentially generate mental states. This concerns those neuronal features that must be present and similar in all individuals, since otherwise they would not share the occurrence of consciousness. Hence, type–type identity theory may apply to neural predispositions while token–token identity theory may hold for neural correlates.

The philosopher may now want to argue that such co-occurrence of both token–token and type–type identity is clearly logically contradictory. There is either token–token or type–type identity between mental and neural states and thus between mind and brain, since otherwise one would assume two different, logically contradictory mind–brain theories at the same time. The neurophilosopher, in contrast, may differentiate the kind of neural features he/she targets, namely neural predisposition and neural correlates, which may stand in different kinds of relationships to mental features. Accordingly, what seems logically-conceptually contradictory and thus implausible from a philosophical point of view may turn out to be empirically plausible in a neurophilosophical framework.

Biography

Saul Kripke

One major objection to identity theory came from the American philosopher Saul Kripke. Kripke was born 1940 in Bay Shore, New York, and as child became keenly interested in logic and mathematics. He would go on subsequently to make essential contributions to these fields. Kripke published several books which became influential. One was on the Austrian philosopher Ludwig Wittgenstein, whose later philosophy on rule-following in language he interpreted in a rather idiosyncratic way. Alluding to Kripke's special way of interpreting Wittgenstein, commentators nicknamed his approach as 'Kripkenstein'.

Let us here briefly describe another landmark book by Kripke, *Naming and Necessity* (Kripke, 1972), in which he discussed the problems of identity and truth. Kripke made complex arguments which cannot be recapitulated here in their entirety because their scope is beyond the purposes of this textbook. Therefore, here the focus is mainly on the concept of the 'rigid designator', which is relevant in the present context of identity theory.

Critical reflection 2

Rigid designators and identity (Kripke)

What does 'rigid designator' mean? Kripke argues that this term refers to the same object in all logically possible worlds – the worlds we can imagine in a logically coherent way independent of the physical constraints of the natural world. Since the object remains the same throughout all logically possible worlds, the term describing an object can be regarded as rigid and unchanging, no matter what.

For instance, heat refers to molecular motion in all possible worlds that are imaginable in a logically coherent way. This means that heat always implies molecular motion. Heat and molecular motion are thus intrinsically related to each other, meaning that their identity is a necessary or logical truth. This means that heat necessarily and unavoidably implies molecular motion, as molecular motion cannot avoid yielding heat. The concept of heat can thus be characterized as a rigid designator, since it describes the same object in all logically possible worlds.

The same holds true in the case of water and H_2O: water refers to the chemical formula H_2O in all logically possible worlds. Water always refers to H_2O in any logically conceivable world so that their relationship is a logically necessary one. The term H_2O is thus a rigid designator, since it refers to one and the same object in all possible worlds.

One may now be inclined to argue that the relationship between water and H_2O was subject to empirical discovery: instead of inferring their identity in a purely logical and thus a priori way (see Chapter 2), the water's chemical formula, H_2O, had to be discovered empirically in an a posteriori way (see Chapter 2 for the distinction between a priori and a posteriori). The identity between water and H_2O is thus a posteriori. Hence, Kripke undermines the traditional linkage of identity with the a priori (see Chapter 2). The example of water signifies that its identity with H_2O can well be associated with empirical discovery and thus with the a posteriori rather than the a priori as it has been traditionally suggested (see Chapter 2 for details of the traditional usage). In short, the example of water demonstrates that a posteriori (rather than a priori) identity is possible.

Critical reflection 3

Non-rigid designators and rejection of identity (Kripke)

How does the a posteriori identity between water and H_2O stand in relation to the identity between mental and neuronal states? There is identity between water and H_2O because both of the concepts imply each other: water is necessarily H_2O, while H_2O necessarily entails water. Can we apply this to the relationship between mental and neuronal states? If mental and neuronal states are indeed identical, neuronal states should imply mental states and vice versa in the same way that water and H_2O imply each other. If yes, one could diagnose identity in an a posteriori sense between mental and neuronal states and thus between mind and brain.

Following Kripke, this, however, is not the case: the experience of pain does not imply any conceptual-logical reference to C-fiber stimulation, nor does C-fiber stimulation imply any reference to the experience of pain. Let us explain this further for both C-fiber and pain. The C-fibers only refer to certain proteins and other molecules, the material out of which they are made. One could therefore imagine C-fibers to be present, even in the absence of pain. Conversely, pain does not imply any reference to C-fibers at all. We experience pain and describe it in various sensory and emotional terms, which conceptually-logically do not imply any reference to C-fibers. For instance, one can imagine that pain could occur without C-fiber stimulation.

The fact that we can purely logically conceive of pain without C-fibers, and C-fibers without pain, means that pain does not imply anything about C-fibers and that the C-fibers do not entail pain. This makes their mutual reference impossible and means that there is no necessary linkage (in conceptual-logical terms) between C-fibers and pain. Therefore the relationship between C-fibers and pain in particular, and thus the one between neural and mental states in general, cannot be compared to the relationship between H_2O and water.

Unlike water and heat, pain therefore cannot be regarded a rigid designator, since it does not necessarily refer to C-fibers in the way that water and heat refer to H_2O

and molecular motion. Therefore pain can at best be considered a non-rigid (rather than rigid) designator. If so there can be no identity between C-fibers and pain and thus between neural and mental states. Identity theory as the answer to the mind–brain problem must therefore be rejected.

Ideas for future research 1

Necessary versus contingent identity in neurophilosophy

Kripke considers the relationship between neuronal and mental states in a purely conceptual-logical way. He investigates whether mental concepts like pain conceptually and logically entail any reference to neuronal concepts like C-fibers and vice versa. This presupposes a purely conceptual-logical approach for which logical conceivability or plausibility is sufficient by itself, independent of whether or not C-fibers are indeed associated with pain in the natural world. Kripke's rejection of identity theory is thus a merely logical and therefore philosophical one.

How about a neurophilosophical approach? One may also investigate the relationship between neuronal and mental states in an empirical way. One would then raise the question of whether the stimulation of the C-fiber is indeed associated with pain. This would test for the natural and thus empirical plausibility of the relationship between C-fibers and pain independently of the purely logical-conceptual plausibility as implied by Kripke.

What would such a neurophilosophical approach look like? If C-fiber stimulation always leads to pain, and if no cases of pain without C-fiber stimulation can be observed empirically, one would provide empirical support to their relationship. One would then assume identity between C-fibers and pain, which, though, is only empirically- rather than logically-based. This means that we can assume some kind of identity between C-fibers and pain on empirical grounds. Even though contingent and thus non-identical in conceptual-logical regard, C-fibers and pain may then be considered identical in empirical and thus natural regard.

The philosopher may frown upon that and argue that this does not provide any support to identity theory. Why? There is simply no identity involved as long as C-fibers and pain do not logically-conceptually entail each other, as, for instance, Kripke demonstrated in the case of water and H_2O. Both terms, C-fibers and pain, would need to be rigid designators to mutually entail each other in a conceptual-logical way. Since this not the case, there is simply no identity between neural and mental states.

The neurophilosopher may, though, not need to be concerned about that. He/she may distinguish between different forms of identity, necessary and contingent. Necessary identity refers to identity in a conceptual-logical sense where both neural and mental states are rigid designators and do therefore logically-conceptually entail each other. Such necessary identity remains by itself independent of any empirical observation in the way the identity between water and H_2O was there before we discovered their identity (which amounts to what Kripke described as 'analytic a posteriori'; see Chapter 2) .

This is different in contingent identity. Contingent identity is based on empirical observation rather than on conceptual-logical grounds. Here both C-fibers and pain do not need to be rigid designators, as, for instance, one can be a rigid and the other a non-rigid designator. If we can observe that C-fiber activation leads to pain, we may be able to postulate identity between them in an empirically-based (rather than conceptual-logically-based) way, entailing contingent rather than necessary identity. This shifts the focus from mere conceptual-logical plausibility to empirical plausibility. Is identity theory in a contingent sense of the term identity empirically plausible? This will be the focus in the next section.

Ideas for future research 2

Empirical implausibility of contingent identity

Let's replace the C-fibers by the different regions in the brain that mediate pain as observed in functional brain imaging. There is indeed empirical evidence supporting both forms of dissociation, neural activity in pain regions without pain and pain without neural activity changes in pain regions. Activation of typical pain regions like the supragenual anterior cingulate cortex, sensorimotor cortex and the periaqueductal gray (PAG), can occur in various instances, including during movements that do not lead to pain. At the same time, pain may occur in, for instance, psychiatric patients with a somatoform disorder (where patients feel abnormal pain sensations), without corresponding activation of the pain regions.

Since the empirical findings suggest dissociation in both directions, the data do not support the assumption of any kind of identity between pain on the one hand and neural activity in pain regions on the other. Both pain and neural activity thus do not seem to be identical in either a contingent or necessary way. This supports Kripke's purely logical objection against identity theory on empirical grounds. Taken together, identity theory seems to be neither conceptually-logically plausible on purely philosophical grounds nor empirically plausible in neurophilosophical regard.

Neurophilosophically, the conceptual-logical and empirical rejection of the mind–brain identity theory implies that we may need to search for other ways. First, one may want to search for neuronal features other than C-fibers or regional activity (see Chapters 8 and 12 as well as Parts IV and V). These other neuronal features may then hold the possibility of showing some kind of identity, contingent or necessary as well as type–type or token–token, to mental features.

Second, one may want to search for features other than neuronal (and physical) altogether that, however, unlike mental features as postulated in dualism, remain nevertheless within the natural world itself (as distinguished from the purely logical world of philosophy). The assumption of informational features, as postulated by Chalmers, can be considered an example (see Chapter 5 for details).

Third, and finally, one may also be more radical and deny the existence and reality of mental features altogether, which makes their assumption of identity with physical states superfluous. This leads to physicalism as metaphysical mind–brain theory, as will be discussed below.

Physical escapes 1

Supervenience of mental states on neuronal states

While identity theory was very popular in the 1960s and 1970s, Kripke's criticism dealt it a serious blow. This necessitated the search for a mind–brain theory that preserved the physical characterization of the mind, while at the same avoiding the problem of identity theory. Hence alternative versions of physicalism were developed which will be described briefly below.

One may distinguish between reductive and non-reductive physicalism. Non-reductive physicalism posits the mind as physical but that its description or conceptual analysis may not be reducible completely to mere physical description. The mental described in mental terms may not be reducible to the physical as described in physical terms because of the difference in concepts. That, though, does not imply that the mental is non-physical in its existence and reality. One may thus assume physicalism in the metaphysical domain while explanatorily one opts for dualism in the conceptual regard (see Part III for details).

One variant of non-reductive materialism is the anomalous monism of the American philosopher Donald Davidson (1917–2003). He combined two different theses: he suggested that our conceptual descriptions of mental features are irreducible to our descriptions of physical features whereas, at the same time, mental and physical features are identical in their existence and reality (in a token–token sense). Metaphysically, mental and physical features are identical. Meanwhile, they differ from each other conceptually when using mental and physical concepts.

The American philosopher, Jaegwon Kim (1934–), suggested another variant of non-reductive physicalism called supervenience. Supervenience claims that the mental can be considered like a pattern or global property, while the physical can be found in the local elements. The global pattern is still physical but operates on the local elements. There is thus a distinction between the mental and the physical as being local and global, while the existence and reality of both is still considered physical.

The distinction between local elements and global patterns can well be related to the brain. The brain consists of neurons that are single cells which can be considered the local elements. The different neurons interact with each other by means of which global patterns of neural activity across different regions in the brain are generated. This leads to what is described as neural networks. These show certain spatial and temporal dynamics on their own (see Part IV for details). One can thus say that the neural networks and their neural activity operate across the single cells.

However, this concerns only neuronal–neuronal supervenience where the networks operate across single neurons. What we want, though, is mental–neuronal supervenience, where the mental can operate across the neuronal. Recent imaging data (see Parts IV and V) suggest that neural activity in these neural networks is indeed related to mental functions like self and consciousness. Different spatial and temporal configurations in the neural network's neural activity seem to be associated with different mental functions. The postulated neuronal–neuronal supervenience thus seems to translate into mental–neuronal supervenience.

Physical escapes 2

Supervenience and epiphenomenalism

What is supervenience? Supervenience aims to keep the causal world of the physical realm closed, meaning that it supports causal closure of the physical world (see Chapter 5). This means that the mental cannot intrude into the physical world.

At the same time, the mental is not supposed to be principally different from the physical; both mental and physical do not reflect different properties because that would lead to dualism. Instead, they reflect only distinct layers of activity like local elements and global patterns, with the former causing the latter.

This distinguishes supervenience from epiphenomenalism. Recall from Chapter 5 that epiphenomenalism assumes mental properties as distinct from physical properties, and denies that the mental can causally impact the physical. Mental properties thus cannot exert any causal impact on the physical. This means that mental causation remains impossible in epiphenomenalism. How is this related to supervenience? Unlike epiphenomenalism, supervenience does not assume any mental properties, not even causally impotent mental properties. Supervenience implies monism, while epiphenomenalism is a dualistic mind–brain theory.

Supervenience claims that there are simply no mental properties. Why? Because the mental is physical. This means that the existence of the mental is sufficiently caused by the physical and thus entirely supervenient on the brain's physical properties. Here, the mental reflects only a different layer, whose existence and reality is principally different from the physical in the context of supervenience. Unlike in reductive theories, consciousness and its various phenomenal features can be acknowledged without implying any mental existence or reality as in dualistic approaches.

Let's compare supervenience to the meal you are cooking. The various ingredients like vegetables, meat and eggs are purely physical. Now put them all together in your pan and you get an omelette. The final omelette is quite different from the initial ingredients. Does this mean that the omelette is no longer physical? No, says the supervenience position: the omelette is sufficiently caused and explained by the physical features of the ingredients and their heating in the pan. In short, the omelette is physical. Nevertheless the omelette is different from the mere addition of its ingredients. In other words, the omelette supervenes the ingredients. Analogously, mental features like consciousness supervene on the physical states of the brain without becoming non-physical.

How does supervenience stand in relation to materialism? Supervenience can be regarded a non-reductive form of materialism. It is non-reductive because mental states are not reduced to and identified with physical states. They are, however, at the same time still physical in their existence and reality. Such non-reductive forms of physicalism must be distinguished from reductive ones. Reductive physicalism claims that consciousness and its supposedly mental features can be completely reduced to physical features. There is no need to maintain mental features as distinct and special even within a physical framework as supervenience does. Instead, the mental features are completely reduced to the physical features.

An even more radical version of materialism is eliminative physicalism or materialism. In this case, mental states are no longer assumed to exist and to be real at all. Since they are not real and existent, there is no need to reduce them as in reductive materialism or to let them supervene on physical states. Instead, the mental states can simply be eliminated. The American philosophers Patricia and Paul Churchland argued for such eliminative materialism, which will be discussed in more detail in the third part of this book (see Chapter 11).

Critical reflection 4

Notion of the 'physical'

There is plenty of discussion about the concept of the mental as the hallmark feature of the mind. The concept of the mental is determined in different ways: physical, non-physical, and in other ways like conceptual and epistemic (see Chapter 3 for more determinations). In contrast, the concept of the physical is often presupposed as given and taken for granted. However, we will see below that the concept of the physical can be defined in different ways, and that this has important implications for how we can logically conceive the relationship between mind and brain.

The concept of the physical can be considered in two different ways. The physical can denote some features as singled out and investigated in physics and thematized in physical theory. For instance, mass, gravity and velocity are physical measures and thus physical features in this sense. I here speak of physical (p) indicating its definition by the science of physics. The physical (p) must be distinguished from the physical that is associated with objects like rocks, trees, chairs, etc. These are physical objects as we, on the basis of our common sense, conceive them. Since they concern physical objects rather than physical measures, I here speak of physical (o).

Let us start with the physical (p). When defining physicalism by the physical (p), one must presuppose that current physics is complete, otherwise one cannot define the physical in the terms of physical (p). However, it cannot be presupposed that physics has attained all possible knowledge. For example, at the end of the nineteenth century, many thought that physics was complete. And then came the theory of relativity and quantum theory. This changed physics and its notion of the physical in a radical way.

How can we be sure that the current physics, in the year 2014, is the physics of, for instance, the year 3012? We cannot, because we cannot know whether physics is ever complete or not. This incompleteness implies that we cannot be sure about the notion of the physical (p). It may turn out that the physical (p) will contain some mental or proto-mental features in the future. Even today some already assume the quanta in quantum theory to be proto-physical and proto-mental. This then makes any exclusive definition of physicalism by the physical (p) impossible.

Critical reflection 5

Notion of the 'physical' and physicalism

What do these considerations imply for the concept of the physical in the context of mind–brain physicalism? Two issues arise. First, the notion of the physical (p) is very much dependent upon the respective historical context. If physicalism is defined by the physical (p), it may change over time, depending on the historical context and its respective presuppositions. Besides referring to different notions of the physical in different times, this introduces an empirical element, historical dependence, into a seemingly purely metaphysical assumption of physicalism. In other words, by presupposing the physical (p), physicalism is transformed from a purely metaphysical thesis into an empirical assumption.

Second, the definition of physicalism would then also depend on whether one operates on the macroscopic level of objects or the microscopic level of quanta. If one operates on the level of objects in the sense of the physical (o), one can maintain a pure concept of the physical. Everything in the world can then be described (more or less) as a physical object in terms of the physical (o). The usage of a pure notion of the physical – physical (o) – also implies a clear-cut boundary between the physical and the non-physical (as in, for instance, the concept of the mental). Such a notion of the physical in the sense of physical (o) seems to be especially presupposed in the earlier concepts of physicalism, including the mind–brain identity theory.

This changes when we revert to the physical (p). The notion of the physical in this model becomes rather blurred once one presupposes quanta. Why? As indicated above, the physical (p) may include the proto-physical and, even more importantly, proto-mental properties. The distinction between the physical and the mental is thus no longer as clear as in the case of physical (o).

Paradoxically, it is the scientific – rather than common-sense-based notion of the physical, physical (p) – that introduces such blurredness. One of the main motivations of the advocates of physicalism is to avoid any kind of blurredness which they attribute to the dualistic theories. In order to avoid the blurredness of dualistic theories, the physicalists turn to science and especially physics. And it is now that very science by itself, quantum physics especially, that comes with a certain degree of blurredness that is then also injected into the respective mind–brain theories. What does this short discussion tell us? We can see here that the scientific results themselves and thus the notion of the physical (p) blur the seemingly clear-cut boundary between the mental and the physical. This implies that the ground upon which mind–brain physicalism is built, the notion of the physical (p) falls down and becomes porous.

Taken together, both conceptions of the physical, physical (o) and physical (p), are problematic because of their definitions' dependence on historical context and the level of description concerning either the macroscopic level of objects or microscopic level of quanta. If however the notion of the physical cannot be defined in a clear-cut and proper way – by excluding any kind of possible mental or proto-mental traces – the assumption of physicalism as based on the definition of the physical, becomes problematic too.

This means that the considerations from physics as science provide some quite strong empirical evidence against physicalism as a metaphysical mind–brain theory. Physicalism may thus turn out to be scientifically- and empirically-grounded as it is often implicitly presupposed in current philosophy of mind. The paradox consists now in the fact that the scientific evidence itself may be taken to argue against a purely physicalistic account of the mind and its relationship to the brain. That means that physicalism as mind–brain theory may not be as empirically plausible as it is often assumed. Hence, even if plausible in conceptual-logical regard, physicalism may fail the test for empirical plausibility.

Critical reflection 6

Physical and phenomenal features

Can physicalism account for the mind and its mental states? If so, we expect physicalism to explain consciousness. Consciousness shows various features that characterize our experience and are therefore described as phenomenal features (see Chapter 1).

What are the core phenomenal features of consciousness? One such feature is subjectivity. Subjectivity describes the individual point of view from which our experience is taken. Another phenomenal feature is intentionality, which concerns the directedness of our experience toward contents and their semantic meaning. Finally, qualia is another phenomenal feature, which describes the phenomenal-qualitative features of our experience (see Part IV for details).

Let me give an example to illustrate these phenomenal features further. When you read these lines, you may be feeling hungry. Your experience of hunger is tied to your first-person perspective as distinguished from observation in third-person perspective. This means that your friend, who is sitting right beside you and reading the same lines, does not have access to your experience of hunger. He/she can have his/her *own* feelings, but he/she cannot know or be a part of yours. As such, your experience of hunger remains subjective.

Finally, your experience of hunger feels a certain way. There is a certain quality to it which current philosophers like Thomas Nagel (1937–) refer to as 'what it is like'. In your case, it would be the 'what it is like to experience hunger' when reading these lines. This quality of your experience is described as phenomenal-qualitative and, in short, as qualia. Qualia are obviously closely related to subjectivity. Without subjectivity and a point of view, we could not experience certain qualities, or qualia, in our experience.

Critical reflection 7

Phenomenal features and physicalism

How do these phenomenal features stand in relation to physicalism as a metaphysical mind–brain theory? The phenomenal features are often considered the central test or litmus test for any physicalistic mind–brain account. If physicalism can account for phenomenal features, it may not only be neuronally plausible, but also phenomenally plausible. And that would make a strong case for physicalism as the metaphysical theory that best describes the mind–brain relationship. In this case, physicalism would be a metaphysical theory not just accounting for the brain itself (and its neuronal working), but would also explain the mind and its relationship to the brain.

If, by way of contrast, physicalism fails to allow for phenomenal features, so physicalism may not be empirically plausible and thus not applicable when it comes to the brain and its neuronal states (depending on the definition of the concept of the physical; see above). In this light, physicalism may just be an empirical theory about the working of the brain, whereas physicalism would then not be a plausible metaphysical theory about the existence and reality of the brain and its relationship to the mind. The failure to accommodate phenomenal features may thus lead to the failure of physicalism as a metaphysical mind–brain theory. In other words, there is much at stake when considering the phenomenal features of consciousness and how they are related to the brain as seemingly purely physical.

Can physicalism account for the phenomenal features of consciousness? This question can be tackled in both conceptual and empirical ways. Conceptually, the philosophers focused especially on the problem of qualia and developed various logical arguments that all aim to show why physicalism can, in principle, not account for qualia (and related) phenomenal features. For instance, qualia may remain absent while the physicalistic features remain the same. If qualia can remain absent in the presence of the same physical features, physical features cannot account for qualia. Physicalism must thus be wrong. This is called the 'absent qualia argument'.

Another argument concerns the possibility of qualia being inverted: each time you experience boredom, your friend experiences excitement, despite the fact that both of you are experiencing similar physical stimuli and have similar neuronal activity and states in your brains. Each time you experience the color red, he/she experiences the color blue. If, however, such inversion in the contents of the qualia is possible, while having similar neuronal states and similar stimuli, qualia and their contents cannot depend on the brain's neuronal, and thus physical, states. Physicalism as a mind–brain theory must thus be wrong because it cannot accommodate the qualia and their contents. This is called the 'inverted qualia argument'.

Several other arguments have been developed to demonstrate that a purely physicalistic-materialistic account of consciousness cannot explain qualia and/or subjectivity. Since these arguments concern consciousness in particular, I will discuss them in detail in Part IV. What is important to note here is that any physical-materialistic approach to the mind–brain problem is confronted by the phenomenal features of consciousness. While physicalism can explain many facets of the mind and its relationship to the brain, the phenomenal features of consciousness like subjectivity, intentionality and qualia seem to remain stubbornly resistant to a physicalistic account.

Taken together, physicalism faces not only empirical problems stemming from the definition of the physical, but also difficulties in accounting for the phenomenal features of consciousness. These difficulties may be manifested empirically in that we are unable to find the neuronal correlates of phenomenal features like subjectivity in the brain's neural activity. Conceptual difficulties also arise with various arguments like the absent and inverted qualia that have been put forward against physicalism.

Computational escapes 1

Functionalism

Recall that we discussed different forms of physicalism as possible solutions to the metaphysical mind–brain theory. These ranged from behaviorism and identity theory to more recent accounts like supervenience. Various objections have been raised against both mind–brain identity theories and physicalistic mind–brain theories. These objections include both conceptual-logical and empirical-neuronal issues.

Both conceptual-logical and empirical-neuronal issues revolve around how mental states as distinct and special can be related to the brain as purely physical. The strategy to resolve this from a position of physicalism is to redefine the concept of the mental in order to integrate and adapt it to the concept of the physical. One may also pursue another strategy, namely to leave out the brain and its physical features altogether. This allows one to account for the concept of the mental independent of the constraints as they are associated with the concept of the physical. This is more or less the strategy functionalism pursues.

Let us detail this strategy. One of the main objections against identity theory consisted in the charge of 'neuronal chauvinism'. The concept of 'neuronal chauvinism' describes the existence and reality of the brain and its neuronal states as basic and foundational for yielding mental states. This is chauvinistic because it excludes the possibility that devices other than the brain might be able to yield mental states.

During the 1970s and 1980s, the charge of neuronal chauvinism led to the downfall of identity theory and the critical view of physicalism. Instead of pursuing mind–brain identity theory, the advocates of 'neuronal chauvinism' developed a novel theory called functionalism. In contrast to mind–brain identity theory, functionalism no longer takes the brain as its starting point. Instead it begins with the computer to make assumptions about the mind.

Why the computer rather than the brain? During the 1970s and 1980s, the computer was starting to enter our private life more and more. Nowadays we can barely imagine a life without computers, cell phones, iPads and all the other fancy electronic devices. However, three or four decades ago these devices were not yet developed and available. At that time the personal computer was not as developed and spread to everybody. Not surprisingly, the introduction of the computer into daily life had a profound impact on people's view of the mind and world. This is visible in functionalism as a possible answer to the mind–brain problem.

What does functionalism mean? In the context of a philosophical mind–brain theory, functionalism compares mind and brain to a computer. We can distinguish between software that runs a particular program and hardware that provides the physical material for running the software. The software is compared to the mind and its mental features, while the hardware that runs the respective program can either be a brain's neurons, the computer's silicon, or some other device's material.

Let us be specific. The brain as hardware is assumed to run a particular program, the software. This software and its program account for mental states as distinguished from the neuronal states of the hardware on which it is running.

How can we characterize the software and its program in general? A program makes certain manipulations of symbols and sets these symbols in specific causal relation to each other. For instance, the input x causes changes in the element y, and that in turn causes changes in the z and so on; x, y and z are causally related to each other but none is related in a necessary or unavoidable way to the others. This means that changes in x could also *not* lead to changes in y but rather to changes in, for instance, a or to no changes at all; x and y (and z) are thus only extrinsically, rather than intrinsically, related to each other. In short, there is extrinsic rather than intrinsic causal relation.

How does this relate to mental states? Mental states are nothing but a program that can be defined by its extrinsic causal relations to sensory input, other mental states and behavioral output. The sensory input causally impacts mental states which in turn causes the behavioral output. However, as in our example, the sensory input could also not cause changes in the mental states. Functionalism thus defines mental states by extrinsic relations between input, internal states and output, which accounts for what is described as the 'functional relation'.

Computational escapes 2

Functionalism, behaviorism and physicalism

How does functionalism stand in relation to behaviorism? Recall the beginning of this chapter. Behaviorism eliminated mental states by considering only the causal relations between sensory input and behavioral output. Functionalism is more careful. As in behaviorism, functionalism acknowledges the central relevance of sensory input and behavioral output. However, unlike behaviorism, functionalism sandwiches mental states between them. Mental states are then defined by the causal relations between sensory input and behavioral output.

Let us return to the earlier example of my belief that it will rain. Behaviorism explained this belief by a behavioral disposition: to take an umbrella (see above). Functionalism, in contrast, acknowledges the belief that it will rain. It does not eliminate our beliefs. Instead, functionalism defines my belief in terms of its causal relation to the sensory input like my perception of seeing dark clouds. And it defines it by the causal relation to other mental states like my desire not to get wet. That in turn is causally related to my behavior of buying an umbrella.

Functionalism thus overcomes two of the central problems of behaviorism. We recall that one charge against behaviorism was that it could not account fully for behavior: it leaves out the mental states associated with our beliefs and desires that are necessary to explain our behavior. This is solved in functionalism by the functional definition of causally relating mental states to the input and other mental states in a causally extrinsic way. Hence, unlike in behaviorism, the association of the belief that it will rain with the desire and urge to not get wet is no problem.

Furthermore, functionalism overcomes the problem of mental causation. Recall that the problem of mental causation consists in assuming the causal impact of a mental state on our behavior and the underlying physical states in a physical world that seems to be causally closed to non-physical causes. Behaviorism fails to account for that. Meanwhile, functionalism can explain the causal impact of mental states on our behavior by including mental states as a sandwiched variable between input–output relations. Mental states like the desire or urge not to get wet can then be assumed to cause physical changes and certain appropriate behaviors, including the act of buying an umbrella.

Functionalism seems to provide answers to some of the major problems associated with behaviorism. At the same time, functionalism also seems to avoid some of the pitfalls of dualism and identity theory. Unlike dualism, functionalism does not encounter the need to introduce any kind of mental feature in order to account for mental states as distinct from physical states. Instead of postulating specific mental features as distinct from physical features, functionalism only introduces the software as distinct from the hardware. The introduction of the software allows for the postulation of causal relations that yield mental states sandwiched between input and output. In a nutshell, functionalism claims that we can have mental states without having to assume specific mental features. The computer-based distinction between software and hardware thus replaces the dualistic distinction between mental and physical as different existences and realities.

Unlike identity theory and other physicalistic accounts, functionalism does not need to assume any kind of physical features that specifically account for mental states. This relieves functionalism of the problems associated with, especially, the type–type identity theories (see above) and the definition of the physical (see above). Hence, functionalism does not only seem to provide an answer to the charges against behaviorism, but also a remedy for the problems of physicalism.

Computational escapes 3

Versions of functionalism

What makes it possible for functionalism to avoid some of the problems associated with behaviorism and physicalism? One central point is the neglect of the brain in functionalism. Especially in physicalism, the brain and its definition as physical plays a central role. This, as we have seen, can lead to many problems. Functionalism no longer attributes a central relevance to the brain and its neuronal states. Instead it is taken over by the concept of software, while the brain itself is considered to be only the hardware.

What is important in functionalism is the software – the program – and the rules according to which it operates. There are, however, different kinds of programs. Hence we need to ask: what kind of program does the software need to run in order to make possible the generation of a mind and its mental features? Strong candidates for the operational features of the program are linguistic-like programs. These will be discussed in the third part of this book (see especially Chapter 10).

What about the hardware? As mentioned above, the same program may run on different computers. This implies that different tokens – distinct hardware – can realize one and the same type, the program. *The same program can run on different hardware and thus be physically realized in different or multiple ways, as neurons in a brain or silicon chips in a computer. There is thus what the functionalists describe as 'multiple realizability', which describes how the same software can be run on different hardware and thus be realized in multiple ways.*

What must the software and its program look like in order to make possible the programming of mental features? Depending on how the software is characterized and used, one may distinguish between different versions of functionalism: machine or computer functionalism focuses on the kind of software that is used to establish causal relations. This distinguishes it from psychofunctionalism, which shifts the emphasis from the software itself to the functions, the psychological functions that are associated with the software. Machine or computer functionalism emphasizes that a particular software or program is needed to yield mental states. Both digital and connectionist versions of machine or computer functionalism will be discussed in further detail in Chapter 11.

The second version of software and programs suggested by functionalism is less oriented toward machines and more on cognitive functions – especially linguistic functions as researched in cognitive psychology. This amounts to what is described as 'psychofunctionalism' (see Chapters 10 and 11 for further details). Here the program is assumed to be structured and organized like the cognitive functions whose psychological features as researched in cognitive psychology are taken as a template and model to suggest the corresponding program and software.

Finally, besides machine/computer functionalism and 'psychofunctionalism', there is a third variety of functionalism called analytic functionalism. Analytic functionalism can be considered as an extension of behaviorism. Analytic functionalism focuses on the conceptual analysis of mental states in terms of their causal relation to input (other mental states) and behavioral output.

This triple constellation allows for more combinations than the simple binary input–output relation in behaviorism, and makes functionalism much richer than logical behaviorism. Functionalism allows us to include mental states like beliefs and desires and, most importantly, to connect them causally to both stimulus input and behavioral output. This makes possible a variety of different causal relations between input, mental states and output. It is also far more complex and exceeds the simple binary input–output relation as postulated in behaviorism.

Ideas for future research 3

'Degradation of the brain' and the brain's extrinsic features

The assumption of 'multiple realizability' implies that the brain is not considered as constituting mental states. Instead, any kind of device – a brain, a computer, or some machine – can, in principle, run the program required to produce mental states. The brain thus no longer occupies a central role in constituting mental states and is degraded in its importance for the mind. One may thus speak of what I describe as the 'degradation of the brain'.

The degradation of the brain in functionalism has important consequences. The brain and its neurons are no longer assumed, in this model, to provide an essential input that would characterize the mind and its mental states. This role is, instead, taken over by software and its specific programs. The degradation of the brain thus shifts the focus from the brain as hardware, to the mind as software, as well as the kinds of programs operated by the software.

Let us go into further detail. What unifies the different versions of functionalism despite their differences is that neither of them considers the brain as a starting point for understanding the constitution of mental states. All three – machine or computer functionalism, psychofunctionalism and analytic functionalism – no longer consider the brain and its neuronal states as the paradigmatic example of how mental states are generated.

Instead of the brain and its neuronal states, machines or computers, cognitive psychological functions, or functional relations, are taken as templates for understanding the constitution of mental features in the different versions of functionalism. There is, to put it baldly, nothing we can learn about how mental states are constituted from the brain and its neuronal states themselves. Instead, we can learn much more from computers, cognitive functions, and functional relations.

Can we really not learn anything from the brain about the constitution of mental states? Let's compare the situation with the heart. The heart is organized and structured as a muscle that pumps blood throughout the whole body in order to provide all the body's cells with oxygen. Once this was discovered, it was possible to construct artificial hearts. The knowledge of the heart itself and its defining or intrinsic features like muscle structure and blood pumping thus paved the way to constructing the heart as a mere machine.

How does this compare to the brain? The brain seems to be involved in constituting mental states in the same way that the heart pumps blood. How does the heart pump blood? By means of its muscle structure that is an intrinsic feature of the heart. How does the brain constitute mental states? We currently do not know. More specifically, we do not know what kind of structure and organization allows the brain and its neuronal states to constitute mental states. In other words, unlike in the case of the heart and its muscle structure, we do not yet know the intrinsic features of the brain that make its purpose possible.

Before associating mental states with machines and computers, we, in analogy to the heart, may thus first need to understand and explain the brain itself and its intrinsic features. In the same way that the knowledge about the heart itself preceded the construction of an artificial heart, we need to gain insight into the brain itself and how its intrinsic features generate mental states before we can associate mental states with computers or other machines.

Ideas for future research 4

'Elevation of the brain' and the brain's intrinsic features

I emphasized the need to understand the brain's intrinsic features and how they are related to mental states. How does this relate to functionalism and its degradation of the brain? Functionalism seems to consider the brain only in an extrinsic way. In this model, the brain only processes the extrinsic causal relations between input, internal or mental states and output. In the same way the software of a computer must run on some kind of hardware, the extrinsic causal relations between input, mental states and output as the software must run on some kind of material, the brain as hardware.

Here, the brain is considered in a merely extrinsic way, namely by itself, it does not possess any intrinsic features that are relevant for the processing of the extrinsic causal relations. The inner design and features of the brain itself do not matter and do not contribute anything to the processing of the extrinsic causal relations. The brain is just the hardware and as such is a mere mediator (rather than the originator). What I referred to as the brain's intrinsic features are replaced in functionalism by the software, the program that is no longer specifically associated with the brain itself. Functionalism thus considers the brain in a merely extrinsic, rather than intrinsic, way. What would a functionalist view of the heart look like? Functionalism would focus only on the purpose of the heart, the pumping of blood. The pumping of blood would be taken as analogous to the causal relations between input, inner states and output. In contrast, functionalism would neglect the intrinsic features of the heart, its structure and organization as a muscle. The detection of the heart's muscle structure allowed us to understand why and how the heart pumps blood. Analogously, we may need to reveal the brain's intrinsic features, its particular structure and organization, in order to understand how the brain can generate mental states.

The consideration of the brain's intrinsic features may thus lead us back to the brain itself and its potentially central role in constituting mental states. The computer-based approach of functionalism and the mind-based approach of dualism may then be complemented if not replaced by a truly brain-based approach to the mind–brain problem. That would reverse the degradation of the brain suggested by functionalism; instead of degrading the brain, the consideration of the brain's intrinsic features would elevate the brain and give it a most basic and fundamental role in generating and explaining mental states. One may therefore want to speak of an 'elevation of the brain'.

Ideas for future research 5

Brain-based approach to the mind–brain problem

How can we explain the brain-based approach to the mind–brain problem in further detail? Rather than taking either the computer (as in functionalism) or the mind (as in dualism) as the starting point to address the mind–brain problem, a brain-based approach considers the brain itself and its intrinsic features as the starting point and template for the constitution of mental states (see Chapter 4 and, especially, Chapters 7 and 8 for details).

This may ultimately lead us back to physicalism, though in a particular version, as brain-based rather than mind–brain physicalism. Unlike in mind-based physicalism, the question here is no longer how the mind can be related to the brain. Instead, a brain-based physicalism takes the opposite standpoint when asking how the brain's intrinsic features can generate mental states. The point of departure here is no longer the mental states themselves, but rather the brain itself and its intrinsic features. The guiding question is: What must the brain's intrinsic features be like in order to make possible the constitution of mental states?

How can we compare that to the heart? Before discovering that the heart shows the structure of a muscle, people knew that the heart was related to the pumping of blood. What they did not know was how and why the heart was related to this action. This is analogous to the current situation in our knowledge of the brain. We know that the brain is involved in and contributes to the constitution of mental states. This has been demonstrated impressively by recent results in neuroscience. However, currently we do not know how and why the brain can contribute to the constitution of mental states.

Critical reflection 8

Functionalism and meaning

Functionalism has been and still is one of the dominating mind–brain theories of the past 30–40 years. Part of its success is that it aims to explain some of the main features of consciousness, like intentionality as the directedness of mental states towards contents, in functionalistic terms. However, problems arise when it comes to the meaning or the semantics of the contents. Opponents argue that functionalism can only account for the constitution of contents in our mental states, whereas it remains unable to explain the meaning or semantics associated with these contents. Accordingly, one of the major problems of functionalism is its inability to account for meaning or semantic properties associated with objects, events, or people as the contents of our mental states.

How can we show that functionalism remains unable to account for meaning, and thus the semantic dimension? In order to illustrate the missing semantic dimension in functionalism, the Californian philosopher John R. Searle (1932–) developed the famous 'Chinese room argument'. Imagine you are sitting in a room with boxes full of Chinese symbols. You have no knowledge of Chinese. Now you receive a rule book that tells you the rules according to which you manipulate the symbols.

Imagine now you are taking part in a Chinese test. Every time the instructor asks you to answer a certain question, you look in your rule book that gives you advice on how to sort and put together the different Chinese symbols. Following the rules well, you create the correct symbol. The instructor gives you a high grade and thinks that you can speak Chinese rather well. In truth, the opposite is true – you do not speak a word of Chinese and do not understand it at all.

How is that possible? By following the rules of the rule book, you performed the same actions as a computer program does: symbol manipulation according to prescribed rules. Most importantly, you were able to complete the test perfectly without understanding a word of Chinese. In other words, you were able to manipulate the syntax (the grammatical structure) without acquiring any comprehension of the semantics, the meaning of the symbols you manipulated so well.

Is our mind and its mental states just a symbol manipulator that does not understand the semantics at all, and thus the meaning of what it manipulates? No, says Searle. We do understand the meaning of terms when we speak a language. In the case of an English exam, for instance, you would not only pass the test in symbol manipulation, but would also pass the semantics portion when the instructor asks you for the meaning of the manipulated terms. This suggests that our mind and its mental states cannot be reduced to symbol manipulation and thus to syntax. The mind and its mental states must be characterized by 'something additional' that goes beyond their mere syntax. And this 'something additional', according to Searle, must consist in the meaning or the semantics of the symbols.

In sum, following the Chinese room argument, functionalism can explain symbol manipulation and thus the syntax (the grammatical structure) of the Chinese room. However, it fails to account for the meaning and thus the semantic dimension associated with that syntactic structure.

Critical reflection 9

Semantic meaning in neurophilosophy

How can we resolve this semantic failure? To solve this, we must shift our starting point back from the computer to the mind. The computer-based approach of functionalism would, in this case, metaphysically be complemented – if not replaced by – a mind-based approach to the mind–brain problem. This may lead us back to dualism (see Chapter 1). Alternatively, one may pursue what I described above as the brain-based approach by revealing and starting from the brain itself and its intrinsic features.

Can this brain-based approach and thus the brain's intrinsic features account for meaning and thus the semantic dimension? Considering the brain in a purely physicalistic way, as in physicalism, will fail, however, as described above. We may thus need to extend our determination of the brain beyond the purely physicalistic account if we want to account for the semantic dimension within the framework of a brain-based approach. This will be discussed in Chapter 7 and especially Chapter 8.

Moreover, we need to account for how semantic content can be generated by the mind and ultimately by the brain. This will be the focus in Chapter 10, where we will focus on the mechanisms of how the mind may eventually generate meaning, while Chapter 12 will go into detail about the brain in this regard.

Take-home message

We discussed mental approaches to the mind–brain problem in Chapter 5. These approaches argued for the special existence and reality of the mind and its mental features when compared to the physical features of brain and body. This is denied in physical approaches to the mind–brain problem. The mind and its mental features are, in this model, no longer considered special and thus as principally different in their existence and reality from brain and body. The mind is thus determined by the physical features of brain and body. This led to the development of behaviorism and physicalism as mind–brain theories. Different versions of reductive and non-reductive physicalism, like identity theory and supervenience, were developed. However, all these theories encountered several conceptual and empirical problems. This fueled the search for theories that avoid the problems of both mental and physical approaches. With the introduction of the computer, the theory of functionalism, which compares the mind–brain relationship to the software–hardware relationship of computers, was developed. Instead of treating the mind as a starting point, as in dualism and physicalism, functionalism starts from the computer. It is thus a computer-based approach. This model also raises several problems, like how to account for meaning (the semantic dimension). This raises the question of whether an approach that focuses on the brain itself and its intrinsic features, a brain-based approach, may provide an escape from these problems.

Summary

The previous chapter discussed mental approaches to the mind–brain problem. This implied the assumption of a mental existence and reality underlying the mind. This led either to dualism or panpsychism as potential solutions to the mind–brain problem. The present chapter takes the opposite position and discusses a physical approach to the mind–brain problem. Here, the mind is considered in purely physical terms – a view that became very popular during the twentieth century. Different versions of physicalism can be distinguished according to time period. In the first half of the twentieth century, behaviorism was the dominating physical mind–brain theory. Behaviorism explained the mind completely in terms of the behavior we can observe. There is no mind above and beyond observable behavior. Mental states, in this model, are consequently reduced to mere stimulus–response patterns. Empirically, methodological behaviorism suggests a methodological strategy of investigating stimulus–response and thus input–output relations in psychology.

Philosophy complemented this with logical behaviorism, claiming that any mental statement can be translated completely into a behavioral statement which no longer uses any kind of mental term. This implied the complete elimination of the mental dimension and thus of consciousness, self, etc., altogether. This idea turned out to be logically inconsistent as well as untenable in common sense. Behaviorism receded and was replaced by mind–brain identity theory. The mind–brain identity theory argues that mental states are identical to neuronal states; this is illustrated by the example of pain, where pain as a mental state is considered identical to the stimulation of C-fibers in the brain. The assumption of identity between mental and neuronal features raises several problems, especially about the concept of identity. This led to the development of other versions of physicalism: reductive and non-reductive. They all, though, suffer from the problem of establishing a clear-cut and proper definition of the concept of the physical. With the development of computers and their introduction into daily life during the 1970s and 1980s, identity theory and physicalism as the dominating mind–brain theories were replaced by what is described as functionalism. Functionalism considers mental states to be like the software and its program running on hardware, which can be either the brain or a computer (or something else). Like a computer program, mental states are then described by their functional relations (causal relations) to the input (other mental states) and the output. By including mental states in the causal relations, functionalism extended behaviorism while at the same time avoiding its pitfalls. Moreover, the analogy to the computer and its software relieved the search for some intrinsic properties in the brain that could generate mental states in a physical way. While still a major mind–brain theory today, functionalism has trouble accounting for the semantic dimension – the meaning we attach to objects, events, or people. This chapter concludes with a critical view of functionalism and highlights its inability to properly account for the semantic dimension and for the brain itself, whose intrinsic features it seems to neglect completely.

Revision notes

- Why was there a shift from mental to physical approaches during the twentieth century?
- How is behaviorism defined?
- Are there different versions of behaviorism, and how can they be distinguished from each other?
- What are the problems that behaviorism faces?
- Give a definition of identity theory.
- What are the problems associated with identity theory?
- Which developments were coupled with the emergence of functionalism?
- What is functionalism?
- How does the distinction between hardware and software map on to the mind–brain distinction?
- What are the problems of functionalism?
- Explain the different versions of physicalism and their respective features.

- What is supervenience?
- How can one define the notion of the physical? Why is it problematic?
- What are the limits of physicalism?
- What are the differences between mind-, computer- and brain-based approaches to the mind–brain problem?

Suggested further reading

- Chomsky, N. (2006) *Philosophy of Mind*, 2nd edn (Boulder, CO: Westview Press).
- Kim, J. (1976) 'Events as property exemplifications', in M. Brand and D. Walton (eds), *Action Theory* (Dordrecht, Netherlands: Reidel).
- Kim, J. (1995) 'Supervenience', in S. Guttenplan (ed.), *A Companion to the Philosophy of Mind* (Oxford: Blackwell), pp. 575–83.
- Kripke, S. (1972) *Naming and Necessity* (Cambridge, MA.: Harvard University Press).
- Putnam, H. (1999) *The Threefold Cord: Mind, Body and World* (New York: Columbia University Press).
- Smart, J. J. C. (1970) 'Sensations and brain processes', in C. V. Borst (ed.), *The Mind/Brain Identity Theory* (Oxford/New York: Oxford University Press).

7
Non-mental and Non-physical Approaches to the Mind–Brain Problem

Overview

The previous chapters discussed mental and physical approaches to the mind–brain problem. Mental approaches characterized the mind's existence and reality as mental and distinguished it from the brain and the body as separate physical entities. Physical approaches, on the other hand, assumed the mind's existence and reality to be physical, which implies that they are essentially identical to the brain and the body. Both mental and physical approaches to the mind–brain problem suffer from various problems, including metaphysical, conceptual and/or empirical implausibility. The main problem is in the definition of the mental and physical as being distinct, different, opposite and mutually exclusive. For example, the concept of the mental is characterized as subjective, whereas the concept of the physical is considered objective. Because of this mutually exclusive difference, the mind–brain problem stubbornly resists any categorically mental or physical solution. How can we escape this dichotomy? This requires alternative approaches where different definitions of both the mental and physical do not posit their mutual exclusivity. This chapter focuses on these non-mental and non-physical alternative approaches.

Objectives

- Understand the background assumptions in the current mind–brain discussion
- Distinguish different problems in the mind–brain discussion
- Explain neutral monism
- Understand biological naturalism
- Understand Kant's diagnosis of epistemic limitation
- Explain why the mind–brain problem is an epistemic problem
- Determine the domain problem

Key concepts

Background assumptions, domain problem, inference problem, neutral monism, definition problem, plausibility problem, biological naturalism, epistemic limitation, epistemic domain.

Background 1

Empirical and phenomenal implausibility of physical and mental approaches

So far we've discussed mental and physical approaches to the mind–brain problem. Mental approaches considered the existence and reality of the mind in mental terms as distinguished from the physical existence and reality of the brain and the body. This led to either dualism of the mind and the brain, or panpsychism, where the mental is inherent in the physical (see Chapter 5). These approaches utilize the mind as their starting point and attribute a special positive existence and reality to it. Accordingly, mental approaches can also be characterized as mind-based.

Physical approaches to the existence and reality of the mind do not consider the mind and the brain as principally different. Various theories that fall within the physicalist approach include behaviorism, identity theory, and reductive and non-reductive physicalism. Functionalism, where the mind is compared to the software of a computer, can also be characterized as a physical approach. In contrast to mentalist approaches, the physicalist approaches deny the special existence and reality of the mind. Since the denial of the special existence and reality of the mind serves as the starting point, physicalist approaches can still be regarded as mind-based, though in a negative sense.

Let us explain in further detail. Though paradoxical, even physicalist approaches can still be considered mind-based. Why? The mind still serves here as the starting point for the mind–brain problem. The mind is considered to be physical rather than mental. This means that the mind is everything that the mental is not, namely physical. The mind and its characterization as mental thus still serve as a template, though a negative one. Compare that to the mentalist approach. Here the mind is considered to be mental rather than physical. This means that the mind is everything that the physical is not, namely mental. The mind and its characterization as mental thus serve as a positive (rather than negative) template or foil for mentalist approaches to the mind–brain problem. Despite their well developed and intricate nature, both mental and physical approaches have encountered serious problems. Mentalist approaches seem able to account for mental states like consciousness, but also appear empirically impossible: no one has ever observed the postulated mental features in either the brain itself, the body, or, most generally, in the world from a third-person perspective. Mental approaches thus seem to be contradicted by the scientific results in physics and neuroscience. They thus suffer from what can be described as 'empirical implausibility'.

The reverse holds for physical approaches. These were developed in close orientation with physics and neuroscience. While this fact makes them (at least intuitively) empirically more plausible, they are unable to explain mental states and, more specifically, the subjective experience of a world that is assumed to be purely physical. As such, physical approaches have major problems accounting for the phenomenal features of subjective experience, including consciousness, qualia, intentionality and subjectivity. Since they cannot properly explain the phenomenal features of consciousness, physical approaches suffer from what can be described as 'phenomenal implausibility'.

Background 2

'Plausibility problem'

The established philosopher may be slightly puzzled by now. For him/her, the most important criterion for any mind–brain solution is 'logical or conceptual plausibility'. 'Logical or conceptual plausibility' states that the suggested mind–brain solution and its arguments are logically (or conceptually) consistent without any logical flaws. Such logical (or conceptual) consistency can, as we saw in Chapters 5 and 6, remain independent of both phenomenal and empirical plausibility. This means that the arguments may be logically consistent while at the same time they may not be in accordance with either consciousness itself, that is, phenomenally implausible, or the data and facts, that is, empirically implausible.

How about the neurophilosopher? In contrast to the philosopher, he/she is not so interested in 'logical or conceptual plausibility'. His/her prime interest is in empirical and phenomenal plausibility rather than logical or conceptual plausibility. A particular mind–brain solution may turn out to be empirically and phenomenally plausible even if logically or conceptually implausible (see, for instance, the example with type–type and token–token identity in Chapter 6). Since we pursue a distinct neurophilosophical rather than purely philosophical approach to the mind–brain problem in this book, we focus on empirical and phenomenal plausibility below (while somehow neglecting logical or conceptual plausibility).

Taken together, the choice between mental and physical approaches presents us with the alternative of either empirical or phenomenal implausibility in neurophilosophy. If we aim to be phenomenally plausible and choose a mental approach, we cannot avoid empirical implausibility. If, in contrast, we opt for empirical plausibility and prefer the physical approach, we are confronted with phenomenal implausibility.

The choice between mental and physical approaches thus presents us with a dilemma, a choice between Scylla and Charybdis: the dilemma consists in the apparent inability to avoid some kind of implausibility in the mind–brain problem that is phenomenal or empirical. This dilemma amounts to what I describe as the 'plausibility problem', which focuses on the mutually exclusive alternative of empirical versus phenomenal plausibility while (more or less) neglecting logical or conceptual plausibility. This makes the 'plausibility problem' a specifically neurophilosophical rather than a philosophical problem. In contrast, from the philosophical perspective, the plausibility problem amounts to nothing but confusion because it confuses empirical/phenomenal plausibility with logical plausibility.

Background 3

Definition of concepts

Why is there a 'plausibility problem'? This question arises from background assumptions out of which the plausibility problem emerges. One of these assumptions consists in the mutually exclusive dichotomy between mental and physical approaches: either you opt for a mental approach or for a physical approach and that's it. Either you are dualist (or panpsychist) or a materialist. There is no alternative.

How can we escape this mutually exclusive dichotomy between mental and physical approaches? One route is functionalism. Recall from the previous chapter that functionalism compared the mind–brain relationship to the software–hardware relationship in the computer. This analogy allows functionalism to sidestep the mutually exclusive dichotomy between mental and physical approaches to the mind–brain problem. However, even functionalism has major problems in accounting for the phenomenal and semantic features of mental states. Moreover, by neglecting the intrinsic features of the brain, it may also turn out to be neuronally (that is empirically) implausible (see Chapters 6 and 8).

The failure of functionalism makes even more urgent the need for methods that can overcome the mutually exclusive characterization of the mind–brain problem as a mental–physical dichotomy. How can we undermine the mutual exclusivity of the concepts mental and physical? We saw in Chapter 6 that the definition of the term 'physical' is not as clear as it is often assumed. The concept of the 'physical' can take on different meanings: the physical can be understood in terms of objects (and thus in terms of common sense), in which case there are clear boundaries between the physical and the mental. Alternatively, the concept of the 'physical' can also be understood in a more scientific sense, in terms of quanta as related to quantum physics. In that case, the boundary between the mental and physical is no longer as clear-cut.

This has major reverberations for the concept of the mental. If the concept of the physical can no longer be distinguished as clearly from the concept of the mental, the latter may also be suffused by the physical. This means that the dichotomy between the concepts of the mental and the physical is undermined and can no longer be considered as mutually exclusive, as we have seen in, for instance, the case of quantum physics (see Chapters 6, 17 and 20).

Background 4

'Definition problem' as a neurophilosophical rather than a philosophical problem

Additional problems also arise out of the concept of the mental. In the same way that the physical can no longer be distinguished as clearly from the mental, the concept of the mental may also by itself no longer be as different. We have, for instance, seen that the concept of the mental can have different meanings even within the various mental approaches. For example, dualists like D. Chalmers seem to determine the concept of the mental in terms of informational capacities (see Chapter 5) while others refer to quanta, as postulated in quantum physics (see Chapter 6).

Taken together, we are confronted by the problem of defining the concepts of the mental and the physical in a categorical and independent way. *The concepts of the mental and the physical seem to resist precise definition by themselves without clear-cut boundaries to the respective other concept – this may be described as the 'definition problem'.* The 'definition problem' concerns the definition of the terms 'mental' and 'physical' by themselves, independent of their supposed characterization of mind and body. Since any suggestion for a mind–brain solution must presuppose (either explicitly or implicitly) some kind of definition of the terms 'mental' and 'physical', the 'definition problem' may be considered a (necessary) presupposition of a possible mind–brain solution.

The traditional philosopher may now be puzzled. His focus is more on arguments than definitions. He tries out different arguments to obtain logical consistency and plausibility for a particular definition and thereby adapts his arguments to the definition in question. This is different in neurophilosophy. Here the focus is on the definition itself rather than on the arguments for or against a particular definition. The philosopher can then the compare the different definitions of his/her concepts like mental and physical features to the data and facts to achieve empirical consistency and plausibility (see Chapter 4 for details). The definition problem as sketched here is thus a genuinely neurophilosophical problem, whereas for the traditional philosopher it is at best a non-problem that, at worst, amounts to nothing but mere confusion.

Background 5

'Definition problem' as a neurophilosophical rather than a neuroscientific problem

Is the 'definition problem' a philosophical or neurophilosophical problem? The philosophical approach to the mind–brain problem focuses on the relationship between mind and brain in predominantly logical and conceptual terms. The prime aim here is to achieve logical or conceptual consistency and thus plausibility. The main problem to be tackled here consists in the question for a logically and conceptually plausible relationship between mind and brain. One may thus be inclined to speak of a 'relationship problem' here.

In order to address the 'relationship problem', the definition of the terms 'mental' and 'physical' may be of secondary importance. Only if the definition of both terms impacts the logical and conceptual characterization of the relationship between mind and brain is their definition relevant. Otherwise the definition of the mental and the physical remains irrelevant in the philosophical approach.

This is different in a neurophilosophical approach. Here the definition of the concepts of the 'mental' and the 'physical' is central. Why? Only if we properly define both terms, we may be able to achieve some kind of empirical and phenomenal plausibility, which in turn is central for the subsequent mind–brain problem. If, for instance, the mental or the physical are defined in an empirically or phenomenally implausible way, any subsequent mind–brain solution will turn out to be empirically or phenomenally implausible too. Hence, an empirically or phenomenally implausible definition of the 'mental' or the 'physical' is detrimental to the neurophilosopher.

In contrast, implausible definitions are not necessarily detrimental to the philosopher: even on the grounds of empirically or phenomenally implausible definitions of the 'mental' or the 'physical', he/she may nevertheless be able to establish some logically and conceptually consistent and thus plausible mind–brain relationship. The 'definition problem' is thus primarily a neurophilosophical rather than a philosophical problem.

Finally, we may also want to distinguish the neurophilosophical approach from a purely neuroscientific approach. The neuroscientific approach will not raise the question of the definition of the concepts of the 'mental' and the 'physical'. Instead, he/she will approach them on purely operational grounds in that he/she searches for some experimental variables that can quantify both terms. The definition of the 'mental' and the 'physical' thus recedes into the background at best and may get lost at worse in a purely neuroscientific approach. This makes it clear that the 'definition problem' is neurophilosophical rather than neuroscientific.

Background 6

Inference from absent to present phenomenal features in mental approaches

Moreover, we cannot infer the validity of the one concept from the deficits of the other. Let us explain this in detail. Many mental approaches argue that the physical characterization of the mind remains insufficient. Why? Because there is a deficit in the physical approaches which makes it impossible for them to account for the phenomenal and mental features of the mind. In order to account for this deficit, the proponents of the mental approaches claim that we need to assume a mind with specific mental features that are (metaphysically) principally different from the physical features of body and brain.

Such assumption of mental features is, however, possible only if one (most often implicitly) presupposes some kind of inference from the explanatory insufficiency of the physical approaches to the existence and reality of the mind. More specifically, the mental approaches presuppose an inference from the absence of the mental features in the physical features of brain and body to their presence in the mind. Put even more succinctly, one infers from something negative, the explanatory insufficiency of the physical approaches, to something positive, the metaphysical existence and reality of mental features.

However, any inference of something positive, the presence of mental features in the mind, from something negative, the absence of mental features in the physical features of brain and body, must be problematic. The assumption of mental features as distinct from physical features relies on the explanatory insufficiency of the latter rather than on some support established by the former itself (independent of the latter).

Mental approaches are thus based ultimately on the deficits of their opponents, the physical approaches, rather than on themselves. Without such inference from the deficits of the physical approaches, the mental approaches may ultimately collapse. This means that they then would no longer be able to account for the kind of phenomenal plausibility they claim to provide. Since it is aimed towards establishing phenomenal plausibility (rather than logical-conceptual plausibility), such inference and its problems may be regarded a neurophilosophical rather than a philosophical problem.

Background 7

'Inference problem' and the inferences between different domains

Do only the mental approaches suffer from what can be described as the 'inference problem'? No, the same applies to the physical approaches, though obviously in a converse way. Here, the missing empirical plausibility of mental features in the observation of the brain and the physical features serves as a starting point. No mental features can be observed in the physical features of the brain in particular, and the world in general. Any assumption of distinct mental features and thus a mind is consequently empirically implausible. This implies that we need to discard the assumption of the mind altogether and to opt for a physical rather than mental approach to the mind–brain problem.

In the same way that the mental approaches are based on a deficit, the lack of phenomenal features in the physical features of brain and body, the physical approaches are also based on a deficit. This deficit consists of the absence of mental features in the physical features. From there it is inferred that only physical approaches are viable, while mental approaches have to be abandoned. Physical approaches thus presuppose an inference from the absence of mental features to the presence of physical features as being explanatorily sufficient for the mind–brain problem.

As in mental approaches, physical approaches are based on an inference from something absent, the absent mental features in the observation of the physical features, to something positive, the explanatory sufficiency of physical features. *I will therefore speak of an 'inference problem' in both mental and physical approaches. The 'inference problem' describes the inference of something positive from a deficit as something negative in the context of the relationship between mental and physical features.*

The 'inference problem' is both a philosophical and neurophilosophical problem. It is philosophical in that one may question the logical-conceptual plausibility of inferring something positive from something negative. Is such kind of inference logically plausible? This is the question the philosopher may want to raise. At the same time, the 'inference problem' is also a neurophilosophical problem, since the inference aims to achieve empirical and phenomenal plausibility in the mind–brain approach.

We will see later in this chapter that the 'inference problem' figures in a major way in various mind–brain theories. Inferences between different domains like empirical, phenomenal, epistemological and metaphysical domains are most often implicitly presupposed in various mind–brain theories. We will demonstrate how such inferences between usually two domains are presupposed in past and present mind–brain theories.

Where do the definition and inference problems leave us with regard to the mind–brain problem? Because both mental and physical concepts seem to resist proper and mutually exclusive definition, the definition problem tells us that both mental and physical approaches may be implausible. This is reflected in the empirical implausibility of mental approaches and the phenomenal implausibility of physical approaches (see above).

The inference problem implies that the presumed dichotomy between the mental and the physical and their negative relation to each other is misguided. Physicalism cannot be inferred from the failure of dualism, nor can dualism be assumed on the basis of the deficits of physicalism. The mental–physical dichotomy in the mind–brain problem and its alleged dichotomy between dualism and monism/physicalism may turn out to be a false positive dichotomy.

How can we escape these problems of definition and inference? To do this, one must revise, undermine, or abandon the conceptual dichotomy between the mental and the physical. This suggestion has been put forward in various different ways (and will be discussed below), which can therefore be described as non-physical and non-mental approaches. The non-physical and non-mental approaches discuss the mind–brain problem in different domains, metaphysical, epistemic, or conceptual domains, which will serve to categorize them in what follows.

Metaphysical escapes 1

Neutral monism (Russell)

Bertrand Russell (1872–1970) was a British philosopher with a rather lively biography. Beyond making seminal contributions to philosophy – especially the philosophy of mathematics and logic – Russell was also politically active and wrote various books and letters to high-level politicians urging liberalism and pacifism. Russell opposed American engagement in the First World War and spent six months in prison for his vocal opposition. Later he strongly and explicitly opposed Hitler, Stalin, US engagement in the Vietnam War, and nuclear weapons.

Russell wrote various colorful political and historical books in the defence of freedom of conscience and humanitarianism, for which he was awarded the Nobel Prize for literature 1950. One of his many books, *The History of Western Philosophy* (1921), became a bestseller whose royalties gave him lifelong financial independence.

Why is Bertrand Russell important in the current debate about the mind and the brain? For one thing, Russell introduced the concept of 'neutral monism'. *Neutral monism describes a position that assumes ultimate reality and existence to be composed of one kind (of what?). In this model, reality and existence is neutral to both the mental and the physical. Both mental and physical realities are posited as reducible to a neutral level of reality. This implies a combination of reductionism and neutralism, where both mental and physical realities are reduced to the neutral reality and existence.*

Neutral monism accounts for observations in physics (like the particle) and quantum physics, as well as for the emergence of psychology as a new scientific discipline. Let me be more specific. Physics at the beginning of the twentieth century raised the metaphysical question of how physical objects and atoms are related to the newly detected particles of quantum physics? The particles and the quanta may describe a deeper reality and existence that underlies physical objects and atoms. If so, the deeper and underlying reality and existence on a more minute or microscopic level may mean that objects and atoms may not as yet be physical of themselves but rather neutral. That neutral reality and existence may thus occur prior to or precede the physical reality in the same way that particles must precede atoms and objects.

An analogous question was raised in psychology. Psychology, as a science of the mind, raises questions about the underlying basis of mental states. This underlying existence and reality might not be mental or physical. Instead, it may be some kind of deeper neutral existence and reality that as such remains indifferent (hence, neutral) and prior to both the mental and the physical. If so, both mental and physical may be subserved by a commonly underlying deeper existence and reality, the neutral, that is prior to and must precede both.

How can we describe this neutral reality in more positive terms? In trying to do this, Russell was inspired by the early American philosopher and psychologist,

William James (1842–1910). As described in Chapter 3, William James spoke of 'pure experience', or 'the stream of consciousness', which describes experience itself as being unmediated by any cognitive (or other) functions. Russell assumed this 'pure experience' to occur prior to any distinction between the mental and the physical. If this is the case, the 'pure experience' must be a more basic and fundamental existence and reality in comparison to both the mental and the physical. This leads to neutral monism as a non-mental and non-physical approach to the mind–brain problem.

Metaphysical escapes 2

Problems in neutral monism

Neutral monism may avoid some of the problems that mental and physical approaches to the mind–brain problem cannot. However, by postulating a third, neutral reality and existence as underlying both the mental and the physical, neutral monism creates a whole new set of problems.

One problem is that a neutral reality and existence may be too mental and thus not sufficiently neutral. Pure experience describes a mental state rather than a neutral state. This implies that the seemingly neutral reality and existence is, instead, weighted toward the mental. Rather than unifying the mental and the physical, the neutral may actually correspond with and mirror pure experience itself, making it more mental than neutral.

The opposite problem has been claimed as well, namely that the concept of the neutral in neutral monism is too physical and thus not sufficiently neutral. One may, for instance, consider particles and quanta as physical, rather than neutral. They describe a different layer of physical reality and existence than atoms and objects. However, the reference to a different layer does put into doubt their physical reality and existence. What is described as neutral thus turns out to be still physical.

While this concept of the neutral can explain the emergence of physical features like atoms and objects, it remains unable to account for the phenomenal features of mental states like subjectivity, intentionality and qualia. This, however, means that the concept of the neutral fails to provide the linkage to the mental and can therefore not really be considered a viable third alternative to mental and physical approaches to the mind–brain problem.

Another problem consists in the assumption of pure experience as mirroring neutral reality and existence. In order to avoid this problem, neutral existence and reality has been characterized in a different way by current philosophers such as David Chalmers (1966–). As detailed in Chapter 5, Chalmers refers to the concept of information as a neutral reality and existence: information remains indifferent, and thus neutral, to both the mental and the physical. At the same time, both the mental and the physical can be reduced to or characterized in terms of information. This makes the notion of information a viable candidate for neutral reality and existence as assumed in neutral monism.

This concept of information also avoids another problem associated with neutral monism – the problem of location, as one might call it. The physical is located inside the brain and the body, while the mental is located outside both brain and body.

How and where must neutral reality and existence be located if it provides the basis for both mental and physical? Considered logically, neutral reality and existence can neither be located inside the brain and the body, because then it would be too physical; nor can it be located completely outside the brain and the body, because then it could not provide the kind of neutral reality and existence that is supposed to underlie the physical.

How does the reference to information address the problem of location? Information can occur in both brain/body and the world; in other words, both inside and outside the brain and the body. Hence information is not tied to a particular location, but instead can occur (or be 'localized') everywhere and nowhere. This may make information a suitable candidate for neutral existence and reality on the basis of which both mental and physical realities can be generated.

Accordingly, the concept of information seems to undermine the inside and outside dichotomy in that information may be everywhere – in the brain, the body, the world, and nowhere (in no specific or particular location). This abstract and metaphorical circumscription makes it rather difficult to describe the location of neutral existence and reality in more specific and concrete detail. This is not only important for conceptual specification, but also for the empirical specification that is necessary to further experimental research.

How can we provide empirical evidence for the concept of information as a third underlying neutral reality? That leads us to the question: how is information generated by itself prior to its association with mental and physical features? How does information come into the world? We currently do not know. The neuroscientist may want to answer that the brain generates information. For that, though, the brain may need to be connected intrinsically to the environment that provides the germs of what we call information. Such an intrinsic relationship between brain and environment will be explored in later chapters in more detail.

Metaphysical–empirical escapes 1

Biological naturalism (Searle)

Neutral monism can be considered a metaphysical assumption that defies the metaphysical dichotomy between the mental and the physical. However, as we have seen, this metaphysical escape from the mental–physical dichotomy raises plenty of problems, as reflected in the definition problem and the plausibility problem. One may thus want to search for a different escape.

John R. Searle (1932–) is a well-known American philosopher who made major contributions to the philosophy of language and later to the philosophy of mind. He is based at the University of California at Berkeley. Searle emphasizes the need to consider the mind and its specific unique mental features as distinct from the merely physical. At the same time, Searle is keenly aware of the latest developments in science, and neuroscience especially.

Searle argues in favor of 'biological naturalism'. Why 'biological' and why 'naturalism'? Searle claims that consciousness is a biological feature just like digestion, photosynthesis, or bile secretion. Consciousness is nothing special when compared to these biological features. In the same way, digestion is a biological process of the stomach, consciousness is a biological feature of the brain. In this model, consciousness is biological, hence the name biological naturalism.

How can biological naturalism be distinguished from the mental and physical approaches to the mind–brain problem? This is where Searle's position on the definition problem comes in. Recall that the definition problem describes the quest for defining mental and physical states. Usually, mental states are characterized by features such as subjectivity, intentionality and qualia (see Part IV for details). Physical states are described by the opposite features, that is objectivity, non-intentionality and no qualia. This mental–physical dichotomy – as we have seen – is the basic foundational assumption for the mind–brain problem and the mental and physical approaches to it.

Where does Searle stand in relation to this mental–physical dichotomy? Searle argues that the mental–physical dichotomy – and thus the definition problem – is simply wrong: physical states can be subjective, qualitative and intentional. If this is the case, we may distinguish between two different kinds of physical states, objective and subjective. Objective physical states do not lead to consciousness and are thus accessible in third-person perspective. Meanwhile, subjective physical states – the ones that presumably lead to consciousness – can only be accessed in first-person perspective. Based on the associated distinction between first- and third-person physical processes, Searle speaks of first- and third-person ontology.

What does the distinction between subjective and objective physical states imply in metaphysical terms? Searle argues that subjective and objective physical states are related to different underlying realities: there is the reality of the first person and the reality of the third person. This means that Searle distinguishes between first- and third-person ontology. Both first- and third-person ontology refer to the physical, but as two distinct aspects, subjective and objective physical states. The traditional dichotomy between the mental and the physical is here replaced by a dichotomy within the physical itself, subjective and objective.

Metaphysical–empirical escapes 2

Biological naturalism versus identity theory

How do first-person physical processes lead to consciousness? Because we are considering this question within the realm of physical reality, this is no longer a metaphysical or ontological question, but an empirical one. Consciousness is the empirical manifestation of the subjective physical states: subjective physical states describe the basic elements from which consciousness is put together and shaped. This means that consciousness is a higher-level function of the whole system that is caused by the combination of the lower-level subjective physical states within that system.

How can we better illustrate this? Searle compares consciousness with photosynthesis. Photosynthesis is the higher-level output of many lower-order physico-chemical processes that cause a particular global output on a higher level, namely photosynthesis. In the same way that physico-chemical processes cause photosynthesis, the subjective physical states as associated with the first-person perspective cause consciousness. This means there is nothing special about consciousness; it is the result of a particular empirical mechanism, the causal effects of subjective physical states. Searle thus combines a metaphysical assumption with an empirical statement.

Let us specify this peculiar linkage between metaphysical and empirical domains. Searle's approach links these two domains. He distinguishes metaphysically between subjective and objective physical states. In order to explain consciousness, he can then refer to the subjective physical states with which he associates consciousness. Consciousness is then nothing but the empirical manifestation of the subjective physical states.

Accordingly, consciousness can be explained in terms of physical features, subjective physical states, which can be described as 'explanatory reduction' within the empirical domain. At the same time, there is no reduction of the reality of mental to physical states in the metaphysical domain as is reflected in the distinction between subjective and objective physical states. In short, ontological dualism is here accompanied by explanatory reduction.

Why doesn't Searle simply assume identity between mental and physical processes, as identity theory does (see Chapter 6)? This does not provide a proper solution either. First, this invokes all the problems of identity theory as discussed in Chapter 6. Second, Searle argues that we cannot compare the case of consciousness with other examples of identity. Consider the case of water and H_2O. Here, the system-level property water and the low-level chemical formula are identical. Most importantly, the low-level property, the chemical formula, is a third-person feature that is objective, non-intentional and quantitative.

One may assume the same kind of identity to hold true in the case of consciousness. Consciousness as a first-person feature, being subjective, intentional and qualitative, may, as a system-level property, be reduced to third-person physical processes that are objective, non-intentional and quantitative. 'No,' says Searle. 'That may work in the case of water but not in consciousness.' For this type of explanatory reduction to be feasible, one needs special physical properties – first-person properties that allow for subjective, intentional and qualitative features. Hence his assumption of subjective physical processes which imply what he calls 'first-person ontology'.

Critical reflection 1

Biological naturalism and the brain

How does Searle's biological naturalism relate to the brain? Extrapolating from Searle's view of the brain, one may be inclined to assume two different types of neuronal processes in the brain: first-person neuronal processes that lead to consciousness, and third-person neuronal processes that do not yield consciousness. How are both first- and third-person neuronal processes related to each other? Following Searle, they seem to operate in parallel to each other. At the same time, they can be distinguished from each other by the presence or absence of consciousness as a system- or high-level property.

One can nevertheless ask how it is possible that there are two types of neuronal processes in the brain. How can they be distinguished from each other? If Searle cannot give an answer to that, his whole approach crumbles. He must assume some kind of neuronal dualism between subjective and objective neuronal processes in the brain to accommodate for his ontological dualism between first- and third-person ontology.

At the same time he must assume some kind of relationship between subjective and objective neuronal processes. Otherwise he ultimately is forced to assume two different types of brain, a first-person physical brain and a third-person physical brain. Such amplification of the brain is, however, rather implausible, both conceptually and empirically. Thus, despite his argument for biological plausibility, where consciousness is a biological phenomenon of the brain, Searle encounters serious problems when it comes to a more concrete and specific account of his position in neuronal terms. Overall, biological naturalism may entail a certain degree of neuronal implausibility on the more specific neuronal grounds of the brain.

Where does this leave us with regard to biological naturalism? Searle provides an interesting answer to the definition problem when shifting the definition of mental states into the realm of the physical. The reality of a subjective, intentional and qualitative nature of consciousness is now no longer associated with a mental reality, but with a special subtype of physical reality: the subjective. The question now is whether the assumption of such subjective physical states as associated with first-person perspective is supported empirically in both physics and neuroscience. Do the empirical data in both physics and neuroscience lend empirical support to the distinction between subjective and objective physical states? This is a matter for future research.

Critical reflection 2

'Phenomenal–physical inference'

Finally, Searle seems to presuppose a special kind of inference. How does he come to the assumption of subjective physical states? The only way for him to reach that conclusion is by inferring from the subjective nature of the phenomenal features of consciousness to the subjective nature of its underlying physical states. If the phenomenal features are subjective, the physical states that cause the phenomenal features must also be subjective. This is so because purely objective physical states cannot account for the subjective nature of physical features. Since, however, consciousness is presupposed as physical by Searle, he must assume subjective physical states.

We can call this kind of inference 'phenomenal–physical inference'. The 'phenomenal-physical inference' describes how the presence of certain specific physical features is inferred from the presence of particular phenomenal features: if particular phenomenal features are present, they must be mediated by certain specific physical features that must be assumed to be also present and be distinguished from other physical features. Searle's argument is based on such 'phenomenal–physical inference', since he assumes first-person phenomenal features to imply first-person physical features as distinct from third-person physical features. By specifying the exact features between which inferences are made, 'phenomenal–physical inference' can be regarded as a specific instance of the more general 'inference problem' discussed above.

Is 'phenomenal–physical inference' plausible and thus valid? One may raise this question on several grounds – logical-conceptual, phenomenal and physical.

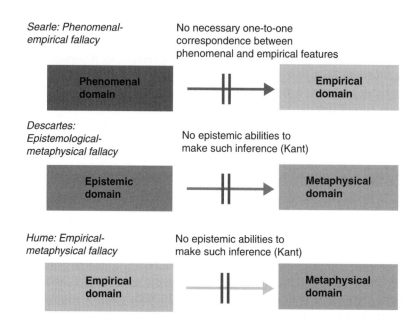

Figure 7.1 'Inference problem' and 'domain problem' in current mind–brain discussion
Figure 7.1a 'Inference problem' as unjustified inferences between different domains
The figure illustrates the 'inference problem' and the 'domain problem' of the mind–brain problem in the current philosophy of mind.

(a) The 'inference problem' describes the inference between different domains in the current discussion about the mind–brain problem. Figure 7.1a represents different varieties of the 'inference problem' concerning inferences between different domains. Figure 7.1b (p. 209) illustrates the 'domain problem' that describes the inclusion of different domains and their relationships in the current mind–brain discussion. Each box and its particular shading in the upper part indicates a particular domain as described in the lower part. The horizontal arrows indicate inferences between different domains.

Upper part in Fig. 7.1a: The figure illustrates the phenomenal–physical inference by Searle, who infers from phenomenal and non-phenomenal features to two different physical features. This is an inference from the phenomenal to the empirical domain. Such inference is, however, problematic because one and the same physical feature may be associated with both phenomenal and non-phenomenal features. There is thus no necessity of a one-to-one correspondence between the two kinds of features.

Middle part in Fig. 7.1a: The figure illustrates the epistemic–metaphysical inference by Descartes, who infers from epistemological features ('I think') to separate metaphysical features ('I am'). This is an inference from the epistemological to the metaphysical domain. Such inference is, however, problematic because according to Kant it transgresses our epistemic abilities.

Lower part in Fig. 7.1a: The figure illustrates the empirical–metaphysical inference by Hume, who infers from empirical features ('bundle of perceptions') the non-existence of a separate metaphysical feature accounting for the mind. This is an inference from the empirical to the metaphysical domain. Such inference is, however, problematic because according to Kant it transgresses our epistemic abilities.

Logically, it seems only plausible to assume that phenomenal features entail physical features that are distinct from those that are not related to phenomenal features at all. While this assumption seems logically plausible by itself, it however presupposes co-occurrence of phenomenal-physical identity and non-identity which is logically implausible.

More specifically, Searle's biological naturalism presupposes identity between phenomenal features and the distinct physical features, the first-person-based physical features. While, at the same time, biological naturalism implies non-identity between phenomenal features and the third-person-based physical features. We are thus confronted here with the co-occurrence of phenomenal–physical identity and non-identity at the same time within the same organism. Even if empirically plausible, logically such co-occurrence between identity and non-identity is obviously contradictory and thus logically implausible.

Critical reflection 3

Empirical and phenomenal plausibility of the 'phenomenal–physical inference' (Searle)

How about empirical plausibility? As mentioned above, there may be severe doubts about two principally different neuronal features in the brain, those related to phenomenal features and those remaining unrelated to them. Such neuronal dichotomy does not seem to gain empirical support in current neuroscience.

How about the phenomenal plausibility of biological naturalism? Phenomenal features as Searle describes them cannot be doubted (though some philosophers, like Patricia Churchland and Daniel Dennett doubted, for instance, the existence of qualia). There thus seems to be some phenomenal plausibility. However, nothing in the phenomenal features themselves entails their modulation by distinct physical states. One could, for instance, also assume that those physical states that, following Searle, do not mediate phenomenal features nevertheless contain some kinds of neuronal features that predispose them to generate phenomenal features in the 'right' circumstances.

What does the assumption of such neural predisposition imply for the mind–brain problem? The strict dichotomy between phenomenal and non-phenomenal features (and thus between first- and third-person perspectives) would then be undermined by the assumption of a continuum between phenomenal and non-phenomenal features. Such a neuronal predisposition could then no longer be described as either phenomenal or non-phenomenal but rather as pre-phenomenal (see Chapters 15 and 16 for details of the concepts of neural predisposition and pre-phenomenal).

Accordingly, the assumption of such neuronal predisposition would undermine Searle's strict dichotomy between phenomenal and non-phenomenal features and its corresponding distinction between different physical features. If so, the strict phenomenal–non-phenomenal distinction may turn out to be neither phenomenally nor empirically plausible. Searle's 'phenomenal–physical inference' may then turn out to be fallacious rather than phenomenally and empirically plausible, as he claims.

Critical reflection 4

Is the mind–brain problem a metaphysical problem?

Neutral monism and biological naturalism share the metaphysical domain as a focus, as well as the aim to undermine and avoid the mental–physical dichotomy. Neutral monism introduces a third reality – the neutral – which supposedly underlies both mental and physical realities. Biological naturalism duplicates physical reality and existence by introducing first- and third-person (subjective and objective) physical processes.

Both positions successfully undermine the mental–physical dichotomy in certain ways. They are, however, confronted by several problems, including the 'definition problem' (how to define the concepts of mental, physical and neutral), the 'plausibility problem' (what is the empirical, conceptual and metaphysical evidence), and the 'inference problem' (can one make an inference from mental to physical and vice versa).

These problems suggest that it is problematic to discuss the mind–brain problem in purely metaphysical terms. Instead of treating the mind–brain problem in exclusively metaphysical terms and thus from within the metaphysical domain, one may consider this problem also in epistemic terms and thus situate it within the epistemological domain. To show such a shift from the metaphysical to the epistemological domain, we should return for a moment to the history of philosophy and introduce the German philosopher Immanuel Kant.

Biography

Immanuel Kant

Immanuel Kant (1724–1804) was a philosopher who lived in Koenigsberg near the Baltic Sea. In the eighteenth century, Koenigsberg was part of Prussia/Germany, but today it is part of Russia and close to the border with Poland. In his entire life Kant never traveled more than 70 miles from the city of Koenigsberg. Hence, Kant never really left Koenigsberg, which is something we can barely imagine today, in the age of global travel. Despite his lack of worldliness, Kant nevertheless had an interesting life – especially as a young man. Despite his relative poverty, he nevertheless was able to access the higher echelons of society with his thoughtful speeches about man and his place in the world. Kant built a reputation for himself as a promising thinker and was known to be a 'charming and intellectual dandy'.

When Kant got older, he withdrew more and more from society in order to focus on developing his philosophy. He performed a strict daily schedule of routines to which he stuck rigidly. For instance, one of his daily routines involved a little walk at 5pm every day; that, according to the legend, was always punctual to the minute so that the Koenigsbergers could set their clocks by him. Besides such physical walks, he performed amazing 'philosophical walks', which led to a revolution in philosophy, as will be explained briefly below.

Epistemic escapes 1

Epistemic limitations in our knowledge of the mind (Kant)

Immanuel Kant was extremely skeptical of metaphysical assumptions about the existence and reality of things and the world itself. Kant argued that any kind of metaphysical inference is merely speculative and extends beyond our cognitive and epistemic capacities. This is the main point in his *The Critique of Pure Reason* (1998 [1781]). Instead of speculating about metaphysical existence and reality, we should investigate our own epistemic abilities and limits. Once we know what we can and cannot know, we might be able to judge whether we can know anything about the metaphysical existence and reality of, for instance, the mind.

Let us illustrate this using the following example. Imagine you want to cook a meal. There are three things you need to check before getting started. First, you need to know the recipe. Otherwise you will not be able to cook what you want. Second, you need the ingredients. And third, you need the right instruments, namely the right pan, pots, knifes, stove, etc.

How does this correspond to the situation of the mind–brain relationship in the context of Kant? We have the ingredients – the mind and the brain. That is what we know. We lack the recipe, which can, in this example, be considered analogous to the 'mind–brain relationship'. We do not know how mind and brain relate to each other and yield consciousness. What about the instruments for realizing and implementing the recipe, the pans, pots, knives, etc.? This corresponds to our epistemic abilities that are required to know anything about how mind and brain are related to each other – the recipe.

Both mental and physical approaches (see Chapters 5 and 6) as well as the alternatives described above, presuppose that we, in principle, have the 'right' 'epistemic instruments' to access and know the existence and reality of the mind and how it is related to that of the brain. Otherwise they could not make the respective metaphysical assumptions. Hence, the various metaphysical assumptions about the existence and reality of the mind presuppose certain epistemic abilities which allow us to access and know the metaphysical domain of existence and reality.

Rather than discussing the various metaphysical assumptions by themselves, as in the current mind–brain discussion, Kant doubted their epistemic presupposition. According to Kant, we do not have the 'right' 'epistemic tools and instruments' in our mind to access and know the mind's underlying existence and reality. We simply lack the 'right' 'epistemic tools' to know anything about the metaphysical existence and reality of the mind and how it relates to that of the brain and body.

Hence, put differently, Kant says that our pots and pans – the 'epistemic instruments' of our mind – are simply not suited to cooking the dish, the 'mind–brain relationship'. Why the pots and pans are not suited is a different question, though. We will come back to this point in Chapters 8 and 12).

Epistemic escapes 2

Epistemic limitations versus metaphysical assumptions (Kant versus Descartes)

How does Kant's epistemic approach relate to the metaphysical approaches of his philosophical predecessors like René Descartes? Recall the French philosopher Descartes from Chapter 5. He assumed two substances, the mental and the physical, to account for mind and body, with both interacting in a tiny structure in the brain, the pineal gland (see Chapter 5). Due to the assumption of two different substances, his position is called substance dualism, which can be specified as interactive dualism because of their interaction in the brain.

How did Descartes come to this conclusion and thus his metaphysical knowledge about mind and brain? As discussed in Chapter 5, he inferred his metaphysical knowledge about the mind's separate existence and reality from his famous sentence 'I think, therefore I am' (cogito ergo sum). Let us explain this in a little more detail. Thinking is an activity which must be performed, e.g. by somebody, a self or an I, which must exist in order to make thinking possible. Without the existence and reality of the self, thinking remains thus impossible. The activity of thinking thus implies and lets us infer the existence and reality of a self, the I. And since thinking is a mental activity, the presupposed self must be mental too and thus be different from the physical existence and reality of the body.

Kant smashed Descartes' line of reasoning. Recall from Chapter 5 that Descartes assumed a mental substance to underlie our 'I'. How did he come to that conclusion? He noticed that we think continuously and that such thinking is possible only if there is somebody who thinks. According to Descartes, our 'I' is possible only if there is an underlying mental substance.

Descartes thus infers from the thinking activity of the 'I think' to a mental substance. It is exactly this inference that Kant deemed false. According to Kant, we simply do not possess the 'right' 'epistemic instruments' to know anything about the existence and reality underlying the 'I' that does the thinking. Thus, we cannot know what, and which kind of, substance underlies the 'I' during the activity of thinking.

According to Kant, Descartes transcends our epistemic abilities when he infers the mental existence and reality of the I from its activity of thinking. In other words, Descartes does not consider our mind's epistemic limitations. That leaves him, Descartes, with nothing but pure speculation about the mind's existence and reality – metaphysical speculation. Following Kant, such metaphysical speculation needs to be rejected, however, because it transgresses our epistemic abilities.

Epistemic escapes 3

'Epistemic–metaphysical inference' and mental approaches (Kant on Descartes)

How can we put this more technically? One can say that Descartes presupposed an inference from the epistemic domain of knowledge – the 'I think' – to the metaphysical domain of existence and reality – the 'therefore I am'. Because this involved inferring assumptions about the mind in the metaphysical domain from the knowledge of ourselves as associated with the epistemic domain, one may speak here of an 'epistemic–metaphysical inference'. The 'epistemic–metaphysical inference' thus describes inferences from the epistemological to the metaphysical domain and thus from knowledge to existence and reality.

Let us detail the 'epistemic–metaphysical inference' in the case of Descartes. The presumed epistemic abilities are manifest in the 'I think' (i.e. 'cogito') and the metaphysical assumption is reflected in the 'I exist' (i.e. 'sum'). The inference itself – the connection between epistemic and metaphysical domains – is expressed in the 'therefore' (i.e. 'ergo') in the 'I think, therefore I am' (cogito ergo sum).

Kant now argues that the inferential component – the 'therefore' – is not justified given our epistemic limitations. Therefore Descartes' epistemic–metaphysical inference is, according to Kant, a false positive one and thus fallacious. We lack the epistemic ability to infer from the epistemic to the metaphysical domain. Descartes' epistemic–metaphysical inference thus turns out to be an 'epistemic–metaphysical fallacy', an inference from the epistemological to the metaphysical domain that is fallacious and thus not plausible because of our lacking epistemic abilities (see the middle part in Figure 7.1a on page 196).

By revealing Descartes' epistemic–metaphysical inference as fallacious, Kant undermines the assumption of dualism in particular, and mental approaches in general, as solutions to the mind–brain problem. If we can no longer infer from our mental activities like thinking (and others like consciousness, self, free will, etc.) to some kind of special mental feature as underlying existence and reality, any kind of mental approach to the mind–brain problem becomes impossible. Following Kant, there is no way that we can determine the mind's existence and reality in a mental way. The only way to do that, according to Kant, is to transgress our epistemic abilities and this leads to pure speculation, metaphysical speculation.

Epistemic escapes 4

Mind as 'bundle' of extrinsic stimuli (Hume)

What about the physical approach to the mind–brain problem? Recall from Chapter 6 that the physical approaches to the mind–brain problem assume the mind's existence and reality to be physical and thus are not principally different from the physical existence and reality of the brain and the body. This position ultimately reduces the mind to the physical existence of the brain and the body: the mind's existence and reality is then considered to consist in nothing but the brain's and body's physical processing of extrinsic stimuli.

This position is roughly the same as that of another philosophical predecessor of Kant, the Scottish philosopher David Hume (1711–76). Not only was Hume a philosopher and economist, he was also a historian. His book on the history of England from the invasion by Roman Emperor Julius Caesar until the seventeenth century became a bestseller. Hume's most famous philosophical work is the *Treatise of Human Nature* (1740), which is still considered one of the most important philosophical texts ever written. Despite its current reputation, though, contemporaneous critics in Great Britain deemed the work to be 'abstract and unintelligible'. This dealt a serious blow to Hume and inspired him to write more clearly in the hopes of becoming a more widely-read philosopher.

Hume's main aim was to rid his philosophy of any speculation. Like Kant, Hume was extremely critical of groundless metaphysical assumptions and argued that we have to search for how the extrinsic stimuli from the environment are processed by our physical brain and body. This entails an understanding of the mind that does not consist of some kind of mental substance. Instead, it consists of nothing but the processing of extrinsic stimuli from the environment.

What is the mind in this model? Following Hume, there is simply no mind with an existence and reality as distinct and separate from brain and body. Instead, the mind consists of nothing but a bundle of psychological processes that process and elaborate the extrinsic environmental stimuli. There are no separate faculties apart from the sensory organs – our five senses – which provide us with information. Anything else does not exist.

In this model, the assumption of a mind with a mental substance is nothing but an illusion. The same applies to mental features like the self. Instead, it is just a collection of experiences of extrinsic stimuli, a psychological bundle. This is often referred to as the 'bundle theory of the self'.

How is Hume's philosophy related to the different solutions of the mind–brain problem? Because of Hume's fundamental reliance on empirical research, his approach to the mind–brain problem rejects all mental solutions. Instead, he characterized the mind as resulting from the processing of extrinsic stimuli in the physical states of brain and body. If forced to choose, Hume definitely would have opted for a physical approach to the mind–brain problem.

Epistemic escapes 5

Intrinsic features versus extrinsic stimuli (Kant versus Hume)

What is the difference between Hume and Kant? Both philosophers were extremely critical of metaphysical assumptions and consider them to be pure speculation. As such, both rejected the assumption of mental substances, as well as Cartesian dualism between the mind and the body. Kant did indeed say that Hume's work and its critical attitude woke him from his 'dogmatic slumber' and made him aware of the groundlessness of metaphysical speculations.

While both Hume and Kant shared a distaste for mental substances, Kant nevertheless rejected Hume's approach. Why? Because for Kant, the mind was more than just a bundle of psychological processes processing extrinsic stimuli from the environment. In Kant's model, the mind is not only determined by extrinsic stimuli, but also has some intrinsic features of its own. The processing of the extrinsic stimuli is thus not completely and exclusively determined by the extrinsic stimuli themselves, but also by the intrinsic features of the mind. One can thus say that the actual processing can be characterized by a 'double input' – the extrinsic stimuli from the environment as well as the intrinsic features of the mind.

What are the mind's intrinsic features? The answer to this question is intricate. Kant rejected the assumption of specific metaphysical features, like a mental substance or other mental properties, as characteristic of the mind. Furthermore, he rejected any metaphysical assumptions about existence and reality in general because that would transgress our epistemic abilities and lead to nothing but 'groundless metaphysical speculation' (see above). This means that the mind's intrinsic features cannot be determined in metaphysical terms.

What, for Kant, are the mind's intrinsic features? Rather than referring to existence and reality, he assumes that the mind's intrinsic features consist in a specific form, organization, or structure. What is this specific form, organization, or structure? It includes spatial and temporal features (and other features; see Chapter 4 for details) which the mind itself imposes upon its own processing of the extrinsic stimuli from the environment. The interaction between the mind's intrinsic form (structure, organization) and the environment's extrinsic stimuli results in mental states like consciousness.

The main difference between Kant and Hume thus consists in the mind's intrinsic features – its form, structure or organization. Hume considers the mind to be a mere bundle or collection of extrinsic stimuli. Kant argues that the mind goes beyond this in that it shows an intrinsic organization which it imposes upon the extrinsic stimuli: the intrinsic organization of the mind structures and organizes the extrinsic stimuli in a particular way that makes their association with consciousness possible.

According to Kant, what we call the mind emerges from an interaction between the mind's intrinsic features and the environment's extrinsic stimuli. This is different from Hume, who ultimately assumes the mind to consist in a bundle of extrinsic stimuli.

Metaphorical comparison

Umbrellas in Edinburgh and Koenigsberg

How can we further illustrate the difference between Hume and Kant? Let's imagine people walking in the often rainy Edinburgh, Scotland, the home of David Hume. They all carry umbrellas with them. Now they enter a house. What do they do? They all close their umbrellas and put them aside. Where and how do they put the umbrellas?

Let us start with the Scottish umbrellas in the case of Hume. In the house of David Hume, they put their umbrellas on to the floor where they all lie chaotically side by side. There is no order at all, small and big umbrellas as well as black and colorful umbrellas are mixed together. Why? Because the house itself does not provide any special way to put the umbrellas in a more orderly and structured way.

This is different in the German-Prussian house of Immanuel Kant (which, though, would be in Koenigsberg, not in Edinburgh). There the house itself would provide a special box or container for the umbrellas. A big container with different holes and with each hole suited for one umbrella of a particular size and a specific color. What would the visitors do? They would look for the 'right' hole for their umbrella, its size and its color, and put it accordingly into the box. The different umbrellas would consequently no longer lie chaotically side by side on the floor as in David Hume's home. Instead, we would find the different umbrellas standing side by side and well structured and organized according to size and color in Kant's house.

Epistemic escapes 6

Intrinsic features and epistemic limitations (Kant)

So far, we have discussed the rejection of mental approaches to the mind by Kant and Hume as well as their differences with regard to the mind's intrinsic features. We must still discuss Kant's relationship to the physical approaches to the mind–brain problem. While Hume's characterization of the mind as a mere collection of extrinsic stimuli puts him close to the physical approaches, Kant's assumption of the mind's intrinsic features and his insistence on the mind's epistemic limitations (see above) sketch a more complicated picture.

Why does our mind suffer from principal epistemic limitations that prevent it from making metaphysical assumptions about its own underlying existence and reality? This is the point where the mind's intrinsic features need to be considered. Any cognition and knowledge of the extrinsic stimuli from the environment is possible only on the basis of the mind's intrinsic features. We can thus know the world and its extrinsic stimuli only through our mind's input – the form it provides. Our knowledge of the world (and of ourselves) is thus dependent upon our mind, which may be described as 'mind-dependence'.

This mind-dependence means that we are unable to know the world independently of our mind's input. However, this makes it impossible for us to access and know how the extrinsic stimuli and the world are by themselves independent of our mind and its form. In other words, we simply cannot differentiate between the contributions of our mind's intrinsic features and the environment's extrinsic stimuli in our cognition and knowledge of the world. Since our knowledge is already always based on the interaction between the mind's intrinsic features and the environment's extrinsic stimuli, we remain unable to know the world (and our own mind) independently of our minds. Instead we can access and know the world (and the mind itself) only through our minds.

Using Kant's more technical terms, this mind-dependence implies that we cannot know the noumenal world, as Kant himself says – that is the world as it is by itself independent of our mind. Instead, we are stuck with the way the world appears to us in our cognition and knowledge. Kant calls this the phenomenal world. As such, we can only access and know the phenomenal world, while the noumenal world remains hidden because of our mind's intrinsic features and limitations.

Epistemic escapes 7

Epistemic limitations and physical approaches

How does this apply to the mind? Our mind's intrinsic features are also involved in accessing our own mind. As in the case of the world and extrinsic stimuli, we can therefore also access the mind and its intrinsic features only through our minds. Hence, while accessing the mind with cognition, the mind's intrinsic features structure and organize our cognition and knowledge of the mind.

This makes it impossible for us to access and know the mind independently of the mind. In the more technical terms of Kant, this means that we can access and know the mind only in a phenomenal, but not a noumenal way (see above). This makes it impossible for us to access and know the mind's existence and reality, and thus its metaphysical characterization.

Why? For any metaphysical characterization of the mind to be possible, we would need the ability to access and know the mind in an independent way without the contribution of the mind itself. Since this is impossible, we remain principally unable to say anything about the mind's underlying existence and reality. Metaphorically put, the mind itself is a blind spot or curtain which makes it impossible for us to discover the mind's 'true and real' reality and existence.

What does this imply for the mind–brain problem? We already demonstrated that Kant, on the basis of the mind's epistemic limitations, rejected mental approaches to the mind–brain problem. Now we can complement that by the rejection of physical approaches. Why would Kant also reject physical approaches to the mind–brain problem?

Precisely because Kant doubts that we can make any kind of metaphysical assumption about the mind's existence and reality. Any kind of metaphysical characterization as either physical or mental (or otherwise) remains impossible for us because of the mind's intrinsic features and their involvement in our cognition and knowledge of the mind. In short, our mind-dependent cognition and knowledge of the world and mind prevents us from making any metaphysical assumption about the mind. As such, any physical approach to the mind–brain problem would ultimately be as speculative as a mental approach.

Epistemic escapes 8

'Epistemic–metaphysical inference' and 'empirical–metaphysical inference' (Kant)

How does Kant judge mental and physical approaches to the mind–brain problem? In mental and physical approaches, Kant would diagnose a false positive and thus a fallacious inference to the metaphysical domain. Let us start with the mental approaches. Descartes suggested a mental approach when assuming mental substances as distinguished from physical substances. This was based on an epistemic ability, the thinking activity, the 'I think'.

Descartes' assumption of mental substances thus presupposes an inference from the epistemic domain of knowledge to the metaphysical domain of existence. One may thus speak of an 'epistemic–metaphysical inference', as a subset of the more general 'inference problem' (see above). Kant now demonstrated that Descartes' 'epistemic–metaphysical inference', that underlies his mental approach to the mind–brain problem, is fallacious (see above).

What is fallacious about the physical approaches? Here the starting point is not a specific mental state and thus an epistemic ability like thinking. Instead, the starting point is the observation of extrinsic stimuli and how they are processed by the mind. The observation of extrinsic stimuli and their processing concerns the empirical domain, the domain of observation of the world (and the mind) in third-person perspective.

The processing of extrinsic stimuli and thus the empirical domain and its physical features provide Hume with the basis upon which he rejects the assumption of any additional reality, like mental features. He thus infers from the empirical domain of the processing of extrinsic stimuli, to the metaphysical domain, and in doing so rejects the existence of anything other than the physical. One can therefore speak of an 'empirical–metaphysical inference' as yet another subset of the more general 'inference problem' (see above) (see lower part in Figure 7.1a on page 196).

One may diagnose what can be described as 'empirical–metaphysical inference' in physical approaches to the mind–brain problem in general. This inference is, however, according to Kant, fallacious and may therefore be discarded as an empirical–metaphysical fallacy. Why? We recall that the epistemic–metaphysical inference leads to a positive assumption of a mental substance as distinct from the physical substance.

This is different in the empirical–metaphysical inference. Rather than inferring a positive assumption of an additional mental reality, the inference leads here to a negative assumption, the rejection of an additional mental reality. Despite these differences that both empirical–metaphysical and epistemic–metaphysical inferences share, they make some kind of metaphysical assumption (whether positive or negative) and they both infer that assumption from assumptions in another domain, empirical or epistemic.

This inference from one domain to another is, however, according to Kant, not justified. Why? Because it transgresses our epistemic abilities. No matter whether the inference leads to a positive or a negative assumption, any kind of inference is problematic and needs to be rejected. Hence, following Kant, inferences between different domains may lie beyond our epistemic abilities and may therefore turn out to be epistemological fallacies. Even though logically-conceptually, inferences between different domains may be consistent and thus plausible, they may turn out to be implausible on epistemological grounds. The above-described fallacies may thus be epistemological fallacies rather than purely logical-conceptual ones.

Critical reflection 5

Current interpretation of Kant's stance on the mind–brain problem

How can we now interpret Kant's position toward the mind–brain problem? Recall that the philosopher expects a metaphysical answer to a metaphysical problem. Nothing less will do. Either the mind is mental or the mind is physical, or something in between as in neutral monism and biological naturalism (see above).

Some, like the American philosopher Karl Ameriks (1947–), assume that Kant opted for an immaterial and thus mental approach to the mind. The mind cannot be physical and material but must rather be mental and immaterial. Why? Because the mind's intrinsic features include unity, spontaneity and self-consciousness (Kant also spoke of apperception in this context) – qualities that by their very nature cannot be physical and material. Hence, by attributing intrinsic features to the mind, Kant introduced some kind of non-physical, and thus immaterial, existence and reality.

Other philosophers like Andrew Brook (1994) and Patricia Kitcher (1992) argue that Kant would opt for functionalism. Recall from Chapter 6 that functionalism compares the mind to the software of a computer, while the brain parallels the hardware. Analogous to the software, the mind can be described by the causal relations between input stimuli, inner or mental states, and behavioral output (see Chapter 6 for details). Following Brooks and Kitcher, Kant focused mainly on the software when he described the structure and organization of the mind and its intrinsic features. In contrast, he did not care much about the hardware, the brain. Kant can therefore be considered a 'functionalist *avant la lettre*' (before the term was created).

Critical reflection 6

Epistemic dualism

Finally, other philosophers like the German Tobias Schlicht (2007) argue that Kant did not make any metaphysical assumptions at all. Kant neither argued for immaterialism (as Ameriks claims) nor for functionalism (as Brook and Kitcher argue). Instead, Kant, so Schlicht argues, refrained from any metaphysical assumption at all including functionalism and immaterialism. Why? Because any such metaphysical assumption transgresses our epistemic abilities and would therefore result in nothing but 'groundless metaphysical speculation' about mind and brain.

How does Schlicht characterize Kant's position on the mind–brain problem? He argues that Kant assumes what can be described as 'epistemic dualism'. Epistemic dualism suggests that there are two different ways of accessing and knowing world and mind. The first is what Kant described as the 'outer sense', which provides us with the perception of extrinsic stimuli and the world. The outer sense thus refers to the sensory system and our five senses. The outer sense is complemented by the 'inner sense', which provides us with introspection and thus awareness about our own mental states.

One may thus roughly parallel the distinction between inner and outer sense (between perception and introspection), with the distinction between first- and third-person perspective. Perception allows for observation that is associated with the third-person perspective. Introspection presupposes experience in first-person perspective which is then accessed as such by introspection in what some authors describe as second-person perspective (which for reasons of simplicity we here subsume under the umbrella of the first-person perspective).

How are the inner and outer senses, and thus perception and introspection, constituted? Kant (and Schlicht) assume that both presuppose the involvement of the mind's intrinsic features. Only by involving the mind's intrinsic features can the material in both inner and outer sense be organized and structured in such a way that it can be accessed and known by us. In other words, both inner and outer senses are dependent on the mind's intrinsic features.

This mind-dependence denies both inner and outer senses access to the existence and reality underlying their respective material. We remain unable to access the existence and reality of extrinsic stimuli as we perceive them in third-person perspective. And we remain equally unable to access the existence and reality of our mental states as we experience and introspect them in first-person perspective. This means that we are stuck and limited to the two ways of epistemic access, inner sense and outer sense, or first- and third-person perspectives. This position is called epistemic dualism.

Ideas for future research 1

Mind–brain problem as epistemological problem

What does epistemic dualism imply for the general characterization of the mind–brain problem? The purely epistemic characterization of the mind–brain problem abstains from any metaphysical assumptions. Why? Because, according to Kant, these cannot be made without transgressing our mind's epistemic limitations.

This means that before starting with any kind of metaphysical characterization of the mind, we need to investigate our epistemic abilities and limitations. One may now come to the conclusion that, unlike Kant, our epistemic abilities are sufficient to make metaphysical assumptions and more specifically empirical–metaphysical and epistemic–metaphysical inferences. Then we can characterize the mind and its relation to brain and body in metaphysical terms and determine their existence and reality. This is only possible on the basis of prior investigation of our mind's epistemic abilities and limitations.

The mind–brain problem can thus be regarded as primarily an epistemic problem rather than a metaphysical one. Any metaphysical assumption about the existence and reality of the mind and the brain needs to be preceded by an investigation of our epistemic abilities. The guiding question here is: what kind of knowledge can we possibly acquire (and what not) given our epistemic abilities? We are then able to see whether those epistemic abilities include the capacity to make metaphysical assumptions about the reality of mind and brain.

Before making metaphysical assumptions, the mind–brain problem is here investigated primarily in epistemic terms; that is, in terms of knowledge. Rather than being a primarily metaphysical problem, the mind–brain problem is now redefined as primarily an epistemic problem.

Ideas for future research 2

Mind–brain problem as 'domain problem'

The proponents of a physical approach to the mind–brain problem can go one step further. For instance, the identity theorists or the advocates of eliminative materialism (see Chapter 6) argue that the mind will either be identified with the brain or completely eliminated on the grounds of future empirical findings about the brain and its neuronal mechanisms. The starting point here is no longer the metaphysical or epistemic domain. Instead, it is the empirical domain, the mechanisms and processes that occur in the brain.

Is the empirical domain the proper home for the mind–brain problem? Kant would argue that even observation in third-person perspective is subject to the epistemic limitation of our mind. Why? Because observation and perception are also possible only on the basis of the involvement of our mind's intrinsic features. Hence, even the empirical domain may not allow us proper access to the metaphysical domain as reflected in the empirical–metaphysical fallacy (see above).

This makes clear that we absolutely need to consider our mind's epistemic abilities and limitations. This remains independent of whether we start from mental activities like thinking, as in Descartes, or from empirical observation, as in current physicalists and neurophilosophers like Patricia Churchland. Without the consideration of our mind's epistemic abilities and limitations, we will not be able to arrive at any sound non-speculative assumptions about the mind.

The epistemic domain rather than the empirical (and the metaphysical) domain may thus be considered the primary domain for the mind–brain problem. And it is from there, the epistemic domain, that one may then want to connect to either the empirical and/or the metaphysical domain. This requires future research into not only the mind–brain problem itself, but also how the different domains – empirical, epistemic and metaphysical – are linked and connected to each other.

One may thus want to speak of a 'domain problem'. The 'domain problem' in general describes the characterization of the different domains, epistemic, metaphysical and empirical, how they are related to each other, and what kinds of inferences we can draw between the different domains. With regard to the mind–brain problem in particular, the domain problem concerns the question of the primary domain for the mind–brain problem and how we can connect and infer from its 'home domain' to the other domains. Both aspects of the domain problem, the domains in general and the domain of the mind–brain problem are, however, subject to future research.

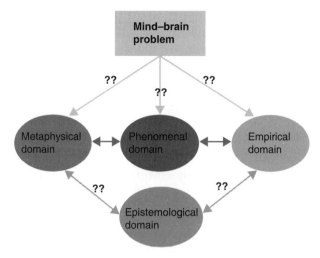

Figure 7.1 'Inference problem' and 'domain problem' in current mind–brain discussion
Figure 7.1b 'Domain problem' as the 'localization' of the mind–brain problem in different domains

(b) The figure illustrates the involvement of different domains and their relationship to each other in the mind–brain problem. This can be described as the 'domain problem'. The question marks indicate the unclear relationship between the different domains. The epistemological domain is placed at the very bottom since, following Kant, it depends on epistemic abilities whether we can draw inferences between the different domains.

Considered in this way, the 'domain problem' must be considered a genuinely neurophilosophical rather than philosophical or neuroscientific problem. Philosophy presupposes one particular domain, like the metaphysical, epistemological, phenomenal or the conceptual domain. Rather than considering the different domains thus presupposing domain pluralism as neurophilosophy does, philosophy usually considers only one particular domain, i.e. domain monism.

The possibility of the 'domain problem' presupposes domain pluralism, while it remains impossible in the context of domain monism. Since only neurophilosophy presupposes domain pluralism, the 'domain problem' must be considered a genuinely neurophilosophical problem rather than being related to either philosophy or neuroscience, which both presuppose domain monism (rather than domain pluralism).

Take-home message

We discussed mental (Chapter 5) and physical (Chapter 6) approaches to the mind–brain problem. Mental approaches assumed the mind to show a specific, e.g. mental, existence and reality as distinct from the physical brain and body. Physical approaches denied the mental character of the mind and argued instead that the mind is physical in its existence and reality. Since both approaches lead to various problems, many researchers and philosophers have searched for alternatives to these approaches. Metaphysical approaches aimed to undermine the mental–physical dichotomy and to provide a non-mental and non-physical characterization of the mind. This is reflected in neutral monism, which argues for a neutral reality and existence that underlies both mental and physical. Biological naturalism opts for two different kinds of physical realities, subjective and objective, as associated with first- and third-person respectively. However, while escaping and undermining the traditional mental–physical dichotomy, these alternative metaphysical approaches nevertheless raised several problems. The solutions so far still consider the mind–brain relationship as primarily a metaphysical problem. A different non-metaphysical approach was suggested by the early German philosopher Immanuel Kant. He argued that we cannot make any metaphysical assumptions about the mind because that simply transgresses our epistemic abilities. This implies that any metaphysical assumption about the mind–brain problem, whether mental or physical, is invalid and must be discarded as 'groundless metaphysical speculation'. This means that the mind–brain problem becomes an epistemic problem that raises questions about our epistemic abilities and the kind of knowledge we can have about the brain and the mind.

Summary

The previous chapters discussed mental and physical approaches to the mind–brain problem. Mental approaches assumed the mind and its existence and reality to be different from the brain and the body and their physical features. This led to dualism and panpsychism. Physical approaches claimed the mind's existence and reality to be ultimately physical, implying reductive or non-reductive physicalism and functionalism. Despite their differences, mental and physical approaches share some tacit background assumptions. One such background assumption is the dichotomy between the mental and the physical, and their mutually exclusive definition; this amounts to what I describe as the 'definition problem'. Closely related is the tendency to infer a mental property or substance from the physical's insufficiency in explaining mental features. I call this the 'inference problem'. Finally, we need to gather evidence to make our assumptions about the mind–brain relationship plausible; this amounts to the 'plausibility problem', which concerns metaphysical, epistemic, conceptual, or empirical evidence.

By calling these background assumptions into question, alternative approaches to the mind–brain problem have been developed. They pursue neither mental nor purely physical approaches and seek alternatives. One such approach is neutral monism (Bertrand Russell). Neutral monism suggests that there is a third reality that is neutral to both mental and physical and precedes them as their commonly underlying basis. Another approach is biological naturalism (John R. Searle). Biological naturalism suggests that we need to replace the dichotomy between the mental and the physical by the dichotomy between subjective and objective physical states. According to Searle, this leads to what he described as first- and third-person ontology. A completely different approach can be traced back to the earlier German philosopher Immanuel Kant. He argued that we have certain limitations in our epistemic abilities and knowledge, which make it impossible for us to make any metaphysical assumptions. Before making any metaphysical assumptions about mind and brain, we must first investigate our epistemic abilities. Kant would have rejected most of the current metaphysical mind–brain assumptions. Rather than making metaphysical assumptions, we should better investigate our epistemic abilities: we need to scrutinize whether our epistemic abilities show the kinds of capacities that are needed to make metaphysical inferences and assumptions. This puts the mind–brain problem into a primarily epistemic context.

Revision notes

- What are the implicit background assumptions in the current mind–brain discussion?
- Name the different kinds of problems.
- How is neutral monism defined? What are the features of this theory and how can it be distinguished from mental and physical approaches?
- What is biological naturalism? Why is it called biological, and why naturalism?
- Why is Immanuel Kant relevant in the current mind–brain discussion?
- What is the stance of Kant himself with regard to the mind–brain problem?
- What are empirical–metaphysical and epistemic–metaphysical fallacies?
- Why can the mind–brain problem be regarded as primarily an epistemic rather than empirical or metaphysical problem?

Suggested further reading

- Chalmers, D. (1997) *The Conscious Mind* (Oxford/New York: Oxford University Press).
- Kant, I. (1998) *The Critique of Pure Reason*, Kessinger Publishing, edited by P. Guyer (Cambridge: Cambridge University Press).
- Kitcher, P. (1992) 'The Naturalists Return', *Philosophical Review,* 101(1), 53–114.
- Russell, B. (1921) *The Analysis of Mind* (London: George Allen & Unwin; New York: The Macmillan Company).
- Russell, B. (1997) *The Problems of Philosophy* (London: Williams and Norgate; New York: Henry Holt and Company. Repr. New York/Oxford: Oxford University Press, 1997).
- Schlicht, T. (2007) *Erkenntnistheoretischer Dualismus. Das Problem der Erklärungslücke in Geist-Gehirn-Theorien* (Paderborn: Mentis).
- Searle. J. (2004) *Mind: A Brief Introduction* (Oxford/New York: Oxford University Press).

8
Brain-based Approaches to the Mind–Brain Problem

Overview

The previous chapter focused on the discussion of alternative approaches to the mind–brain problem. These approaches were described as alternative because they do not adhere to the strict dichotomy between the mental and the physical by pursuing non-physical and non-mental approaches to the mind-brain problem. While these approaches were able to avoid some of the problems plaguing mental and physical approaches, they also generated novel problems. In yet another effort to side-step the problems associated with mental and physical approaches to the mind–brain problem, others have pursued different methodological strategies. For example, rather than starting with the mind as a point of departure for mind–brain investigation, philosophers have also taken the brain itself as a point of departure. Approaches that begin with the brain can be described as 'brain-based' and follow specific methodological strategies to approach the mind–brain problem that allow consideration of the brain in different domains, as well as in its relation to the body and the world. The discussion of brain-based approaches to the mind–brain problem is the focus of this chapter.

Objectives

- Understand the difference between mind- and brain-based approaches to the mind–brain problem
- Define the property P and why we cannot know it
- Distinguish McGinn's and Nagel's approaches to the mind–brain problem
- Explain the concept of the 'deep interior of the brain'
- Understand how Descartes considered the body
- Explain Merleau-Ponty's various concepts of the body
- Understand the implications of embodiment for the mind–brain problem
- Explain the thought experiments of the brain in the vat
- Know what semantic or content externalism amount to, including the concept of the extended mind
- Distinguish the brain as subject from the brain as object
- Distinguish between different concepts of the brain and understand their rationales
- Explain and define different concepts of the brain
- Explain what a philosophy of brain is

Key concepts

Property P, 'deep interior of the brain', mental and physical concepts, lived body, Merleau-Ponty, brain in the vat, extended mind, semantic or content externalism, Schopenhauer, brain as subject of cognition, brain as object of cognition, brain paradox, philosophy of brain.

Methodological background 1

'Domain monism' versus 'domain pluralism' in the mind–brain problem

Let us briefly recapitulate the first three chapters of this Part. Chapter 5 focused on mental approaches to the mind. The mind was determined by a mental existence as distinguished from the physical brain and body. This led the French philosopher René Descartes to assume mental and physical substances that were supposed to interact with each other in a small structure in the brain called the pineal gland. Descartes' position can be described as interactive substance dualism. While this position is rarely advocated today, the assumption of the mind's special mental existence and reality as distinguished from the brain and the body is still maintained in different variations (see Chapter 1).

The assumption of the mind's special existence and reality as mental was criticized by those who advocated physical approaches to the mind–brain problem. Physical approaches, the focus of Chapter 6, assume the mind's existence and reality to be as physical as the brain and the body. In this model, the mind is considered identical, or reducible to, the brain and the body. Functionalism – one type of physical approach – compares the relationship between the mind and the brain to the relationship between the software and hardware of a computer. Since both physical and mental approaches lead to various problems, alternative non-mental and non-physical approaches have been suggested. These either assume a reality and existence separate from the mental and the physical (as in neutral monism), or they extend the concept of the physical by distinguishing between first- and third-person physical processes (as in biological naturalism by John Searle).

Finally, in Chapter 7, based on the philosophy of German philosopher Immanuel Kant, the metaphysical focus was relinquished in favor of more epistemic considerations of the mind–brain problem. There was increasing concern with the kinds of epistemic abilities that are necessary to access and know the mind's underlying existence and reality.

What exactly did Kant do? He shifted the investigation of the mind from the metaphysical to the epistemic domain. This implied a different methodological strategy and raised different types of questions. Most importantly, it tells us that even if the metaphysical domain is usually considered the 'home domain' of the mind–brain problem, one may nevertheless discuss the mind–brain problem in the context of a different domain, including the epistemic or even the empirical domain. One thus speaks of domain pluralism rather than domain monism of the mind–brain problem (see Chapter 4 for the distinction between domain monism and domain pluralism).

Methodological background 2

'Fixed domain location' versus 'flexible domain location' of the mind–brain problem

This has important implications. Such domain pluralism means that the traditional association of the mind–brain problem with the metaphysical domain is not 'written in stone'. We can refer to this as the 'flexible domain location' of the mind–brain problem. The concept of 'flexible domain location' means that, methodologically, the mind–brain problem can be considered in different domains, including metaphysical, epistemic, or empirical domains thus entailing domain pluralism rather than domain monism. Such 'flexible domain location' must be distinguished from what can be described as 'fixed domain location', where the mind–brain problem is situated in only one fixed domain. This 'fixed domain' is usually the metaphysical domain as it is presupposed most often in the current philosophical discussion.

Both the philosophers and the neuroscientists among you may now be puzzled. The philosopher usually takes the mind–brain problem as a metaphysical problem; this is the way it is approached especially in current philosophy of mind. In contrast, the neuroscientist often considers the mind–brain problem to be an empirical problem that can be resolved by brain research. Both thus presuppose domain monism rather than domain pluralism, and fixed rather than flexible domain location. The main difference in their methodological approach consists in the fact that they associate the mind–brain problem with different domains, metaphysical versus empirical.

Such 'domain monism' and 'fixed domain location' (in reductive neurophilosophy and traditional philosophy) must be distinguished from a (non-reductive) neurophilosophical approach that includes multiple domains entailing domain pluralism (as in non-reductive neurophilosophy; see Chapter 4). Here the focus on a single domain is replaced by the consideration of several or multiple domains. Domain pluralism replaces domain monism as methodological strategy.

Domain pluralism implies that the mind–brain problem is no longer situated in one fixed domain but rather flexibly in several domains. Hence the suggested domain pluralism and flexible domain location can be regarded as a methodological strategy that is characteristic of a neurophilosophical approach to the mind–brain problem as distinguished from both philosophical and neuroscientific approaches.

Methodological background 3

'Mind-based approach' to the mind–brain problem

We pointed to one feature of the methodological strategy in mind–brain discussion, the 'domain location'. This allowed us to characterize the current mind–brain discussion by 'domain pluralism' and 'flexible domain location'. There is yet another methodological feature that needs to be considered, namely the distinction between a mind-versus a brain-based approach. Let us start by explaining the mind-based approach.

What is a mind-based approach? Despite their differences, physical, mental and non-mental and non-physical approaches nevertheless share a point of departure. They all take the mind and its mental features as their starting point when tacitly presupposing the following question: how is the mind and its mental features related to the brain and its physical features? This question serves as a guide for the investigation of the mind–brain problem. Since the starting point here is the mind, these approaches can be characterized methodologically as 'mind-based'.

Let us differentiate between mental and physical approaches. The mind and its mental features – including consciousness, self, free will, etc. – serve as the point of departure for any metaphysical assumptions about the existence and reality of mind and brain in mental approaches. The mind serves here as a positive template for any subsequent metaphysical assumption. These approaches are thus mind-based in a positive way: the mind and its mental features are taken as the starting point to infer the mind's existence and reality as well as its relationship to brain and body.

Even if the mind's special existence and reality is denied, as in physical approaches, the mind nevertheless still serves as the methodological starting point even though in this case as a negative one. The same holds true for non-mental and non-physical approaches, where the concept of mind still serves as a negative template for the characterization of both the mental and the physical. These approaches are mind-based in a negative way: the concept of mind still serves as a template or foil, though as a negative one to infer the metaphysical relationship of mental features to the physical features of brain and body.

One may now want to interject that a physical approach to the mind–brain problem cannot be mind-based. Why? The physical approach rejects the mind and its mental features as a separate reality and claims them to be identical or reduced to the brain and its physical features. The physical approach must thus be brain-based rather than mind-based. This, however, is to confuse metaphysical and methodological domains. Metaphysically, the physical approach is indeed brain-based rather than mind-based, if not brain-reductive (meaning that it reduces mental features to the physical features of the brain; see Chapter 6).

Such a metaphysical claim of a brain-reductive approach, however, needs to be distinguished from a methodological claim. A methodological claim describes the strategy, how an approach proceeds in its investigation of the mind–brain problem independently of whether it characterizes the mind as physical or not in metaphysical terms. One feature of the methodological strategy is the point of departure and the direction of the investigation: that starting point is, as pointed out, the mind and it is from the mind to the brain that the physical approaches proceed methodologically. Methodologically, the physical approach to the mind–brain problem can therefore be characterized as mind-based even if metaphysically it is brain-reductive.

Methodological background 4

'Mind-based versus brain-based approaches' to the mind–brain problem

Most current philosophical approaches to the mind–brain problem, including both dualism (Chapter 5) and materialism (Chapter 6), can be characterized as mind-based. They all presuppose the mind as a positive or negative template or foil in their methodological approach to investigating the metaphysical relationship between mental and physical features. An exception to the mind-based treatment of the mind–brain problem is functionalism. Functionalism uses the computer – along with the relationship between hardware and software – as its starting point in considering the mind–brain relationship (Chapter 6). In this sense, functionalism can be considered computer-based: the computer and its characterization by software and hardware is taken as a model or template on the basis of which the mind–brain problem is discussed.

What does the example of functionalism tell us? In the same way that Kant told us that the 'domain location' of the mind–brain problem is not fixed in the metaphysical domain, functionalism shows that the mind-based methodological strategy is not 'written in stone' either. Functionalism replaces the mind as the methodological starting point with the computer. Functionalism can thus be characterized methodologically as computer-based, rather than mind-based.

You may now be puzzled. While the mind–brain problem is about the mind and the brain, only the first, the mind, is usually taken as the methodological starting point for most approaches (with the exception of functionalism). The current discussion about the mind–brain problem is thus mostly mind-based (or computer-based as in functionalism). This is reflected in the question that guides any investigation of the mind–brain relationship in current philosophy: 'How are the mind and its mental features related to the brain and its neuronal features?'

One could, however, also imagine the opposite methodological strategy. Rather than starting with the mind and using its template or foil, one could start with the brain and proceed from there. Here, the guiding question would be: 'How are the brain and its neuronal features related to the mind and its mental features?' Instead of basing one's methodological strategy on the mind, one would then take the brain as the methodological starting point. Instead of a mind-based approach, one would then need to speak of a 'brain-based approach'.

You may now want to argue that it makes no difference whether one goes from mind to brain or from brain to mind, the outcome will always be the same. More specifically, one would think that the methodological approach, the starting point, should make no difference and lead to the same solution of the mind–brain problem in either case. Does the methodological difference between brain-based and mind-based approaches ultimately make no difference for the metaphysical characterization of the mind and the brain? For the time being, we leave open the answer to this question.

While the philosopher traditionally approaches the mind–brain problem in a mind-based way, the neurophilosopher may want to take the reverse stance, approaching the mind–brain problem in a brain-based way. This, as we will see, raises novel questions. One such question concerns the characterization of the brain itself and in which domain we can investigate it. The neuroscientist and the philosopher alike may now want to say that the answer to that question is easy. The brain must be considered exclusively in the empirical domain. This, however, as we will see, is not as clear as it seems, especially given some philosophers who shift the brain into, for instance, metaphysical or epistemological domains (see below).

Methodological background 5

Brain-based approach versus physicalism

What would a brain-based approach to the mind–brain problem look like? Rather than starting with the mind and mental features, one would take the brain and its specific features as a starting point. In order to accomplish this, the brain and its intrinsic features need to be revealed and defined. Based on the above-mentioned flexibility of the presupposed 'domain location', the brain, when conceived of as a starting point, may be investigated in either the metaphysical, the empirical, or the epistemological domain. How, then, should we define the concept of the brain and its various neural features? The answer to this question requires an extensive and rich discussion.

That the answer is complicated might be puzzling. After all, a brain-based approach seems to imply that the mind is not more than the brain, a physical organ with no special mental substances. Is a brain-based approach thus nothing but a metaphor for a brain-reductive approach and hence ultimately for physicalism or materialism?

Not necessarily. A brain-based approach describes a methodological strategy – more specifically the point of departure from which one considers the mind–brain problem. This methodological strategy should not be confused with a metaphysical claim, like assuming that the mind can be reduced to the brain and its physical features, as in physicalism or materialism. Hence, to characterize a brain-based approach as physicalistic or materialistic is to conflate method and metaphysics and thus the strategy to approach a question with the answer to that question.

Methodological background 6

Neurophilosophical versus philosophical approaches to the mind–brain problem

Let us summarize what we have discussed so far. We pointed out some methodological features in the current philosophical approaches to the mind–brain problem and indicated some possible neurophilosophical alternatives. One methodological hallmark feature in the current mind–brain discussion is the presupposition of one particular domain, domain monism, and the fixed association of the mind–brain problem with that domain, usually the metaphysical domain, entailing 'fixed domain location'. This is accompanied by the (implicit or explicit) presupposition of the concept of mind as a (positive or negative) template as it is reflected in the guiding question, namely how mental features are related to the physical features of brain and body.

Such a philosophical approach is contrasted with a neurophilosophical approach to the mind–brain problem. Here the domain monism and the fixed domain location are replaced by domain pluralism and flexible domain location. In addition, the brain rather than the mind serves as the template to approach the mind–brain problem: rather than asking how mental features are related to physical features, one reverses the direction and asks how physical and neuronal features are related to mental features. This implies a shift in the methodological strategy from a mind-based to a brain-based approach.

The aim of this chapter is to pursue such a neurophilosophical (rather than philosophical) approach to the mind–brain problem. The focus is thus on brain-based approaches to the mind–brain problem that consider the brain in either metaphysical, empirical, or epistemological domains as the starting point for any subsequent discussion about the brain's relationship to the mind. This chapter will reveal different conceptualizations and definitions of the brain which will lead us beyond the brain itself to the body and the environment. These can be called embodiment and embeddedness, respectively.

Biography 1

Colin McGinn

The brain as a topic for discussion can be considered using the particular lenses of different domains, including the empirical, metaphysical and epistemic domains. Most frequently, however, the brain is associated with the empirical domain, the domain of observation in third-person perspective. The brain in the empirical domain – observable in terms of physical features – is then taken to be identical with the brain in the metaphysical domain, where its existence and reality is determined as physical. In other words, there is no difference between the brain in the empirical domain – the brain as we observe it – and the brain in the metaphysical domain.

One may, however, also suggest a discrepancy between the empirical and metaphysical characterization of the brain. Empirically, the brain is characterized by its physical features, like its neuronal states. This does not, however, imply that metaphysically the brain must also be determined by physical existence and reality. The existence and reality of the brain in the metaphysical domain can differ from the way we can observe the brain in the empirical domain. In other words, empirical and metaphysical characterizations of the brain do not need to be co-extensive. This discrepancy between the brain's characterizations in the empirical and metaphysical domain is, for instance, suggested by the British American philosopher Colin McGinn.

Colin McGinn (1950–) is a philosopher, originally from England, who now teaches at the University of Miami in Florida. He is best known for his work in philosophy of mind, as well as his intellectual autobiography and a novel called *The Space Trap*. Most recently his has written a book with the interesting and attention-grabbing title, *Mindfucking*.

Brain in the metaphysical domain 1

Property P and the brain (McGinn)

How does McGinn approach the mind–brain problem? McGinn assumes that consciousness can be characterized by two properties, which he has termed surface and hidden properties. The surface properties are the phenomenal properties accessible via introspection. An example of this is when we become aware of the main phenomenal features of our experience: subjectivity, intentionality and quality. Surface properties are thus those that characterize our experience and consciousness. They will be discussed in detail in Part IV.

In addition to surface properties, McGinn assumes a 'hidden property'– the so-called 'property P'. Property P is not a physical property. Instead, he assumes property P to be a property that is sandwiched between phenomenal and physical properties. As such, property P allows for the transformation of the physical states of the brain into the phenomenal states of consciousness. Why does McGinn assume property P, rather than reducing mental states to physical states? Because mental states, unlike physical states, are essentially non-spatial and therefore not extended into space. Since a spatial or physical state cannot account for something non-spatial like the mental state, McGinn argues for the need to assume a separate property, property P, that can account for the non-spatial nature of mental states.

What does property P look like? McGinn compares the hidden property P with other hidden structures in other disciplines. Examples include the atomic and subatomic structure of matter, and the curved structure of space–time in relativity theory. Without these hidden structures we cannot understand the surface properties of physical objects. Another example of a hidden property is the molecular structure of DNA, our genome. Without the molecular structure of the DNA, we cannot understand macroscopic physiological mechanisms. The syntactic structure of our language is another example of a hidden structure that lies beneath the surface of our sentences and linguistic expressions.

As in his analogous examples from other disciplines, McGinn suggests property P to be sandwiched between physical and phenomenal states. As such, property P can transform one state into another, for example, the physical into the phenomenal. McGinn says: 'it would be situated somewhere in between them. Neither phenomenological nor physical, this mediating level would not (by definition) be fashioned on the model of either side of the divide, and hence would not find itself unable to reach out to the other side ... The operative properties would be neither at the phenomenal surface nor right down there with the physical hardware; they would be genuinely deep and yet they would not simply coincide with physical properties of the brain. Somehow they would make perfect sense of the psychophysiological nexus, releasing us from the impasse that seems endemic to the topic. They really would explain how it is that chunks of matter can develop an inner life' (McGinn, 1991, pp. 103–4; see also p. 100).

What does McGinn's approach presuppose in methodological regard? On the whole McGinn claims that we need to assume a hidden structure in the brain itself to explain the transformation of the brain's physical states into the phenomenal states of consciousness. His postulated need to start with the brain itself and to search for specific properties in the brain from which we could infer the mind and its mental features puts McGinn's approach close to a brain-based approach. He searches in the brain itself and discusses its neuronal (and other non-neuronal) features. This distinguishes him from mind-based approaches that determine and define the mind and its mental features first in order to then either distinguish it from or accommodate the brain and body and their physical features.

Brain in the metaphysical domain 2

Access to and knowledge of property P (McGinn)

The hidden character of property P as located and sandwiched between phenomenal and physical states raises an important epistemological question: how can we access and know anything about this property P? McGinn argues that we cannot know anything about it. Why? Because we simply lack the 'right' 'epistemic instruments and tools' to acquire any knowledge about it.

Let us be more specific. We can perceive the world. To perceive the world means that we can access the physical properties of objects in the natural world. I can perceive the table as a physical object. I can observe the tree standing in front of my balcony as a physical object. But I cannot perceive anything else beyond that. From this perspective, my perception is limited to the realm of the physical, thereby excluding mental (phenomenal) features.

In addition to our perceptual abilities, we also have introspective abilities. I can become aware of my own mental states as mental states. To do this provides access to what are described as phenomenal features (another way to say mental states?). The phenomenal features of mental states include subjectivity, point of view, intentionality and qualia (see Part IV for more detailed discussion).

How, though, can we access property P, which, according to McGinn, is vital in transforming the physical into phenomenal features? McGinn argues that we have available only two kinds of different concepts, physical concepts and mental concepts. We use physical concepts to describe the physical properties in our perceptions. Since our perceptions are spatial, the physical concepts are intrinsically spatial (otherwise they could not describe the perceptions and their spatial nature).

What about the mental concepts we use to describe our introspections? In contrast to physical concepts, mental concepts are non-spatial. Because we lack any positive characterization of non-spatial nature by itself, our mental concepts refer to the non-spatial nature of mental features only in a negative way. A positive characterization of the non-spatial nature of mental features remains elusive to us, however.

Taken together, we have two kinds of different concepts – mental and physical – which seem mutually exclusive. Why? Because mental concepts are non-spatial (in a negative sense) and physical concepts are spatial (in a positive sense). What is lacking are concepts that describe the non-spatial nature of mental features in a positive sense. This means that our differentiation between mental and physical concepts does not provide the kind of conceptual equipment necessary to properly describe property P and its role in transforming spatial physical states into non-spatial mental states.

Let us explain this further. To define property P and its transformative role we need a concept that integrates and links spatial and non-spatial features. Because we remain unable to move beyond perception and introspection as the dominant modes of inquiry, however, we remain unable to develop integrative concepts that bridge the gap between spatial and non-spatial descriptions. In short, a bridge concept is lacking in our 'epistemic tools and instruments'.

McGinn argues that this lack of a bridge concept means that property P will remain inaccessible to our understanding through introspection or perception. And because we have no other epistemic abilities beyond perception and introspection, property P cannot be known in principle. We therefore have no way of acquiring any knowledge about how property P works and what its features are. McGinn calls this being 'cognitively closed', which describes a basic epistemic inability on our side, meaning that we remain principally unable to access and know property P. The assumption of cognitive closure illustrates the degree to which property P and its features remain mysterious.

Moreover, even the method McGinn uses to assume and describe property P remains opaque, as any assumption of property P already presupposes some kind of access and knowledge of it, which McGinn claims to be impossible. It is in this way that McGinn seems to contradict himself. One might ask, however, how McGinn can assume that property P is cognitively closed without having any knowledge of it?

In sum, the positive (rather than negative) characterization of property P, as well as the methodological strategy to obtain that knowledge, remain unclear in McGinn's account. Because of the unclear definition of property P, his position has therefore been described and classified as mysterious, or as mysterianism in current philosophy of mind.

Critical reflection 1

McGinn versus Kant I

How can we describe McGinn's position in further methodological detail? McGinn combines a metaphysical claim with an epistemological claim. The metaphysical claim consists in the assumption that a special property, property P, exists, is real, and is sandwiched between physical and phenomenal features. How about the epistemological claim?

The epistemological claim is the assumption that property P is inaccessible and thus unknowable because of our own epistemic limitations. This focus on epistemic limitation is similar in some ways to the philosophy of Immanuel Kant, who also postulated basic epistemic limitations. Kant refrained from making metaphysical assumptions (see Chapter 7) about the existence and reality of mind and world precisely because he believed that it was impossible to access or know anything beyond the world as we can experience and observe it on the basis of our epistemic capacities. McGinn, in contrast, makes a metaphysical assumption when he postulates property P, on the basis of postulated epistemic limitations. Unlike Kant, McGinn presupposes that we are able to draw metaphysical inferences from epistemic claims and thus from our knowledge of the world. Accordingly, what Kant would have claimed to be an epistemic–metaphysical fallacy, McGinn considers as valid epistemic–metaphysical inference.

An imaginary dialogue between McGinn and Kant would be very interesting. Let's imagine what that might sound like in the context of the twentieth century.

Kant: You seem to have an interesting theory about the mind. You argue that neither perception nor introspection allows us any access and knowledge of the mind.

McGinn: Yes, Mr Kant, that is indeed so. In that respect I learned a lot from you and your *Critique of Pure Reason* and its insistence on our principal epistemic limitations.

Kant: I can see that. But what I do not see is how you can postulate this strange property you call property P. What are you talking about?

McGinn: It's very simple. There must be some deeper property in our brain that allows the transformation of physical states to phenomenal states. Metaphorically speaking, this deeper property must serve as a bridge between physical and phenomenal states. And because physical states do not entail phenomenal states and phenomenal states do not refer to physical states, that bridge must be described as a third property that is non-physical and non-phenomenal. Hence, the only way this type of transformation is possible is to assume a third property, property P, that as an intrinsic feature could be compared, if one wishes, with what you, Mr Kant, described as categories.

Critical reflection 2

McGinn versus Kant II

Kant: Intuitively, your approach is appealing since it seems to solve the question how mental states can be generated on the basis of physical states. However, there is a serious flaw to your approach. How can you

	assume the existence and reality of property P, if we cannot know anything about it?
McGinn:	I do not understand what you are asking me, Mr Kant.
Kant:	You argue in favor of the existence and reality of a special property called property P. This assumption presupposes some kind of knowledge about its existence and reality, namely metaphysical knowledge. How can you acquire this knowledge if, at the same time, you suffer from an epistemic limitation that prevents you from acquiring any knowledge about it?
McGinn:	You mean that even the knowledge about the bare existence and reality of property P is already too much and should remain impossible given our principal epistemic limitations?
Kant:	Yes, exactly that. You cannot argue in favor of the existence and reality of property P because you are unable to know anything about it. You cannot know if it really exists, or if it is even independent of your own mind.
McGinn:	What you say means that we cannot know anything at all about the mind's existence and reality. It means it is impossible to make any metaphysical claims whatsoever.
Kant:	Exactly. That is why I refrained from making metaphysical assumptions. It is of my opinion that they are completely speculative.
McGinn:	What is that?
Kant:	You also presuppose that property P is located in the brain. How can you know that? You claim that we cannot know property P due to our epistemic limitations. Now you claim to know property P when you say that it is located in the brain. You must thus have some knowledge about property P. How, however, is that possible, given your claim that we cannot know property P?
McGinn:	Why is that a problem?
Kant:	How can you know that property P is 'located' in the brain? You claim that our 'epistemic instruments and tools' include nothing but perception and introspection. Perception is unable to demonstrate any hidden property in the brain and only reveals surface properties like neuronal states and activity. Introspection is likewise unable to provide you any insight into the brain at all, since you do not introspect your brain and its neuronal states.
McGinn:	My goodness, I knew that you were radical Mr Kant. But that radical?
Kant:	This has nothing to do with being radical. Nothing but mere logic.
McGinn:	Wait, wait. If you are this radical about the epistemic limits, I have to put forward an argument against you. You claim not to know anything about existence and reality, given our epistemic limitations. And you claim that you know that you do not know anything about existence and reality, correct? You thus have some knowledge about your own epistemic limitations. Doesn't the knowledge of the epistemic limitations already presuppose some kind of knowledge about an underlying existence and reality? If you claim that you cannot know anything about the property P, you must at least presuppose that it exists and is real. Otherwise you remain unable even to make your claim of an epistemic limitation. And that is exactly what I did in the case of property P.

Kant: Bad, bad, bad, that is what happens when metaphysics comes in and takes over and hijacks the epistemic domain. You will end with nothing but pure and groundless metaphysical speculation. Now I understand why they describe your position as mysterious.

Biography 2

Thomas Nagel

How can we overcome the principal difference between mental and physical concepts? Immanuel Kant and Colin McGinn would probably answer that we remain unable to do so because of our epistemic limitations. This skeptical stance is not supported by the American philosopher, Thomas Nagel (1937–). Nagel was born in Belgrade in the former Yugoslavia but moved early in his life to the United States. He was educated there and later took a position at New York University in New York.

Nagel covered many philosophical topics in his writings, ranging from mind–brain issues and consciousness to ethical and political questions. All revolve around the assumption of a basic subjectivity (as opposed to mere objectivity) that provides the guiding thread throughout the different philosophical issues. How can we describe such basic subjectivity, and how is it, for instance, manifest in consciousness? Nagel argues that such basic subjectivity is manifested in a certain quality, the so-called qualia that characterize our consciousness (see Chapters 13 and 16 for details). That quality can be described as 'what it is like', which led Nagel to title one of his most famous articles 'What is it like to be a bat?'

Brain at the border between metaphysical and empirical domains 1

The 'deep interior of the brain' (Nagel)

How does Nagel associate his insistence on a basic subjectivity with the brain and its relationship to the mind? Let's recall that mental concepts are intrinsically subjective because they refer to subjective experience from a particular point of view. Physical concepts, on the other hand, are objective because they refer to observation and do not imply any particular point of view.

Most importantly, the mental concept does not include any reference to the objective nor, conversely, do physical concepts imply any link to the subjective. Hence, mental and physical concepts are mutually exclusive and are therefore only contingently related. A concept that bridges the difference between mental and physical concepts must consequently be able to explain the link between the subjective – the mental – and the objective – the physical.

Nagel assumes that this subjective–objective link is not merely contingent. Instead, he assumes this link between the subjective and the objective to be necessary and unavoidable. What this means here is that the purely objective physical features of the brain cannot avoid yielding the mental features of the mind, including their subjective nature. In the same way that your heart cannot but produce a heartbeat, your brain and its physical features cannot avoid generating mental states. Why this might be the case currently remains unclear.

Where can we search for this necessary truth? Nagel points to the 'deep interior of the brain', where we may detect some as yet unknown properties. The exact methodological strategy for investigating the as yet unknown properties in the 'deep interior of the brain' remains unclear at this point in time.

Brain at the border between metaphysical and empirical domains 2

Characterization of the 'deep interior of the brain' (Nagel)

How can we characterize the 'deep interior of the brain' in further detail? One could characterize it in empirical terms, as well as search for specific neuronal mechanisms and features that currently remain hidden to us. We will see that the brain's intrinsic activity – its resting state activity (as distinguished from its stimulus-induced activity as related to extrinsic stimuli) – may be a viable empirical (neuronal) candidate for what Nagel describes as the 'deep interior of the brain'.

The brain's intrinsic activity is the basis of any extrinsic stimulus-induced activity. No extrinsic stimulus can induce any stimulus-induced activity in the brain without interacting with the brain's intrinsic activity. The brain's intrinsic activity may then indeed be considered an empirical candidate for what, as pointed out above, Nagel describes as the 'deep interior of the brain' and its central role in yielding mental states. For that, however, we need to describe the brain's intrinsic activity in much more detail. This will be done in Parts III–V.

In addition to its empirical characterization, the 'deep interior of the brain' can also be characterized conceptually. How can we account for the 'deep interior of the brain' in conceptual terms? The necessary generation of mental features by the brain's physical features stands in contrast to the mere contingent linkage between mental and physical concepts. We thus may need to revise our concepts in a revolutionary way to allow for a necessary link between subjective and objective features in mental and physical concepts. Only when we can account for the necessary subjective–objective linkage in our concepts, will we be able to begin to understand the necessary generation of mental states by the brain's physical features.

How does Nagel's position compare to McGinn's? Like McGinn, he starts with the brain and searches for features in the brain itself that could account for the mind and its mental features. The methodological starting point is thus the brain. Given our earlier discussion, we can therefore characterize Nagel's approach methodologically as brain-based, rather than mind-based.

Unlike McGinn, Nagel does not postulate a special, inaccessible property. Instead, Nagel leaves open the possibility that certain mechanisms or features in the 'deep interior of the brain' will be discovered through future research. This means that Nagel seems to localize the main problem right at the border between metaphysical and empirical domains. More specifically, the assumption of specific features in the 'deep interior of the brain' seem to characterize the brain as brain and may thus signify its existence and reality as brain. One could thus speak of a metaphysical (or, better, ontological) characterization of the brain. Since, however, Nagel does not exclude the possibility that these features in the 'deep interior of the brain' can be observed empirically, his characterization of the brain must be located methodologically right at the border between the metaphysical and empirical domains.

What does this 'location' of the brain and the mind–brain problem at the border between metaphysical and empirical domains imply for our future methodological strategy? First and foremost, it opens the 'deep interior of the brain' to empirical-experimental investigation. Even if mysterious at this point in time, the 'deep interior of the brain' may become less mysterious once the 'right' empirical discoveries are made. Hence, metaphorically speaking, Nagel offers some hope that, in the future, we may find the 'right' key to unlock the brain and open the door to how it relates to mental features. This is different in the case of McGinn's property P, which will always and thus principally remain mysterious by default. Accordingly, unlike Nagel, McGinn does not offer any hope of us ever being able to find the 'right' key to unlock the brain and open the door to its relationship with mental features.

There is, however, more to Nagel's characterization of the brain by features in the 'deep interior of the brain'. In addition to empirical investigation, we may also need to discuss the implications of the 'deep interior of the brain' for our concepts like the mental and the physical. If we can indeed find some empirical support for the 'deep interior of the brain' in the sense of Nagel, our traditional definitions of the concepts of the mental and the physical need to be revised in a rather radical way. The discoveries in the 'deep interior of the brain' may then serve as a methodological starting point for conceptual discussion of the mind–brain relation. We will see later in this chapter and in Chapter 12 how we can further characterize the 'deep interior of the brain' in both empirical and conceptual regard.

Biography 3

Arthur Schopenhauer

So far we have discussed brain-based approaches to the mind–brain problem in the metaphysical domain (McGinn) and at the border between metaphysical and empirical domains (Nagel). Are there any approaches that start with the brain in the epistemic domain? Yes. German philosopher, Arthur Schopenhauer, did precisely that.

Arthur Schopenhauer (1788–1860) was born in Gdansk, then part of German-speaking Prussia. Today this region is part of Poland. Schopenhauer became a famous philosopher thanks especially to his book, *The World as Will and Idea* (1818/1819). However, his university career in Berlin never took off – in part because his lectures were scheduled at the same time as G. W. Hegel's. Schopenhauer was from a wealthy merchant family, so he left the university. He lived and wrote on philosophy without a teaching job for the rest of his life, in Frankfurt, Germany.

His outlook on the world was rather pessimistic: he characterized the human condition as painful and full of suffering. For Schopenhauer, the only way to get some relief was through art. It is for this reason that he is often especially popular among artists and authors.

Brain in the epistemological domain 1

Schopenhauer and the brain

In order to understand Schopenhauer, we must return briefly to one of his inspirations, Immanuel Kant. Kant, as we recall from the previous chapter, was a German philosopher who postulated that it is impossible for us to access and know the underlying existence and reality of our minds because of our epistemic limitations. Kant therefore rejected the 'location' of the mind–brain problem in the metaphysical domain, and instead localized it in the epistemic domain (see Chapter 7).

Why do we have these epistemic limitations? Because, according to Kant, the mind has intrinsic features, including a specific form, structure and organization. This makes it possible for the mind to impact and impose itself upon the processing of extrinsic stimuli from the environment, which the mind structures and organizes. Because the mind does this structuring and organizing, the extrinsic stimuli can be associated with consciousness and a self. Accordingly, Kant argued that the mind's form, structure, or organization is manifest in various functions or faculties like consciousness and self-consciousness, which otherwise – without the mind's input – would remain impossible.

How does Schopenhauer relate to this position? He took Kant's point that there must be some kind of form, structure and organization that is necessary and thus indispensable to generate consciousness and self. However, unlike Kant, he did not 'locate' this form with the mind, but associated it with the brain itself. Rather than revealing the mind and its various faculties, Kant investigated the 'share of the brain functions' (though not under this name) (Schopenhauer, 1966, I, p. 418).

Schopenhauer's shift from mind to brain has serious consequences. Kant attributed various faculties to the mind. These are now associated with the brain and the faculties of the brain. Today, the assumption that faculties like consciousness and self-consciousness are attributable to the brain is nearly commonplace. However, at the time of Schopenhauer, it was a daring statement, since scientific investigation of the brain had just started (see Chapter 3). Schopenhauer can be credited as the first person to introduce the brain into philosophy and its associated epistemological domain. One might also consider him the first neurophilosopher *'avant la lettre'* ('neurophilosopher prior to the invention of the term').

Brain in the epistemological domain 2

Brain as subject and brain as object of cognition

Kant associated the mind's own input – its faculties – with our ability to cognize the mind and the world. The mind, in this model, is thus the subject of cognition. This idea changed once Schopenhauer shifted these faculties – our cognition of the world – from the mind to the brain. Now the brain itself becomes the subject of cognition: rather than the mind, it is the brain that perceives and cognizes the world. In this model, the brain is the subject of cognition. One may thus speak of the 'brain as subject of cognition'.

How can we further characterize the 'brain as subject of cognition'? One may at first glance be puzzled. The brain as we observe it is an object. All we observe is a mere 'gray pulpy mass', as Schopenhauer himself said. In his time, the brain could only be investigated in anatomy. Today, we have brain imaging that allows us to investigate the brain electronically as it functions. This produces colorful images of the brain, which will be discussed further in Parts IV and V. However, neither in the gray pulpy mass nor in the colorful images of the brain can we detect any mental states; no consciousness and no self-consciousness. As in Schopenhauer's day, we still remain unable to locate any mental states, their contents and their respective meaning in the brain we observe and investigate in neuroscience, the 'brain as object of cognition'.

How does the 'brain as object of cognition' relate to the 'brain as subject of cognition'? The impossibility of discovering mental states in the 'brain as object of cognition' implies that it must be different from the 'brain as subject of cognition'. The 'brain as subject of cognition' should contain mental states and all the faculties that Kant attributed to the mind and that Schopenhauer shifted to the brain. However, neither of these faculties can be observed in the 'brain as object of cognition'. This leaves open the question: how do we characterize the 'brain as subject of cognition' and distinguish it from the 'brain as object of cognition'?

Do we have two different brains, the 'brain as subject of cognition' and the 'brain as object of cognition'? No, says Schopenhauer, we have only one brain. But our one brain can be considered in different ways, or what one might call different epistemic modes. One mode of consideration is the mode of third-person perspective or observation. This mode reflects the 'brain as object of cognition' and is 'located' in the empirical domain. This is the brain we research in neuroscience, where it is visualized as a gray pulpy mass or in terms of colorful images as described above.

The alternative mode is the mode of experience in first-person perspective that, following Schopenhauer, corresponds to the 'brain as subject of cognition'. Since the 'brain as subject of cognition' is associated with the acquisition of knowledge, it may be associated with the epistemological domain – the domain of knowledge. The brain is distinguished from the empirical domain of the 'brain as object of cognition'. What does the 'brain as subject' look like? We currently do not know.

Ideas for future research 1

'Brain-paradox'

Do we have two different brains? One brain as the subject of cognition and one as the object of cognition? Is the brain that we observe, the brain as object of cognition, thus different from the brain on whose basis we are able to experience and have consciousness, the brain as subject of cognition? How is this possible? Critics argue that it is logically impossible. One and the same brain cannot simultaneously be both subject and object of cognition. Either the brain is the 'subject of cognition' or the brain is the 'object of cognition'. The double characterization of the brain as both the subject and the object of cognition is logically contradictory and thus paradoxical, implying what is described as the 'brain paradox'.

How can we resolve the 'brain paradox'? One possible solution is to refrain from using the double characterization of the brain as both the subject and the object of cognition. The brain is just the object of cognition; it is no longer the subject of cognition. What, then, is the subject of cognition? Here, it is useful to introduce the concept of mind. The mind is the subject of cognition. This is the tacit and usually implicit presupposition of Kant and most present-day approaches in current philosophy of mind. In contrast, the body and the brain can only be the objects of cognition, and can only be investigated in the sciences, including neuroscience.

The introduction of the mind also solves another puzzle. If the brain is the subject of cognition, it should be able to feel, think, etc. That, however, according to the proponents of the mind, is not the case. It is the person, not his/her brain, that feels, thinks, etc. The British philosopher Peter Hacker and the Australian neuroscientist, Max Bennett, highlight this point in their book, *Philosophical Foundations of Neuroscience* (Bennett and Hacker, 2003). They speak of a 'mereological fallacy', since the assumption of the brain feeling, thinking, etc., confuses a part (the brain itself) with a whole (the subject). The concept of a 'mereological fallacy' is based on the concept of mereology, which describes the theory of the relation between parts and wholes. In other words, we should not assume the brain to be the subject of our thoughts, feelings, perceptions, etc., since that confuses the person as a whole with the brain as one of its parts.

These considerations show that Schopenhauer's problem with replacing Kant's concept of mind by the brain leads us back to what he wanted to escape from: the mind as distinguished from the brain. This in turn invariably raises the question of how such a mind is related to the brain. Accordingly, the introduction of the mind solves one problem, the brain-paradox, but introduces another, the mind–brain problem.

Ideas for future research 2

Shift from the 'brain-paradox' and the mind–brain problem

Why is the shift from the brain-paradox to the mind–brain problem important for us to consider? First and foremost, it tells us that the mind–brain problem is not 'written in stone'. Rather, it depends on our background assumptions, including that the brain itself cannot be the 'subject of cognition' since it does not feel, think, etc. by itself. Second, it tells us that Schopenhauer's solution of 'locating' the Kantian concept of form in the brain rather than the mind leads us back to the mind itself. If we want to escape from the mind and the mind–brain problem, we thus have to find other solutions rather than merely shifting Kant's concept of form from the mind to the brain, as Schopenhauer suggested.

Third, it raises the question of whether the brain-paradox cannot be solved in other ways beyond reverting back to the mind–brain problem. Since we have seen that the mind–brain problem is of itself rather tricky, if not unsolvable, as postulated by some philosophers like Colin McGinn (see above), we may want to refrain from introducing the concept of mind. We may thus need an alternative strategy to escape from the 'brain paradox'.

Doubling the brain and assuming two different brains, one brain as subject and the other as object, is no option either because there is simply only one brain. One could, for instance, associate the brain's right and left hemispheres with the brain as subject and object of cognition. The left hemisphere as the subject of cognition would then cognize the right hemisphere as the object of cognition. That, however, sounds absurd in itself. And it is also not empirically plausible since both hemispheres act closely together as one functional unity. Hence, doubling the brain in its existence and reality, i.e. in a metaphysical way, is no option either to resolve the 'brain-paradox'.

Ideas for future research 3

Concept of the 'brain as observed'

How can we resolve the brain-paradox? One possible solution is to associate the 'brain as subject of cognition' and the 'brain as object of cognition' with two different domains: the empirical domain, the domain of science and the epistemological domain, the domain of knowledge. The two brains, the brain as subject and the brain as object, then reflect our different ways of accessing and knowing the brain, rather than suggesting that they are in fact two different brains.

The distinction between the brain as the subject and the brain as the object is thus an epistemological distinction – a distinction between different kinds of knowledge – rather than a metaphysical distinction between different existences and realities. There is thus one brain that can be described, conceptualized, accessed and known in different ways, including both concepts, the brain as the subject and as the brain as the object of cognition.

Where does this leave us? We did not double the brain itself. Instead, we only doubled our access to and knowledge of the brain by associating it with different domains – the empirical and the epistemological. Hence, rather than doubling the brain itself, we only doubled our methodological approach to it by considering the brain in both epistemological and empirical domains. This amounts to a 'doubling of domains' rather than a 'doubling of the brain'. To confuse the 'doubling of domains' with the 'doubling of the brain' is thus to conflate method and brain; that is, our methodological strategy of approaching the brain with the brain and its functions themselves.

What do I mean by 'doubling of domains'? The different domains allow us to consider and investigate the brain in different ways. Considering the brain in the empirical domain makes it possible for us to observe the brain and investigate it using the scientific method. This is different once one considers the brain in the epistemological domain. Now we can investigate the brain with regard to knowledge and whether, for example, it is relevant to our epistemic abilities. For that, we may require a different methodological strategy, like the rational-argumentative approach or phenomenal analysis. Finally, like Nagel and McGinn, we may also approach the brain in the metaphysical domain, which may open the door for novel, hitherto unknown features of the brain that, being intrinsic, may define and thus signify the brain as brain.

What does this investigation of the brain in different domains look like? Let us return to Schopenhauer once more. Schopenhauer spoke of the 'brain as the subject' and the 'brain as the object of cognition'. If one now associates the 'brain as the object of cognition' with the empirical domain, one may want to reformulate his original concept. Since the empirical domain is the domain of observation, one may want to re-conceptualize what Schopenhauer described as the 'brain as object of cognition' by the concept of the 'brain as observed'. *The concept of the 'brain as observed' describes the brain as we can observe it in third-person perspective. This is the* brain we target in neuroscience and investigate by using the various techniques to visualize its neuronal states.

Ideas for future research 4

Concept of the 'brain as functioning'

How can we investigate the brain in the epistemological domain? The epistemological domain is about knowledge and that is based on our epistemic abilities and limitations. Considering the brain in the epistemic domain will thus make it possible for us to link the brain and its various neural operations to our epistemic abilities and limitations.

The association of the brain with our epistemic abilities and limitations should reflect how the brain really works, independent of our cognition and knowledge of it. In other words, the brain as subject of cognition must be associated with mind-independent cognition rather than mind-dependent cognition. Mind-independent cognition reflects the world as it is, independent of ourselves and our particular epistemic abilities and limitations.

Considering the brain in the epistemological domain thus means assuming that mind-independent cognition is based on our brain and its neural functions. Those neural functions must remain independent of our observation and thus our epistemic abilities, since otherwise they could not yield mind-independent cognition. Accordingly, mind-independent cognition may provide a glimpse into how the brain actually functions by itself. *One may thus want to introduce the concept 'brain as functioning' to describe the brain as it functions independently of our observation and cognition.*

The neuroscientist may now be puzzled. Doesn't he/she investigate the brain as it functions by itself, independent of his/her observation? For example, the neuronal activity he/she observes in the motor cortex during a movement may also occur in exactly the same way when he/she does not scan and observe the subject's brain. Hence, the kind of neuronal functions he/she investigates when observing the brain in his/her mind-dependent way may be identical to the way the brain functions independent of his/her observation. If so, there should be no difference between the 'brain as observed' and the 'brain as functioning'.

This assumption remains to be proven. If future empirical discoveries do not lend any support to the distinction between the 'brain as observed' and the 'brain as functioning', the conceptual distinction may turn out to be empirically moot. What is clear, though, is that the 'brain as functioning' must be able to account for consciousness and other mental features of the mind. Moreover, the 'brain as functioning' would need to be related to specific neuronal mechanisms associated with our epistemic abilities and limitations. In other words, the 'brain as functioning' would need to provide insight into what the philosopher Thomas Nagel described conceptually as the 'deep interior of the brain' (see above).

However, whether we will ever be able to really tap into the 'brain as functioning' in a mind-independent way remains unknown at this point. Immanuel Kant and his modern successor Colin McGinn would deny any possibility for this, while others like Thomas Nagel are more optimistic (see above).

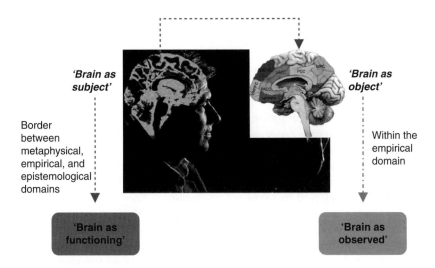

Figure 8.1 Different concepts of the brain

Upper part: The upper part of the figure describes the brain as subject of cognition, which concerns the brain that contributes to our perception and cognition (left part) and hence to our acquisition of knowledge about the world including its various objects like bodies and brains (--- arrows). Among those objects may also be brains which are then considered as objects, i.e. brain as object (right part).

Lower part: The lower part of the figure distinguishes between different concepts of the brain. The brain as functioning is the brain as it functions independent of our observation of it; this is supposed to reflect the brain as subject of cognition as indicated by the --- vertical arrow (and the situation of this concept at the border between metaphysical, empirical and epistemological domains). We have access to the brain, though, only via our observation in third-person perspective, which concerns the brain as observed. The brain as observed is considered to reflect what Schopenhauer described as 'brain as object of cognition', as indicated by the vertical ----- arrow (and the situation of this concept within the empirical domain).

However, even if McGinn is right and Nagel wrong, this will not yet be the end of our neurophilosophical investigations. One may then, for instance, search for those neuronal mechanisms that supposedly underlie the kind of epistemic inabilities or limitations McGinn postulates. One may, for instance, investigate the neuronal mechanisms underlying our apparent inability to describe the non-spatial features of our mental states in positive terms: what neuronal mechanisms prevent us from describing the non-spatial features of our mental states in a positive way? This would amount to a truly neurophilosophical and ultimately brain-based account of our mental features, including their epistemic inabilities, and thus of the mind–brain relationship in general.

Ideas for future research 5

'Brain as observed' and the brain as object of cognition

The philosophers and neuroscientists alike may now want to ask, first, why we need to introduce different concepts of the brain, and, second, why the distinction between the 'brain as functioning' and the 'brain as observed' is advantageous over the one between the 'brain as subject of cognition' and the 'brain as object of cognition'. Let us start with the first issue, the need for introducing different concepts of the brain.

Both philosophers and neuroscientists usually associate the brain with only one particular domain, the empirical domain. This implies that conceptually we need to characterize the brain in only one particular way. Since the empirical domain is associated with the methodological strategy of observation, the brain in the empirical domain can be characterized as 'brain as observed'. This is the traditional way of conceiving the brain.

However, the issue becomes more complicated once one raises the epistemological question of whether we can know everything about the brain as it is (implicitly) presupposed by, for instance, McGinn and Nagel. If there is something about the brain that we may in principle (McGinn) or in the current moment in time (Nagel) not know, the brain as we can observe it (currently) may not reveal everything there is about the brain itself and its existence and reality.

This means that the intrinsic features of the brain, those that define the brain as brain and its existence and reality as brain, may be different from those that we can (currently or in principle) observe. The brain in the empirical domain, that is, as we can observe it, may thus be different from the brain in the metaphysical (or ontological) domain, that is, the brain as it functions by itself, independent of our observation. We thus need to distinguish the concept of the 'brain as observed' from that of the 'brain as functioning'.

How does this distinction between the brain as observed and the brain as functioning stand in relation to the one between the brain as subject and object of cognition? This touches upon our second question as indicated above. Let us begin with the easy side of things, the brain as observed. The concept of the brain as observed is more or less analogous to that of the brain as the object of cognition. Both presuppose the empirical domain and its focus on observation as the main methodological strategy.

Ideas for future research 6

'Brain as functioning' and the brain as subject of cognition

Now let's turn to the more difficult side of things. The brain as the subject of cognition clearly presupposes the epistemological domain. Cognition is related to knowledge and for us to acquire and have knowledge we need a subject. Schopenhauer determined the brain as the subject of knowledge, hence his notion of the brain as the subject of cognition (he does not distinguish between cognition and knowledge). The concept of the brain as the subject of cognition can thus be situated in the epistemological domain.

This is different in the concept of the brain as functioning. The concept of the brain as functioning presupposes the metaphysical (or, better, ontological) domain since it refers to specific features that characterize the brain by itself, that is, as brain and its existence and reality as brain. This is, for instance, reflected in McGinn's assumption of the property P and Nagel's description of specific features in the 'deep interior of the brain'.

At the same time, the concept of the brain as functioning borders the empirical domain since it implies concrete functioning and thus specific neuronal processes and mechanisms. The concept of the brain as functioning can thus be situated right at the border between the metaphysical and empirical domains. This is especially clear in Nagel.

Adding Schopenhauer, one may now want to argue that the brain as functioning is also closely related to the brain as the subject of cognition. The brain of functioning and its particular intrinsic features and neuronal processes should then be central in making possible knowledge as such and the acquisition of knowledge. If so, the brain as functioning may be situated right at the interface or border between the metaphysical, empirical and epistemological domains. Rather than being an exclusively metaphysical, epistemological, or empirical concept, the concept of the brain as functioning may then be considered a border concept. This marks it as a truly neurophilosophical concept distinguished from both philosophical and neuroscientific concepts that are located within a particular domain, e.g. the metaphysical (or epistemological) or empirical domain, rather than between domains.

Biography 4

Maurice Merleau-Ponty

How can we access the 'brain as functioning'? On the basis of our observation, the neuroscientist can provide an indirect answer. Most importantly, that answer itself depends on our own brain and mind. Thus we cannot exclude that our observations may be traced back to the our mind's input. Furthermore, we remain unable to experience our brain in our consciousness. We experience objects, people, or events as the content of our consciousness. In contrast, we remain unable to experience our own brain directly as the content of our consciousness.

In contrast to the brain, we are able to experience the body in our consciousness. I experience sensations and tickling in my arm; I feel my heart pounding when it races; and I perceive the rumbling of my digestion after a heavy meal of too many hamburgers. Unlike my brain that has no propioceptors, I am able subjectively to experience my body in consciousness with all its phenomenal features: subjectivity (with the body in first-person perspective and point of view), intentionality (as directedness towards the body as content) and qualia (qualitative features of my body).

The body may thus be central for constituting consciousness and other mental features of the mind. This pivotal insight was introduced into the philosophical discussion by the French philosopher Maurice Merleau-Ponty (1908–61). He studied the science of his time, psychology, and the clinical field of neurology in patients with brain lesions.

Based on this background, he argued that the body played a central role in constituting consciousness and other mental features. The body provides the sensorimotor functions which make it possible to 'locate' us in the world. And that 'location in the world' make consciousness and other mental features possible. In order to understand the mind and its mental features, we have to go back to the body and its sensorimotor functions, and how they relate and link us to the world. Consciousness in particular and the mind in general are thus embodied and embedded, that is, integrated in both body and environment.

Brain and body 1

Different concepts of the body (Merleau-Ponty)

Why is the body so central for consciousness? Merleau-Ponty argues that we can consider the body in different ways. We can observe our body in third-person perspective in an objective way; this concerns what can be described conceptually as the 'objective body'. The 'objective body' is the body that we investigate in science. And it is the body that the medical doctor investigates in his/her medical examination when he/she listens to the heart, tests for reflexes, etc.

This is different when we conceive our own body in first-person perspective. Now we no longer observe our own body as an 'objective body'. Instead, we experience our own body from the inside. This is what Merleau-Ponty described as the 'lived body'. Most importantly, the 'lived body' is the body that provides the basis for our consciousness. It is through the 'lived body' that we can become conscious of the events, objects, and people in the environment.

Let us be more specific. Merleau-Ponty bases his argument on the lessons he learned from the early phenomenologically-oriented neurologists like Kurt Goldstein and Erwin Strauss (see Chapter 3). They investigated the subjective experiences and thus the consciousness of patients with selective lesions in the brain from the First World War.

What do these patients show exactly that inclines Merleau-Ponty to attribute the body with such a central role for consciousness? In addition to specific isolated deficits in particular contents of consciousness, the patients experienced both their body and the world in a fundamentally different way. The fundamental changes in their consciousness could be related to the alterations in the sensorimotor functions of their body. If so, the body and its sensorimotor functions must be central for consciousness.

In this model, the 'lived body' is thus seen as the very basis of consciousness in general. Changes in the experience of the 'lived body' are often accompanied by alterations in consciousness of the world. The 'lived body' seems to anchor us at a specific location or point within the world. That location or point can be described by the concept of the 'point of view'. The concept of point of view refers to a particular stance within the world that makes the experience of the objects, events, or people within that world possible. Since Merleau-Ponty associated point of view with the 'lived body', the lived body can be regarded as the very site, or even the origin, of first-person perspective. This in turn makes the 'lived body' the very basis and ground upon which the mind and consciousness stand.

Brain and body 2

Embodiment and sensorimotor functions

Current philosophers like Andy Clark (1997), Alva Noe (2004), Shan Gallagher (2005), Evan Thompson (2007) and Mark Rowlands (2010) have appropriated and adapted certain aspects of Merleau-Ponty's embodiment theory.

How can we specify the embodiment of the mind and its mental features? Merleau-Ponty assumed the body to be central in providing our stance and thus our point of view within the world. The body connects us and makes it possible for us to perceive the world via its sensory functions and acts upon the world using its motor functions. Consequently, sensorimotor functions are central in providing the linkage between body and world as manifest in our point of view.

The emphasis on the central role of sensorimotor function in constituting consciousness and other mental features provides an empirical counterweight to the predominantly cognitive approaches. Much research in both philosophy and neuroscience associates consciousness and other mental features with predominantly higher-order integrative and cognitive functions like attention, memory, executive functions, etc. (see Part III and especially Part IV for details). The proponents of embodiment now suggest that we need to focus on sensorimotor functions rather than cognitive functions in explaining consciousness.

Why the focus on sensorimotor functions rather than cognitive functions? The sensorimotor functions of the body connect us directly to and anchor us in the world. This makes it possible to constitute a point of view which is essential for consciousness. Cognitive functions, in contrast, are based and dependent on sensorimotor functions and their linkage to the world. In order to understand consciousness and other mental features fully, we therefore have to go back to the body and its sensorimotor functions.

Because we operate in the world using our sensorimotor functions, we cannot avoid being always already connected to our environment. This is what Merleau-Ponty calls 'being in the world'. Since such 'being in the world' is tied to the constitution of a point of view, it is closely associated with consciousness and other mental features. Conceptually, this intimate connection between the body and the world as the very basis of consciousness has been described by various terms like enactment, embeddedness and situatedness. Therefore, the concept of mind is described by corresponding terms like 'embedded mind', 'enacted mind', or 'situated mind' (or 'extended mind' as we will see below).

Critical reflection 3

How are the brain and its sensorimotor functions related to the body during consciousness?

Merleau-Ponty and his contemporary followers emphasize the body and its sensorimotor functions in constituting the kind of intimate body–world connection that they suppose to be necessary for consciousness. How is this intimate body–world connection possible? Sensory functions rely on sensory input, while motor functions reflect motor output. Moreover, to provide an intimate linkage between the body and the world, the merely extrinsic sensory input needs to be linked to the body and its intrinsic features. Conversely, the body's intrinsic motor output needs to be linked to the extrinsic world in order to provide the kind of intimate body–world connection Merleau-Ponty and his followers assume.

How can the extrinsic sensory input be intimately related to the body's intrinsic features? How can its intrinsic motor output be intimately linked to the extrinsic world? It is the brain itself and its intrinsic features that must provide some kind of special yet unknown input. Why and for what is such special input needed? The special input must provide the linkage between the extrinsic sensory input and the body and its intrinsic features. Only if the extrinsic sensory input and the body's intrinsic features are processed in a common format, can they be linked to each other. Thus there must be some kind of translation between extrinsic sensory stimuli and intrinsic bodily feature.

In addition, the brain's input seems to endow its intrinsic motor output with some features that relate it to the extrinsic world. However, the exact empirical, neuronal mechanisms of processes with regard to both sensory input and motor output remain unclear at this stage. Further research is required in order to understand these processes more fully.

What does this imply for the role of the brain and the body in consciousness? It may ultimately be the brain and its own intrinsic features that organize and structure sensory inputs and motor outputs in a specific way that makes possible their direct sensorimotor connection with each other. And that sensorimotor connection could in turn make the integration of the body and the world possible. If so, the emphasis on the body and its sensorimotor functions as constitutive of consciousness leads us back to the brain itself and the kinds of neuronal processes by means of which the brain becomes embodied and embedded.

Critical reflection 4

Concept of 'embodied and embedded brain'

One may want to speak of what can be described conceptually as an 'embodied and embedded brain'. *The concept of an 'embodied and embedded brain' describes the linkage and integration of the brain with the body and environment. This means that because of its particular way of processing, the brain itself may make the intrinsic integration of the sensory and motor stimuli from body and environment possible.*

However, how the brain intrinsically integrates its own neural activity with the sensory and motor functions of the body remains unclear at this point. The exact neural mechanisms are yet to be detected. Even if the empirical mechanisms of the integration between brain, body, and environment are currently unknown, they still have major conceptual implications. If the integration between the body and the environment can indeed be traced back to the brain itself and its specific mode of neural operations, we need to trace back Merleau-Ponty's body-based approach to a brain-based approach.

The brain-based approach suggested here would thus be considered an extension of (rather than a contradiction to) the body-based approach and its emphasis on embodiment and embeddedness. How can we trace back embodiment and embeddedness to the brain? Conceptually we may assume that the brain shows certain neuronal features that predispose it to link brain, body, and environment as described by the concepts of embodiment and embeddedness. What do such predisposing neuronal features look like, and how are they realized in detail? This will be discussed further in Part IV.

Brain and world 1

'Evil demon' and 'brain in the vat'

The intimate connection between the body and the world has important implications for the 'location' of mental states. Traditionally, mental states are associated with and localized in the mind (see Chapter 5). This position changed with the growing association of consciousness with the brain. Mental states were no longer associated with the mind, but 'located' in the brain (see Part III for further details when discussing the concept of representation).

The 'location' of mental states in some kind of mind or in the brain itself is doubted in the approaches that presuppose an embodied and embedded mind. Because of the intimate connection between the body and the world, mental states can no longer be 'located' either in some kind of mind, brain, body or world. Instead, mental states operate across the boundaries of all of these entities. They are thus 'extended', implying the concept of an 'extended mind'.

How can we further illustrate the mind's extension and its embodiment and embeddedness? To do this, we might want to briefly return to the French philosopher René Descartes. He assumed an interesting thought experiment called an evil demon. Imagine that there is an evil demon in your mind who is clever and deceitful and aims at nothing but to mislead you. He presents you with an illusion of an external world including other people as well as an illusion of your own body. You consequently perceive an external world and your body, a 'lived-body' as Merleau-Ponty would say.

Despite how convincing it might be, these are only deceptions created by the evil demon. In fact, there is neither an external world, nor a body in reality. This means that you will not be able to figure out whether the body and world you experience are real or illusory. You cannot exclude that the contents of your experience and perceptions are implanted by the evil demon. Since we cannot rule out the possibility of the evil demon, Descartes infers that we have to be skeptical about our knowledge of an external world: we cannot know for sure that what we experience and perceive in our consciousness really exists independently of ourselves and our mind.

One may now assume that the evil demon is purely mental and not anchored in the physical world. This assumption does not, however, hold true since we as humans are anchored in the physical world. In order to demonstrate the skepticism about our knowledge of the external world, one would need an example that is based on our physical and thus material brain rather than invoking a purely mental evil demon. In other words, one needs a physical rather than mental version of the evil demon. A physical version of the evil demon was provided by the American philosopher Hilary Putnam (1926–). It is known as the 'brain in the vat' thought experiment.

What does the 'brain in the vat' thought experiment look like? Imagine a mad scientist who completely separates and removes a person's brain from its original body. The mad scientist maintains the brain by suspending it in a vat of life-sustaining liquid. The brain is thus disconnected completely from both body and world. In other words, the brain is no longer embodied and embedded but isolated. In short, it is disembodied and disembedded.

The neurons of the brain in the vat are connected via wires to a supercomputer that provides the kinds of electrical impulses the brain normally receives from its body and the environment. The input that the isolated brain in the vat receives from the supercomputer thus does not differ from the input an embodied and embedded brain receives. At the same time the isolated brain in the vat sends out exactly the same output to the supercomputer as an embodied and embedded brain would send out to the body and world. Accordingly, on the whole then there is no difference in either input or output. What does this imply for the experience and the perception of the brain in the vat?

Brain and world 2

Perceptions of the 'brain in the vat'

What does the brain in the vat perceive: does it experience and perceive itself as embodied and embedded with a real body and environment? Or does the brain in the vat experience and perceive itself rather as isolated, as disembodied and disembedded? These are not easy questions but they are of high importance. If the brain in the vat experiences itself as embodied and embedded, it would experience itself walking down the street, hearing the voices of other people, and eating ice cream in much the same way that we do.

This sounds paradoxical because the brain in the vat as an isolated brain is not connected to a body or an environment. How can the brain in the vat walk down the street, eat ice cream and hear voices? This requires a body and an environment to which the brain is intrinsically connected. Perceptions thus remain impossible without the intrinsic connection between brain, body and environment. This implies that any kind of perception in the case of the isolated brain in the vat should remain impossible.

Is the brain in the vat thus able to decipher that the voices it hears, the steps it walks, and the ice cream it eats are not real and existent? Is it able to know that these are all simulations induced by its wiring to the supercomputer? This implies the question: is the brain in the vat able to recognize itself as such, and to distinguish itself as an isolated brain rather than an embodied and embedded brain? The isolated brain in the vat should remain unable to recognize itself as brain in the vat and to distinguish itself from an embodied and embedded brain. If so, the brain in the vat may experience and perceive the voices as real, the ice cream it eats as real, and the steps it takes as real.

This has serious epistemological consequences, that is, for our knowledge and its ability to discriminate between reality and illusion. More generally, the brain in the vat seems unable to make the distinction between the simulated and the real. It remains unable to distinguish whether the input is simulated and provided by the supercomputer or by a real body and world. This implies that the brain in the vat will not make a distinction as to whether the voices are 'real' or hallucinatory: it will simply hear voices and take them for real. The same applies also to the motor function. The brain in the vat will experience itself as walking down the street and eating ice cream. Whether the movements and the actions are real, the brain in the vat will not be able to tell.

Brain and world 3

Semantic or content internalism versus externalism

What does this imply for the 'location' of the semantic content of perceptions and actions? If the brain in the vat cannot distinguish between real and simulated perceptions/actions, their meaning must be 'located' in the brain itself – the brain in the vat. Let's be more specific. Despite the fact that it has no connection to the body and the world, the brain in the vat is nevertheless able to experience and perceive the same kinds of meaning as an embodied and embedded brain.

Most importantly, the brain in the vat remains unable to distinguish between simulated and real semantic contents. This implies that the meaning, and thus the semantic content, must be 'located' in the brain itself rather than in the body and the world. Meaning or semantic content is thus not embodied and embedded, but rather isolated in the brain itself and remains independent of both the body and the world. *Semantic content is thus located internally in the brain, rather than externally in the body and world. This is described as 'semantic or content internalism'.*

Content internalism is contested by the American philosopher Hilary Putnam (1926–). He argues that the brain in the vat is lacking something. What does the brain in the vat lack? Putnam argues that the brain in the vat is isolated and lacks the kinds of interaction the brain needs to have with body and world in order to generate beliefs, thoughts and words about the world it is in. Without this type of interaction with its environment, the brain is not be able to generate beliefs like 'I am the brain in the vat'. Putnam provides a complex argument that can be summarized by the two following alternatives, both of which lead to contradictions and therefore make the assumption of a brain in the vat futile.

In the first case, the brain is able to make the statement 'I am the brain in the vat'. While this is possible in the case of an embodied and embedded brain, it remains impossible for the isolated brain in the vat. Since it remains impossible for the brain in the vat to make such a statement, the brain making this statement cannot be a brain in the vat itself. Accordingly, the statement, 'I am the brain in the vat' can only be made by an embodied and embedded (and thus a) real brain, which then, by making this statement, would obviously contradict itself.

What about the second case? The second case is that the brain is not thinking about a real brain when making the statement, 'I am the brain in the vat'. Instead it is only making statements about images conveyed to it by the impulses from the supercomputer; however the impulses from the supercomputer are associated with the statement that they provide images of a real brain. In this version, the brain in the vat does not make a statement about itself but rather about an image of the brain as provided by the supercomputer. Hence, the statement 'I am the brain in the vat' would then not apply to the brain in the vat itself but only to the image of the brain. This would be equally as false and contradictory as the first case.

Taken together, both cases show that the brain in the vat remains unable to make the statement, 'I am the brain in the vat' without any logical contradiction. This implies that the content, and thus the semantic meaning, can neither be generated inside the brain itself nor be characterized as purely internal. Instead, content and semantic meaning must be generated in interaction of the brain with the body and environment. *This means that meaning, i.e. semantic content, is generated externally to the brain in its relation to the body and world which amounts to what is described as 'semantic or content externalism'.*

What holds true – semantic internalism or externalism? There has been much debate in recent philosophy of mind about whether semantic contents are generated inside the brain itself, i.e. semantic internalism, or in its relation to body and world, i.e. semantic externalism? The present-day proponents of an embodied and embedded mind clearly side with semantic content externalism. In order to generate meaning, and thus semantic content, we need a relationship to the body and world.

Ideas for future research 7

Concept of 'extended brain'

What does semantic externalism imply for the concept of the brain? It implies that *the brain can no longer be regarded as distinct and isolated from the body and the world. Rather than being only extrinsically connected to body and world in a contingent way, the brain in this model is intrinsically (necessarily) linked and connected to the body and the environment. Analogous to the concept of the 'extended mind' (see above), one can also speak of an 'extended brain'.*

The concept of the 'extended brain' describes the brain's neural operation as extended in a virtual way beyond itself and its own physical boundaries into both body and environment. This means that the concept of the 'extended brain' implies the embodiment and the embeddedness of the brain into the body and environment. The concept of the 'extended brain' finds its opposite in the concept of the 'isolated brain', whose neural operations are not integrated and dependent on the body and environment. This means that the 'isolated brain' is disembodied and disembedded.

How can we further specify the concept of the 'extended brain'? The 'extended brain' is characterized by an intrinsic relationship between the brain and both the body and the environment. Here, the concept of intrinsic denotes that the relationship of the brain to both the body and environment constitutes the brain as brain. Without such a relationship to the body and environment, the brain would no longer be a brain. This is analogous to the heart: without the pumping of blood, the heart would no longer be a heart.

The characterization of the brain by an intrinsic relationship to the body and environment has major empirical and conceptual implications. Empirically, one needs to search for the constitution of the brain's intrinsic activity, and more specifically, how the activity is encoded. If the brain is indeed constitutionally dependent upon body and environment, one would expect the intero- and exteroceptive stimuli from the body and environment to be encoded into the brain's intrinsic activity. The brain's intrinsic activity would then no longer be based exclusively on the brain itself, but would reflect an amalgam between the brain, the body, and the environment. One would thus assume that the brain encodes its relationship to the body and environment into its neural activity.

Conceptually, then, one can assume what can be described as an intrinsic body–brain and environment–brain relationship. This means that the concept of the brain cannot be defined independently of the body and the environment. The assumption of the concept of a 'pure' brain that is independent of the body and the environment would not correspond to any empirical reality and thus remain empirically implausible. In contrast, the characterization of the brain by an intrinsic relationship to body and environment may be considered empirically plausible. Taken together, the consideration of the relationship between brain, body and environment has major reverberations for our definition of the brain.

Ideas for future research 8

'Extended brain' and the 'brain as functioning'

How does the concept of the 'extended brain' relate to the concepts 'brain as observed' and 'brain as functioning'? The 'brain as observed' abstracts the brain from both the body and the world. In this model, the body and the world are considered in terms of input, as some kind of stimulus whose origin does not matter.

For example, for the 'brain as observed', whether an interoceptive stimulus – a stimulus from the internal organs of the body – stems from the real body or is simulated by some kind of supercomputer, does not matter. We only observe what the brain does and how it processes the interoceptive stimulus. The concept of the 'brain as observed' may consequently correspond to what can be described conceptually as the 'isolated brain' that is disembodied and disembedded.

Does the 'brain as functioning' therefore correspond to the concept of an 'extended brain'? This depends on whether one presupposes semantic internalism or externalism. If one presupposes semantic externalism, meaning is 'located' externally to the brain. It could be that the brain's intrinsic structure makes these relationships possible by means of specific neuronal mechanisms, and intrinsically links and integrates the brain and its intrinsic activity to body and environment. Those neuronal mechanisms should reflect the functioning of the brain itself, independent of our observation.

Let's talk specifically about these neuronal mechanisms. The brain's input, its intrinsic activity, must somehow link the extrinsic sensory input to itself and the body. At the same time it must endow the intrinsic motor output with some special features that allow it to be linked to the extrinsic world. The neuronal mechanisms underlying the extension of the brain to the body and environment would thus reflect the brain's neural functions, independent of our observation. The concept of the 'extended brain' would then indeed correspond to that of the 'brain as functioning'.

What if one presupposes semantic internalism, rather than semantic externalism? In that case the distinction between 'extended brain' and 'isolated brain' would remain impossible. If defined in a semantically internal way, the brain could no longer be extended to the body and environment. Why? Because such a definition would violate the internal definition of the brain and would thus no longer concern the brain. That makes even the assumption of the concept of an 'extended brain' logically impossible. If, however, the concept of an 'extended brain' is impossible, its opposite, the concept of an 'isolated brain' also remains impossible.

The conceptual distinction between 'extended brain' and 'isolated brain' remains consequently impossible in the case of semantic internalism. This implies that the corresponding conceptual distinction between the 'brain as observed' and the 'brain as functioning' might also collapse at this point. The 'brain as observed' and 'the brain as functioning' would then be assumed to be identical rather than (at least conceptually, even if not empirically) different.

On the whole, we discussed different concepts of the brain and how they are related to each other. Moreover, when considering the whole chapter, we investigated the brain in different domains – metaphysical, empirical and epistemic – and how this in turn impacts the description and thus the concept of the brain. As such, the brain may be considered in different domains and described in different concepts. This may lead to the future development of a philosophy of brain as analogous to the philosophy of mind.

Ideas for future research 9

'Philosophy of brain' and the mind–brain problem

What is a philosophy of brain? *A philosophy of brain discusses the brain in different domains and investigates different concepts of the brain as they are related to the investigation of the brain in different domains.* Why is a philosophy of brain important? It is important because it may provide a new, brain-based methodological approach to the mind–brain problem. As explained earlier, a brain-based methodological approach takes the brain itself as a starting point for the investigation of the mind and its mental features.

The direction of the methodological strategy would then be reversed; rather than proceeding from the mind to the brain, as in almost all current approaches, one would take the brain as the point of departure and proceed from there. To approach the mind–brain problem from this methodological vantage point may provide new solutions to a problem that so far has remained stubbornly resistant.

The traditional philosopher may now want to argue that such a philosophy of brain remains redundant if not contradictory. The brain can only be investigated in the empirical domain entailing a neuroscience of brain. In contrast, a philosophy of the brain remains impossible, since for that we would need to investigate the brain in metaphysical, epistemological or conceptual domains as the domains traditionally associated with philosophy. Hence, philosophy would consider a philosophy of brain to be superfluous. The same applies to neuroscience, which analogously considers the brain only in the empirical domain.

This, however, changes once one reverts from a philosophical or neuroscientific approach to a neurophilosophical approach. The brain is then considered in different domains, metaphysical, epistemological, empirical and conceptual, as I demonstrated in this chapter. The investigation of the brain is then no longer limited exclusively to neuroscience and its observational-experimental method. Instead, the brain can also enter philosophy and its traditional domains as the subject of investigation by using a rational-argumentative approach. This results in a philosophy of brain as a core theme of any future neurophilosophy.

What does such a philosophy of brain imply for the mind–brain problem? The discussion of different concepts of the brain could lead to a reformulation of the mind–brain problem. Traditionally, the mind–brain problem presupposes the brain as observed and takes it to be identical with the brain as functioning. Thereby it seems as if there is no difference between the extended and the isolated brain. It is simply asked whether the existence and reality of the mind is identical or not with the one of the brain as observed and the isolated brain. This is the strategy and question implied by the mind-based approaches as physical and mental approaches in current philosophy of mind.

This changes, however, once one takes a truly neurophilosophical (rather than philosophical) approach. One then differentiates between the different concepts of the brain and, at the same, takes the brain rather than the mind as the starting point, thus presupposing a brain-based rather than mind-based approach (see above). The main question is then no longer how the mind relates to the brain. Instead, the guiding question is: how is it possible for what we call mind and mental features to differentiate from the brain and its purely neuronal features?

Because of its reverse methodological strategy, brain-based approaches to the mind–brain problem may reformulate certain questions. It may be a problem of how the mind and its mental features can be differentiated from the brain as observed on the basis of the brain as functioning. The mind–brain problem may consequently be transformed into what can be described as the 'brain–mind differentiation problem'. The 'brain–mind differentiation problem' describes the question: how is it possible that the brain as observed and what we describe as mental features and associate with a mind can differentiate from each other on the basis of the brain as functioning? We will come back to this question in Parts IV and V, where we will describe the neural predispositions and thus the brain-based nature of mental features like consciousness and self.

Take-home message

The previous chapter discussed non-physical and non-mental approaches to the mind–brain problem. These approaches searched for alternative strategies within the metaphysical domain when suggesting non-mental and non-physical solutions. Alternatively, they shifted the mind–brain problem from the metaphysical domain to the epistemic or empirical domains. However, while addressing some of the problems plaguing physical and mental approaches, these alternative solutions generated different issues of their own. How can we escape these problems? One method is to start any investigation of the mind–brain problem with the brain itself. This type of approach is brain-based rather than mind-based. This brain-based approach can consider the brain in the empirical, the epistemic or the metaphysical domain. In addition to the methodological strategy, we also considered the relationship between the brain and the body and the world. The consideration of the different brain-based approaches in the current debate on the mind–brain relationship has led to the development of new and different concepts of the brain. These new concepts are central for a brain-based and thus truly neurophilosophical investigation of the mind–brain problem, as well as the development of a philosophy of the brain in the future.

Summary

The previous chapter focused on non-mental and non-physical alternative approaches to the mind–brain problem. Alternative strategies proposed the brain, rather than the mind, as the starting point for the discussion of the mind–brain problem. This method is brain-based rather than mind-based. The concept of a brain-based approach describes only a methodological strategy as distinguished from any metaphysical claim about the existence and reality of the brain and its relation to the mind. A brain-based approach can be accompanied by considering the brain in different domains, including the metaphysical, empirical and epistemic. Colin McGinn's approach to the brain localizes the discussion in the metaphysical domain, and assumes the existence of a specific property in the brain called property P. Property P, according to McGinn, is responsible for producing consciousness, despite the fact that we cannot know property P because of principal epistemic limitations in our own brains. Thomas Nagel suggests the possibility of specific properties in the 'deep interior of the brain' that have yet to be discovered in future empirical investigation. This implies that Nagel, unlike McGinn, considers the brain in the empirical, rather than metaphysical, domain. Arthur Schopenhauer, a nineteenth-century German philosopher, localized investigation of the brain in the epistemic domain when he distinguished the 'brain as subject of cognition' from the 'brain as object of cognition.' As we have seen above, this differentiation leads to a logical contradiction called 'the brain-paradox' because the same brain cannot be both a subject and an object simultaneously. However, according to Schopenhauer's analysis, one can introduce the conceptual distinction between the 'brain as observed' and the 'brain as functioning'. Finally, we needed to consider how the brain is related to the body and the world. For that we turned to the French philosopher Maurice Merleau-Ponty and the contemporary concepts of 'embodiment' and 'embeddedness'. Here, the body and its sensorimotor functions are considered to be central for constituting consciousness and other mental features. The implications of embodiment and embeddedness for the 'location' of meaning or semantic content are discussed by imagining a brain isolated from both the body and environment. This is called the 'brain in the vat' thought experiment. This experiment asks whether meaning is 'located' internally or externally to the mind and the brain. This in turn is accompanied by the need to introduce different novel concepts of the brain like the 'extended brain' and the 'embodied and embedded brain' that may be central in future neurophilosophical investigations of the mind–brain problem.

Revision notes

- Can you determine what the property P is? Why is it mysterious? And how is it related to Kant and his emphasis on epistemic limitations?
- What are the features of Nagel's approach to the mind–brain problem? How can you distinguish Nagel's and McGinn's accounts? Why are both subsumed under the camp of the 'mysterians'?
- Distinguish a brain-based approach from a mind- and computer-based approach to the mind–brain problem.
- What is the difference between a brain-based and a brain-reductive approach?
- What is the evil demon?
- What is the difference between the brain as subject of cognition and the brain as object of cognition?
- How can you define different concepts of the body and why these are important?
- Can you explain the thought experiments of an evil demon and a brain in the vat? What purpose do they serve and how can they be distinguished from each other?

- What is semantic or content externalism and its opposite positions?
- What is the extended mind?
- What is the extended brain?
- Can you explain the different concepts of the brain as suggested by Schopenhauer?
- Why do we need to distinguish between different concepts of the brain?
- What is philosophy of brain?

Suggested further reading

- Bennett, M. and Hacker, P. (2003) *Philosophical Foundations of Neuroscience* (Oxford/New York: Blackwell).
- McGinn, C. (1991) *The Problem of Consciousness* (London/New York: Basil Blackwell).
- Merleau-Ponty, M. (1962 [1945]) *Phenomenology of Perception* (London: Routledge).
- Nagel, T. (1998) 'Conceiving the impossible and the mind–body problem', *Philosophy*, 73(285), 337–52.
- Schopenhauer, A. (1818/1819) *The World as Will and Idea. Vols I and II*, Dover edition 1966 (London: Dover).

Part III
Philosophy of Psychology and Neuroscience: From Explanation of Mind to Explanation of Brain

In Part II, I discussed the various possible solutions to the mind–brain problem and the ways in which the brain can be considered. In addition to a metaphysical discussion about the existence and reality of the mind, there has been much empirical research in both psychology and neuroscience. Part III considers these developments in psychology and neuroscience, shifting from the metaphysical domain to the empirical domain, the domain of observation. However, rather than discussing concrete empirical research, I will focus on theoretical problems that arise in the context of the empirical investigation of the mind and the brain. These problems are subsumed under the umbrella terms philosophy of psychology and philosophy of neuroscience.

The first two chapters in this part, Chapters 9 and 10, will focus on topics in the philosophy of psychology, like personal versus subpersonal levels of explanation (Chapter 9) and the question of meaning and semantic content (Chapter 10). This will be followed by two chapters focusing on topics in the philosophy of neuroscience, like neuroscientific explanations and concepts (Chapter 11) and the characterization of the brain (Chapter 12).

The approach taken here is a theoretical one. Such a theoretical approach includes models of mental features for instance, of meaning and semantic content (Chapter 10), as well as models of the brain (see Chapter 12). Such theoretical models aim to put the various experimental data into a larger framework. They thus remain within the empirical domain where they complement the observational-experimental data from the theoretical side. One can therefore speak of theoretical psychology and theoretical neuroscience, which are usually subsumed under the umbrella terms philosophy of psychology and philosophy of neuroscience.

In addition to such theoretical models, philosophy of psychology and philosophy of neuroscience also include a strong methodological line of research. This concerns the explication of methodological strategies like the relation between subpersonal and personal levels in psychology (Chapter 9) and the kind of explanations and concepts in neuroscience (Chapter 11). The concepts of philosophy of psychology and philosophy of neuroscience can here be regarded as extensions of the more general philosophy of science, one of the main fields in philosophy.

How about a philosophy of neurophilosophy? We have already discussed different concepts of neurophilosophy when comparing reductive and non-reductive approaches to neurophilosophy in Chapter 4. There we also discussed the methodological strategy that is specific to non-reductive neurophilosophy and distinguishes it from both philosophy and neuroscience. Such a philosophy of neurophilosophy is now, in this part, complemented by discussing the main issues and topics in philosophy of psychology (Chapters 9 and 10) and philosophy of neuroscience (Chapters 11 and 12).

Part III
Principles of Psychology and Neuroscience: From Key Lessons of Animal Brains to Humans

9
Philosophy of Psychology: Personal versus Subpersonal Levels of Explanation

Overview

Part II focused on exploring the existence and reality of the mind and how it is related to that of the brain. These problems were investigated primarily within the context of the metaphysical domain – the domain of existence and reality. However, this approach ignored the exact mechanisms, functions and processes underlying the mind and its mental features. To get a better understanding of these mechanisms etc. it is necessary to shift the investigation from the metaphysical domain to the empirical domain – the domain of observation. This shift in domain is accompanied by a shift from philosophy of mind to psychology. At the root is the question: how can we explain the mind and its mental features in empirical terms? In our daily life, we observe other people's behavior and from this infer and predict their mental states. The private and individual psychology of mental states is called common sense or folk psychology. Folk psychology needs to be distinguished from scientific psychology. The latter relies on publicly observable mental states, rather than private mental states. How are folk psychology and scientific psychology related? This is a central topic in contemporary philosophy of psychology, which we will discuss further below.

Objectives

- To understand the difference between metaphysical and empirical accounts of the mind and their related disciplines
- To grasp the distinction between vertical and horizontal explanations
- To understand the difference between personal and subpersonal levels of explanation
- To delineate the criteria for reduction, like commensurability, and derivability
- To understand the difference between folk and scientific psychology
- To account for the different positions concerning the relationship between folk and scientific psychology, like irreducibility, continuity, and elimination
- To understand the implications of this discussion for the concept of neurophilosophy and its relationship to neuroscience and philosophy

Key concepts

Mind–brain problem, mind problem, interface problem, vertical and horizontal explanations, philosophy of psychology, philosophy of mind, folk and scientific psychology, reduction, functionalism, philosophical and psychological functionalism, elimination, eliminative materialism.

Background 1

From philosophy of mind to psychology

The previous part discussed various relationships between the mind and the brain within the context of the mind–brain problem. The focus was on the existence and reality of the mind and how it is related to the existence and reality of the brain. If someone believes that the mind's existence and reality is different from that of the brain, we can characterize that person as a dualist. If, in contrast, someone maintains the mind's existence and reality is traceable to the brain, we can characterize that person as a materialist, or a physicalist. Both dualism and materialism, as well as the various alternative positions like functionalism, neutral monism, etc. discussed in Part II are metaphysical assumptions about the mind and its existence and reality.

The mind–brain problem is a metaphysical problem that seeks to explain the existence and reality of the mind, and its relationship to the brain. Because philosophy is traditionally the discipline that deals with the metaphysical domain – the domain of existence and reality – the mind–brain problem is considered a philosophical problem that, more specifically, belongs to philosophy of mind.

In addition to the mind–brain problem, philosophy of mind is also concerned with features of the mind. The mental features often attributed to the mind include consciousness (subjective experience), subjectivity (point of view and first-person perspective), self (sense of self and identity), intersubjectivity (relations to other people as subjects) and others like free will, emotions, etc. Some of these core features of the mind will be dealt with within the context of the brain in subsequent parts (see Part IV about consciousness, and Part V about subjectivity and intersubjectivity). For reasons of space, other features associated with the mind, like free will, emotions and mental causation, will not be elaborated in further detail.

Who and which discipline is in charge of these mental features? Traditionally, these mental features were discussed in philosophy, which focuses on the mind and its existence and reality. Since these various mental features are attributed to the mind, they are also dealt with in philosophy and more specifically in philosophy of mind. There has thus been much discussion among philosophers of mind about these various mental features.

However, we already saw in Part I that the mind's mental features can also be investigated empirically. An example of this is psychology. Rather than discussing possible metaphysical assumptions about the existence and reality of the mental features, psychology investigates them experimentally by observation. This presupposes a shift from the metaphysical domain of philosophy to the empirical domain of psychology.

The shift from the metaphysical to the empirical domain is accompanied by a focus on processes, functions and mechanisms associated with the mind and its mental features. Rather than merely describing the mind's existence and reality as in philosophy of mind, the focus is now on explaining the mind and its mental features. What are the mechanisms, functions and processes that underlie the kinds of mental features we attribute to the mind? How can we explain the occurrence of the mental features and their specific properties?

Unlike philosophy in general – and philosophy of mind in particular – psychology no longer considers the existence and reality of the mind as a problem within the metaphysical domain. Instead, it tackles the mind as a problem fit for empirical terms by observing it and its mental features during experimental manipulation. There is thus a shift from the metaphysical to the empirical domain, where the search for existence and reality is replaced by observation.

Background 2

Folk versus scientific psychology

How do we access the mind in empirical terms? One way to access the mind is to rely on what is described as common sense. Every one of us has a certain way of deciphering other people's minds. For instance, by observing another person's behavior, we infer that he/she might possess particular beliefs, ideas, or desires, etc. These assumptions are called propositional attitudes. For instance, our experience tells us that every time we observe the behavior x, it is coupled with a belief about the content w. When we observe another person smiling and laughing (the behavior x), we assume that the person believes the story I just told (the content w) to be funny and amusing.

This example demonstrates how we make law-like generalizations and predictions in the social world about behaviors and their associated mental states. This allows us to decode and interpret others' speech and the contents of their thoughts. In short, we are all able – to some degree – to act as 'hobby psychologists'. 'Hobby psychology', though often faulty, has been described in more technical terms as 'folk psychology'. *In the following chapters, I will use the term 'folk psychology' to describe the kinds of law-like generalization everybody makes on a daily basis to predict behavioral and mental regularities.*

There is, however, much more to psychology than 'folk psychology'. Folk psychology is complemented by a more rigorous and disciplined psychology. Psychology as a discipline relies on empirical observation in third-person perspective. Unlike the private and individual convictions of folk psychology, the empirical observations of psychology are accessible to everybody. This means that they utilize publicly accessible data, rather than private data. *The third-person data yielded by observation are not based on an individuals' experience and learning, but rather on experimental paradigms that follow strict rules of scientific validity (internal coherence) and reliability (repeatability without changes). This is what can be described as 'scientific psychology'.*

Background 3

Interface problem

How are folk and scientific psychology related? This raises the question: how is the scientific explanation of the mind and its mental features related to our common sense explanations in folk psychology? The American philosopher Jose Luis

Bermudez (2005) describes this as the *'interface problem'*. *The interface problem is a problem concerning the different forms or levels in explanations of the mind. This is most prominently visible in the tension between common-sense and scientific approaches to the mind. How are the different psychological approaches, and thus the mind's different explanations (common sense and scientific), related to each other? This is the question that best describes the central concern of the interface problem.* The interface problem is consequently dealt with in *philosophy of psychology – the discussion of theoretical issues within the empirical domain of psychology.*

How is the interface problem related to the mind–brain problem, as discussed in Part II? The interface problem must be distinguished from the 'mind–brain problem'. The mind–brain problem is not a problem about different levels, ways, or approaches of explanation. Instead, it is a problem about how the existence and reality of the mind, along with its mental features, can be described and related to that of the brain and its physical features (see Chapter 5 for a definition). The mind–brain problem is thus concerned with the description of the mind's metaphysical features, rather than its different forms or levels of explanation. The mind–brain problem thus presupposes the metaphysical domain, the domain of existence and reality.

This distinguishes it from the interface problem, which is 'located' in the empirical domain, the domain of observation. The empirical domain presupposes different forms of explanation of the mind and its mental features – common-sense and scientific. The interface problem raises a question about the relationship of these different explanations of the mind within the empirical domain. As such, the interface problem concerns a theoretical issue within the empirical domain. Therefore, the interface problem is not a problem within the philosophy of mind, but one within the philosophy of psychology that focuses on theoretical issues in the empirical domain.

Biography

Daniel Dennett

The interface problem concerns the relationship between folk and scientific psychology. Possible solutions to this problem may range from the strict distinction between folk and scientific psychology to the assumption of continuity to elimination. More specifically, one may consider folk psychology as irreducible to scientific psychology. Or one may assume continuity between both. And finally one may also suggest the elimination of folk psychology completely in favor of scientific psychology. Below, all three positions will be discussed. Let's start with the first one: the assumption that folk psychology cannot be reduced to scientific psychology. To further explain this position, I want to introduce the American philosopher, Daniel Dennett.

Daniel Dennett was born 1942 but spent some years of his childhood in Beirut, Lebanon, where his father was a covert agent in service for the USA during the Second World War. He then returned with his father to Boston, USA, where he began to study philosophy, his lifelong dedication. Daniel Dennett is currently a Professor at Tufts University in Boston.

Dennett made major contributions to the description and explanation of the mind, with consciousness as its hallmark feature. He aimed to show that many of the mental concepts we use to describe the mind are nothing but illusions. Beyond revealing the mysteries of the mind as conceptual illusions, he is also strongly engaged in revealing religion as a myth based on evolutionary development, rather than inspiration by God.

Explanatory irreducibility 1

Different levels of explanation (Dennett)

What does Daniel Dennett uncover about the explanation of the mind? Dennett (1987, 1993) distinguishes between different hierarchical levels of explanation. At the top we can find the human agent, a person, who acts in the world and is directed towards specific contents, demonstrating intentionality. Consequently, Dennett speaks of an 'intentional stance', which implies the personal level.

The 'intentional stance' needs to be distinguished from the 'design stance'. The design stance describes the structure and organization of the system that underlies the person as agent. The design level of the underlying machinery must be distinguished from a person's intentional stance. Consider the experience of perceiving a tree in front of you. The tree is the intentional content of your perception; your perception is directed toward the tree. This is called the intentional stance. In contrast, the physical organization of your eyes, brain, and their specific structure provide the necessary mechanisms and design that make perception possible. We call this the design stance (with the terms 'stance' and 'level' being used more or less interchangeably).

More specifically, the design level concerns the algorithms that allow for the processing of information in the system. Explanations of the mind from the design stance no longer target the person within the world – the personal level – but rather the system and design underlying the person – the subpersonal level. Hence, the explanation of the design stance does not presuppose the personal, but rather the subpersonal level. In addition to the intentional and the design stance, Dennett assumes another stance that is useful when explaining the mind. This third stance concerns the physical features that realize and implement the design stance. Dennett calls this the 'physical stance'. How can we characterize the 'physical stance'? The physical stance concerns the physical realization and implementation of the system operating at the design level. There must be particular physical features and structures present that allow for the physical implementation and realization of the system's design features. For instance, this would concern the chemical and physical properties of the single cells in your retina and visual cortex.

Explanatory irreducibility 2

Irreducibility of folk psychology (Dennett)

How do these different stances relate to the distinction between folk and scientific psychology? The level of the intentional stance more or less reflects folk psychology, while the other levels, the design level and the physical level, are more related to scientific psychology and ultimately neuroscience. How does Dennett consider their relationship? He takes the intentional stance, and thus folk psychology, very seriously and thinks that there is some truth in our everyday law-like generalization about others' behavior and mental states.

This means that folk psychology and its explanation of the intentional stance on the personal level cannot be reduced to scientific psychology (and ultimately neuroscience) and its explanation of the design and physical stance on the subpersonal level. More specifically, the directedness toward content (intentionality) can neither be found within the design itself, the structure and organization, nor the physical features. In other words, the intentional level can be reduced to neither the design level nor the physical level.

Accordingly, the personal level and folk psychology in general cannot be reduced to the subpersonal level of scientific psychology. The assumption of such irreducibility distinguishes Dennett from other philosophers like the American couple, Patricia and Paul Churchland, who deny any truth to folk psychology (see subsequent sections on the elimination of folk psychology in favor of scientific psychology). As we will see below, they consider our everyday psychological explanation as simply wrong. This position leads the Churchlands to insist that we reject folk psychology altogether, in favor of scientific psychology.

Dennett does not share their opinion. He assumes that our everyday psychological assumptions carry some truth about the personal level and its hallmark feature, intentionality. As such, these folk psychological explanations on the personal level must be distinguished from the subpersonal-level explanations of scientific psychology. There is consequently more to our psychological life, and thus to our mind and its mental features, than what scientific psychology tells us about the subpersonal level. And this 'more' consists in the intentionality that characterizes the personal level, the level of folk psychology. In short, folk psychology cannot be reduced to or eliminated in favor of scientific psychology.

Explanatory irreducibility 3

Horizontal explanations

Why can the personal level of folk psychology not be reduced to the subpersonal level of scientific psychology? Dennett argues that the personal level is characterized by a specific feature, namely intentionality. Because of intentionality, the personal level cannot be reduced to, nor be found in, the subpersonal level – the level of the design and physical stance. In short, the personal level is non-reducible to the subpersonal level.

Since the different features imply personal and subpersonal levels, any explanation of the personal in terms of the subpersonal implies an explanation between different levels. *This type of explanation between different levels is described as a 'vertical explanation'. Vertical explanation must be distinguished from explanations that remain within a single level, which are known as 'horizontal explanations'.* What are 'horizontal and vertical explanations'? Let us start with explaining horizontal explanations. A horizontal explanation occurs within one level of the explanatory hierarchy. It remains, for instance, within the level of the intentional stance and thus within folk psychology. A horizontal explanation does not recruit any other level, such as the design level or the physical level.

Metaphorical comparison 1

Soccer with Lionel Messi

Let me give you an example. Imagine you are playing soccer and your role model is Lionel Messi of Barcelona. When you try to kick the ball from the midfield's outer edge into the goal, unlike Lionel Messi, you fail. Instead of the ball going in the net, it ends up smashing into the camera of a photographer on the sideline of the field. Why did the ball destroy the poor photographer's camera? 'Very simple,' you will say, 'because I kicked the ball the wrong way.' And you may want to add: 'Since I am not Lionel Messi, I failed and should not try to do the kinds of things he does so easily.' So far, so good.

Your explanation of the incident remains purely within the intentional level of explanation and thus on the personal level. The description of you as an agent and the consequences of your behavior figure prominently in all stages. There is no reference at all to any subpersonal level – the design or physical level – in your explanation. This is what is described as 'horizontal explanation'.

Such horizontal explanation may also apply to other levels, like the physical-neuronal level. You may, for instance, explain the action of a neuron solely by its electrical features. The single neuron was activated (showing an action potential) because it received input from other neurons that also exhibited action potentials. Here, the explanation remains solely within the electrical, physical level of the neuron, and does not refer to either the design or the physical stance. It is thus a horizontal explanation.

Explanatory irreducibility 4

Vertical explanations

However, you are eager for explanations for other 'why' questions. Why did the camera break when it was hit by the ball? How could the ball break the camera's glass? These examples raise questions about the physical features of the ball and the camera's glass. Maybe the ball you kicked was manufactured with particular leather which made it very hard. 'Yes,' you say, 'my foot really hurts.' This is again a horizontal explanation. But how could the hard leather of the ball destroy the camera's glass and make my foot hurt? By kicking the ball very hard, you gave it an enormous speed and velocity, which increased the likelihood of destroying the camera. Moreover, the photographer forgot to use his special camera lens with strong glass.

Now you have combined the ball's velocity, its hard leather, and the brittleness of the camera's glass in explaining the incident. The explanation no longer remains within the personal level of the intentional agent. Instead, the physical features of both the ball and the camera's glass are used in explaining the breaking of the latter. You thus recruit the physical level – the subpersonal level – to explain what happened on the intentional level – the personal level. In other words, the horizontal explanation is now combined and complemented with a vertical explanation.

You may also do the same in the case of the neuron. How could the excitation of the two other neurons lead to an action potential of your target neuron? You may search for how the action potential of the two neurons was mediated by biochemical substances that carried forth the excitation of the former to your target neuron. The latter's electrical activity is thus no longer explained solely in electrical but also in biochemical terms as yet another level or layer of explanation.

You may want to investigate things even further by predicting the personal level using the subpersonal level. Do the physical and biological features of the subpersonal level predict the behavior and intentions you observe on the personal level? Rather than remaining within the subpersonal level – neuronal and biochemical levels of explanation – you aim to show the relevance of the latter for the personal level. You have thus crossed different levels in your explanation, making it a vertical, rather than a horizontal explanation.

Explanatory irreducibility 5

Incommensurability and non-derivability

Why is the distinction between horizontal and vertical explanation important for the relationship between folk and scientific psychology? Proponents of their irreducibility do allow for horizontal explanation, but not for vertical explanation. They argue that explanations on the personal level, as in folk psychology, cannot be reduced to those on the subpersonal level, as in scientific psychology. This makes vertical explanation impossible. The impossibility of such vertical explanation implies, however, that folk psychology, in describing the personal level, cannot be reduced to scientific psychology, which is more concerned with the subpersonal level.

How, though, can we further specify the irreducibility of a personal to a subpersonal explanation? To do so, one may want to apply specific criteria. In order for vertical explanation to be possible, subpersonal and personal levels of explanation must be commensurable and derivable. *Commensurability describes the possible connection between the concepts and terms of the two different levels, implying that they can be translated into each other. Derivability concerns the degree to which the higher level's features are already present in the lower one such that the former can be inferred and thus derived from the latter.*

How do both criteria – commensurability and derivability – relate to the irreducibility of the personal to the subpersonal? Proponents of irreducibility reject both criteria, commensurability and derivability. According to them, personal and subpersonal levels of explanation are incommensurable with each other because their concepts and terms cannot be translated into each other.

Why can't personal and subpersonal terms be translated into each other? Because the concepts of the personal level are normative or prescriptive, and thus law-like; they describe inferences from behavior to mental states on the basis of law-like regularities mirroring 'how one ought to behave and think'. This is different in explanations of the subpersonal level, which concern descriptive concepts. Subpersonal level explanations concern the actual mental state and the description of its various features. Rather than 'how the mental states of other persons ought to be on the basis of their behavior', the guiding question here is 'how the mental states actually are on the basis of their careful observation'.

Metaphorical comparison 2

Soccer and the photographer's camera

Let's illustrate this using our soccer example. If you kick the ball at the camera, it ought to break the camera's glass. You know that if you break the photographer's camera, he will be angry. You know this on the basis of law-like generalizations of behavior, and hence folk psychology on the personal level. Most importantly, your explanation of the whole scene treats the anger of the photographer as more or less an unavoidable consequence. Why? Because you know that his behavior is the norm. You are thus implying a normative or prescriptive explanation.

This is different when you explain the same incident, but focusing on the physical properties of the ball and the lens. When trying to explain the photographer's action, you consider the psychological mechanisms. The anger he feels is induced by the breaking of his camera, which is mediated by psychological, neuronal and hormonal changes that cause anger. None of these things, not the psychological, neuronal or hormonal changes, implies anything about the photographer's reaction: for example, his purple face or loud shouting. The scientific explanation on the subpersonal level thus does not contain any normative or prescriptive elements. Instead it remains purely descriptive.

Explanatory irreducibility 6

Are normative-prescriptive and descriptive explanations incommensurable?

How are both normative-prescriptive and descriptive explanations related to each other? The proponents of irreducibility now argue that both personal and subpersonal levels of explanation are incommensurable. Why? Both kinds of explanations use principally different concepts. Personal levels of explanations, as in folk psychology, concern normative and prescriptive concepts. Subpersonal levels of explanation, as in scientific psychology, remain purely descriptive in that they use only physical concepts that are neither normative nor prescriptive.

Due to their use of different concepts, normative-prescriptive and descriptive, subpersonal and personal levels of explanation cannot be translated into each other. And if they cannot be translated into each other, they are incommensurable. There is thus incommensurability between subpersonal and personal levels of explanation, and thus between scientific and folk psychology.

Moreover, none of the features associated with the normative-prescriptive concepts on the personal level can be found on, and related to, the subpersonal level and its descriptive concepts. If there is no connection at all between both levels, the personal level and its normative-prescriptive concepts do not correspond to anything on the subpersonal level and its descriptive concepts. This makes it impossible to derive the former from the latter. Because neither is commensurable to the other, and neither can be derived from the other, vertical explanations between personal and subpersonal levels remain impossible. This means that folk and scientific psychology – as characterized by personal and subpersonal levels – cannot be reduced to each other either. In other words, folk psychology remains irreducible to scientific psychology.

Critical reflection 1

Explanatory models and mind–brain models: metaphysics versus explanation

The irreducibility view implies a characterization of the mind and its personal level that is radically different from that of its underlying design and system. As suggested by Dennett, the mind is characterized as an intentional agent. Since there are no explanatory connections between personal and subpersonal levels, the mind must be not only principally different but also autonomous when compared to the subpersonal level.

Jose Luis Bermudez (2005) therefore speaks of an 'autonomous mind' that is autonomous on the personal level when compared to the psychological and neuronal processes characterizing the subpersonal level. This implies not only a dichotomy between the two levels of explanations, personal and subpersonal, but also a dichotomy between the two different explanations themselves, folk and scientific psychology.

Does the explanatory dichotomy between folk and scientific psychology imply a metaphysical dichotomy between mind and brain? Recall from Part II that the mind–brain problem is concerned with explaining the existence and reality of the mind, and thus presupposes the metaphysical domain. This is different in the explanatory dichotomy. Both scientific and folk psychological explanations presuppose the empirical domain: they describe two different forms of explanation – personal and subpersonal – within the same domain, the empirical domain.

This difference in domains, metaphysical versus empirical, is important. Why? Because it tells us that the explanatory dichotomy between folk and scientific psychological explanations is not to be confused with the metaphysical dichotomy between mind and brain. If one were to confuse the metaphysical and explanatory dichotomy, one would also conflate and identify false positively the metaphysical and empirical domains.

Critical reflection 2

Explanatory models and mind–brain models: method versus subject of investigation

Why is all this relevant? Because it tells us that the mind–brain problem as a metaphysical problem needs to be distinguished from the interface problem. Recall from the beginning of this chapter that the interface problem concerns the relationship between explanations on different levels, and ultimately the relationship between folk and scientific psychology. If the proponents now argue for the irreducibility of folk psychology to scientific psychology, they only imply explanatory dualism. This does not imply any preference for, or necessity of, metaphysical dualism. Instead, it opens the door for explanatory dualism to be associated with different metaphysical positions about the mind–brain problem ranging from dualism through materialism to alternative solutions.

One could, for instance, be an explanatory dualist who claims the irreducibility of folk to scientific psychology, while at the same time advocating materialism with regard to the mind–brain problem. This shows that both explanatory models and metaphysical assumptions can dissociate from each other. We consequently need to distinguish clearly between our explanatory models on the one hand and metaphysical issues like the mind–brain relationship on the other. In other words, the answer to the interface problem does not imply a specific answer to the mind–brain problem.

Put in even more general terms, the distinction between explanatory models and metaphysical issues makes it clear that we need to distinguish between method and subject of investigation. Both folk and scientific psychology describe specific methodological strategies in investigating psychological states and mental features. In contrast, the metaphysical question of the existence and reality of those very same mental features is the subject of investigation which can be investigated by relying on different methodological strategies.

In sum, the possible dissociation between explanatory dualism and metaphysical mind–brain dualism makes it clear that we should not confuse method and subject of investigation as we may, for instance, project false positively from the former to the latter. That, however, needs to be detailed further, as we will do in the next section.

Critical reflection 3

Explanatory dualism versus metaphysical dualism

Disregarding the distinction between method and subject of investigation, one could claim that explanatory dualism between personal and subpersonal levels may force one to assume metaphysical dualism between the mind and the brain. Why? Personal and subpersonal levels can be characterized not only by different kinds of concepts (see above), but also by principally different features like intentionality and non-intentionality. What is the existence and reality of these features?

The assumption of different features on subpersonal and personal levels may thus lead one from the empirical to the metaphysical domain. The explanatory dualism between subpersonal and personal explanations in the empirical domain may then resurface as dualism between the brain and the mind in the metaphysical domain. In short, personal–subpersonal explanatory dualism might go hand in hand with mind–brain metaphysical dualism.

Does the explanatory dualism between personal and subpersonal levels within the empirical domain imply metaphysical dualism, and thus a different existence and reality of subpersonal and personal features? This is a tricky question! In order for explanatory dualism *not* to imply metaphysical dualism, subpersonal and personal features cannot be principally different from each other. This position may endanger the assumed irreducibility of personal to subpersonal explanations and thus of folk to scientific psychology.

How can we thus avoid metaphysical dualism while at the same time preserving explanatory dualism? Recall from above that the proponents of explanatory dualism assume non-commensurability and non-derivability between personal and subpersonal levels of explanation. This means that both subpersonal and personal explanations are not related to each other, since otherwise they would no longer be incommensurable or non-derivable. In other words, there is no 'explanatory relationship' between subpersonal and personal levels.

Critical reflection 4

Explanatory relevance

Does the absence of any explanatory relationship imply metaphysical dualism? The contemporary American philosophers, John McDowell and Jennifer Hornsby, argue that this is not the case. They assume that subpersonal and personal levels of explanations are indeed not related to each other, signifying the absence of any explanatory relationship between them.

However, the absence of an explanatory relationship between the personal and subpersonal does not imply that the subpersonal is irrelevant for the personal. McDowell and Hornsby argue that 'enabling conditions' provide the necessary non-sufficient conditions or 'conditions of the possibility of', i.e. 'capacities' (Hornsby, 1997) (or 'predispositions'; see Part IV about consciousness for an explanation of the term 'predispositions') for the realization of the personal level. In other words, without the underlying subpersonal level, no personal level could be generated at all. The subpersonal level is thus relevant in explaining the possibility of the personal level. This is called 'explanatory relevance' (as distinguished from explanatory relation; see below).

However, we still need to distinguish between the possibility and the actual realization of the personal level. Consider the chicken. The hen has the possibility of laying an egg, whereas the rooster does not. This does not yet imply that the hen lays an egg. For that to happen, additional conditions, different from those that provide the hen with the possibility of laying an egg, are needed. Now, is the subpersonal level as explanatorily relevant for the actual realization of the personal level as it is for its possibility? Even if the actual realization of the personal level remains independent of the subpersonal level, both levels are still connected to each other by the explanatory relevance of the subpersonal level for the possibility of the personal one.

Why is the assumption of explanatory relevance so important? Because it indicates that subpersonal and personal levels are somehow linked to each other. If so, their underlying existence and reality cannot be principally different. If it were, the subpersonal level could not provide the enabling conditions necessary for the personal level. This, however, excludes the assumption of metaphysical dualism between mind and brain as metaphysical placeholders for personal and subpersonal explanations within the empirical domain. In other words, the assumption of explanatory relevance and its enabling conditions excludes the possibility that explanatory dualism implies metaphysical dualism.

Critical reflection 5

Philosophical versus neurophilosophical approach

One may now be eager to know more about the necessary non-sufficient or enabling conditions of the subpersonal level. What are these 'enabling conditions', 'capacities', or 'predispositions'? We currently do not know. For the philosophical proponents of irreducibility, this is not the main problem. Their focus is on the explanation of the mind and its personal level, and how that is principally different from the subpersonal explanations provided by science.

The conceptually- and logically-minded philosophers simply need to account for explanatory dualism with the irreducibility of personal to subpersonal levels, without implying metaphysical dualism between the mind and the brain. To do this, they need to logically and coherently demonstrate how explanatory dualism does not imply metaphysical dualism. With this aim, they have introduced the concept of 'capacities' or 'enabling conditions', which signify explanatory relevance as distinguished from explanatory relations.

But there are more questions to ask. What about the exact mechanisms, functions, and processes that signify and characterize the 'enabling conditions' or 'capacities' on the subpersonal level? What kind of neuronal mechanisms must the brain and its neuronal states show on the subpersonal level in order to enable mental states on the personal level?

The philosophers arguing for explanatory relevance and irreducibility may not be interested in these questions. For them it is sufficient to provide a logically and conceptually coherent claim to argue for explanatory dualism without metaphysical dualism. The exact empirical realization and implementation of these 'enabling conditions' or 'capacities' on the subpersonal level doesn't matter for the philosopher. These are questions for science to answer.

In contrast, the neurophilosopher may be interested in these questions. She or he may want to reveal and investigate the exact subpersonal, neuronal, mechanisms that enable, predispose or create the capacity of the brain to generate mental states on the personal level. If that is possible, the neurophilosopher may be able to directly link personal and subpersonal levels of explanation in terms of explanatory relevance. However, the assumption of such explanatory relevance excludes both explanatory reduction of personal to subpersonal levels as well as metaphysical dualism.

Explanatory continuity 1

Continuity between personal and subpersonal explanations

Functionalism was introduced in Chapter 6 on the mind–brain relationship. Roughly, functionalism describes the characterization of the mind in orientation to its functional role – the relationship between input, internal states and output – rather than in terms of either a mental or physical substance or property. By invoking the concept of the functional role, mental states are compared to a computer, and more specifically to computer software that runs a specific program.

As comparable to software, mental states must be distinguished from the underlying hardware, the material. In this metaphor, the hardware may consist either of neurons, in the case of the brain, or of silicon, in computers (or otherwise). The hardware – the brain or the computer itself – may then be regarded as what is described as a realizer of mental states.

How does the assumption of the functional mind relate to the distinction between personal and subpersonal levels? Recall that the personal level of explanation was characterized by the law-like generalization typical of folk psychology. The subpersonal level relies on causal generalization as used in scientific psychology.

How are both law-like and causal generalizations related to each other? The irreducibility proponent claims that they are principally different and not related to each other at all. The functionalist, however, denies this. For the functionalist, law-like generalizations on the personal level are a species or subtype of causal generalization as made on the subpersonal level. The causal, physical mechanisms are the concrete manifestations on the subpersonal level. On the personal level, these are simply more abstract.

Explanatory continuity 2

Explanations of anxiety at different levels

Let me give an example of an abstract specification. The explanation on the personal level is that I was afraid and angry when, on waking up, I noticed a thief in my apartment, who wanted to steal my money, credit cards and cell phone. To prevent him from leaving, I threw the alarm clock next to me on my night table towards him, which resulted in the breaking of a window.

How can we describe this scene and thus my mental states in functional terms? The input was the thief having entered my apartment, which led to my anxiety and anger. That in turn was coupled with another internal state, the intention to not let him escape and to get back my valuables, which caused my action of grabbing and throwing the alarm clock towards him.

What would be a subpersonal explanation of the same scene? A subpersonal explanation would explain the same in terms of certain biochemicals and neuronal states underlying my anxiety and anger, and the subsequent intention to throw the alarm clock. Specific biochemicals may have increased, which led to my anxiety and anger. Then my premotor cortex became active, signifying my intention to throw the alarm clock, which induced activity in motor cortex and the subsequent action, e.g. the throwing of the alarm clock. Finally, since my movement was quick, the velocity of the alarm clock was extremely high, which caused the already brittle window to break.

How is the subpersonal explanation related to the personal explanation? Functionalism argues that the personal explanation is just an abstraction of the subpersonal. The personal explanation of the functional role of mental states in terms of input, internal states and output implies the subpersonal explanation, the physical and neuronal processes. Those are left out of the personal explanation, making it an abstraction from the subpersonal explanation. These abstractions allow one to filter out from the subpersonal level what is relevant and thus specific to explain the personal level. One may thus speak of the abstract specification of subpersonal explanations on the personal level.

What does the characterization of personal explanations as abstract specification imply for its relation to the subpersonal level? Once subpersonal causal explanations and laws are acknowledged to characterize the personal level via their abstract specification, the principal distinction between both levels fades away. Instead of being different, personal and subpersonal explanations can be related to each other, thus allowing for an explanatory relationship between them (rather than mere explanatory relevance as the proponent of irreducibility argues; see above). There may thus be continuity, rather than irreducibility, between personal and subpersonal levels of explanation.

Explanatory continuity 3

Commensurability and derivability

Recall from above the criteria for the successful reduction of personal to subpersonal explanations. Commensurability describes the possibility of translating personal into subpersonal concepts, while derivability concerns the inference of personal from subpersonal explanations. Now, if personal explanations are regarded as abstract specifications of subpersonal explanations, both can be translated into each other. The subpersonal neuronal and physical explanation of the example above – the biochemicals and the activation of the motor and premotor cortex – can be translated into the personal explanation of intentions and actions as we illustrated. If so, there is commensurability rather than incommensurability between subpersonal and personal explanations.

This commensurability is accompanied by derivability. The subpersonal, biochemical and neuronal processes are assumed to correspond not only to the intentions and actions on the personal level. Even more importantly, the latter may be inferred from the former, thus allowing for the derivability of personal level explanations from subpersonal explanations. The personal level explanations are thus 'echoed' and mirrored in the explanations of the subpersonal level.

One may now want to argue that the same software, as in the same functional relations, may be realized and implemented by different hardware or realizers. Since the different realizers in the subpersonal level may show different causal relations, they may affect the personal level explanations. This position would undermine the assumption of functionalism that defines mental states by their functional roles on the personal level, rather than by causal relations on the subpersonal level.

Do different realizers on the subpersonal level imply different explanations on the personal level? The functionalists say no, since the realizer only realizes and implements the causal relation whereas it does not determine and specify the kind of causal relations that occur. The same causal relationship on the subpersonal level can be realized in multiple ways by different realizers. This is called 'multiple realizability' (see Chapter 6 for details).

How can we define the concept of multiple realizability? Multiple realizability means that the functionalist characterization of mental states can be realized on different levels, both subpersonal and personal. This implies explanatory continuity between subpersonal and personal levels. Subpersonal and personal explanations can consequently be commensurable and derivative, without giving up the functionalist assumption of determining mental states by their functional roles on the personal level.

Explanatory continuity 4

Philosophical functionalism

Let us recall an important distinction from Part II. There we distinguished between philosophical functionalism and psychological functionalism. Philosophical functionalism describes the characterization of mental states by their functional roles in a conceptual way, independent of any empirical observation. Psychological functionalism refers to the empirical observation of mental states' functional roles in psychology, thus presupposing an a posteriori, rather than an a priori strategy.

Why is this distinction between philosophical and psychological functionalism relevant within the current context of the relationship between subpersonal and personal explanations? Both forms of functionalism imply different strategies for getting to the explanations on the personal level. This in turn implies different ways of characterizing folk psychology. Let us detail this further.

How can we find and detect the causal generalizations and laws on the personal level as discussed in folk psychology? Advocates of philosophical functionalism argue that they can literally be read from our common-sense or everyday psychology. Everybody has at least an implicit understanding of the fundamental principles and causal laws that guide the personal level of folk psychology. Folk psychology can thus be thought of as theory that provides us with knowledge about causal explanations and causal laws.

Most importantly, the knowledge about causal generalizations and laws acquired on the personal level can be used to guide our investigation on the subpersonal level of the realizers. This may occur in two different ways. The first is by observing the behavior and its associated mental states, sometimes called 'folk functionalism' (Rey, 1997). This occurs by mere observation without the prior recruitment of concepts, implying an a posteriori rather than an a priori strategy.

Alternatively, rather than relying on observation, one may analyze the concepts themselves used in the causal generalization of folk psychology. Unlike folk functionalism, analysis of concepts can occur a priori (prior to and thus independent of the observation of the behavior and mental states they describe). This implies an a priori strategy and is called 'a priori or conceptual functionalism' (see Shoemaker, 1984).

Despite their different ways of accessing the causal generalizations and laws on the personal level, both varieties of philosophical functionalism share the same basic presupposition. Namely that the causal generalization and laws on the personal level can be accessed and brought to light without empirical or scientific investigation of the subpersonal level. This is possible on the basis of assuming isomorphism between the personal and subpersonal levels of explanation, where the latter is mirrored in the former.

Accordingly, everything is already at least implicitly present on the personal level itself. Folk psychology provides the key to the causal laws and generalizations on the personal level. The personal level can thus be explained completely by folk psychology and its concepts. Scientific psychology is not necessary. This is the main claim of philosophical functionalism, which argues for the prime importance of philosophy and folk psychology in explaining the personal level. It therefore denies the relevance of scientific psychology and its subpersonal explanations for explaining the personal level.

Explanatory continuity 5

Psychological functionalism

Is scientific psychology really irrelevant, as claimed in philosophical functionalism? This is denied in yet another version of functionalism, psychological functionalism (e.g. psychofunctionalism). Why? While the personal level of folk psychology provides us with predictions, it does not imply that these predictions are causal generalizations and laws. According to psychological functionalism, laws are useful for predictions and thus for the personal level of folk psychology. Such predictions should not be confused with explanations, which can be made only on the subpersonal level.

How can we further specify the distinction between predictions and explanations? Laws operating on the personal level do not account for the explanation themselves, i.e. the explanatia (things that do the explaining). Instead, they are what needs to be explained, the explananda. And this explaining of the predictions and laws characterizing the personal level can only be done by the subpersonal level and thus by scientific psychology.

Psychological functionalism thus claims that we need scientific psychology and its causal explanations on the subpersonal level in order to explain and account for the predictions made by the personal level and by folk psychology. In other words, the predictions on the personal level do not provide any explanation. As such, folk psychology does not yield explanations, but only predictions on the personal level. Hence, there could be no explanations on the personal level without the subpersonal level and the work of scientific psychology.

How does psychological functionalism and the emphasized role of scientific psychology relate to philosophical functionalism? Philosophical and psychological functionalism can be distinguished by the role of the personal level in acquiring explanations of the mind. Philosophical functionalism considers the personal level to be sufficient to discover the causal laws and generalizations by either observation of behavior ('folk functionalism') or analysis of concepts ('a priori or conceptual functionalism').

Psychological functionalism, in contrast, argues that we need to consider the subpersonal level too in order to explain the causal laws and generalizations. The personal level is thus by itself not sufficient for yielding personal level explanations. For that we need to recruit the subpersonal level, which is only possible on the basis of scientific psychology. Hence, unlike in philosophical functionalism, psychological functionalism considers the recruitment of scientific psychology necessary for explaining the causal generalizations and predictions on the personal level.

Let me detail the relationship between the personal and subpersonal levels in both forms of functionalism. The philosophical functionalist presupposes isomorphism between both levels with the personal level providing access to the subpersonal. In contrast, the psychological functionalist emphasizes the necessity of the subpersonal level and its multilayered structure in explaining the personal level. More specifically, he/she, unlike the philosophical functionalist, considers the investigation of the multilayered subpersonal level by scientific psychology, cognitive psychology, and neuroscience to be necessary to understand the personal level.

This is different in the philosophical functionalist. The philosophical functionalist does not really take the subpersonal level itself into account. Rather than making a contribution to the personal level, the subpersonal level is more or less enslaved by the personal level, to which it is assumed to be isomorphic. Everything, including any causal law and generalization characterizing the personal level, is supposed to also apply on the subpersonal level. Unlike in psychological functionalism, the subpersonal level does not 'stand on its own feet' in philosophical functionalism.

Critical reflection 6

Explanatory levels and neurophilosophy

Why is all this important in the current neurophilosophical context of the brain and its relationship to philosophy? Today scientific psychology is highly concerned with the brain and neuroscience. The way one determines the relationship between folk and scientific psychology may thus impact the role one could attribute to the brain and its investigation in neuroscience.

Let us be specific. If folk psychology is assumed to be sufficient and thus able to account for personal-level explanations, as claimed in philosophical functionalism, the brain and its subpersonal-level investigations in neuroscience may remain irrelevant when it comes to mental states on the personal level. Neuroscientific explanations of the brain's neuronal mechanisms might then be considered unrelated (and irrelevant) to the personal level explanations of mental states.

More generally, neurophilosophy would then not only be superfluous and redundant but impossible. If there is no relationship between subpersonal and personal levels, neuroscience has to remain within the confines of the subpersonal, purely neural, level. Any linkage of neuroscience to mental states on the personal level as they are dealt with originally in philosophy then remains illusory.

The assumption of such impossible linkages between mental and neural states would also make the linkage between neuroscience and philosophy, and thus neurophilosophy, impossible. The disciplinary triangle between neuroscience, philosophy and neurophilosophy postulated here would then be replaced by mere parallelism (and sègregation) between philosophy and neuroscience without any neurophilosophical border stations. The situation changes, however, once one considers scientific psychology and its subpersonal characterization for explaining the personal level in folk psychology. This may open the door to include the brain and its neuronal mechanisms that may underlie the subpersonal explanations in scientific psychology. This in turn makes it possible to attribute a direct explanatory relationship between the subpersonal neuroscientific explanations and the personal explanations of mental states.

Critical reflection 7

Neurophilosophy and philosophical and psychological functionalism

What could such an explanatory relationship between subpersonal and personal, that is, neural and mental levels, look like? This is the moment where neurophilosophy comes in. As we have seen in Chapter 4, one may opt for different kinds of models for the explanatory relationship between neural and mental levels. One may pursue either a reductive, and more specifically brain-reductive, explanation that remains within the confines of the empirical domain; or, one may opt for a non-reductive and more specifically brain-based model of explanation that includes and links the different domains, including metaphysical, empirical, epistemological, phenomenal and conceptual (see Chapter 4 for details).

How does this stand in relation to the distinction between philosophical and psychological functionalism? Philosophical functionalism argues for the irrelevance of scientific psychology for personal-level explanations. By that, philosophical functionalism excludes any possible explanatory relationship between the subpersonal-level and the personal-level explanations. Going beyond scientific psychology, this also implies that the brain and its neuronal mechanisms as part of the subpersonal level will not show any explanatory relationship (and relevance) to the personal-level explanation of mental states. This implies that neurophilosophy is basically irrelevant, if not impossible. In other words, any possible field of neurophilosophy is impossible.

This is different in psychological functionalism. Here the assumption of the explanatory relationship between the subpersonal-level and the personal-level explanations opens the door not only for scientific psychology, but also neuroscience. The brain and its subpersonal neuroscientific explanations might then be related to (and relevant for) the personal-level explanations. In this model, neurophilosophy as distinct from both scientific psychology and philosophy becomes possible.

Explanatory elimination 1

Theoretical posits and other properties in folk psychology

So far we have investigated two main positions on the relationship between folk and scientific psychology. The first position argued for irreducibility, meaning that folk psychology cannot be reduced to scientific psychology. The second position, functionalism, assumed continuity between personal- and subpersonal-level explanations. While there are many intermediate positions discussed in current philosophy of psychology, I focused here on only the main contrasting suggestions. Now we shall complement this picture by considering another extreme, namely the elimination of folk psychology altogether.

Let us characterize folk psychology in further detail. Folk psychology describes our common-sense explanations of our everyday behavior and its mental states. Such common-sense explanations tacitly presuppose some kind of theory, Theory-Theory as it was called (first by Morton, 1980). For instance, if somebody is crying, we infer that the respective person must be sad. We start speculating that she or he must have experienced some bad event or just received some bad news.

We make law-like generalizations from our observations of the person's behavior. This leads to assumptions about mental states in the respective person. These generalizations describe certain relations and regularities, so that when a person shows behavior x, he or she must show the mental state y. Though not observable, the mental state y is assumed to be a 'theoretical posit'– the assumption of a particular existing and real mental entity corresponding to and underlying the mental state y.

In addition to the theoretical posit, mental states are assumed to show causal, semantic and qualitative properties. Causal properties are manifest, in that they impact subsequent behavior. For instance, if you have the desire to eat a hamburger, you will go to McDonald's to get one. You may experience certain qualities while desiring and eating the hamburger, certain phenomenal-qualitative features, which are manifest in what is described when you answer the question: 'What is it like to desire and eat a hamburger?'

What about semantic properties? Semantic properties concern the meaning of the hamburger, what it means to me and why it is so important to me. Semantic properties are manifest in the content; the hamburger itself as mere object is not what characterizes your mental state. Rather it is the meaning of the hamburger to you, as traced back to previous experiences of the wonderful taste, that accounts for your mental state. Your mental state has hamburgers as its contents – it is directed toward hamburgers, which is described by the term intentionality (or directedness towards; see Part IV for details).

Explanatory elimination 2

Rejection of folk psychology (Churchland)

On the whole, folk psychology can be characterized by theoretical posits and their specific causal relations, their semantic properties and their qualitative features. Folk psychologists seem to presuppose the theoretical posits and their respective features to be real and existent. This is denied by the American philosophical couple Patricia and Paul Churchland, who teach at the University of California, San Diego.

We got to know a little bit about Patricia and Paul Churchland in the first part of this book when I introduced the concept of neurophilosophy (see Chapter 4). There we discussed the Churchlands' reductive definition of neurophilosophy as the investigation of originally philosophical concepts using the empirical methods of scientific psychology and neuroscience.

The Churchlands' strong emphasis on the empirical domain is also true in their opinion of folk psychology. They reject folk psychology altogether for its non-scientific basis and call for its complete replacement by scientific psychology. The explanations of folk psychology simply do not hold true because they assume theoretical posits with causal, semantic and qualitative features to be real and existent. However, according to the Churchlands, nothing can be found in reality and existence that corresponds to such theoretical posits and their respective features.

The assumption of such theoretical posits must thus be discarded completely. This in turn implies the elimination of folk psychology and its replacement by scientific psychology, and ultimately neuroscience. Personal-level explanations will thus be completely replaced by subpersonal-level explanations, which are in turn therefore considered both necessary and sufficient to account for mental states on the personal level.

Explanatory elimination 3

Arguments against folk psychology (Churchland)

Why do the Churchlands reject and eliminate folk psychology and its assumption of theoretical posits? They discuss various arguments that will be considered briefly below. First, they argue that folk psychology has not yielded any progress. Instead, it remains stagnant. For progress to take place, folk psychology would need to develop a clear research program with considerable explanatory power and experimentally testable hypotheses. However, there is neither a clear research program nor any foreseeable strong explanatory power in folk psychology, let alone any hypotheses that are amenable to rigorous experimental testing.

Following the Churchlands, folk psychology simply cannot explain nor test experimentally various mental states and phenomena like dreaming, mental disorders, consciousness, etc., that all remain mysterious. Hence, according to the Churchlands, folk psychology is futile and needs to be eliminated.

Second, previous folk theories have not been very successful. Common-sense explanations are not only prevalent in psychology but also in other subject matter that is now described by scientific disciplines. For instance, scientific astronomy told us that the earth revolves around the sun, whereas before this folk astronomy believed the sun to revolve around the earth, and that the earth was the center of the universe. The same applies to folk physics, folk biology and others. All these were fundamentally mistaken and replaced by their respective scientific disciplines. The same fate, according to the Churchlands, will befall folk psychology, because its assumptions and concepts will eventually be completely replaced by the insights of scientific psychology and ultimately by neuroscientific concepts.

Finally, the methodological strategy of folk psychology is problematic. It predominantly relies on introspection, the inner observation and perception of mental states and their contents. Since introspection relies on private access in first/second-order perspective, rather than public access as in third-person-based observation, folk psychology remains unable to give valid and reliable accounts of mental states.

Due to its inherently private nature, folk psychology may also be loaded with personal and private theories that affect observation and perception. One may consequently introspect what one wants to introspect, rather than what there is, independent of one's own introspection. This distinguishes folk psychology from scientific psychology. Unlike folk psychology, scientific psychology relies on publicly accessible observation in third-person perspective. Therefore scientific psychological explanations can be verified, as well as falsified, which remains impossible in folk psychological explanations.

Critical reflection 8

Matching between folk and scientific psychological concepts

The Churchlands are quite radical in considering folk psychology a complete failure. They do not believe it has any place at all in future explanations of the personal level. The Churchlands' approach, however, is not the only possibility. One may opt for a more differentiated approach. Let us consider some possible arguments against the complete elimination of folk psychology.

There are many instances where we are quite successful in predicting other people's behavior. This is, for instance, the case when a person stares at a cake on a table. We may then assume that that person is not only hungry but has a strong longing for something sweet. Most of the time, we are not deceived, and are thus usually correct in our folk psychological explanations.

Instead of eliminating folk psychology altogether, we may, therefore, include personal-level explanations. This would allow us to test single folk psychological concepts and explanations against the explanations of the subpersonal level using psychological and neuroscientific investigations. In other words, folk psychological concepts and explanations can then be compared and matched with the scientific psychological and neuroscientific concepts and explanations.

What are the possible outcomes of such a matching procedure? If both correspond and match, folk psychological concepts might sometimes be empirically sound. If, in contrast, there is no matching, the respective folk psychological concept may be diagnosed as empirically implausible and thus not applicable in our natural world. However, we need to be careful. Such empirical implausibility only applies to the natural world, whereas the very same constellation may be conceivable and thus conceptually plausible in a purely logical world. Accordingly, empirical implausibility (in the natural world) does not exclude conceptual plausibility (in the logical world).

Critical reflection 9

Criteria for success and failure of matching (Stich)

The question is, of course, what kinds of criteria does one apply for successful versus unsuccessful matching of folk psychological concepts to scientific psychological and neuroscientific concepts? The American philosopher Steven Stich (1943–) speaks here of 'reference failure' and 'reference success'. Stich argues that it is difficult to distinguish between 'reference failure' and 'reference success', and that there may be plenty of intermediates. The boundary between folk and scientific psychological concepts may thus turn out to be rather blurred when it comes to either linking or distinguishing between them. We consequently remain unable to draw a clear demarking line between folk and scientific psychological explanations. In turn, this makes it impossible to completely eliminate folk psychology and replace it completely by scientific psychology.

What are the criteria upon which we can say that the theoretical posits of folk psychology refer to nothing real? How can we distinguish these cases from those that refer to something real? The distinction between reference failure and success may be rather blurred and not as clear-cut as eliminativists of folk psychology assume. However, we then no longer know whether folk psychology refers to something real or not; this implies that the ground upon which eliminativism stands, the possible distinction between reality and illusion, is eliminated. If, however, the ground or the presuppositions of eliminativism are no longer valid, eliminativism itself is eliminated.

Explanatory elimination 4

Elimination of mental concepts

What does the elimination of the folk psychological conception of mental states described above imply for the metaphysical status of the mind? As described in Part II in detail, some, mostly Anglo-American, philosophers in the early and mid-twentieth century, like C. Broad, P. Feyerabend, W. Quine and R. Rorty, doubted that mental notions like belief, fear, hope and sensation are useful in describing what happens. Instead, the mental concepts we use to describe our subjective experience should be replaced by neuroscientific concepts describing the brain processes. The paradigmatic example is that of the mental concept of pain, and the neuronal concept of the C-fiber. The concept of the C-fiber as an underlying neuronal concept is assumed to replace the former mental concept, the one of pain.

Why are mental concepts not useful? In general, advocates of materialism argue that mental terms may be reduced to neural terms describing brain processes. This amounts to identity theory and other forms of reductive materialism as discussed in Chapter 6. Reductive materialism still presupposes some kind of mind, which is then reduced to the brain. This tacit presupposition, the assumption of mind, is denied in an even more radical version, eliminative materialism. Rather than assuming the existence and reality of the mind as reducible to that of the brain, one may deny any kind of existence and reality to the mind whatsoever.

This denial results in what is described as eliminative materialism. The position of eliminative materialism is a metaphysical assumption about the existence and reality of the mind and its relationship to the brain. Let me explain in more detail. The eliminative materialist goes one step further. By considering the reduction of mental to neural terms, the materialist still acknowledges that the mental terms refer to something real and existent – the mind, even if it is more properly described by neural and materialist terms. This assumption of something real and existent is doubted and rejected by the eliminative materialist.

The eliminative materialist argues that mental terms fail because they do not refer to anything that is actually real or existent. As such, according to the eliminative materialist, there is no mind or mental states at all. Any assumption about the existence and reality of a mind and mental states is nothing but an illusion. If so, even the question of any kind of reduction of the mind's existence and reality to that of the brain is superfluous and redundant. Nothing but the physical brain and body exist. This is what our philosophical couple, Patricia and Paul Churchland, assume.

Explanatory elimination 5

Empirical versus metaphysical elimination

How does eliminative materialism relate to the elimination of folk psychology? The metaphysical elimination as suggested by eliminative materialism must be distinguished from the explanatory elimination of folk psychology. As pointed out at the beginning of this chapter, both folk and scientific psychology presuppose the empirical domain. As such, the difference between folk and scientific psychology consists not so much in their presupposed domain, but rather in their methodological strategies and how they yield explanations of the personal level.

The elimination of folk psychology should thus not be confused with the metaphysical elimination of the mind in eliminative materialism. Both forms of elimination presuppose different domains: the metaphysical elimination of the mind presupposes the metaphysical domain, while the elimination of folk psychology is 'located' in the empirical domain. One may therefore contrast them as empirical and metaphysical eliminations.

How do metaphysical and empirical eliminations relate to each other? Since they concern two different domains, both eliminations are distinct issues. This means that the elimination of one does not imply the elimination of the other. For instance, one may well eliminate the mind and postulate eliminative materialism, while at the same time rejecting empirical elimination of folk psychology in favor of scientific psychology. How is that possible? Folk and scientific psychology do not imply any metaphysical commitments about the existence and reality of the mind. Instead, they reflect only different methodological strategies and conceptual explanatory frameworks to account for the observation of psychological states. We now come back to the distinction between method and subject of investigation, and thus method/explanation and metaphysical assumptions as pointed out above. This tell us that one needs to be careful in delineating the input of the observer or the investigator, namely the method he/she applies, from the subject matter itself, i.e. the psychological states and their mental features, he/she aims to investigate with that very method.

Explanatory elimination 6

Phenomenal elimination

So far we have discussed different forms of elimination – empirical and metaphysical. Metaphysical elimination concerns the elimination of the mind, which in this model is not even given a physical existence and reality. Empirical elimination concerns the elimination of a particular conceptual and thus explanatory framework – folk psychology. Both metaphysical and empirical eliminations are complemented by yet another form of elimination called phenomenal elimination.

Let us first describe the concept of the phenomenal. Today, consciousness is often considered the paradigmatic example of mental states upon which the question of mind is to be decided. Consciousness describes the subjective experience of ourselves and the environment in first-person perspective. As such, subjective experience is characterized by various features like intentionality (the directedness towards contents), unity (the linkage of different objects and contents into one unified field of experience) and subjectivity (point of view and first-person perspective).

How can we characterize these concepts, i.e. unity, subjectivity, intentionality? Since these concepts describe the features and structure of our subjective experience or consciousness, they are conceptualized as phenomenal features (see Chapters 1 and 14 for details). Another phenomenal feature is the qualitative character of our subjective experience, the 'what it is like' one feels when experiencing certain events. This qualitative character of our subjective experience is described by the term 'qualia'. Besides subjectivity and intentionality, qualia are often considered to be one of the hallmark features of consciousness.

Explanatory elimination 7

Do we need to eliminate qualia?

Are qualia real and existent? Can they, as phenomena on the personal level, be reduced to psychological and neuronal mechanisms on the subpersonal level? There has been much discussion about this issue, which will be recounted in Chapter 13.

One might also eliminate the problem of qualia altogether by simply eliminating them. Unsurprisingly, the Churchlands also opt for this strategy. The same is also advocated by Daniel Dennett, whose arguments will be discussed briefly below.

In his well-known article 'Quining Qualia' (Dennett, 1988), Dennett goes one step further and confronts the basic phenomenal features of consciousness, namely subjectivity and qualia. Dennett argues that subjectivity and qualia are simply intuitions. He considers qualia and subjectivity to be private, meaning that they can be accessed only by the experiencing subject itself. This is called 'auto-phenomenology' – a term that highlights how phenomenal features refer to and are only accessible by the experiencing subject itself. According to Dennett, we are, however, able to access another person's experience. Empathy is an example of this. In other words, we have some (but limited) intersubjective access to other people's consciousness.

The description of phenomenal features such as subjectivity and qualia as private is consequently flawed and can no longer be maintained. In other words, subjectivity and qualia are not as private as they are presumed to be. One may therefore replace subjective 'auto-phenomenology' with 'hetero-phenomenology' – a term that describes consciousness as intersubjective, rather than in terms of concepts like qualia that refer only to intra-subjective (rather than inter-subjective) states.

Does this mean that all phenomenal features are eliminated, and consequently consciousness is eliminated as well? No, Dennett and others only argue that a certain conception of phenomenal features are private and that auto-phenomenology needs to be eliminated. According to Dennett, we conceptualize phenomenal features the wrong way: rather than characterizing them as private and auto-phenomenological, we should describe them as public and hetero-phenomenological. Accordingly, we only need to eliminate a particular conception or description of phenomenal states rather than the phenomenal states themselves. In other words, we need to change our conception of phenomenal states and thus the methodological strategy by means of which we describe and thus conceptualize phenomenal states. To reject phenomenal states altogether would thus be to conflate our conceptions of phenomenal states with what we aim to conceptualize, the phenomenal states.

Critical reflection 10

Do we need to eliminate mental concepts in neurophilosophy?

Why is all this relevant within the present context of neurophilosophy? Different forms of elimination entail different kinds of neurophilosophy. Recall Chapter 4, where we distinguished between reductive and non-reductive forms of neurophilosophy. Reductive neurophilosophy focused only on the brain and the empirical-experimental investigation of original philosophical concepts. Meanwhile, non-reductive neurophilosophy also made room for philosophical investigation in different domains. This allows us to compare and match the conceptual determinations, as suggested in philosophy, with the results from the empirical-experimental investigation in psychology and neuroscience.

How do these different forms of neurophilosophy relate to the different forms of elimination discussed above? The empirical elimination of folk psychology implies that no concepts other than scientific psychological (and neuroscientific) ones can be used to explain psychological phenomena. Not only does that eliminate folk psychological concepts but it eliminates mental concepts as well, including those used in philosophy and philosophy of mind. One may thus want to speak of conceptual elimination.

This has major reverberations on neurophilosophy. If philosophical and folk psychological concepts are eliminated, there is no longer a possibility for non-reductive neurophilosophy. Why? Non-reductive neurophilosophy presupposes concepts other than scientific psychological and neuroscientific concepts when assuming the comparison and matching between philosophical and neuroscientific concepts. Philosophical concepts are, most often, mental concepts. If the mental concepts are eliminated, there are no longer any concepts with which the neuroscientific concepts can be matched and compared. Non-reductive neurophilosophy, relying on such matching and comparison, would consequently become impossible.

Let us put this in a slightly different way. If mental concepts are eliminated, including the elimination of folk psychology in favor of scientific psychology, all domains other than the empirical domain (like the metaphysical and epistemological domains that rely on mental concepts) are also eliminated. There would thus be domain monism rather than domain pluralism. That, in turn, restricts any subsequent neurophilosophy to its reductive formulation, which confines itself to the empirical domain exclusively and thus to domain monism (see Chapter 4). In contrast, non-reductive neurophilosophy that presupposes domain pluralism rather than domain monism then remains impossible.

Critical reflection 11

Are phenomenal and metaphysical elimination relevant in neurophilosophy?

Such an exclusive focus on the scientific psychological and neuroscientific concepts is further sharpened and condensed when considering phenomenal elimination. If, for instance, the concept of qualia is eliminated because of its subjective and auto-phenomenological nature, any non-objective and thus non-scientific concepts are also excluded. Subjectivity as expressed in the concept of qualia would, in this case, remain barred from any future neurophilosophy. There would simply be no phenomenal features any more.

What does such phenomenal elimination imply for this kind of neurophilosophy? Neurophilosophy would become purely objective in the case of such phenomenal elimination. This implies the reductive model of neurophilosophy.

More specifically, neurophilosophy would then, in the best of all cases, be realized as an appendix to neuroscience, as a theory of neuroscience that addresses theoretical models and issues about explanation and concepts in neuroscience. This is exactly the way reductive neurophilosophy is set up and therefore is often considered to be closely related to, if not identical with, philosophy of neuroscience. Meanwhile, in the worst of all cases, neurophilosophy would be eliminated and be completely replaced by neuroscience in the same way that the phenomenal features are replaced by the neural features of the brain. What about the metaphysical elimination of the mind? Metaphysical elimination of the mind means that the mind is no longer attributed any existence and reality. Does such elimination of the mind eliminate neurophilosophy? We should be careful about what exactly is eliminated in such a case. The metaphysical elimination of the mind only considers the existence and reality of the mind as such, whereas it does not concern the elimination of the mental and phenomenal features themselves.

What does this imply for neurophilosophy? Neurophilosophy is about how the brain and its neural features relate to the phenomenal and mental features (that philosophy attributes to a mind; see Chapters 5–7). This is what I described as brain-based neurophilosophy as distinguished from mind-based philosophy (see Chapters 7 and 8). The metaphysical elimination of the mind would consequently make any kind of philosophy impossible because of the latter's mind-based approach.

In contrast, a brain-based approach, as in non-reductive neurophilosophy, remains untouched by the metaphysical elimination of the mind. Why? Brain-based neurophilosophy does not presuppose any kind of metaphysical existence and reality of the mind at all. Instead, brain-based neurophilosophy only presupposes the existence and reality of the brain. This means that the metaphysical elimination of the mind as, for instance, in eliminative materialism does not eliminate non-reductive and thus brain-based neurophilosophy but rather is highly compatible with it.

Take-home message

The previous part discussed the existence and reality of the mind and its relationship to the brain and the body. This is described as the mind–brain problem and is a metaphysical problem concerning the existence and reality of the mind. This problem has traditionally been discussed within the realm of philosophy and especially the philosophy of mind. In addition to the metaphysical domain of existence and reality, the mind has also been investigated in the empirical domain. The empirical investigation of the mind takes place in psychology and, most recently, also in neuroscience. The investigative shift of the mind from the metaphysical to the empirical domain raises important theoretical issues, which are dealt with in the philosophy of psychology. One central topic in the philosophy of psychology is the relationship between our common-sense, or folk, conceptions, and the scientific psychological explanations of our mental states and behavior. The present chapter discusses different possible relationships between folk psychology and scientific psychology that range from irreducibility and continuity to elimination. Most importantly, the suggested relationship between folk and scientific psychology is not only relevant within philosophy of psychology, but also for the future of neurophilosophy and its relationship to both philosophy and neuroscience.

Summary

The mind–brain problem is a metaphysical problem that explores the existence and reality of the mind and its relationship to the brain. The mind–brain problem is a core feature of philosophy of mind, a subfield of philosophy that also considers the mind's mental features, including self, subjectivity, consciousness, etc. The metaphysical mind–brain problem is complemented by the empirical investigation of psychology, and most recently neuroscience. This raises the question: how are the everyday and common-sense explanations of our mental states and behavior, so-called folk psychology, related to their explanations in scientific psychology (and neuroscience)?

How are both forms of psychology related to each other? This question is called the 'interface problem', exemplified by its concern with the interface between folk and scientific psychology. Different relationships between both forms of psychology have been discussed as part of a theory of psychology, or philosophy of psychology.

Proponents of irreducibility claim that folk and scientific psychological explanations rely on principally different and mutually exclusive forms of explanation. While horizontal explanations are possible within each level, personal and subpersonal, vertical explanations, as explanations between both levels, are denied. Reduction of the personal level to the subpersonal thus remains impossible, implying incommensurability and non-derivability between both levels.

This is different in functionalism. Functionalism characterizes the mind by its functional roles between input, internal states and output on the personal level. These functional roles are regarded as abstract specifications of the causal relations that apply on the subpersonal level. The relationship between personal and subpersonal levels is thus one between realizer and functional role, implying continuity between both levels. This continuity between subpersonal and personal levels may be either conceptual or empirical: philosophical functionalism assumes concepts on the personal level to correspond to isomorphic structures on the subpersonal level. Psychological functionalism does not presuppose such isomorphism but argues instead for the subpersonal psychological functions themselves to provide continuity to the personal level explanations.

Finally, in addition to irreducibility and continuity, one may also argue in favor of eliminating folk psychology and its personal-level explanations altogether. In its place, scientific psychology and its subpersonal-level explanations would be considered sufficient. This is called empirical elimination. Empirical elimination must be distinguished from metaphysical elimination, the elimination of the existence and reality of the mind, and phenomenal elimination, the elimination of phenomenal features of consciousness, like subjectivity and qualia. These different forms of elimination – metaphysical, empirical and phenomenal – carry important implications for the kind of neurophilosophy that might be possible, including neurophilosophy's relationship to both philosophy and neuroscience.

Revision notes

- What is the difference between vertical and horizontal explanations?
- Explain the difference between explanation and description of the mind.
- Define the mind–body problem and the mind problem.
- What is the interface problem?
- How can we distinguish between metaphysical and empirical accounts of the mind?
- How is philosophy of mind different from psychology and philosophy of psychology?

- Why is there is a distinction between folk and scientific psychology, and how can one characterize both?
- What are the criteria upon which one might reduce folk to scientific psychology?
- What is the argument of functionalism with regard to the relationship between folk and scientific psychology? How does this relate to the interface problem?
- How can we distinguish between different varieties of functionalism?
- Why do some authors argue in favor of eliminating folk psychology?
- Can you distinguish between different forms of elimination?
- Why is the relationship between folk and scientific psychology relevant for neurophilosophy?

Suggested further reading

- Bermudez, J. L. (2005) *Philosophy of Psychology* (London: Routledge).
- Dennett, D. C. (1993) *Consciousness Explained* (Boston, MA: Back Bay Books).
- Hornsby, J. (1997) *Simple Mindedness: A Defence of Native Naturalism in the Philosophy of Mind* (Cambridge, MA: Harvard University Press).
- McDowell, J. (1994) *Mind and World* (Cambridge, MA: Harvard University Press).
- Nagel, E. (1961) *The Structure of Science: Problems in the Logic of Scientific Explanation* (New York/Burlingame: Harcourt, Brace & World).

10
Philosophy of Psychology: Mind and Meaning

Overview

The previous chapter discussed different suggestions as to how the personal level of our mental explanations is related to the subpersonal level Three different possible relationships between personal- and subpersonal-level explanations, which reflect the relationship between folk and scientific psychology, were discussed, ranging from irreducibility and continuity to elimination. This discussion, however, left out one of the central hallmarks of the personal level, the semantic meaning. Mental states as experienced on the personal level refer to certain objects, events, or people. And most importantly, these objects, events, or people are attributed certain meanings that determine our mental state and associated behavior. By attributing meaning, the objects, people, or events become semantic contents that characterize the respective mental state. How does our mind generate these meanings and thus the semantic contents? This is the focus of this chapter.

Objectives

- To understand the difference between functional and representational roles
- To grasp the distinction between propositional attitudes and contents
- To reveal the insufficiencies of functionalism and the necessity of a representational account
- To learn what is meant by structural isomorphism
- To characterize the concept of the language of thought
- To understand what is meant by music of thought
- To understand the distinction between representation and consciousness
- To discuss the formal-syntactic structure of the brain's neural activity
- To differentiate between mind-based top-down and brain-based bottom-up approaches

Key concepts

Content, semantic properties, propositional attitudes, functional roles, representational roles, structural isomorphism, vehicle-content distinction, formal language, consciousness, phenomenal features, brain and its formal-syntactic properties, brain-based approach, bottom-up approach.

Background 1

Mind and semantic content

Recall the previous chapter, which described functionalism as a theory about the mind that postulates that the mind's mental states are determined by their functional roles – the relation between input, internal states and output. For instance, my intention to write this book (a mental state in itself) has a certain functional role, namely that it determines my subsequent behavior. Having seen another successful book on the subject served as input, which in turn yielded my desire to become famous as an author. The desire to become famous is then coupled to another internal state – the intention to write a book – and induces a certain output: I take up my laptop and start writing.

Functionalism claims the functional role as a necessary and sufficient condition of mental states. Mental states are then defined exclusively by their functional roles – the specific relation between input, internal states and output. There is, however, more to my mental state than its mere functional role. In addition to its functional role, my mental state has a certain meaning.

Let me describe this using our initial example. My desire is not pure desire, but also the desire to become famous. My intention is not just an intention, but the intention to write a book. This makes my subsequent action meaningful: my movement of picking up the laptop is one part of an overall action that is tailored towards realizing my intention. The movement can thus be understood only in the context of my intention, whereas by itself, considered in an isolated way, it does not make sense.

Taken together, my mental state includes certain contents – the desire to become famous and the desire to write a book. These contents provide the kinds of meaning that define my mental states. How are these contents, 'semantic contents' as described in philosophy, generated? Where do semantic contents come from?

The mental state bears a specific semantic relation to the world. If it did not, it could not produce meaning and meaningful action. Let's get back to our example. The physical book itself, the mere pages and the material, does not cause my desire to become famous. Instead, my desire to become famous results from the importance and relevance the rest of the world attributes to the other book. In other words, it is the meaning or semantic content associated with the other author's book that causes my mental state, the desire to become famous.

In order to understand mental states, we thus need to investigate how semantic content or meaning is generated. To do this we need to understand how the mental states and their respective semantic contents are related to the world. The mental state seems to bear a semantic relation to the world. This semantic relation to the world and how it is associated with semantic content is the focus in the present chapter.

Background 2

Semantic content and representation

How can the semantic meaning in my mental state be associated with its contents, like the book? Though the book itself is not physically present in my mental state, its associated semantic features, i.e. its contents, are nevertheless present. Furthermore, despite the physical absence of the object itself, the associated meaning and thus the semantic content cause my subsequent action: to pick up the laptop and begin writing. Hence it is meaning that instigates my behavior. This is called 'causation by content'.

The associated content must thus stand in some kind of semantic relation to the environment, otherwise the semantic content could not cause my actions, which are targeted toward and occur in that very same environment. There is thus a semantic relationship to the world or environment. How is this semantic relationship to the world possible when the object itself remains physically absent? This is where the notion of representation becomes important. The object's meaning, and thus its semantic relationship to the environment, must be somehow represented in the mental state. And through that representation, the physically absent object is associated with meaning and becomes the kind of semantic content that determines our mental states.

This point seems to suggest that our mental states are determined by the representation of semantic content, rather than by their mere functional roles as defined by the causal relations between input, internal states and output. In short, the mental state is signified by what can be described as its representational role – the representation of semantic contents. Before defining the concept of representation (see below), we will briefly shed light on the difference between the representational account of the mind and its functionalist counterpart.

How can we describe the concepts of functional and representational roles in further detail? The concept of functional roles is concerned with 'what the mental state does': my mental state is wanting to write this book. Its representational role consists in 'what the mental state stands for': my mental state of wanting to write this book implies that a book is characterized by text that must be written down, and that those sentences might appeal to many people, making both author and book famous.

By inferring the mental state's semantic content from its functional role – the input–internal states–output relationship – the functionalist more or less infers 'what the mental states stand for', from 'what the mental states does'. This contrasts with a representational account of the mind. A representational account makes a clear distinction between 'what the mental state does' and 'what the mental state stands for': 'what the mental states stands for' and its semantic contents are considered to be different from 'what the mental states does', and are therefore accounted for on separate grounds.

Unlike in functionalism, the focus of the representational account is on semantic content and its special relation to the world, also known as the semantic relationship. The representational account of the mind is thus about semantic content, or how the objects in our mental states can represent meaning, including their special semantic relationships to the world. How is semantic content and its special semantic relationship to the world generated by, and represented in, our mind and its mental states? This question will be discussed below.

Content 1

Mental representation

What does the term representation mean? This is a question about content, or the semantic meaning, of the sequence of letters r-e-p-r-e-s-e-n-t-a-t-i-o-n. In the most basic and abstract sense, representation describes an entity that (i) is about another entity, while the latter is not about it; (ii) stands in for the other entity, which remains physically absent; and (iii) can initiate and guide subsequent behavior in the same way that the other entity would. In order to understand this rather abstract definition, we will provide the following example.

Let us consider the case of a painting that fulfills all three criteria of representation. The painting of a horse stands for, and thus represents, the horse. In this example, the painting is the 'something', while the horse is the 'other entity'. The painting represents the horse, while the horse does not represent the painting. There is thus an asymmetric relationship between the 'something' – the painting – and the 'other entity' – the horse. The horse remains absent; there is no horse around and you do not receive any perceptual input from a real horse. However, the painting nevertheless represents it. And the painting of the horse may initiate and cause subsequent action. For instance, I might want to ride a real horse or to paint a similar scene in a new painting.

This is the broadest meaning of the term representation. As such, it can apply to all sort of things, mental states among them. Mental representation can thus be considered a subset of representation in general. In the following we focus only on mental representation, leaving the general notion of representation aside.

How can we determine the concept of mental representation? The content of a mental state represents or stands in for another entity that remains physically absent. However, the other entity is mentally present even though it remains physically absent. What is mental presence during physical absence? The original entity is substituted for by something else, the mental content, which therefore represents the entity.

Let's come back to my initial example of wanting to write a book. Though the book itself is not physically present, it nevertheless figures prominently in my mental state, implying that it is represented as content. In other words, the book is only mentally, but not physically, present. This is only possible if my mental state stands for and thus represents the book as its semantic content.

The reverse does not hold, though. The physical book itself does not represent the semantic content associated with the mental state of the book. There is thus an asymmetric relationship between the semantic contents in my mental states and the (physically absent) objects it represents: the semantic content refers to the (physically absent) objects while the latter does not imply the former. This fulfills the second criterion of representation as discussed above.

Finally, the representation of the book in my mental state triggers and initiates, and thus causes, my subsequent action, like picking up the laptop, etc. The example of me wanting to write a book thus fulfills the third criterion of a representation. The case of mental representation can thus be considered an instance of representation in general.

Content 2

Propositional attitudes

Is mental representation just another example of representation in general? It should be noted that there is something specific about mental representation. It is not the content itself that causes the subsequent action. Instead, it is rather the attitude towards the content that causes the behavior. In the case of my example, it is my desire, my wanting, that reflects my attitude. It is desire that initiates my writing of the book. Since they are associated with propositional contents, like 'writing a book', our attitudes towards semantic contents are called propositional attitudes.

Imagine another example of a propositional attitude, the one of social anxiety. In that case it is not the social situation itself, its meaning or content, that causes your anxiety before starting your public speech. Instead, it is rather your attitude towards the social situation, the conference with many people in the audience, that causes your anxiety. That attitude will lead to and be manifest in your belief that you may stutter at the beginning of your speech and may consequently embarrass yourself in front of the audience.

We therefore need to distinguish between the content itself, the proposition, and the attitudes towards that content, or the propositional attitude. Propositional attitudes consist of beliefs, desires, hopes, etc. that are usually associated with particular semantic contents. The propositional attitude causes the subsequent behavior associated with the respective semantic content. In my case of wanting to write a book, it is not the mental representation of the book, but my desire and will (and thus my propositional attitude) that causes my subsequent behavior.

The distinction between propositional attitude and semantic content is further supported by the fact that both can be dissociated from each other. Such dissociation may be manifest in the association of one and the same propositional attitude with different semantic contents. For instance, your belief about failure may be directed toward situations when you have to give a speech in public. This is called social anxiety.

Another example is examination anxiety. You may be well prepared for an exam, which there should be no problem in passing, given your knowledge and previous grades; however, there is some irresistible belief deep inside you that you may fail. The first time that belief occurred was after one test in primary school when the teacher failed you. Since then you pre-empt failure for any examination you have to undergo.

What do we take away from these examples? Despite their different contents, social situation and examination, both examples share the belief of failure and thus the propositional attitude. They show us that the same propositional attitude, the belief of failure, can be associated with different contents like public speeches and written examinations. This means that attitudes and contents can dissociate from each other, which makes it even more important to distinguish them.

Content 3

Can we find semantic contents on the subpersonal level?

We pointed out that semantic content is present in mental states even if the related object is physically absent. How is it possible that something as physically absent as a horse surfaces as semantic content and is thus mentally present in my mental states? The semantic content must have an origin that is different from the actual physical content. There must thus be something special about semantic contents.

Semantic content occurs on the personal level. It is on this level that we experience mental states. The personal level of semantic contents must be distinguished from the subpersonal level. The subpersonal level can be characterized by physical processes that are supposed to underlie mental states and their semantic contents on the personal level. The physical states are thus supposed to provide the vehicle of the personal level and its semantic content.

How is this possible? One way is to assume that the subpersonal level and its physical processes are structured in a way that somehow corresponds to the semantic contents on the personal level. The subpersonal level may, in this case, demonstrate an internal structure that allows it to generate and yield the kinds of semantic contents we experience in our mental states on the personal level. In other words, the subpersonal physical structure itself shows a particular structure – an internal structure that serves as a vehicle for constituting semantic contents on the personal level.

Biography

Jerry Fodor

What does the internal structure of the subpersonal level, the vehicle, look like? Here it would be useful to introduce the American philosopher Jerry Fodor. Fodor is a philosopher born in 1935 in New York City. He studied philosophy and later held a teaching position at Massachusetts Institute of Technology (MIT) in Cambridge, Massachusetts. However, his urge to go back to the New York area must have been stronger than the fame of being at MIT. He subsequently returned to the city, where first he taught at New York University, and now teaches at Rutgers University in New Jersey.

Fodor seems to have asked himself the question: where does the semantic content on the personal level come from? We express semantic content in terms of concepts and most generally in our language. Semantic content, from this point of view, can thus be considered as linguistic content, where our language shows a specific, linguistic structure. That linguistic structure is reflected in our sentences, in propositions like 'I believe that...', 'I hope that...', etc. These linguistic structures, of propositional attitudes like 'I believe' and 'I hope', are always followed by a semantic content, that is a propositional content, as indicated by the 'that...'.

Content 4

Language of thought (LOT) on the subpersonal level (Fodor)

These examples make it clear that a propositional attitude is always followed by a specific content – semantic or propositional content. Propositional attitudes and contents are thus distinguishable elements on the personal level. Where, though, does this linguistic structure with its distinction between propositional attitudes and semantic contents come from?

Fodor now assumes the same kind of structure to be true of the subpersonal level of the physical structure. There must be distinguishable subpersonal physical elements that correspond to the distinguishable linguistic elements on the personal level – propositional attitude and semantic contents. In other words, the basic elements of our sentences, like propositional attitudes and semantic contents and their structure – or how they relate to each other, must correspond to the analogous elements of the subpersonal physical structure. Basic elements on the personal level of propositional attitudes and semantic contents must then be traced back to corresponding physical elements on the subpersonal level.

This correspondence between the linguistic structure on the personal level and the physical structure on the subpersonal level is what Fodor describes as 'structural isomorphism'. The concept of structural isomorphism describes that the subpersonal level and its physical structure are structured and organized in a way that is analogous to the linguistic structure on the personal level.

The subpersonal level must thus be organized and structured like a language, showing the same kind of linguistic structure as the personal level. In other words, the subpersonal level and its physical structure must show some kind of internal language that determines the operations, i.e. the composition and combination of its different basic elements. There is thus what Fodor describes as 'language of thought' (LOT) on the subpersonal level.

Content 5

'Structural isomorphism' between personal and subpersonal levels (Fodor)

What does the language of thought (LOT) look like? Should one imagine the LOT to be a natural language like Spanish, English, or otherwise? No, because no such language can be found in the subpersonal level and its physical structures. Instead, the LOT is a rather formal language that, analogous to the syntax of a natural language, emphasizes the logical form, its formal structure like the sequence of propositional attitudes and semantic contents.

This linguistic structure is manifest on the personal level of language. Where, though, does it come from? Jerry Fodor assumes that it is already ingrained at the subpersonal level, the vehicle. The subpersonal vehicle shows a structure that is analogous or isomorphic to that of the personal level. The vehicle shows the same distinguishable elements, with the same composition and combination, as the contents and their linguistic structure on the personal level.

The subpersonal level of the vehicle can thus be characterized by a formal-syntactic structure, a language of thought (LOT) as Fodor says, that corresponds to our language, including its semantic features on the personal level of contents. Accordingly, the LOT on the subpersonal level may show a form or structure that is analogous and thus corresponds to the linguistic structure on the personal level. For instance, the linguistic sequence of a propositional attitude followed by semantic content may correspond to some analogous structure that is ingrained and encoded into the subpersonal level and its physical structure.

However, we have to be careful. The semantic content itself is not encoded and ingrained into the subpersonal level and its physical structure. Instead, the subpersonal level and its physical structure provide only the vehicle for generating and yielding the semantic content on the personal level. One may thus want to distinguish between vehicle and content. The vehicle is the subpersonal level, while the personal level provides the content.

How is this distinction between vehicle and content related to the assumption of structural isomorphism? Structural isomorphism argues for analogous structures on subpersonal and personal levels. At the same time it argues that the semantic contents are not 'located' by themselves on the subpersonal level; the semantic contents can thus not be identified with the physical structures on the subpersonal level. For instance, we cannot find and observe any such semantic contents in the neural activity of the brain in the subpersonal vehicle of our mental states.

Content 6

Syntax and semantics are associated with subpersonal and personal levels (Fodor)

How are both assumptions, structural isomorphism and non-location of semantic contents on the subpersonal level, compatible with each other? Fodor and others argue that the relationship between vehicle and content can be compared to the relationship between syntax and semantics in a formal system. A language can be considered in terms of both syntax and semantics: sentences can be analyzed in terms of their semantic properties, as well as in terms of their syntactical, grammatical structure.

The same holds true for vehicles and their contents, whose relationship corresponds (more or less) to the one between syntax and semantics. The subpersonal level provides the vehicle and thus the syntax for the semantics of the semantic content that operates on the personal level. Hence, subpersonal and personal levels relate to each other like syntax and semantics, with both showing an analogous structure: structural isomorphism. At the same time the characterization as syntax, rather than semantics, does not necessitate that the semantic contents themselves be located in the subpersonal level and its physical structure.

To put it differently, each distinguishable element on the subpersonal level of the vehicle stands for a particular content and its meaning on the personal level. The structure of the subpersonal level of the vehicle is thus determined by the content and its semantic features on the personal level. This presupposes a top-down approach, from the personal to the subpersonal level.

Thus far we have focused mainly on semantic contents and how they are related to the subpersonal level and its physical structure. What about propositional attitudes – how are they generated? To answer this question, it may be important to consider perceptual input. Perceptual input is processed on the subpersonal level and its physical structure. Because it is processed subpersonally, the perceptual input is associated with the subpersonal syntactic structure and organization; this, so Fodor says, reflects the encoding of the semantic content on the subpersonal level. The semantic content is thus manifest on two different levels: personally in the gestalt of propositional attitudes, and subpersonally in terms of the structure and organization.

How is it possible that the semantic content is manifest on both subpersonal and personal levels at the same time? Both subpersonal and personal levels show a more or less analogous formal structure and organization. The propositional attitudes on the personal level find their analog in the syntactic structure of the subpersonal level. Subpersonal level syntactic structure and personal level semantic content can thus be considered to belong together, like yin and yang.

Critical reflection 1

Localized versus distributed representation

The representational account characterizes contents, the semantics or meaning associated with objects, events, or people, by propositional contents and attitudes. These are linguistically determined, as reflected in the linguistic-grammatical structure of propositional attitudes like beliefs, desires, etc. The linguistic structure of our sentences represents the contents in terms of propositional contents and attitudes on the personal level.

This linguistic structure on the personal level of content is supposed to be reflected in the structure on the subpersonal level, the vehicle. As we recall from above, the subpersonal level is supposed to be characterized by a structure corresponding to the linguistic one on the personal level. This was described as *'language of thought' by Fodor. The concept of the 'language of thought' describes a syntactic-like formal structure that determines the operations and processing of the subpersonal level, the vehicle.*

How does all this apply to the brain, which may be regarded as the subpersonal vehicle of the mind on the personal level? Does the brain have an internal formal-syntactic structure with distinguishable elements that symbolize contents? Here we can return to the work of the American philosophers, Patricia and Paul Churchland. Is the brain a mere digital computer? According to Patricia and Paul Churchland, the answer is no: 'We need to consider the brain, including its subpersonal level, as independent of the semantic contents on the personal level.' More specifically, we need to consider the design of the brain, as independent from the contents and their meaning.

How do the Churchlands characterize the brain? The brain is not a digital computer, as suggested by Fodor and his LOT. Instead, the brain is a parallel-distributed processing computer (PDP) with various circuits that process the same input and information in different ways at the same time. There are various circuits where the different inputs are processed in parallel and distributed ways. Imagine a group of people who want to enter a palace. If there is only one entrance gate, all the people must go through there and can only enter the palace sequentially – this corresponds to a digital computer. If, in contrast, there may be more than one entrance gate to the palace, different people can enter the castle in parallel but distributed ways at the same time. This obviously corresponds to parallel-distributed computing as suggested by the Churchlands.

Critical reflection 2

Parallel-distributed processing

How is such parallel-distributed processing possible? It is possible by including three layers with different kinds of units. There are input units in the artificial neural network that process all inputs: for instance, sensory inputs. There are output units that are in charge of the output, like motor action and behavior. And there are hidden units sandwiched between input and output units, which allow for the parallel and distributed processing of the same inputs.

All three units, input, output and hidden, show a certain level of activation, called an activation weight, and are connected to each other via their synapses, which results in connection strength. How can this system represent any kind of content at all? There are no longer any single units that correspond to one specific content. Instead, each unit may participate in the processing of various contents.

How do the three units, input, output and hidden, and thus the PDP network, represent semantic content if not in single units? The activation pattern, the spatial and temporal distribution of the activity across the networks and its different layers, stands for a particular content. Instead of in single units as distinguishable elements, contents are now represented in the activation pattern holding across the different units.

What does this imply for the representation of contents? Contents then are no longer represented in single units. Instead, they are represented in the activation pattern across different single units entailing a distributed, rather than localized, way. Unlike in the LOT, where each unit (that is, concept) represents a specific content, one unit can now represent several contents. Conversely, one and the same content can now be represented by several units at the same time.

Critical reflection 3

Linguistic versus vectorial representation

Can a formal-syntactic-like structure be found in the brain? Patricia and Paul Churchland do not think so. They argue that there is simply nothing in the brain, and its neural activity, that resembles or corresponds to the linguistic structure of propositional contents and attitudes. All we can see in the brain are nerve cells that fire and show action potentials and spiking frequencies. We do not see any linguistic, sentence-like structures in the brain.

What do we see in the brain? As indicated above, the Churchlands assume that we see different units, input, output and hidden, across a particular pattern of activity that can be observed. Let us specify this account of the brain further.

The brain can be characterized by input units or vectors, neurons that mediate inputs from the outer world. These concern mainly sensory neurons that receive and process the sensory input from the external world. In addition to such input vectors, there are also output vectors; these process the activation from the input vectors – the sensory inputs – beyond those to the output units, mainly motor units, that cause behavior and motor action.

Between input and output vectors, one can find intermediate layers, processing layers, that help transform the input vector into an output vector that corresponds to the motor-behavioral output. The action of the brain can thus be described as vector-to-vector transformation and certain activity patterns, the activity vectors.

How does such vector-to-vector transformation relate to contents? The Churchlands assume that the activity vector, the activity distribution across input and output vectors and processing layers, allow for representation of content. Contents are then no longer represented in propositional contents, but rather in the activity vectors. Propositional contents are thus replaced by what one can describe as 'vectorial contents'.

These vectorial contents are represented in the activity pattern in the three- (or four-)dimensional space of the brain. The activity pattern can be characterized by various points in the three- (four-)state space of the brain. And how these different points in space are related to each other determines the kinds of content they represent. For instance, different colors may be represented by activity at different points and how they relate to each other.

Content is consequently represented as vectorial content in the activity vector that describes the actual state-space of the brain. Therefore one may speak here of 'state-space vectorial semantics' to replace the linguistically-based propositional semantics and the 'language of thought' of the representational approach. One should be careful, though. The vectorial approach still assumes representation to apply, which it shares with the linguistic approach. The difference lies more in the means or the way representation is carried out: while it is primarily linguistic in the linguistic approach, it is assumed to be vectorial and state-space-based in the vectorial approach.

Ideas for future research 1

'Language of thought' (LOT) and the brain

The focus on the personal level of content and consciousness neglected the underlying vehicle and its specific characterization by the language of thought (LOT). The assumption of the LOT is based on the comparison between the mind and a computer. The distinction between vehicle and content as mirroring the relationship between syntax and semantics is analogous to a computer, where purely syntactic calculations operate without any semantic consideration. According to this model, mind in the representational approach is thought of as a computer – a computer whose underlying physical operations are implemented and realized by the brain.

Does the brain really exhibit an internal formal-linguistic(-syntactic)-like structure in the sense of Fodor's language of thought? American philosopher Dan Lloyd (1953–) asks this question. Lloyd was a philosopher at Trinity College in Hartford, Connecticut. Beyond philosophy, Lloyd is also interested in how the brain yields what we call consciousness and mind. He is not afraid of tackling empirical issues and approaching them from a sometimes unusual angle.

One question he investigated is the coding of the brain's neural activity. The brain shows neural activity during both a resting state (i.e. without any specific external stimuli, as for instance with eyes closed or during sleep) and stimulus-induced activity (i.e. during specific external stimuli). How is such neural activity generated and encoded? To address this question, Lloyd (2011) compared the coding schemes of the brain's neural activity with those of music and language. He was thus interested in the formal syntactic properties of all three: brain, music and language.

Language and the brain may share similar formal syntactic properties, for example. This, considered within philosophy, would amount to a thesis of a 'language of thought'. Despite their different semantic and thus content features, language and the brain may nevertheless share formal-syntactic features: our thoughts that arise from the brain may be structured like the language we use to express these thoughts. This amounts to what is described as 'language of thought' in philosophy. What are formal-syntactic properties? Formal-syntactic properties may, for instance, consist in a particular coding of the different elements and their relationship to each other. The elements may be tones in music, words in language, and neural activity patterns in the brain. How are these basic elements related and combined with each other? Certain tones may appear together often, while others hardly ever occur in conjunction with one another. The same holds true of the sequences of words in language, neural activity patterns in the brain, the combination of tones in music, and their respective combinations.

Now, if certain combinations occur more often than others, the full range of possible combinations between the different elements (i.e. neural activity patterns, tones, words) may not be fully exhausted by the respective system (i.e. brain, language, or music). In this instance, Lloyd aimed to compare different combinations of the brain's spatiotemporal activity patterns across time with a constellation of the basic elements in music (i.e. tones) and language (i.e. words).

Ideas for future research 2

Formal-syntactic properties in language, music and the brain

What exactly did Lloyd do in this investigation? He investigated how words are combined in different languages (English, French, Spanish, Finnish, Chinese), and how tones are related and combined in different pieces of music (Schubert, Gershwin, Chinese folk, British folk, African folk). In addition, he conducted an analysis of different imaging studies, during either resting state activity (e.g. absence of any specific stimuli or task), or the presentation of repeated and deviating auditory stimuli (e.g. an auditory oddball paradigm: presentation of repeating standard stimuli with occasional deviant stimuli).

For the analysis of the brain imaging studies, he used a specific statistical technique, an independent component analysis. The independent component analysis can detect different patterns, constellations or combinations between the single elements across time. This analysis revealed one particular spatiotemporal neural activity pattern at each point in time, and allowed Lloyd to observe how, and how many, patterns change and are yielded across time.

Following this experiment, Lloyd was interested in seeing how the different activity patterns in the brain are related to, and combined with, each other across time. He was interested specifically in seeing whether they all occurred across time, or whether particular constellations of specific activity patterns were favored, while others occurred less often.

He did the same for the constellations or combinations of tones in the different music types and words in the different languages. As in the case of the brain and its activity patterns, he calculated distinct variants of the standard deviations, i.e. the variations across time and space, thereby indicating the degree of sparseness.

Sparseness means that certain combinations of elements (e.g. the activity patterns, the tones, the words) occur more often while others occur rather rarely. There is thus a low or sparse number of actually occurring constellations or combinations when compared to the number that is possible (based on pure mathematics). Alternatively, the number of possible combinations may be high, in which case the discrepancy between actually and possibly occurring constellations is low. That implies a low degree of sparseness.

Ideas for future research 3

Resemblance between brain and music

What did Lloyd observe in his results? He observed that music was sparser than language. Certain constellations of notes occurred more frequently, and others less often, whereas all words and all possible constellations of words were used quite often in language. Language thus shows a higher number of constellations that are therefore less sparse when compared to music.

How does the brain and its neural activity pattern compare to the constellations observed in language? The results show that, overall, the brain's spatiotemporal activity patterns exhibited a rather high degree of sparseness (i.e. the number of actually occurring sequences (or constellations or combinations) relative to the number of possible sequences).

Most importantly, the number of the brain's actually occurring spatiotemporal activity patterns proved to be more or less similar, or closer to, the number of constellations observed among the tones of music. Both patterns, that is, in the brain and music, differed from the constellations between the words in language: the number of actual constellations between the elements, that is, the words, relative to all possible constellations, was much higher in language than in both music and the brain. This means that the brain's activity patterns across time are coded in a sparser way when compared to the coding of words in language, while the brain and its sequence of spatiotemporal activity patterns seem to be more or less similar to the formal constellations and sequences between tones in music.

What do these results imply more concretely? Let's start with the example of music and the structure between the different tones. In addition to the key, one particular note or tone in music strongly determines and predicts which note comes next. The same is true in the case of the brain. Here, the occurrence of one actual spatiotemporal activity pattern seems to entail a high likelihood of a particular activity pattern following. At the same time, the previous activity pattern decreases the probability of others as its possible successors. Hence, the actual spatiotemporal patterns have a strong say in what will happen next. This implies a certain predictive quality.

This is different in language, where the use of a particular word does not really predict which word or words will follow. Language seems to use all possible constellations of words, which makes this type of prediction very difficult. As a result, language seems to differ from both music and the brain, where the following sequence can be better predicted based on those that preceded it.

Ideas for future research 4

'Music of thought' (MOT) and the brain

What do these results imply for the characterization of the brain? The tones in music and the spatiotemporal activity patterns in the brain's resting state activity may better fulfill the expectations. In other words, they are more predictable than the sequences and constellations of words in language. To put it differently, we may be caught by surprise more often by the use of words in our language than when listening to a piece of music, or when considering the patterns of our brain's spatiotemporal activity.

On the whole, Lloyd clearly observes that the brain's neural activity during both resting state and stimulus-induced activity is characterized by the sparseness of its neural activity pattern across time and space. This means that one neural activity pattern predisposes and predicts the subsequent one. Thereby the degree of sparseness of the brain's neural activity pattern seems to be more or less similar to that of notes in music. In other words, the formal structure and organization of their respective elements across time seem to be more or less similar between the brain and music. Meanwhile, both seem to differ in their formal structure from language.

The similarity of the brain's formal structure to that of music inclines Lloyd to speak of a 'music of thought' (MOT) as distinguished from a 'language of thought' (LOT). One can extend this analogy even further and speak of a 'music of the brain' (MOB). The concept of 'music of the brain' may thus indicate metaphorically the similarity of the brain's formal structure and the organization of its elements to music.

What does this imply for the assumption of a language of thought with regard to the brain? As already pointed out by Lloyd himself, the assumption of a LOT does not seem to be plausible when it comes to the neural activity of the brain. Even if it is logically, conceptually and linguistically plausible, the LOT does not seem to be in accordance with the empirical data about the brain and its neuronal states. Accordingly, the LOT must be regarded as empirically implausible when it comes to the brain.

Thus, while a representational account including the LOT may work well for the mind and its linguistic functions, it may not be plausible when it comes to the brain and its purely neuronal functions. This shifts the focus from the concept of mind and how it can be accounted for to the brain itself. More specifically, we need to understand the structure and organization of the brain independently of the mind. This will be further detailed in the following chapters. Before addressing these issues, we have to go back to the representational models of the mind as discussed in current philosophy of mind and philosophy of psychology, and see how they account for consciousness.

Representation 1

Higher-order representation and consciousness

The representational approach focuses on semantic contents and thereby aims to explain meaning as a central component of consciousness in particular, and our mind in general. However, the representation of semantic contents does not necessarily need to go along with consciousness. Not all semantic contents are conscious. In addition to conscious contents, as in, for instance, my desire to write this book, there are also numerous unconscious contents.

The Austrian psychologist and psychoanalyst Sigmund Freud (1856–1939) observed that there may be numerous semantic contents in our unconscious that exert an impact on our behavior. For instance, my conscious desire to become famous as a writer may be overshadowed by some unconscious anxiety of becoming famous and being in the public eye as a famous writer. My unconscious anxiety may then affect my subsequent behavior in different ways. For instance, it might prevent me from writing the kind of book that would make me famous. My unconscious may propel me to write a report about my little hometown, which unlike a story about, for instance, New York, doesn't have much potential to attract a broad audience.

How can we distinguish between unconscious and conscious representations of semantic contents and propositional attitudes? So far we have only described the mechanisms of how semantic content is yielded and generated. To do this we referred to structural isomorphism between subpersonal and personal levels, as discussed above. This does not yet carry any distinction between conscious and unconscious representations of semantic content. For that, an additional mechanism needs to be postulated that allows us to distinguish between conscious and unconscious representation.

What would this additional mechanism look like? One possible theory is what is described as 'higher-order representation'. Higher-order theorists, like the Anglo-American philosophers David Rosenthal (2005) and Peter Carruthers (2005), assume that the representation of semantic content on its own remains unconscious. The semantic content is thus represented in an unconscious way; this is what is described as first-order representation.

In order for the semantic contents in first-order representation to become conscious, 'something else' – an additional mechanism – must occur. This 'something else' or the additional mechanism is now assumed to consist in further elaboration of the semantic content as represented in the first-order representation.

More specifically, the first-order representational semantic content must be re-represented as such. This results in second- (and third-)order representations of semantic content. A second-order representation of the semantic content is supposed to allow the association of consciousness to the originally unconscious semantic content as represented in the first-order. Hence it is the second-order representation, as an 'additional mechanism', that makes the generation of conscious semantic content, as distinguished from the unconscious first-order semantic content, possible.

Representation 2

Awareness and higher-order representation

How is such second-order representation possible? The higher-order representation theorists assume that cognitive functions like awareness and attention are most likely central. By focusing the awareness and attention on the first-order representational content, the latter shifts into awareness and attention. This shift in awareness and attention makes the association of consciousness, to the otherwise unconscious first-order representational content, possible. Awareness and attention thus mediate the transition from unconscious first-order to conscious second-order representations. How can we illustrate the role of awareness and attention by our previous example? I have a desire to write this textbook. This desire remains unconscious and can be characterized by first-order representation. Now my awareness and attention shift to the textbook, and thus also to my desire to write the book. The textbook consequently becomes second-order represented, which makes its association with consciousness possible. Consciousness is thus no longer explained by the semantic content, as represented in first-order representation, but rather by its association with awareness and attention that turn the first-order represented content into a second-order representation.

But what is awareness? Awareness is a kind of inner perception or thought. If taken to be an inner perception, like an inner monitoring, it must be conceived of as a higher-order perception that complements the original lower-order perception. These are the higher-order perception models (HOP). They claim that in addition to my lower-order perception of outer stimuli from the environment, we also have some kind of higher-order perception of the inner processes in our mind that, for instance, represent the semantic content in first-order. By recruiting higher-order processes, awareness is shifted to the first-order represented semantic contents, which in turn makes their representation in second-order representation possible, and thus their association with consciousness.

Representation 3

Additional mechanism

Alternatively, one may assume that certain kinds of internal thoughts may be directed toward the first-order representation of the semantic content. These special thoughts may then represent the first-order semantic content in a second-order representation and the subsequent association with consciousness. Since they are based on higher-order internal thought, these theories are described as higher-order thought theories (HOT).

Whether HOP or HOT, many challenges have been put forward against higher-order representational theories. Therefore different variants of both have been developed which will not be discussed in detail here. One variant is called 'self-representation'. Like higher-order representational theories, self-representational theories (see Kriegel and Williord, 2006) claim that simple representation as outlined in the LOT is not sufficient to associate the semantic content and its first-order representation with consciousness. As in HOT and HOP, there is thus a need to assume an additional mechanism.

The nature of this additional mechanism is different in self-representational approaches. Both HOT and HOP suggest such an additional mechanism remains completely independent of the represented content. In self-representational approaches, it is assumed that the semantic content itself may bear certain features that induce the recruitment of awareness and attention and its subsequent association with consciousness. Unlike in HOT and HOP, the semantic content and awareness/attention are intrinsically related to each other, with the former 'determining' the degree of recruitment of awareness and attention. Consciousness is thus represented in the content itself. It is self-representational.

Self-representational theories thus link awareness and attention to the content itself. Unlike in the higher-order theory, awareness is intrinsically linked to the originally represented content. The represented content and the awareness are internally, and thus constitutively and logically, related to each other. Metaphorically speaking, the content itself 'decides' whether 'it wants to become conscious or not'. This stands in contrast to the higher-order view, where the represented semantic content and awareness remain more or less independent of each other without any internal, logical and constitutive relation.

Representation 4

Higher-order representation

Consciousness signifies our ability to have subjective experience. Subjective experience can be characterized by what is called phenomenal features (see Chapter 1). What are the phenomenal features of our experience, and thus of consciousness? Most authors assume that consciousness can be characterized by a qualitative dimension, a 'what it is like' dimension, which is subsumed under the umbrella term 'qualia' (see Part IV for details). Qualia are experienced only from a particular point of view and are therefore subjective. This is what the philosopher Thomas Nagel (1974) described in his famous 'what is it like to be a bat?' (see Chapter 14 for phenomenal details). Consciousness can thus be characterized by qualitative and subjective features as its core phenomenal features.

Any representational account of the mind that aims to account for consciousness must, therefore, be able to explain the qualitative and subjective features of consciousness. However, nothing in our account of representation so far has hinted at an explanation for these qualitative and subjective features.

Let's be more specific. The representational approach can account for propositional attitudes, content, and the linkage between subpersonal and personal levels via structural isomorphism. However, none of these implies qualitative and subjective features. The structural isomorphism can account for contents and their linkage to propositional attitudes. The assumption of additional mechanisms, like awareness or self-representation, may make the difference between unconscious first-order representation and conscious higher-order representation.

Representation 5

Higher-order representation and phenomenal features

How, though, can higher-order representation account for the subjective and qualitative features of our subjective experience? There is nothing subjective or qualitative in either the structural isomorphism or the higher-order representation. How is it possible that the purely objective content and its propositional attitudes are assigned subjectivity and a specific quality? To answer this question one may pursue different strategies.

One such strategy is to simply deny that consciousness can be characterized by qualitative and subjective features of consciousness. In this strategy, consciousness is characterized in a purely quantitative and objective way that is compatible with the assumption of a representational account of the mind. In this model, qualia and subjectivity are understood as mere illusions. This eliminative position has been advanced by philosophers like Daniel Dennett and Patricia Churchland, as already discussed in Chapter 9.

Another strategy is to acknowledge the subjective and qualitative features of consciousness in their own right, while at the same time distinguishing them from the represented contents. One may then distinguish between qualitative features and representational content (see, for instance, Block, 1995, 2007). If so, both qualitative features and representational contents should be able to dissociate from each other, or occur independently.

For instance, in the case of pain there is a qualitative and a subjective experience. There is, however, no specific content with which the pain is associated. There is thus a pure qualitative subjective experience without any represented content. Pain can thus be considered an example where the qualitative features of subjective experience occur without a corresponding represented content.

The converse can also hold true. In the case of the unconscious, there is unconscious content – like a certain event from the past – as in, for instance, the scene when I was forced to read a textbook by my teacher. I experienced that scene at the time with a feeling of horror because of all the complicated material that made me feel inadequate and unable. Now, years later, that scene is engraved in my unconscious, while at the same time I do not experience it in my consciousness. The scene of the teacher forcing me to read the book is thus still represented as content in my unconscious. Meanwhile, it is detached and thus dissociated from any subjective experience and thus from qualitative features.

Due to the possibility of such dissociation between represented content and qualitative features, both must be accounted for in separate ways. This means that represented content and qualitative features must be related to different underlying mechanisms. If, in contrast, they were related to one and the same underlying mechanism, they would not be able to dissociate from each other.

Representation 6

Non-conceptual representation of content

So far we have discussed two main strategies in the quest to reconcile the representational account with the qualitative and subjective features of consciousness. One strategy is simply to deny and discard subjectivity and qualia. Instead, these features are seen as conceptual illusions. Another strategy is to acknowledge subjectivity and qualia, but to distinguish them from the represented content. One may also assume that qualitative and subjective features can be associated with the representation of content, if the latter occurs in a special way. Various suggestions for such a special form of representation have been made recently by different (and mainly American) philosophers. While a discussion of the various suggestions would be beyond the scope of this book, let us at least discuss one of them, the one advanced by the American philosopher, Michael Tye.

Tye argues that the qualitative features, the qualia, are identical to a particular kind of represented content. Such content can be experienced in a qualitative and thus conscious way if it is represented in a specific manner. What does this specific representation look like? The original content may already be represented, namely as content. In contrast, the concept that is used to describe that content in linguistic terms may not yet be represented. There is thus a dissociation between content and concept in the representation.

For instance, there may be a certain feeling, like some kind of feeling of boredom, before I start watching a movie that my friend told me was boring, even though I am not yet able to put that feeling into a concept; it is just a feeling that cannot be expressed linguistically. Such representation of content without a corresponding concept can be characterized as non-conceptual. Following Tye, the representation of content in a non-conceptual way is central to associating the represented content with subjective and qualitative features.

Representation 7

Attribution of phenomenal features to non-conceptual content

In addition, the represented content must also be directed toward certain objects that remain physically absent but are mentally present, or that – in other words – show intentionality. In our example, I experience the feeling of some kind of boredom when I merely think about the movie, without even having watched it. The mentally imagined book (that the movie is based on) is thus the content or object of my representation that remains purely mental. This purely mental (but not physical) content is necessary, according to Tye, for associating represented content with qualitative and subjective features.

In order to become qualitative and subjective, another feature that the represented content needs to show is its direct linkage to belief and desire. The represented content must be poised, as Tye says. This means that the represented content must be immediately ready for belief and desire. In our case, the book, as the represented content, and its associated feeling of some kind of boredom are ready and available for direct association with specific beliefs and desires: the belief that the movie is bad, and the desire to avoid reading the actual physical book.

Another issue is that the representation of specific contents as non-conceptual, intentional, and associated with beliefs and desires needs to be repeatable, i.e. I can have the same feeling of boredom about the movie every time I want to start watching it. If all four criteria – non-conceptual, intentional, poised (e.g. association with belief and desire), and repeatability – are fulfilled in the representation, the respective content – according to Tye – will be represented as subjective and qualitative. If so, representation is compatible with not only consciousness but also its subjective and qualitative features.

Finally, one may also raise the issue of how to naturalize contents in general, and phenomenal content in particular. This means to develop naturalistic theories of content and content determination, as has been discussed extensively by philosophers like Dretske, Millikan, Block, Kriegel, Papineau and Fodor. This is a truly psychological issue that extends beyond the neurophilosophical focus of this book. Therefore I refer to sources that focus especially on philosophy of psychology, including the question of naturalistic contents (see, for instance, Botteril and Carruthers (1999) as well as Bermudez (2005) for excellent overviews).

Critical reflection 4

Mind-based methodological strategy

Why is all this important in the present context of philosophy and the brain? One may want to argue that the suggestion of representation and its linkage (or non-linkage) to the subjective and qualitative features of consciousness is just a philosophical issue or, more specifically, a theoretical issue that is dealt with in philosophy of psychology. In this case, it addresses the question of how the mind represents content, while at the same time allowing this content to be conscious.

However, we have already seen that psychology, and especially cognitive psychology, has extended toward the brain and neuroscience, resulting in the development of cognitive neuroscience (see Chapter 3). The mechanisms, processes and functions discussed in cognitive psychology are extended toward the brain, where the related neuronal mechanisms are investigated. In this case, we have already seen the language of thought that has been 'taken to the brain', which was made possible by searching for analogous formal or syntactic-like structures in the neural activity pattern of the brain (see above).

The same thing happens in the case of representation. The assumption of awareness as a necessary prerequisite of higher-order representation, and subsequently consciousness, has been associated with higher-order cognitive functions like attention, working memory, etc., for which neural activity in specific regions like the prefrontal cortex is central (see Part IV for details).

The specific qualitative features are now related to specific psychological functions like a particular short-term memory, e.g. iconic memory. Since such iconic memory is assumed already to occur in the sensory cortex, like the visual cortex, our sensory experiences are already associated with subjectivity and quality.

What kind of methodological strategy is presupposed here? One can think of this methodological strategy as top-down because it starts with the mind and 'moves down' to the brain. It can also be considered as mind-based. It is top-down because it takes its models of representation and consciousness as starting points and then aims to find corresponding neuronal mechanisms on the level of the brain. More specifically, the model of the mind serves as a template to characterize its current descendants, like representation and consciousness, and, even more importantly, how they ought to be implemented in the brain.

Critical reflection 5

Elimination of the mind and the methodological strategy

Does the elimination of the mind lead us closer to a brain-based approach? Even if a particular method denies any subjective and qualitative features of consciousness, as the eliminativists do (see above), their approach still remains mind-based. Why? The elimination still presupposes the characterization of the mind even though in a negative way. Such negative characterization of the mind is then taken as a basis for understanding the brain. So even though the mind is only taken as a negative template, which is to be eliminated, it nevertheless serves methodologically as the basis and starting point for the characterization of the brain (see Chapters 7 and 8 for more details). Therefore, even the eliminativist of the mind still presupposes a mind-based approach.

Wouldn't one nevertheless assume that the elimination of the mind turns the eliminativist's approach into a brain-based rather than a mind-based approach? To think this would be to confuse the methodological strategy with its outcome or result, though: the eliminativists' assumption of locating and reducing the mind and its various features in the brain and its neural activity is the outcome, or the result. Following Paul M. Churchland, this result, the location of the mental features in the neural activity of the brain, can be characterized as brain-reductive or 'brain-friendly rather than brain-aversive'.

Despite being brain-reductive or brain-friendly, the result must nevertheless be distinguished from the methodological strategy and its starting point. The starting point is still the mind and its supposed mental features, and how they are supposed to organize and represent contents (in a vectorial rather than linguistic way; see Chapter 9). The methodological point of departure for the eliminativist is consequently still the mind, rather than the brain. Therefore even the eliminativist (and other less radical but reductive) approaches can still be characterized by a mind-based methodological strategy.

Critical reflection 6

Brain-based methodological strategy

What does the alternative, a truly brain-based methodological strategy, look like? To answer this question, we need to understand the structure and organization of the brain independently of the assumptions made on the level of mind, as in philosophy of psychology. This raises the issue of whether or not we might be able to develop a theory of brain function: how is the brain's neural activity structured and organized? Are there specific principles of neural organization? How can we investigate and access them? These questions touch upon what can be subsumed under the umbrella of philosophy of neuroscience (see Chapter 11) and philosophy of brain (see Chapter 8, and, especially, Chapter 12).

Why is philosophy of brain important in the present context? A future philosophy of brain may open the door to understanding the brain itself – its structure and organization, independent of our models of the mind. Any such future philosophy of brain may provide the foundation for a brain-based rather than mind-based methodological strategy in our approach to the mind and its mental features.

How about philosophy of neuroscience? Philosophy of neuroscience may develop the specifics of a brain-based methodological approach. By considering both the brain's characterization, as developed in philosophy of brain, and the know-how from philosophy of science, a future philosophy of neuroscience may provide us with new insights for developing a methodological strategy based on the brain's capacities and predisposition (rather than on the mind in either a positive or negative way), while at the same time such philosophy of neuroscience may allow us to pinpoint the epistemological limitations of the brain, which should be considered and thus incorporated in the development of a brain-based methodological strategy.

By developing such a brain-based methodological strategy, neurophilosophy would distinguish itself from both philosophy and neuroscience. As explained above, philosophy relies on a mind-based methodological strategy, whereas that of neuroscience is brain-reductive. Applying a brain-based rather than either a mind-based or a brain-reductive methodological strategy distinguishes neurophilosophy from both philosophy and neuroscience in methodological regard. Such brain-based neurophilosophy may, in turn, open new doors to themes and topics like representation and semantic content, as well as the mind–brain relationship, that are heavily discussed in current philosophy of psychology and philosophy of mind.

Take-home message

The previous chapter investigated the relationship between the personal and subpersonal levels. This led us into an investigation of the philosophy of psychology, where different models for this relationship were suggested. These ranged from irreducibility and continuity to elimination. However, these various possible relationships left open the question: how can meaning be assigned to the objects, events, or people that feature as semantic contents in our mental states. How must our mind be organized and structured in order to allow for the assignment of meaning to the contents in our mental states? This raises the question: how are semantic contents generated? One of the main suggestions is that the meaning is structured and organized in a linguistic way, as reflected in the structure and organization of our language. This led to the assumption that the subpersonal level itself is structured and organized in a language-like way, which resulted in the hypothesis of a 'language of thought'. The 'language of thought' as formal structure and organization is considered to realize the semantic content on the subpersonal level. How can such representation account for consciousness, and specifically its subjective and qualitative features? Various suggestions are discussed in philosophy of mind. However, all proposals raise serious problems – so much so that a fully satisfying representational account of consciousness remains unattained. In addition to the problem of representation, there are also methodological issues. One more methodological problem may be that the strategy starts with higher-order functions, like language and consciousness, as features of the mind. The methodological strategy presupposed in the current philosophical discussion may thus be characterized as top-down and mind-based. This may be complemented by a more bottom-up and brain-based strategy in neurophilosophy. Such a bottom-up and brain-based approach might shed new light on the hitherto unresolved issues in both philosophy of mind and philosophy of psychology.

Summary

The previous chapter focused on the explanation of the mind on the personal level and how that relates to the subpersonal level. Proponents of irreducibility denied any connection between personal and subpersonal levels. In contrast, functionalist accounts of the mind linked both levels by suggesting corresponding structure and organization on subpersonal and personal levels. Meanwhile, eliminativist approaches suggested the elimination of the personal level as distinct from the subpersonal level. However, all approaches have left open the question: how can meaningful or semantic content be generated by the mind? This was the topic of the present chapter.

How does the subpersonal level generate semantic meaning on the personal level? Contents describe the meaning of a particular object in relation to the world and thus its semantic properties. Functionalism explains semantic contents that are represented in mental states ('what a mental state stands for') in terms of their functional role ('what a mental state does'). The functional role is here clearly distinguished from the semantic contents themselves. What does this functional role consist of? Propositional attitudes like belief, desire, or hope, determine the functional role.

One therefore has to distinguish between propositional content and propositional attitude. Propositional attitudes can be determined by their functional roles on the personal level. In contrast, the generation of propositional contents remains unclear. This is where the representational account of the mind comes in. The representational account argues that the propositional content is generated by its underlying subpersonal vehicle, e.g. brain or computer, and its particular physical structure. Thereby, the subpersonal vehicle shows an internal physical structure.

Most importantly, this internal physical structure on the subpersonal level is supposed to correspond to the linguistic structure that expresses the relation between propositional attitudes and content on the personal level. How can we further describe the subpersonal physical structure? Such internal structure is described as 'language of thought' (LOT), which refers to a formal structure. The language of thought describes the syntax of a language that determines and encodes the operations and processing in the physical structure. Due to the alleged correspondence between the personal and subpersonal levels, there is what is called 'structural isomorphism' between the subpersonal vehicle's internal structure and the linguistic structure of the content on the personal level.

The assumption of such mental representation characterized by LOT is often compared to the digital computer. That, though, is contested by the vectorial approach, which takes a parallel-processing computer, rather than a digital computer, as the starting point. The vectorial approach assumes that the representation of content is based on activity vectors (as modelled on the basis of the brain and its neural activity patterns) rather than linguistic-like structures on the subpersonal level. A recent investigation of the brain does indeed lend support to such an assumption. It shows that the formal structure of the brain's neural activity pattern does not correspond to the formal-syntactic structure of language but rather to one that is more similar to music and its constellations of tones. The exact details remain open, though, and subject to future research.

One may return to mental representation: can mental representation explain consciousness and its subjective and qualitative features? The representational approach primarily concerns contents that by themselves remain unconscious. Hence, a special additional mechanism for transforming unconscious representational contents into conscious ones must be assumed. Whether such additional mechanisms can also account for the subjective and qualitative, i.e. phenomenal, features of consciousness, remains open at this point. Several suggestions for the relationship between a representational account of the mind, and the subjective and qualitative features of consciousness, are discussed in current philosophy of mind. One suggestion is to eliminate the subjective and qualitative features altogether and to focus only on the objective and quantitative features, as they can be investigated in science. Alternatively, special forms of representation are suggested to account for the subjective and qualitative features.

Finally, the relevance of these philosophical discussions are put into the context of the current framework of philosophy and the brain. Since they start with higher-order mental features like language and consciousness as hallmarks of the mind, the methodological strategy presupposed in these philosophical discussions can be characterized as top-down and mind-based, which can be distinguished from a bottom-up and brain-based methodological strategy, as suggested by neurophilosophy.

Revision notes

- How is content defined?
- What is the difference between representational and functional roles?
- Is there a difference between propositional attitudes and propositional contents? If so, what does this difference consist of?
- How can representational and functional accounts of the mind be distinguished?
- Explain the concept of structure. Where and how is it realized?
- Why must content and vehicle be distinguished from each other?
- What is structural isomorphism?
- How can the concept of the 'language of thought' be explained?
- What is the difference between a linguistically- and vectorially-based account of mental representation?
- What problems does a representational account of the mind present when it comes to consciousness? How can these problems be solved?
- Why does a representational account of the mind have difficulties in explaining the phenomenal features of consciousness?
- Is the internal formal-syntactic structure of the brain's neural activity compatible with the assumption of a 'language of thought'? Are there any formal-syntactic alternatives?

Suggested further reading

- Bermudez, J. L. (2005) *Philosophy of Psychology* (London: Routledge).
- Block, N. (1980) 'Troubles with functionalism', in Block, N. (ed.), *Readings in the Philosophy of Psychology* (Cambridge, MA: MIT Press).
- Dretske, F. (1988) *Explaining Behavior. Reasons in a World of Causes* (Cambridge, MA/London: MIT Press).
- Kriegel, U. and Williord, K. (eds) (2006) *Self-Representational Approaches to Consciousness* (Cambridge, MA: MIT Press).
- Lloyd, D. (2011) 'Mind as Music', *Frontiers in Psychology*, 2(63).

11
Philosophy of Neuroscience: Explanations, Concepts and Observer in Neuroscience

Overview

The preceding chapters focused on the theoretical issues of psychology, and more specifically on the relationship between folk and scientific psychology. This discussion raised questions about a particular theoretical issue in psychology, namely the relationship between the different levels of explanation, the subpersonal and personal. The subsequent question that arises is whether there are similar theoretical and foundational issues in neuroscience that might benefit from a philosophy of neuroscience. Such issues concern the relationship between the different levels of explanation in current neuroscience – ranging from genetic, molecular, regional and network levels of explanation – and raise questions about the kinds of concepts used in neuroscience. The problems and issues surrounding a philosophy of neuroscience is the focus of this chapter.

Objectives

- To understand the distinction between philosophy of psychology and philosophy of neuroscience
- To determine the difference between philosophy of neuroscience and neurophilosophy
- To highlight the difference between incorporation and contextualization models as explanatory strategies in neuroscience
- To distinguish between brain-reductive and brain-based concepts
- To define brain-based versus observer-based concepts
- To distinguish between intrinsic and extrinsic observer-related intrusions

Key concepts

Philosophy of neuroscience, neurophilosophy, incorporation model, contextualization model, critical neuroscience, brain-based versus brain-reductive concepts, brain-based versus observer-based concepts, intrinsic versus extrinsic observer-related intrusions.

Background 1

From psychology to neuroscience

The previous two chapters focused on theoretical issues in psychology, as subsumed under the umbrella of philosophy of psychology. Philosophy of psychology describes the discussion of theoretical and foundational issues in psychology. One such issue consists of the relationship between folk and scientific psychology. Folk psychology describes our private common-sense assumptions about our own person and others' mental states on a personal level. This is different from scientific psychology. In scientific psychology, specific psychological functions are investigated, as observed publicly in third-person perspective. These psychological functions, like attention, memory, etc. are associated with the subpersonal rather than the personal level.

How are personal and subpersonal levels of explanation (and thus folk and scientific psychology) related to each other? This is a major debate in current philosophy of psychology. Different models ranging from the irreducibility, continuity, or elimination of folk psychology in favor of scientific psychology were discussed in earlier sections (see Chapter 9). The guiding question here is whether one explanation – the folk psychological explanation – can be reduced to the scientific psychological explanation.

In order to address the question of reduction, philosophy of psychology reaches out to philosophy of science, where the issue of explanatory reduction has long been discussed. Here the focus is centered on two criteria that must be met in order to reduce one explanation to the other, as, for instance, between two theories. The first criterion consists of commensurability, as determined by the possibility of translating the concepts of the one explanation into those of the other. The second criterion, derivability, describes how the concepts of the one theory as, for instance, folk psychology, can be inferred from those of a theory like scientific psychology.

Why is this relevant for neuroscience? Let us recall Chapter 3, where we described the development of neuroscience as a discipline. In this chapter we pointed out that historically the scientific examination of the mind took place in psychology as a distinct discipline from philosophy. However, with the increasing knowledge of the brain, psychology, and especially cognitive psychology, extended and reached out more and more to the brain and to neuroscience. At the same time, the introduction of better technologies, like functional brain imaging, made it possible to investigate the neuronal states associated with the mind and its mental features, including consciousness, self, free will, emotions, etc.

This illustrates a direct encounter between neuroscience and psychology, while at the same time highlighting that each discipline uses different concepts (neural versus psychological). For example, psychology uses psychological concepts like working memory, attention, reward, episodic memory, etc., whereas neuroscience employs neural concepts like the prefrontal cortex, visual cortex, action potentials, etc. The direct encounter between psychology and neuroscience in our current investigation of the mind and its mental features raises the question: how are psychological and neural concepts (and their respective explanations) related?

Background 2

Relation between psychology and neuroscience

Can we reduce psychological concepts to neural concepts? If so, one would expect both psychological and neural concepts to be commensurable with each other: can we translate psychological concepts into neural concepts? And can psychological concepts be derived from neural ones accounting for derivability? If commensurability and derivability as the criteria of reduction are met, one may reduce psychological concepts to neural concepts and thus ultimately reduce scientific psychology to neuroscience.

As in the case of folk and scientific psychology, different positions are debated concerning the relationship between scientific psychology and neuroscience. Some opt for the irreducibility of psychology to neuroscience, primarily because of the principal differences in their subject matter, and in their concepts and explanations. Others suggest continuity between both disciplines. What this means is that both psychology and neuroscience are considered to be essentially the same processes despite the fact that they are manifested on different levels of explanation. Finally, others argue in favor of the reduction (or even elimination) of psychology to neuroscience, ultimately defending the position that neuroscience should replace psychology.

How are scientific psychology and neuroscience related to each other? This question is not only about different scientific disciplines and their relation to each other, but also a question about different levels of explanation. The psychological level is concerned with psychological functions like attention, working memory and executive functions. Meanwhile, neuroscience is concerned with the physiological level, which also includes different levels of explanation like genetic, molecular, cellular, electrical, biochemical, regional and network levels. Hence, any question about the relationship between scientific psychology and neuroscience is a question about how to deal with these different levels of explanation.

Finally, the concept of level will be explained briefly. The concept of level denotes here a particular framework of theories and concepts to describe certain empirical findings. For instance, the genetic level of investigation uses genetic concepts and theories about genes, which is different from the level of neural networks. And the concept of levels reflects this, using different theories and concepts to describe respective findings.

Background 3

Philosophy of neuroscience

How are these different levels of explanation – ranging from the genetic, the cellular and the psychological levels – related to each other? This question is central to neuroscience. As we recall from Part I, neuroscience originated in different disciplines, ranging from anatomy and physiology to clinical neurology and psychiatry. These different disciplines brought with them a range of methods that targeted different levels of explanation.

The heterogeneity of different methods and levels in neuroscience is amplified and extended even further by recent developments in the field. Neuroscience, as discussed in Part I, has extended its investigatory reach into the territory of the social sciences and humanities, like anthropology, philosophy, theology, religion, politics, law, ethics, economics, etc. Many new disciplines are now characterized by the prefix 'neuro'. This change brings along with it new levels of explanation. Some examples include: the cultural level as in neuroanthropology, the moral level as in neuroethics, the political-social levels as in social sciences and politics, the economical level as in neuroeconomics, the legal level as neurolaw, etc.

How are all these different levels of explanation related to each other? The success of neuroscience, and its extension and outreach to other disciplines beyond itself, makes it even more important for us to understand the relations between these various levels of explanation. For example, how are the genetic or regional levels of the brain's neural activity related to the social and economic levels of explanation? This is not only a question about different levels of explanation, but also one about different disciplines, like neuroscience, social science and economy. This issue will be discussed in the first part of this chapter.

The question of the explanatory relationship between neuroscience and other disciplines does not concern the practice of neuroscience by itself; it is not about designing and conducting specific experiments that measure the various kinds of neuronal and psychological variables. Instead, *the debate surrounding the explanatory relationship between neuroscience and other disciplines is more a theoretical and methodological issue that is generally subsumed under the umbrella term 'philosophy of neuroscience'. The concept philosophy of neuroscience describes the discussion of theoretical, methodological and foundational issues in neuroscience.*

In this sense, the philosophy of neuroscience must be distinguished from neurophilosophy. Neurophilosophy is often considered as the application and relevance of neuroscientific data and studies to the investigation of originally philosophical questions (see Chapter 4 for different definitions). For instance, the question of consciousness has long been a topic dealt with in philosophy, but today it is also tackled in neuroscience. In neuroscience, it is manifested in the search for the neural correlates of consciousness (see Part IV for details). Both this chapter and the next focus not so much on neurophilosophy, but rather on methodological and theoretical issues in neuroscience, and thus on the philosophy of neuroscience.

Background 4

Philosophy of brain

We have already touched upon one major methodological issue in the philosophy of neuroscience: what are the relationships between different levels of explanation? This central issue will be discussed first. Other issues within the philosophy of neuroscience include thinking about the ways in which we can access and investigate the brain – a process that is possible only in an indirect way – with the help of technology. This is called the 'indirectedness of studies of mind and brain', which is also of central importance and will be discussed in the second section of this chapter.

Finally, there are other issues in philosophy of neuroscience that are more about the brain itself than the scientific investigation of the brain or neuroscience. These questions concern mainly the characterization of the brain as a whole rather than its specific neuronal, psychological and mental functions as investigated in neuroscience. *One may thus speak of a 'philosophy of brain' that deals with the empirical and theoretical (as well as metaphysical and epistemic; see Chapter 8) characterization of the brain as a whole and as distinguished from its various empirical functions.* These issues will be discussed in the next chapter.

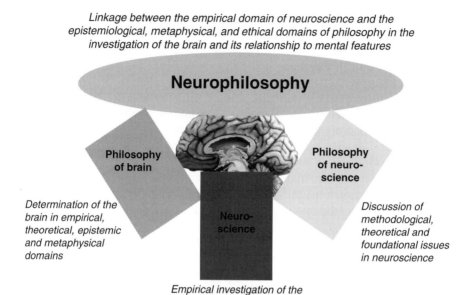

Figure 11.1 Relationship between neuroscience and other disciplines
Figure 11.1a Neuroscience, philosophy of neuroscience, philosophy of brain and neurophilosophy
The figure illustrates related disciplines to neurophilosophy (a) as well as two different models of the relationship between neuroscience and other disciplines (b) and (c) on pages 309 and 311.

The figure illustrates that neurophilosophy is based on three different fields: neuroscience, philosophy of brain and philosophy of neuroscience. All four are related to each other and overlap with regard to the brain. However, each has its particular focus on the brain. Philosophy of brain considers the brain in different domains, neuroscience investigates the brain in the empirical domain, philosophy of neuroscience focuses on theoretical and methodological issues in neuroscience, and neurophilosophy aims to link the different domains in the investigation of the brain and its relationship to mental features.

Explanation 1

Different levels of explanation in neuroscience

How are the different explanatory levels, and thus the different concepts and theories in neuroscience, related to each other? At the turn of the nineteenth century, neuroscience originated as a discipline with roots in different disciplines, including anatomy, physiology, and clinical neurology and psychology. What is important to remember is that each discipline considers the brain in a different way. Anatomy focuses on different regions in the brain whose neural activity is related to certain psychological functions. This is also true of clinical neurology and psychiatry. Physiology focuses more on the cellular level of neural activity by investigating the neuronal cells' electrical and biochemical activity, and their action potential.

Since its origins, neuroscience has reached out to different levels. The cellular level was complemented by the biochemical, the molecular and the genetic levels. The biochemical level focuses on different proteins, known as biochemical substances, and their action in the brain. The molecular level is complemented by the genetic level, where the underlying genes are investigated. Neuroscience has assembled an enormous amount of knowledge about the genetic, molecular, biochemical and cellular mechanisms in the brain in the past 20–30 years.

In addition, the cellular level has also been investigated extensively. Thanks to new technology, including brain imaging, regions of the brain can now be characterized in different ways: for example, by their pure anatomical structure, as well as by their functional activity during specific tasks. This has been complemented by the search for neural networks, as characterized by the connections and relationships between different regions.

Even more recently, neuroscience has been extending its investigatory reach to include theories and concepts, and thus levels of investigation, outside the brain. For example, the social level is concerned with the processes in our society, including relationships between different subjects and groups; this is addressed in a subfield called social neuroscience, which investigates the neuronal mechanisms underlying social functions like empathy, self, mindreading, etc.

Closely related is the investigation of cultural factors on the brain's neural activity: different cultural contexts may lead to different patterns of neural activity while perceiving the same objects, as has been shown in the novel fields of cultural neuroscience and neuroanthropology, while neuroeconomics investigates the neuronal mechanisms of economically relevant processes like buying and selling.

In addition to these levels, various other levels, like the level of religious belief ('neurotheology'), the legal level ('neurolaw'), the political level ('neuropolitics'), and the pedagogic level ('neuropedagogy') have been investigated by neuroscience. In the following, I will therefore distinguish between the neuronal levels that lie within the brain itself – genetic, molecular, biochemical, cellular, regional and network levels – from those that are located outside the brain, like the social, political, etc. These I describe as non-neuronal.

Explanation 2

Relation between different levels of explanation in neuroscience

How are these different levels of explanation – neuronal and non-neuronal – related to each other? Recall Chapter 2, in which we discussed the issue of naturalization. Naturalization describes the assumption that different domains, like the metaphysical and epistemic (or ethical) domains, are part of the empirical domain, and thus the world as we observe it in third-person perspective. How are the different domains related and connected in naturalism?

According to the relationship between the different domains, one may distinguish between different forms of naturalization. One possibility is that all other domains are replaced by, and thus incorporated into, the empirical domain (known as replacement or incorporation naturalism). This ultimately leads to the reduction of the metaphysical, epistemic and ethical domains to the empirical domain.

In an alternative model, the metaphysical, epistemic and ethical domains become linked and related to the empirical domain, without being completely reduced to it.

This leads to non-reductive or cooperative naturalism, as distinguished from replacement or incorporation naturalism.

How can we characterize such non-reductive or cooperative naturalism? On the one hand, non-reductive or cooperative naturalism needs to be distinguished from its more radical siblings, which opt for replacement or incorporation naturalism. The main difference here is that metaphysical or epistemic concepts and theories, describing the metaphysical and epistemic domains, are preserved and not replaced by or integrated into empirical concepts. In other words, metaphysical/epistemic and empirical concepts here stand side by side.

On the other hand, non-reductive or cooperative naturalism needs to be distinguished from non-naturalized approaches too. Non-naturalized approaches claim independence between the different domains with no relationship at all between the metaphysical, epistemic and empirical domains. This contrasts with non-reductive or cooperative naturalism, where the different domains are assumed to be related to and dependent on each other. There is a certain overlap and dependence between epistemic and empirical domains, even though they cannot be replaced by each other or taken as identical.

Explanation 3

Explanatory reduction and the 'incorporation model'

How do the different forms of naturalism relate to the different levels of explanation in neuroscience? To answer this question, one may first want to make the distinction between domains and levels because it is an issue that concerns the relationship between philosophy and neuroscience.

Naturalism is about the relationship between different domains. In contrast, the explanation of the different levels is supposed to remain within the empirical domain itself – the world as we can observe it in third-person perspective. As such, the problem of explanation is one within neuroscience, rather than one about its relationship to philosophy. This highlights a problem within the empirical domain itself, rather than its relationship to other domains.

The principal difference between levels and domains means that we cannot make any direct inferences between naturalization and the explanatory relations between different levels. This should not distract us, however, from learning models of possible relationships in general – independently of whether they apply to the relations between different domains or levels. In other words, we may be able to shift the two models discussed in the context of naturalism (incorporation and cooperation) to the context of the explanatory relations between different levels.

How does the model of incorporation resurface in the context of the explanatory relation problem? The explanatory relation problem concerns the relationship between different levels of explanation. If, for instance, the social and cultural levels of explanation are incorporated into the neuronal levels of explanation – the molecular, cellular, regional and network levels – the former are supposed to be necessarily and sufficiently explained by the latter.

The explanation of the social and political levels is here reduced to, and ultimately incorporated by, the explanations of the various neuronal levels. If so, one

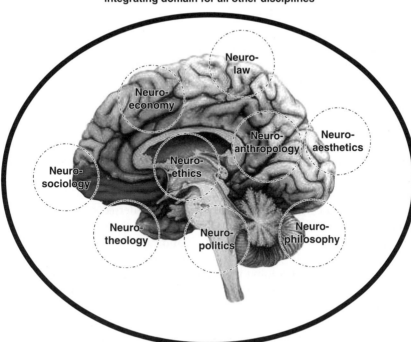

Figure 11.1 Relationship between neuroscience and other disciplines
Figure 11.1b *Incorporation model*: incorporation of other disciplines into neuroscience

The figure illustrates the incorporation model in the relationship between neuroscience and other disciplines. All other disciplines are incorporated within the empirical domain (bold ellipse) and are then directly related to and thus incorporated into neuroscience and its investigation of the brain. The other disciplines thus become part of neuroscience, hence the various subfields in neuroscience that start with the prefix 'neuro' (like neuroanthropology, neurotheology, etc.). This incorporation into neuroscience implies that the originally different framework of these disciplines is more or less replaced by the empirical framework of neuroscience, as indicated by the small dotted circles.

would assume that both social-political and neuronal levels can be translated into each other with both being commensurable. And that the former can ultimately be derived from the latter, thus accounting for derivability as the second criterion of reductive explanation.

The reductive model in the explanation of the different levels of neuroscience is assumed in particular by the advocates of neurophilosophy in the Anglo-American world, who presuppose a reductive neurophilosophy (see Chapter 4). Here, the neurophilosophers stick to domain monism by focusing on one particular domain, namely the empirical domain, which replaces and incorporates other domains. Moreover, the methodological arsenal is sparse, and focuses almost exclusively on the experimental-observational method. Finally, there is a brain-reductive approach, in that existence and reality, knowledge, etc. are supposed to be reduced to and fully accounted for by the brain and its neuronal processes.

Explanation 4

Context and critical neuroscience

The exclusive and complete explanation, and thus the incorporation, of the non-neuronal levels by the neuronal levels has recently come under attack. Scholars who advocate what they call 'critical neuroscience' argue that the explanations of the neuronal levels of the brain cannot necessarily and sufficiently account for those on the non-neuronal levels. Instead, the explanations of the neuronal levels need to be considered by themselves within their respective social, economic, ethical, political, etc. contexts.

For example, images from functional brain imaging seem to suggest the localization of our mental states (like one's sense of self) in particular brain regions. If the self cannot be found in our observations of the brain's neural activity, one cannot infer that there is no such thing as self and that the self does not exist. That would be an over-interpretation of the data, meaning that we've inferred some statement from the data that the data itself does not contain. What is the information that is not contained in the data? The additional information might consist of specific social and cultural (and possibly political) assumptions that are implied by the current socio-cultural and political context in which the investigation takes place.

How can we avoid this type of over-interpretation? While the acquisition of the data itself may remain independent of the respective social, economic, ethical, political, etc. contexts, their interpretation and conceptualization (their association with specific concepts), may be highly dependent upon these respective contexts. In other words, we need to assess critically and evaluate the neuroscientific data about the various neuronal levels by putting them into their respective non-neuronal contexts. This is called 'critical neuroscience'.

Critical neuroscience is a recent movement that calls for a critical and contextual evaluation of the findings in current neuroscience, where the social, political, historical and economic contexts within the current neuroscientific findings will be considered. Critical neuroscience aims to address scholars in the humanities, as well as, most importantly, neuroscientists themselves, policy makers and the public at large.

Explanation 5

Explanatory context and the 'contextualization model'

What does the claim of critical neuroscience imply for the relationship between the different explanatory levels, neuronal and non-neuronal? Rather than incorporating non-neuronal into neuronal levels, the latter's explanations should be considered in the context of the former. In other words, we need to contextualize the neuronal levels of explanations with regard to their respective social, political, economic, legal, etc. contexts.

This has major implications for the explanatory relations between the different levels. If context needs to be considered, the explanations of the neuronal levels can no longer be regarded as independently sufficient in the explanation of the non-neuronal levels. Instead, we need to consider the non-neuronal contexts themselves when explaining and interpreting the neuronal levels. Hence, we need literally to contextualize the explanations of the neuronal levels within the brain by using the non-neuronal levels outside the brain. Even if we want to explain the non-neuronal levels outside the brain by the neuronal ones within the brain, as in various disciplines with the prefix 'neuro', we still must presuppose the respective non-neuronal context in explaining those neuronal levels that are supposed to explain the non-neuronal levels.

How, then, can we characterize the relationship between the different levels of explanation? Rather than incorporating non-neuronal into neuronal levels, we may contextualize the latter by the former. One may consequently speak of a *'contextualization model'* for the relationship between the different explanatory levels. The contextualization model assumes the neuronal level explanations to be only necessary (but not sufficient) for non-neuronal level explanations.

In sum, one may distinguish between two different models in the relationship between different explanatory levels in neuroscience. The incorporation model argues in favor of the replacement and incorporation of the non-neuronal levels

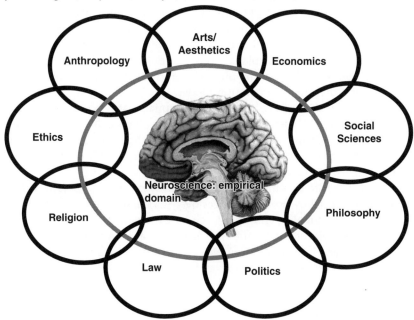

Figure 11.1 Relationship between neuroscience and other disciplines
Figure 11.1c Contextualization model: bilateral contextualization between neuroscience and other disciplines

The figure illustrates the contextualization model in the relationship between neuroscience and other disciplines. Rather than being incorporated, the different disciplines here form the context for neuroscience and its investigation of the brain. Instead of being situated inside the empirical domain of neuroscience, the different disciplines are then situated at the border of neuroscience where they overlap with its empirical domain. This means that the original framework and domain of the respective disciplines is somehow preserved, as indicated by the ellipses around each discipline.

of explanation into the neuronal explanations. Meanwhile, the contextualization model denies the possibility of replacement and incorporation. Instead, it argues in favor of the need to consider the respective non-neuronal contexts for explanations of the different neuronal levels. Hence, one may distinguish between explanatory incorporation and explanatory contextualization.

Critical reflection 1

Multi-level causal explanations in neuroscience

One might want to argue that any explanation in neuroscience involves multiple levels and is thus contextualized. For example, the contemporary American philosopher, Carl Craver, assumes that explanations in neuroscience are both causal and multi-level. Rather than being reduced to each other, he suggests that the different levels of explanation can be causally related to each other. For instance, the molecular level describing different proteins might cause the generation of action potentials on the cellular level. In this case the cellular level is not reduced to the molecular level but rather is causally related to it. Taking the different levels of investigation together, this leads to what Craver describes as 'mosaic unity' of causal explanations: including different mechanisms, multiple levels and various fields.

One may say that the different levels serve here as contexts for each other. Contextualization in this sense – i.e. as part of a multi-level causal explanation – does not imply that the lower-level explanation is insufficient at the higher level. If so, the contextualization of lower levels by higher levels does not imply the contextualization model but rather the incorporation model.

Let's consider the example of the action potential that describes an electric discharge of a single cell. An action potential is induced by small electrical changes in the cell's actual state. This is related to changes in biochemical substances (proteins), like the excitatory glutamate or the inhibitory GABA, whose balance, the 'excitation–inhibition balance' (EIB), may change.

How does that relate to a multi-level causal explanation? The causal explanation of the action potential by protein changes over EIB changes, and electrical changes to cellular changes, involves multiple levels of explanation: the cellular level, the biochemical and molecular level, the electrical level, and ultimately also the atomic level. All these levels interact in a causal way with the next higher level, providing the context for the respectively lower level, while at the same time the latter causally affects the former.

For instance, the electrical level of the EIB provides the context for the biochemical-molecular level, e.g. GABA and glutamate: GABA and glutamate act and change within the context of a certain actual degree of the EIB, the electrical level, while at the same time they also change, and thus causally impact, the degree of the EIB itself.

This example demonstrates that a multi-level contextualization and a causal explanation can go hand in hand. A causal explanation implies that the lower level explanation is necessary and sufficient for the explanation of the next higher level. If so, a causal explanation of higher levels by lower ones does not exclude contextualization. This means that contextualization does not necessarily imply the contextualization model, but rather the incorporation model. This is the kind of reduction that, according to Craver and neurophilosophers like John Bickle, is common in current neuroscience. Bickle calls this 'reductionism in practice'.

Critical reflection 2

Conceptualization in neuroscience

How does such contextualization relate to the claim of critical neuroscience that the need for contextualization makes any sufficient explanation of higher, non-neuronal level, by lower, neuronal level, explanations impossible? On the one hand, Craver and others could argue that there is no principal difference between neuronal and non-neuronal level explanations and, on the other, say there are explanations of different levels within the brain itself, e.g. the neuronal levels.

One may, for instance, argue that political phenomena outside the brain, like certain voting behavior, can be linked to the neuronal explanations inside the brain in the same way that two different, purely neural, explanations are related to each other. The neuronal activity in particular regions or networks in the brain may then necessarily and sufficiently explain the voting behavior.

The advocates of critical neuroscience may want to argue that the data about the brain regions involved in racism do not need to be interpreted and put into concepts, as these concepts are used to describe multi-level causal explanations, and the choice of these concepts strongly affects the interpretation of the data on whether respective brain regions are indeed considered to be the neural correlates of racism. Thereby, the choice of the concepts is strongly dependent upon the respective social, political, etc. context.

One could thus imagine the same data to be conceptualized in different ways, which would also lead to different interpretations. If, for instance, an imaging study presents personally relevant trait adjectives to subjects, one may associate the respective brain regions with the self. The brain region x causes the self, and is therefore sufficient for the neural correlate of the self. This is different once one associates the very same neural activity with a particular process.

One may, for instance, postulate self-related processing that describes the process by means of which stimuli are related to the self of a specific person. Such self-related processing needs, however, to be distinguished from the self itself, and more specifically from the experience of a self. In this case, the observed neural activity in the respective regions can at best be considered a necessary, but no longer a sufficient, condition of the self. The region's neural activity is thus no longer a neural correlate of the self but only its neural predisposition (see Chapters 15 and 16).

Critical reflection 3

Conceptualization and contextualization

Furthermore, the advocates of critical neuroscience argue that the choice of our concepts strongly depends upon the respective social, political, etc. context. In a very materialistic and naturalistic context, where everybody speaks of a 'neurorevolution', one may be more inclined to speak of the concept of self, rather than of self-related processing. Moreover, the stronger the claims, the more the actual data seem to carry major findings, and the more likely the respective scientist or philosopher can acquire grant funding, attract media attention, and become a major player in the field. Hence, it is the context, the non-neuronal context lying outside the brain, that strongly affects our choice of concepts to describe, interpret, and explain our data about the different neuronal levels and their relationship to non-neuronal levels.

This makes it clear that even the multi-level causal account is prone to contextualization in the sense of critical neuroscience. Since contextualization affects our selection and the choice of concepts we use to describe, interpret, and explain our data, one may want to say that contextualization goes hand in hand with conceptualization. Accordingly, contextualization is necessary even in the multi-level causal explanation, since otherwise the respective data could not be described, interpreted and explained.

Such need for contextualization implies the contextualization model rather than the incorporation model in the relationship between neuroscience and the other disciplines. The incorporation model eliminates all other contexts and thus any context in general by incorporating the other disciplines into its empirical domain. Contextualization is thus eliminated and replaced by conceptualization, that is, conceptualization in exclusively neural terms, implying that the concepts are interpreted and explained in an exclusively neural context. Conceptualization is thus decoupled from contextualization.

This is different in the contextualization model. Here the different disciplines and their respective contexts are preserved. The conceptualization of neuroscience and thus the explanation and interpretation of its concepts can here occur in the context of the different disciplines in the same way that the concepts of the latter are contextualized by the neural concepts of neuroscience. Hence, contextualization and conceptualization go hand in hand in the contextualization model.

Explanation 6

Brain-reductive explanations

Two models in the relationship between the different explanatory levels can be distinguished from each other. The incorporation model replaces and incorporates non-neuronal explanations with neuronal explanations. The contextualization model suggests the non-neuronal explanatory levels to provide the explanatory context for the explanations of the neuronal levels.

Since the different levels of explanations are associated with different disciplines, the explanatory models imply different forms of relationships of neuroscience to other disciplines. In the case of the incorporation model, the other disciplines are integrated into and ultimately replaced by neuroscience. They are part of neuroscience itself and form a specific partition within the empirical domain of neuroscience.

Neuronal explanations from neuroscience are here considered to be both necessary and sufficient for explaining the originally non-neuronal levels that are now incorporated and translated into neuronal-level explanations. This implies that the explanations of the originally non-neuronal levels are reduced to the neuronal-level explanations and ultimately to the brain. The explanations within the incorporation model can thus be characterized as brain-reductive explanations. *Brain-reductive explanations are explanations where neuronal level explanations, as explanations about levels inside the brain itself, are considered necessary and sufficient for the explanations of non-neuronal levels, which concern phenomena outside the brain.*

Explanation 7

Brain-based explanations

This is different in the contextualization model. Here, the different disciplines are linked and related to neuroscience, demonstrating overlap, but there is no superiority of one over the other. In other words, this overlap is not exhaustive. There are areas within the respective disciplines that are not covered by neuroscience and its explanations of the neuronal levels. This makes it possible for those disciplines to provide the context for neuroscience and its explanations of the neuronal levels. Disciplines like anthropology, social science, etc., thus provide the context for neuroscience.

How is it possible that disciplines like anthropology, theology, etc., simultaneously both overlap with, and are yet distinguished from, neuroscience? Explanations of the neuronal levels provide only a necessary, but not sufficient, explanation of non-neuronal levels. The necessary conditions of explanations may be well reflected in the various disciplines with the prefix 'neuro', which basically signify the neuroscience of anthropology, sociology, politics, ethics, etc.

How about the sufficient conditions for explanations? They can only be acquired by considering the explanatory levels outside the brain as the context for the explanations of the neuronal levels inside the brain itself. This means that everything outside the brain serves as the context for what lies inside it. The brain and our neuronal explanations of its processes inside are thus put in a larger context and are therefore contextualized.

What about the role of the brain in such contextualized explanations? Since the neuronal-level explanations are both necessary and sufficient, the explanations of the non-neuronal level can no longer be reduced to the brain. At the same time, they cannot be detached completely from the brain, since that would confuse sufficient and necessary conditions. One can say, then, that the explanations of the non-neuronal levels are brain-based, rather than brain-reductive (which is the case in incorporated explanations). *Brain-based explanations arise where neuronal-level explanations are considered necessary but not sufficient for the explanations of non-neuronal levels.*

Explanation 8

From philosophy of neuroscience to philosophy of neurophilosophy

Why is the distinction between brain-reductive and brain-based explanation important in the present context of philosophy and the brain? The kind of explanation, that is, brain-based versus brain-reductive, sets the explanatory framework for the relationship between the brain and philosophy. If one assumes brain-reductive explanations, one ultimately admits that the reduction of philosophy to the brain and neuroscience is possible. Philosophy as a discipline is then no longer considered as principally different from neuroscience; it more or less becomes part of neuroscience in the form of a 'meta-science' and/or as philosophy of neuroscience. This implies a reductive, that is, brain-reductive concept of neurophilosophy.

This is different if one presupposes a brain-based explanation. With a brain-based theory, the relationship between philosophy and the brain is supposed to be bilateral rather than unilateral. In other words, philosophy may then provide the context for the brain and neuroscience, while at the same time being based on it. In this case, philosophy has its own domains but, unlike in the historical tradition, no longer remains completely independent of the empirical domain in general, and neuroscience in particular. That requires a non-reductive – that is, brain-based rather brain-reductive form of neurophilosophy.

Accordingly, the distinction between brain-reductive and brain-based explanations does not only carry important implications for neuroscience, but also for its relationship to philosophy. More specifically, different forms of explanations, brain-based versus brain-reductive, entail different concepts of neurophilosophy, reductive versus non-reductive. Hence, what is originally discussed in philosophy of neuroscience, different forms of explanations, has important ramifications for the concept of neurophilosophy. There is thus direct transition from philosophy of neuroscience to what may be described as 'philosophy of neurophilosophy'.

What do I mean by 'philosophy of neurophilosophy'? Philosophy of neurophilosophy is about methodological, theoretical and foundational issues in neurophilosophy. These issues concern, for instance, different kinds of possible methodological approaches in neurophilosophy, types of neurophilosophical explanations, and the relationship between neurophilosophy and philosophy. The different concepts of neurophilosophy, like reductive and non-reductive and their characterization, may be considered one of the core topics of a philosophy of neurophilosophy.

Brain and observer 1

Indirectedness in the investigation of mind and brain

We have already discussed one particular methodological issue – the explanatory relation of neuroscience to other disciplines. Another methodological problem includes the kinds of inferences and conclusions we can (or cannot) draw from our empirical data. For instance, the recent advent of functional brain imaging raises the question: can we make inferences about specific psychological, and ultimately mental, states from the colorful visualizations of neural activity in specific regions and neural networks of the brain?

This touches upon a more basic theoretical and ultimately epistemic problem, namely, the problem of knowledge. We can access and observe the brain only in an indirect way, using different technologies like cell recording devices and various brain imaging techniques. This is described as 'indirectedness of studies of mind and brain'. *Indirectedness, as a concept concerning the studies of mind and brain, points out that we can investigate the brain and the neuronal activity that supposedly underlies our mental states only in an indirect way and with the help of technological devices.*

Meanwhile, direct access to the brain without the help of additional technological devices remains impossible.

The indirectness of studying the brain and the mind makes it necessary to disentangle the contributions of the brain to our data from those related to our observations via the various technological devices. In other words, we need to distinguish between the brain's input and the observer's (and his/her technologies') input to our data. To do this, we clearly need to distinguish between concepts that are more related to the observer him/herself, and his/her technological input, than to the brain itself and its neuronal activity (as independent of our technological observation).

Let us explicate this point. Our data and concepts may include two components. One component refers to the brain itself, and its actual neuronal processes, as they are independent of our observation and our technological devices. This component needs to be clearly segregated from the one that is more related to our observation, and its technological devices, and reflects technological or observational artifacts. Only if we can clearly segregate the first from the second component, will our data be valid. Valid data means that the data we obtain should reflect the brain itself as it operates – independently of our observation of it. It is also important that the data is not confounded by the brains and minds of us as observers – the data should remain mind- (and brain-)independent.

The exclusion of technological artifacts is, of course, an empirical and, more specifically, a technological issue. As a result, it is not a topic of philosophy of neuroscience. This is different, though, in the case of the alleged mind-independence of our data that is not only an empirical issue, but also a conceptual one. It is conceptual in that the concepts we use to describe our data should reflect the brain itself, rather than the observer and his/her mind. In other words, our data should be based on the brain rather than the observer, so that it is brain-based rather than observer-based. This will be discussed below.

Brain and observer 2

Encoding of multiple data and facts into one concept

Neuroscience acquires data and ultimately facts to describe the brain (the distinction between data and facts may itself be worth discussing from a philosophical point of view (see also Northoff, 2011)). Neuroscience uses concepts to describe data and facts. Usually, one would expect the concepts to correspond and thus match the data and facts in a one-to-one way: where one particular datum or fact corresponds to one specific content as referred to in the respective concept.

Things are not quite so simple, however. Concepts are usually more general and vague than data and facts. This means that concepts usually include more than one particular content, and thus, by definition, are general. The concepts neuroscientists (and any scientists) use (or must use) remain consequently unable to completely match and correspond to the data and facts in a one-to-one way. Instead, the concepts may also refer to contents other than the ones from the particular data and facts in question.

Rather than a one-to-one relationship, such association of several contents with one concept implies a many-to-one relationship, where one concept stands for and can be associated with many different sets of data/facts. Let us express this relationship between contents and concepts in terms of coding. Rather than coding contents and their related data and facts in a local way – that is, one-to-one – concepts seem to encode contents, i.e. data and facts, in a rather dense way – that is, in a many-to-one way.

Such dense encoding of several contents into one concept has major ramifications. We can never be completely sure whether the concept we use to describe our data and facts really matches and corresponds to the single concept. Why does there remain such uncertainty? The same concept could also refer to some other data and facts that are associated with and encoded by the same concept. Since this applies to concepts in general, both neurophilosophical and neuroscientific concepts are affected by such dense encoding and the respective uncertainty in the conceptualization of our contents, including our data and facts.

Brain and observer 3

Brain-based concepts

What does the dense encoding of data and facts into our concepts imply for the role of the observer in his/her own observation? The observer introduces the concepts in order to describe the data he/she observed. By choosing and selecting particular concepts, the observer intrudes him/herself into his/her own observation of the brain. Since the observer cannot avoid using concepts, there seems to be almost no way of the observer not intruding, even in the seemingly most objective investigation (that, being objective, claims to remain independent of the observer). The observer thus intrudes into his/her own observation via the use and definition of concepts which, due to the necessity of conceptualization, is inherent in our neuroscientific and neurophilosophical methodology.

For instance, the concepts the neuroscientist uses to describe data/facts are the same or similar concepts he/she uses to interpret images from brain scans. These images may be visualizations of statistical analyses based on a large number of images and scientific constructs, rather than pictures of what is actually occurring in the brain. It seems that there will always be some degree of observer-related intrusion in interpreting and analyzing imaging.

Let us specify the implications for the observer and his/her concepts. Concepts are generated by the observer. The very same observer who conducts the experiments also needs to generate concepts to describe his/her data/facts and to formulate his/her hypothesis. Yielding hypotheses and data/facts is possible only when considering certain requirements that need to be fulfilled within the experimental context. One such experimental requirement is the careful distinction between different experimental variables, each of which needs to be treated in a segregated and independent way. This makes the introduction of concepts to describe these segregated and independent variables necessary.

So far, so good. The problem starts once the very same concepts that describe these segregated and independent experimental variables are also assumed to describe the brain itself. More specifically, based on the experimental data/facts, the concepts describing the respective experimental variables are assumed to describe one-to-one the processes and mechanisms in the brain itself. Due to such a perfect match between the concept and the data one would assume that the concept is brain-based rather than observer-based.

Brain and observer 4

Observer-based concepts

However, one could also imagine instances where the concept does not match or correspond to the brain's neuronal processes and mechanisms. In this case, the concept is more related to the observer and his/her experimental requirements than the brain, and its neuronal processes, and mechanisms independent of our observation. This means the concept is more observer-based than brain-based. I distinguish between what I describe as 'observer- and brain-based concepts' below.

The distinction between observer- and brain-based concepts is not an 'all-or-nothing' distinction, but rather a 'more-or-less' distinction. This means that a particular concept may be based on both the observer's experimental requirements and the brain's neuronal processes. This reflects a continuum between brain- and observed-based concepts. It may be just a matter of degree and balance between observer and brain that determines the concept in question. And thus, a concept is more or less strongly based on either the observer and his/her experimental requirements, or the brain's neuronal processes.

One would then say there is a continuum, with its two extreme poles describing purely observer- and brain-based concepts. The neuroscientist seeks, of course, concepts where the balance is tilted strongly toward the brain-based pole of the continuum and away from the observer-based pole. While it is the task of the neurophilosopher to reveal the degrees to which the observer himself intruded on the concept, the neuroscientist claims to describe his/her data and facts. Hence, the neurophilosopher must reveal the degree to which neuroscientific concepts are observer-based rather than brain-based.

Brain and observer 5

Examples of observer-based versus brain-based concepts: GABA and glutamate I

In neuroscience, we encounter several examples of suspicious concepts where the balance is seemingly more strongly tilted toward the observer than the brain itself. In the following, I want to briefly mention some of them.

Two central transmitters in the brain are called glutamate and GABA. Experimentally, we need to segregate glutamate and GABA, and correspondingly, neural excitation and inhibition, from each other. In order to measure, for instance, glutamate and neural excitation, we need to experimentally parse both variables from any traces of GABA and neural inhibition, otherwise we cannot be sure whether our data really tells us about glutamate and neural excitation. This means ultimately that GABA, glutamate, and neural inhibition and excitation, are treated as segregated and independent experimental variables.

The designation of GABA and glutamate as segregated and independent variables occurs on purely experimental grounds and is therefore strongly observer-based. Based on data whose acquisition presupposes such experimental segregation and independence, one would assume GABA and glutamate also to act as segregated and independent in the brain itself. One further assumes that certain levels of GABA and neural inhibition are necessary for a specific neuronal process. While these levels may be open to (secondary) modulation by glutamate and neural excitation, they are considered (primarily) as independent and segregated.

Brain and observer 6

Examples of observer-based versus brain-based concepts: GABA and glutamate II

What does this imply for our distinction between brain- and observer-based concepts? This means that the observer's concepts are transferred to the brain itself. The initially observer-based characterization of GABA and glutamate as independent and segregated experimental variables is now projected on to the brain and assumed to accurately describe its neuronal processes. In short, it is no longer regarded as observer-based, but rather as brain-based.

Does such experimentally-based segregation and independence between GABA/neural inhibition and glutamate/neural excitation really correspond to the empirical data? I deny this to be the case. Instead, I assume that both GABA and glutamate can be characterized by difference-based coding: rather than the absolute levels of GABA and glutamate, it is their relation that is encoded into neural activity (see Northoff, 2013a). Their relation is empirically manifest in the excitation–inhibition balance (EIB) that is supposed to provide the measure for the subsequent generation of neural activity. But such coding of the difference to the respective other puts the assumption of (primary and constitutive) segregation and independence between glutamate and GABA into doubt.

The characterization of GABA and glutamate by (primary and constitutive) segregation and independence may be relevant (and even required) for the observer (and his/her experimental approach). It may not, however, apply to the brain as it is by itself independent of the observer. This means that such a characterization is more strongly related to the observer and his/her experimental requirements, than to the brain's neuronal processes, independent of the observer's observation. In other words, assuming segregation and independence between GABA and glutamate may turn out to be more observer-based than brain-based.

Brain and observer 7

Examples of observer-based versus brain-based concepts: stimuli and their origins

Another example is the distinction between different types of stimuli according to their origin in either the brain (i.e. neuronal stimuli), the body (i.e. interoceptive stimuli), or the world (i.e. exteroceptive stimuli). Based on these distinct origins, different anatomical structures and pathways have been assumed as well as reflected in the radial-concentric threefold anatomical organization (see Northoff 2011).

However, on a functional level, the distinction between the different origins of the stimuli and their respective anatomical structures seems to be blurred. This was apparent in the observed neural activity (functional connectivity and low–high frequency fluctuations), and the coding strategy (difference-based rather than stimulus-based coding) that operated across and superseded the underlying anatomical structure (see Northoff, 2013a). Yet this means that the distinction of stimuli according to their origin may not be as relevant for the brain and its neuronal processes as it is for us as observers with our experimental requirements. The experimental requirement is not to confuse stimuli of different origins, because otherwise we cannot say anything about, for instance, exteroceptive stimuli and their underlying neuronal processes.

However, as relevant as the distinction between the different stimuli origins may be for us as observers, it does not seem to be as relevant for the brain itself. Rather than the origin of the different stimuli, the statistically-based differences between different stimuli are relevant for the brain and its encoding of stimuli into neural activity. It is from this that my characterization of the brain's neural operation by 'matters of degrees and differences' stems, rather than 'matters of origins and stimuli'. However, this means that the characterization of the brain's neural processing by 'matters of origin and stimuli' may be more strongly related to the observer him/herself than the brain itself: that is, it is observer-based rather than brain-based.

This is different in the case where one describes the brain's neural processing by 'matters of degrees and differences'. That characterization seems to be closer to the way the brain itself operates and processes its neural activity; it thus seems to be brain-based rather than observer-based.

Brain and observer 8

Examples of observer-based versus brain-based concepts: resting state and stimulus-induced activity

Let us provide a final example where brain- and observer-based concepts may be confused with each other: the distinction between resting state and stimulus-induced activity. Experimentally, we clearly need to segregate and delineate both, otherwise we would never be able to know the component in the brain's neural activity that the stimulus itself induces in the brain and the one that stems from within the brain itself: that is, its resting state activity (independent of the stimulus). One may consequently assume segregation between resting state and stimulus-induced activity. However, as the empirical data suggest, both cannot principally be distinguished from each other, let alone segregated.

Instead of a principal difference and segregation, resting state activity and stimulus-induced activity can only be distinguished from each other on the basis of degrees. This, though, means that the principal distinction between resting state and stimulus-induced activity is more strongly based on the observer than the brain itself. One may thus formulate what can be described as a 'continuity hypothesis', which assumes a neuronal continuum and discontinuum between resting state and stimulus-induced activity. If the stimulus-induced activity does not alter the pre-existing resting-state activity very much, the latter may shift toward the brain end of the continuum between brain- and observer-based concepts.

If, in contrast, the extrinsic stimulus and its stimulus-induced activity exert major changes to the brain's intrinsic activity – its resting state activity – the increasing discontinuum between the two may be accompanied by a shift towards the observer-end of the continuum. Why? Because the observer is always already involved in the extrinsic stimuli he/she applies, be it directly as part of that stimulus itself, or indirectly as the applicant or cause of that stimulus. In other words, the investigation of extrinsic stimulus-induced activity may be more prone to intrusions by the observer and thus to the consecutive development of observer-based rather than brain-based concepts.

How can we escape the possible confusion between brain- and observer-based concepts? In order to shift concepts away from the observer-based pole towards the brain-based pole we need also to shift our perspective. More specifically, we need to abandon our observer-based perspective and imagine how it is for the brain to generate the kind of neuronal processes we observe. We should aim to move from the observer's perspective to the brain's perspective (taken in a figurative sense because the brain itself has no perspective). Metaphorically, one may consequently say that we need to replace 'what it is like for the observer' by 'what it is like for the brain'.

On the whole, I have demonstrated various examples of concepts that seem to be more strongly based on the observer him/herself and his/her experimental requirements, than on the brain's neuronal processes independently of any observation. We may want to sharpen the point even further: is an observer-free and thus truly objective (in an absolute rather than relative sense) investigation of the brain possible at all? Or will there always already be some intrusions that we cannot avoid (e.g. intrinsic observer-related intrusions)? One may argue that raising this question is possible only on the basis of an observer. Accordingly, we can in principle not avoid the contamination of our data by the observer and his/her intrusions.

Ideas for future research 1

Observer-related intrusions

How can we be sure that the concepts we apply are more brain- than observer-based? The only way for us to know is to develop corresponding hypotheses and conduct appropriate experiments. If the data are in accordance with the characterization implied by these concepts, the assumption that they might be more brain-based could be justified. If, in contrast, the data does not support these hypotheses, the concepts may turn out to be as observer-based as the ones they replaced.

If the data are in accordance with the concepts, the concepts are empirically plausible. They are thus to a higher degree based on the brain rather than the observer. Meanwhile, the opposite case of no empirical support would suggest that they are more based on the observer than the brain. Hence, the degree of empirical plausibility, the accordance of the concept with the empirical data, may be regarded as a measure of the degree to which the concept is more strongly brain- or observer-based. This also implies that purely brain-based concepts are more of an ideal than a reality. It may in fact be that purely brain-based concepts are, in principle, impossible.

However, alternative experimental designs, using different experimental variables, should also be applied. If the different experimental variables yield the same or analogous results, the likelihood that both data sets are affected by the observer and his/her experimental variables is rather low. The data may then provide an excellent basis for being associated with a concept and this would show a high probability of being brain-based rather than observer-based.

How about the reverse case, with low empirical plausibility and a more strongly observer-based concept? In this case the observer and his/her experimental requirements seem to intrude too much into the concept and the subsequent experimental design. The observer thus intrudes into the brain and imposes him/herself, thereby manipulating what he/she can observe from the brain's neuronal processes according to his/her own stance and needs. I call this type of intrusion 'observer-related intrusion'.

The concept of 'observer-related intrusion' refers to the intrusion or imposition of the observer him/herself into/on to his/her own observation of the brain's neuronal processes. In short, observer-related intrusions describe how the observer him/herself confuses his/her own observations. Observer-related intrusions consequently lead to low degrees of empirical plausibility and more observer-based than brain-based concepts.

Ideas for future research 2

Extrinsic observer-related intrusions

How can we deal with observer-related intrusions? We can try out alternative concepts and conduct respective experimental designs. This allows us to compare the results from both experimental lines with the two different concepts. The concept that shows a higher degree of empirical plausibility may then be the one that is more brain-based, while the concept that shows a lower degree may be more observer-based.

In other words, we have to try out different alternative concepts and put them to rigorous experimental testing (see Chapter 4 about concept–fact iterativity). This means that we are not at the mercy of observer-related intrusions, but that we have the (methodological) tools to minimize and ultimately avoid them. In turn, we can minimize the degrees to which an observer intrudes and imposes him/herself on to our concepts. In the best case, we can avoid observer-related intrusions altogether. In this case, respective concepts are strongly brain-based.

Since we are in principle able to minimize the degree of observer-related intrusion, I characterize these kinds of intrusions as extrinsic. The concept of 'extrinsic observer-related intrusion' describes how the observer's intrusion and imposition can, in principle, be minimized and at best be avoided, thus remaining extrinsic to both observation and concept.

Biography

Georgy Buszáki

I assume that extrinsic observer-related intrusions can, in principle, be minimized and at best be avoided altogether. This is possible by refining our concepts as described, and developing better and more precise technological tools for measuring and acquiring data (for instance, higher resolution brain scanning). However, there may be instances where we remain principally unable to minimize observer-related intrusion. This, to make it clear, does not concern the individual observer as distinct from other individual observers; it rather pertains to all possible observers.

Let me start with the Hungarian-American neuroscientist Georgy Buszáki and his emphasis on rhythms and oscillations. He argues in his excellent book *Rhythms of the Brain* (Buszáki, 2006) that rhythms and oscillations are a hallmark feature of the brain. To prove his point he would need to investigate experimentally a brain without oscillations and see whether it does not show the kinds of effects for which he assumes oscillations to be necessary. This remains impossible, since we cannot even imagine a brain without oscillations, let alone test it experimentally, as Buszáki himself remarks (see Buszáki, 2006, p. 360).

Ideas for future research 3

Intrinsic features of the brain

Even pathological cases like schizophrenia, depression, or vegetative states that may help in overcoming extrinsic observer-related intrusions do not provide an option here. Why? Because they still show rhythms and oscillations, which, despite being distorted, are none the less present, and thus not completely absent, as experimentally required. There is thus a principal limit to the possible experimental testing that can, in principle, not be overcome and avoided. More specifically, to gain experimental proof of the causal role of oscillations in specific psychological processes, and even consciousness, we would need to eliminate them completely.

This is not the case even in neuropsychiatric disorders. There, oscillations are still present, though in a distorted way. From these abnormalities, we may gain some clues about the possible role of oscillations for particular psychological processes, especially if the latter are also altered in the psychiatric patients. However, this is not sufficient to demonstrate a causal role of oscillations for the respective psychological processes, but only a modulatory or correlational role.

Mere modulation or correlation is not to be equated with causality, since the alterations in the patients' psychological processes may be modulated by the oscillations, but caused by other neuronal processes, completely different from the oscillations. Hence, neuropsychiatric disorders can help in our understanding of the brain, as distinct from the observer, but cannot help us in overcoming our principal epistemological limits, e.g. intrinsic observer-related intrusions, as I will call them below.

How can we describe these principal limits in further detail? The limit consists in the fact that we remain, in principle, unable to prove whether our concepts of rhythms and oscillations are ultimately based on the independent brain, or whether they are more related to us as observers. We are thus stuck in our own intrusion, unable to ever free ourselves from ourselves and our own brain in our observations. I therefore speak of an 'intrinsic observer-related intrusion' as distinct from an 'extrinsic observer-related intrusion'.

How is it possible that 'observer-related intrusions' are intrinsic rather than extrinsic? This is a question about how the different concepts deal with intrinsic and extrinsic observer-related intrusions. The concepts of rhythms and oscillations refer to a feature that characterizes the brain's design and, even more importantly, defines the brain as brain.

Buszáki cannot even imagine a brain without oscillations, because otherwise he would no longer be talking about a brain that is recognizable to us. In short, such a brain would be senseless and meaningless. He thus considers rhythms and oscillations to be what I describe as 'design features' of the brain that define it as brain. In short, rhythms and oscillations are design features that are intrinsic to the brain and define it as brain.

Ideas for future research 4

Intrinsic features of the brain: examples

How can we describe the brain's 'intrinsic design features' in further empirical detail? The brain's intrinsic activity seems to show an elaborate temporal and spatial structure as its design feature. This temporal structure seems to consist in the fluctuations of the intrinsic activity level in different frequency ranges (from 0.001 to 60 Hz). Thereby, the phases, e.g. their onsets and peaks, in the different frequency ranges are somehow linked together. This seems to provide a template with a quite elaborate (but not yet fully understood) temporal structure.

The spatial structure of the intrinsic activity seems to consist in the distribution of the activity levels across different regions, yielding trans-regional balances. This provides a certain spatial pattern or structure of intrinsic activity. Moreover, it seems that temporal and spatial structures are somehow related to each other, though the exact mechanisms of such spatiotemporal couplings currently remain unclear.

How is the neural activity in this spatial and temporal structure coded? Empirical data suggest that it is not a single peak at some discrete point in time and space that is coded. Instead, what is coded in neural activity in the resting state (and the stimulus-induced state) are the differences in neural activity between two different discrete temporal and spatial points. This means that the neural activity is not based on a single stimulus or single activity change at one discrete point in time and space, but rather on spatial and temporal differences. This is what can be called difference-based coding and should be distinguished from stimulus-based coding.

In addition to the coding strategy, other more specific design features of the brain concern the high and low frequency fluctuations of neural activity in both resting state and stimulus-induced activity. This mirrors Buszáki's assumption of rhythms and oscillations. Functional connectivity between different regions during both forms of neural activity is yet another design feature.

Finally, and most importantly, the brain's intrinsic activity, its resting state activity, and its consecutive constitution of a spatiotemporal structure, must also be regarded as a design feature of the brain without which the brain would not be a brain (at least not a human brain). 'Intrinsic feature' means here that it cannot be changed in principle by any extrinsic stimulus in the same way that the intrinsic muscle structure of the heart cannot be abolished by the extrinsic blood flow. Though different, all these features share the trait that their absence could not even be imagined without abandoning all that we know of the brain. They must therefore be assumed to define (at least the human) brain as brain and are thus what I describe as the brain's 'design features'.

Ideas for future research 5

Intrinsic features of the brain: intrinsic observer-related intrusions

We now face a serious problem. One may focus on these design features, and they may be more brain-based than other rivaling concepts that I regard as being more observer-based, but in order to show that that these concepts are brain-based, I would need to put them to experimental testing. That means that I would need not only to show that the presence of the intrinsic design features induces neuronal and phenomenal/mental effects, but also that their absence makes the neuronal and phenomenal/mental effects impossible. If I were able to show the latter, I could assume that the brain's intrinsic design features are a necessary condition of possible consciousness (see Part IV).

This is where the problems start. While I can test the effects of the presence of these design features, the experimental testing of their absence remains principally impossible. In the same way that Buszáki cannot even imagine a brain without oscillations, let alone test its effects experimentally, we cannot imagine a brain without difference-based and sparse coding, a brain without functional connectivity, a brain without intrinsic activity, and a brain without spatiotemporal structure. Why? Because these are design features of the brain that are intrinsic to the brain and therefore define it as brain.

Let me make this clear. There are principal constraints (and ultimately limits) to how far we can go experimentally. Since these principal constraints (and ultimately limits) can be traced back to the brain itself and its particular design features, I here speak of 'neuro-experimental constraints'. *And since they are intrinsic features of the brain of the observer that define his/her brain as brain, he/she, the observer, remains in principle unable to overcome them and avoid their intrusion into his/her observations. One can therefore speak of an 'intrinsic observer-related intrusion'.*

These 'neuro-experimental constraints' constrain (and ultimately limit) the knowledge we can acquire about the brain; these epistemological constraints may therefore be described as 'neuro-epistemological constraints'. And since they concern the brain's design features and their neuronal and phenomenal/mental effects, any hypothesis is prone to both 'neuro-experimental and neuro-epistemological constraints'.

Does this invalidate future hypotheses? Yes and no. Yes, because ultimately one remains principally unable to know whether one is right or wrong, unable to overcome intrinsic observer-related intrusion. One will thus remain principally unable to know whether the brain indeed operates on, for instance, the basis of difference-based coding independent of my concept and hypothesis of it. And no, because one can at least work on minimizing the extrinsic observer-related intrusions and develop novel experimental designs in order to put hypotheses on more secure empirical grounds. Hence, empirically one may move forward while epistemologically one remains stuck between what can and cannot possibly be known in principle about the brain.

Take-home message

The mind and its various functions are scientifically investigated in psychology. This raises several theoretical and methodological problems that are subsumed under the heading of philosophy of psychology. These were discussed in the previous chapters. Here we shifted to the brain and its scientific investigation in neuroscience. Here, too, several theoretical and methodological problems were raised, which can be subsumed under the umbrella of philosophy of neuroscience. One central problem is how to link different levels of explanation, ranging from the genetic, molecular, cellular and regional, to social, political and economic levels of explanation. Different models of their explanatory relationships – incorporation versus contextualization – as well as different kinds of concepts – brain-based versus brain-reductive – were discussed. In addition, we needed to consider the role of the observer in our experimental investigation of the brain and our descriptions of the data in terms of concepts. Concepts may either be more related to the brain itself, i.e. brain-based, or the observer, i.e. observer-based. The observer can thus intrude into his/her own observations and confuse the data he/she acquires. This leads to observer-related intrusions. On the whole, this shows us that neuroscience is confronted with several methodological and theoretical problems, which require us to be careful in taking the data at face value. In short, the data and observations may not be as directly related to the brain as we often presume them to be.

Summary

The previous chapters discussed theoretical and methodological problems in psychology. Now we shift our focus to the methodological and theoretical problems in neuroscience, also known as the philosophy of neuroscience. One central problem in neuroscience is how different levels of explanation are related to each other. Examples of different explanatory levels include: economic, social, and cultural, as well as molecular, cellular and regional. If they can be reduced and incorporated into each other, the various associated disciplines will ultimately be reduced to and incorporated by neuroscience. This is called an incorporation model and needs to be distinguished from a contextualization model, where the different disciplines form the context for interpreting the data from neuroscience. Both models go along with different types of concepts, brain-based versus brain-reductive. Brain-based concepts signify necessary but not sufficient conditions for concepts on other levels outside the brain, e.g. social, political, economic, etc. Such brain-based concepts signify the contextualization model. In contrast, brain-reductive concepts are both necessary and sufficient to explain levels of explanation outside the brain, e.g. non-neuronal levels.

In addition to different levels of explanations, we also need to consider the role of the observer and his/her impact on the observations and data of the brain. We can investigate and observe the brain only in an indirect way, with the help of technological devices like electrodes, scanners, etc. This introduces the possibility that the observer and his/her technological devices may confuse observations and the data. Therefore the resulting concepts the observer uses to describe his/her data may be related more to either the observer himself, or the brain itself. If the concept is more related to the observer, one may want to speak of an observer-based concept. If, in contrast, the concept is more related to the brain, one can speak of a brain-based concept. Finally, one may want to distinguish between the ways the observer can intrude into, and thus confuse, his/her own observations and data. The observer may impact his/her own observations by the kinds of experimental and technological settings he/she chooses. These kinds of intrusions can in principle be avoided; they are thus extrinsic to the observation, for which reason I refer to them as 'extrinsic observer-related intrusions'. However, there may also be other kinds of intrusions which the observer can, in principle, not avoid, as they are related to the way his/her own brain functions during observation. In this case the intrusions are intrinsic to his/her observation and can be referred to as 'intrinsic observer-related intrusions'.

Revision notes

- Define philosophy of neuroscience.
- Describe the different levels of explanations in neuroscience and make the distinction between neuronal and non-neuronal levels.
- What are the explanatory reduction and incorporation models?
- Define explanatory context and the contextualization model.
- What is critical neuroscience?
- Distinguish between conceptualization and contextualization.
- Why is multi-level causation relevant?
- What is the difference between brain-reductive and brain-based concepts?
- What is philosophy of neurophilosophy?
- Describe indirectness of studies of mind and brain.
- What is the difference between brain-based and observer-based concepts? Give examples.
- Why is the distinction between intrinsic- and extrinsic observer-related intrusions relevant?

Suggested further reading

- Bechtel, W., Mandik, P., Mundale, J. and Stufflebeam, R. (eds) (2001) *Philosophy and the Neurosciences: A Reader* (New York: Basil Blackwell).
- Bickle, J. (2009) *The Oxford Handbook of Philosophy and Neuroscience* (Oxford/New York: Oxford University Press).
- Churchland, P. and Sejnowski, T. (1994) *The Computational Brain* (Cambridge, MA: Bradford Books).
- Craver, C. (2007) *Explaining the Brain: Mechanisms and the Mosaic Unity of Neuroscience* (Oxford: Clarendon Press).
- Northoff, G. (2013a) *Unlocking the Brain. Vol. I: Coding* (Oxford/New York: Oxford University Press).

12
Philosophy of Brain: Characterization of the Brain

Overview

The previous chapter discussed various theoretical and methodological problems that exist within the field of neuroscience and can be subsumed under the heading of a philosophy of neuroscience. These problems included the different kinds of explanations neuroscience utilizes; how to link the different levels of explanation within the brain itself (e.g. genetic, molecular, cellular, regional, networks, etc.) to the levels of explanation outside the brain (e.g. religious, social, political, etc.). The previous chapter also questioned the kinds of concepts we use in neuroscience, whether they are more related to the brain independently of us, i.e. brain-based, or more related to us as the observer, i.e. observer-based. While this described the explanations, methods and concepts applied in neuroscience, it left open any discussion of how the brain itself can be characterized. More specifically, we need to characterize the brain as a whole. This topic may be subsumed under the umbrella concept of philosophy of brain. This characterization of the brain as the central core of a future philosophy of brain is the focus in this chapter.

Objectives

- To understand the difference between philosophy of neuroscience and philosophy of brain
- To determine the concept of modules and its various features
- To grasp the difference between holism and localization
- To understand the concept of intrinsic activity of the brain and how it differs from extrinsic activity
- To understand the possible implications of intrinsic activity for our philosophical accounts and methodological strategies
- To understand how Kant and his view of the mind can be related to the intrinsic view of the brain

Key concepts

Philosophy of neuroscience, philosophy of brain, modules, encapsulation, domain-specificity, holism, localization, intrinsic activity, intrinsic versus extrinsic view, Kant.

Background 1

Theoretical and methodological issues in neuroscience: philosophy of neuroscience

The previous chapter focused on neuroscience's methodological problems, including how we can link the different levels of explanation that range from molecular, cellular and regional, to social, political and cultural. Different models to help answer this question included incorporation versus contextualization. The incorporation model assumed that non-neuronal levels of explanation can be reduced to and thus incorporated into neuronal-level explanations. Contextualization denies this possibility and instead considers non-neuronal levels of explanation as contextual support for the neuronal level explanation.

The question for the explanatory relations between different levels went along with the distinction between different types of concepts: brain-based concepts versus brain-reductive concepts. Brain-reductive concepts are concepts that describe the brain's neuronal function as necessary and sufficient for explaining non-neuronal levels of explanation. On the other hand, brain-based concepts claim only to be necessary but not sufficient in detailing non-neuronal levels of explanation.

In addition to the relationships between different levels of explanation, considering the role of the observer is central in neuroscience. The observer's role is important because we can only approach and investigate the brain as an observer: in an indirect way and with the help of technological devices. This is called indirectedness. Indirectness, in the study of brain and mind, makes it necessary to distinguish conceptually between the brain and the observer. In doing so, it becomes clear that our observations and data about the brain and its neuronal states may actually stem from two different sources: from the brain itself, independent of our observation, and from our observation, including the respective technological devices.

The concepts we use to describe our neuroscientific observations and data may thus reflect an amalgam between both brain and observer. Therefore one needs to distinguish between brain-based and observer-based concepts, while still recognizing the possibility for a continuum. Concepts are predominantly observer-based if they reflect the observer and his/her methodological, experimental and technological means of observation. Since the observer him/herself can avoid this impact on his/her own concepts, one can speak here of extrinsic observer-related intrusion. We must also consider the fact that the observer him/herself has a brain that imposes certain constraints upon observation and, most importantly, might interfere with observation. Since without a brain he/she could not observe at all, the intrusion of one's own brain cannot be avoided. This is called intrinsic observer-related intrusion.

Taken together, the discussion of these methodological and technological issues in neuroscience can be subsumed under the umbrella of philosophy of neuroscience. *Philosophy of neuroscience describes the investigation of methodological, theoretical and foundational issues in neuroscience. As a result, the philosophy of neuroscience is not so much concerned with how to practice neuroscience as a discipline, but is, instead, more about theory.*

Background 2

From philosophy of neuroscience to philosophy of brain

What does philosophy of neuroscience do? It theorizes about the explanations, methods and approaches we apply when investigating the brain in neuroscience. This leaves out something essential, namely what neuroscience is about: the brain. So far, we have talked about how neuroscience explains, investigates and approaches the brain, but we have not yet discussed what the brain itself is like. How can the brain be characterized? What is the purpose of the brain? What makes the brain so special in comparison to other organs, like the heart and kidney?

The neuroscientist may now be surprised. He/she assumes that he/she already characterizes the brain. He/she investigates the brain and its various functions and mechanisms using different levels of explanation. Doesn't that characterize the brain? Yes and no. Yes, it does characterize the brain to a certain extent, but only specific functions, regions, networks, biochemicals, etc. No, because it does not characterize the brain as a whole, as an organ.

How can we characterize the brain as a whole? This is a question neuroscience does not address in its search for specific mechanisms and functions. The characterization of the brain is more a theoretical issue that draws upon the various data neuroscience assembles. One may thus speak of a theory of the brain. A theory of the brain would provide a cornerstone in what we called 'philosophy of brain' in Chapter 11 as well as in Chapter 8.

Metaphorical comparison

Is our brain nothing but a car?

What is a philosophy of brain? Philosophy of brain aims to characterize the brain as a whole, distinguished from its various and distinct functions and mechanisms. Let's invoke an analogous comparison: a car. The car mechanic is interested in single functions of the car. He distinguishes various parts of the car, all of which have specific mechanisms that contribute a particular function to the car as a whole. From the mechanic's point of view, the car is thus nothing but a collection of different parts.

However, that collection of different parts does not yield any information about the car as a whole. Imagine the mechanic puts all the different parts of the car side by side on the floor. He will tell you that this is a car and that it is in no way different from what you perceive as a car. 'No', you will say, 'the car is more than the mere sum of its different parts. The parts themselves and their collection do not move at all. Whereas put together in the right way they amount to a car that moves.' The same is true of the brain. The single functions themselves and their collection do not tell us about the brain as a totality, nor how we can characterize it and its purpose. In other words, neuroscience is like a car mechanic who only cares about the different parts and puts them side by side. As the car is just a mere collection of parts to this mechanic, so the brain is nothing but a collection of different mechanisms and functions to the neuroscientist. And in the same way that the car mechanic misses the purpose of the car as a whole – its ability to move and provide transportation – the neuroscientist may miss the brain's characterization and purpose as a whole.

One issue concerning the brain as a whole are the features that define the brain as brain, its intrinsic design features (see Chapter 11 for details). The intrinsic design features must provide a particular purpose for the brain. What is the purpose of the brain?

Background 3

Philosophy of brain

Let's invoke another comparison and compare the brain with other organs of the body. We know the purpose of the heart: pumping blood. This is made possible by the muscle structure of the heart. We know the purpose of the kidneys: to filter and rid the blood of toxic substances. This is also made possible by the kidney's structure. What, though, is the purpose of the brain? Many have argued that the brain's purpose consists in yielding consciousness and other mental features like emotions, free will, etc. What kind of structure and organization does the brain use for that?

These considerations are empirical in nature. They presuppose the investigation of the brain using observation in third-person perspective.

What now is philosophy of brain? Philosophy of brain aims to investigate and characterize the brain as a whole, in all its different roles. What is the brain? What is the purpose of the brain? What defines the brain as brain? What is the role of the brain in our cognition, knowledge, and existence and reality? Philosophy of brain thus raises 'what' questions about the brain as distinguished from the 'how' questions neuroscience addresses when investigating specific functions and mechanisms ('How does something function?').

In addition to the kinds of questions, what versus how, philosophy of brain can also be distinguished from neuroscience with regard to the domains. Neuroscience as third-person observation of the brain presupposes the empirical domain, the domain of observation. This is different in philosophy of brain. As we saw above, the brain can be considered in different domains, empirical, epistemic and metaphysical, and be characterized in each domain in different ways according to the respective context.

We have already discussed the epistemic and metaphysical characterization of the brain in Chapter 8. There we investigated the brain and its role and purpose in metaphysical and epistemic domains. The brain in the empirical domain is the focus in the present chapter.

Brain and function 1

Modularity and modules: definition

How can we empirically characterize the brain as a whole? One central issue is how the brain and its different regions and networks are related to the various functions, sensory, motor, affective, and cognitive, as well as the different mental states. Can a particular function or a specific mental state be related to the neural activity in one particular region or network within the brain? Or is the respective function or mental state a result of the interplay between the different regions and networks in the brain? This touches upon the problem of localization of function in the brain.

Can we localize different mental functions in different regions and networks in the brain? The question of the localization of mental functions has been discussed in both psychology and neuroscience. As we saw in the historical overview in Part I, psychology focused on the scientific investigation of the mind and its different mental functions. This was later extended to cognitive functions, and resulted in the birth of cognitive psychology. At that time it was, and still is, much discussed as to whether different sensory functions as well as cognitive functions like working memory, attention, executive functions, etc. are processed by different units or not. Such units of processing are described as modules.

What is a module, and how can we characterize it? There is much discussion about the criteria for determining a module. Let us rely here on the original criteria as formulated by the philosopher Jerry Fodor, whom we already encountered in the context of the language of thought in Chapter 10. According to Fodor, a module can be characterized by various features that imply certain design features:

(i) A module has a special purpose that accounts for one particular function, but not others. For instance, a special region or network in the brain may only mediate one particular function like the visual cortex, which presumably mediates only visual, and not auditory, function. There is a one-to-one relationship between function/purpose and module.
(ii) A module receives one particular kind of input, but no other inputs. For instance, the visual cortex may be assumed to receive visual stimuli, but not input from auditory stimuli or cognitive input from attention, or working memory, as it may, for instance, come from the prefrontal cortex.
(iii) Such a characterization of a specific input implies for the respective module that its processing can account for only one particular domain, like the visual domain in the case of the visual cortex. There is thus a specific type of processing tailored to the respective input, which is described as 'domain-specificity'.
(iv) Finally, since the modules receive only one particular type of input and perform only the respective type of processing, but no others, the module is like a capsule isolated from other modules and their respective inputs and types of processing. The module is thus what is described as 'encapsulated'. Due to the restriction of one input and one type of processing, such encapsulation allows for fast processing of the respective input.

Where do such modules come from? According to Fodor, modules in this sense are biologically ingrained; they are innate in the structure of the mind and ultimately encoded in the brain's structure and function. The different sensory systems (auditory, visual, olfactory, tactile, gustatory) and the motor system can be regarded as typical modules in this sense. They can be traced back ultimately to evolutionary development, as manifested genetically and in previous gene–environment interactions.

Brain and function 2

Modularity and modules: sensory functions

Are there such sensory and motor modules ingrained and implemented in the structure of the brain? This raises questions about the empirical (e.g. neuronal) plausibility of the assumption of sensory and motor modules in the brain. Let us focus on the sensory system, and especially the visual cortex, which is often used as a paradigmatic example of modularity.

The visual cortex seems indeed to be highly specialized for visual function. It contains different subregions (V1–V5) that are all assumed to process distinct aspects of visual function like shape, motion, orientation, etc. And most importantly the visual cortex is only in charge of visual function, but not of any other sensory function, e.g. auditory, tactile, etc. This strongly suggests that the visual cortex and its visual function are a module that is characterized by a particular purpose: a specific input, domain-specificity and encapsulation.

However, empirical reality paints a different picture. The visual cortex does not only receive visual input. In addition, the visual cortex also receives inputs from other sensory modalities like the auditory cortex. There are direct pathways or connections from the visual to the auditory cortex and vice versa. The visual cortex thus also receives auditory input. And that auditory input can modulate the neural activity in the visual cortex elicited by the visual input. This is called 'cross-modal interaction', the direct interaction between inputs from different sensory modalities.

How does such cross-modal interaction manifest itself? For instance, concomitant auditory input can enhance or decrease the neural activity induced by visual stimuli. 'That is perfectly clear,' you may say, 'that is why what you hear heavily influences what you see.' Hence, the findings are very much in line with our experience.

Even in the absence of visual stimuli, the visual cortex may change its neural activity level with auditory input. Conversely, activity in the auditory cortex has also been shown to change in the presence of visual stimuli, even if auditory stimuli remain absent. And the cross-talk goes even further. The auditory cortex does not only receive auditory input, it also receives direct tactile input from the somatosensory cortex. Hence, the same degree of tactile stimulation may, for instance, feel ticklish or not, depending on the concurrent auditory input. 'My goodness,' you may remark, 'that is why I do not feel tickled by you when you turn on nice music at the same time.'

Brain and function 3

Modularity and modules: sensory functions as modules

Can the different sensory cortices then still be considered modules according to the criteria listed above? The observation of cross-modal interaction clearly demonstrates that the visual cortex and the other sensory cortices receive multiple inputs from different sensory modalities. In addition, the sensory cortices also receive inputs from other non-sensory functions like cognitive and affective functions, and their respective regions and networks. Hence, the assumption of a specific input characterizing the sensory cortex must be repudiated. This sheds some doubt on its characterization as a module.

The proponent of the module, however, could make the following argument. Even if receiving different inputs, the sensory cortex may nevertheless subserve one particular function and thus be domain-specific. That, however, may turn out to be problematic too. For example, investigations in congenitally blind patients (who are blind from birth) demonstrated that the visual cortex can take over other functions, including tactile and auditory functions. Auditory and tactile stimuli then strongly recruit not only their respective sensory cortex, but also the visual cortex. This means that the visual function is not innate or ingrained into the visual cortex. In turn, this would suggest that the visual cortex has no single function and calls into question the idea that it has both a special purpose and domain-specificity.

Does this mean that there are no modules in the brain that fit the criteria mentioned above? The defenders of modularity may want to revert to different strategies to uphold their arguments. Either they could argue that the criteria for modularity as formulated above are too strict and need to be more flexible. If one, for instance, no longer requires one particular input as defining criteria, one may also be able to assume modularity in the case of the visual cortex. Hence, the characterization of the brain by modularity may then very much depend on the criteria and the definition of modularity.

Or one might argue that certain regions of the brain, like the subcortical regions, the regions beneath the cortex as the outer surface of the brain, show modularity. However, even these regions are strongly, and most often reciprocally, connected with each other. The same holds true for the various regions on the cortical level that are strongly connected to each other and can therefore interact only in conjunction.

Brain and function 4

Modularity and modules: cognitive functions

So far we have discussed only sensory functions. But we still need to discuss whether modularity applies to other functions, like cognitive functions. Cognitive functions concern, for instance, processes like attention, working memory (e.g. short-term memory), episodic memory as long-term memory, and executive functions, e.g. the elaboration of goals and plans for subsequent action and behavior.

In contrast to sensory functions, Fodor and others no longer assume modularity in the case of cognitive functions. Why? Because cognitive functions no longer rely on one particular input like visual stimuli. Instead, cognitive functions are based on and require multiple inputs from different sources, like different sensory inputs, motor inputs, inputs from emotional functions, etc. Therefore, unlike sensory and motor functions, which are often assumed to be encapsulated, cognitive functions do not show any sign of encapsulation.

The absence of encapsulation also makes the assumption of one specific type of processing, and thus domain-specificity, impossible. Instead, there is domain-generality with several inputs and stimuli processed in different ways. Rather than relying on a particular input, different inputs from the external environment (exteroceptive input), the body (interoceptive input) and inputs from one's own memories are combined and processed in a domain-general, rather than a domain-specific, way. Cognitive functions are thus constructed by the processing of different inputs in a domain-general way. This implies what is described as a constructionist approach.

How now do cognitive functions stand in relation to sensory functions? Some proponents of modularity (see below) suggest that sensory functions may indeed be organized in terms of modules, whereas cognitive functions seem to defy such modularity. This is also assumed in the organization of the brain and its cortex, i.e. the outer surface: sensory (and motor) cortices are distinguished from other regions like the prefrontal, parietal and temporal cortex. Unlike in the case of the sensory (and motor) cortex, where one particular stimulus type seems to dominate, the other cortical regions can be characterized by the confluence and convergence of different types of stimuli and inputs. These regions are consequently described as association cortex.

This type of double organization is debated in current philosophy. Some, like the British-American philosopher, Peter Carruthers, support modularity in the case of cognitive functions. He suggests, for instance, that the different cognitive functions like attention, working memory, episodic memory, etc. represent different modulates that are then based on different (cognitive) contents rather than different (sensory) stimuli and inputs. The brain, according to him, is characterized by what he describes as 'massive modularity', ranging from sensory to cognitive modules.

Brain and function 5

Localization versus holism: cognitive functions in specific regions

Is the assumption of cognitive modules plausible when considering the brain? This raises the question: can cognitive, and ultimately mental functions, be localized in specific regions or networks of the brain? This is a basic question that neuroscientists raised, and these past observations are complemented by the current findings.

One of the main methodological approaches in neuroscience at the beginning of the twentieth century was the investigation of patients with brain lesions. These patients could reveal how their higher-order cognitive functions like consciousness, memory, attention, learning, and so on were affected by lesions in particular regions. This was, for instance, the way Paul Broca found out about the specific region in the brain that is in charge of language comprehension, now known as the Broca region. He observed that patients with a lesion in the left lateral prefrontal cortex show major deficits in uttering words and language, a symptom referred to as aphasia.

Observing patients with brain lesions and understanding their corresponding mental disturbances has since been a major tool deployed to learn about brain function. From the exact localization of the lesion and the corresponding mental disturbances, scientists may infer which region in the brain mediates the respective underlying higher-order cognitive function. Many other higher-order cognitive functions, including consciousness and self, are currently being investigated in this way. This is called the 'localization-based approach' to the brain.

The concept of the 'localization-based approach' can be defined in two ways. First, it implies the assumption that a particular function can be related to the neural activity in a specific brain region, meaning that the former can be localized precisely in the latter. Second, the concept of the 'localization-based approach' refers to a particular methodological strategy: the brain is here approached in terms of regions rather than in terms of, for instance, processes.

The methodological approach to the brain in terms of regions is not restricted to patients with local brain lesions. It may also extend to healthy subjects as, for instance, investigated in functional magnetic resonance imaging (fMRI). The use of techniques like fMRI is indeed guided by the search for the localization of particular functions in specific regions of the brain, which it therefore approaches in terms of regions (as distinguished from processes). For instance, many imaging studies aim to associate specific regions or networks in the brain with particular social, affective, or cognitive functions. Even functions like consciousness, self and empathy are currently often associated with specific regions or networks in the brain. This will be discussed in detail in Parts IV and V.

The search for the localization of higher-order cognitive functions in patients with brain lesions and functional brain imaging converges with the assumption of modules in cognitive psychology. Cognitive psychology assumes specific functional unities that are in charge of processing and operating a specific cognitive content as, for instance, attentional content, working memory content, conscious content, self-specific content and so on. A recent development in cognitive neuroscience put the concept of modules into the context of the brain and considered questions about localization in the brain.

The concept of modules as described in cognitive psychology was then easily transferred to the brain, and more specifically to particular brain regions and their connections. Hence, the localization-based view of brain function was intimately coupled with the module-based view of psychological function. This resulted in the assumption of the localization of specific cognitive modules in particular regions (or networks of regions) in the brain.

Brain and function 6

Localization versus holism: holism in the history of neuroscience

However, the opposite position can also be argued. A strictly localization-based approach was doubted early on by neurologist, Hughlings Jackson. Jackson suggested a more complex and systematic neural organization with multiple interdependencies between different regions. This paved the way for a more holistic view of brain function, one that relates higher-order cognitive functions to the neural operations in the whole brain and its multiple regions.

Interestingly, Sigmund Freud, the founder of psychoanalysis who initially was a neuroanatomist, also rejected a localization-based approach to the brain. His reason was that more complex psychological disorders like hysteria or depression could not be confined to alterations in specific brain regions. He instead regarded these disorders as more complex system disorders, where the organization of the psychic apparatus, as he called it, is abnormal.

There is one essay, *Project for a Scientific Psychology*, from 1895 where Freud applies this systems view to the brain and its neurons. He himself considered this attempt a failure and therefore never allowed it to be published during his lifetime. Based on this posthumous publication, one may consider Freud a forerunner of a more holistic view of brain function.

Later, the American neuroscientist Karl Lashley observed in his post-mortem dissections that the extent of brain lesion predicts the degree to which higher-order cognitive functions and mental states are disturbed. This led him to develop what he called the 'Law of Equipotentiality' and the 'Law of Mass Action'. Both laws describe the distribution of neural processing across the whole brain during higher order cognitive functions like consciousness and memory.

Different regions were assumed to contribute equally to the generation of complex functions, which therefore must be considered the result of 'mass action' in the brain. Higher-order cognitive functions like memory and consciousness were assumed to result from the neural processing throughout the whole brain rather than being localized in particular regions or modules within the brain (see also other authors like Wolfgang Koehler and Kurt Goldstein).

Analogous observations were made by the Russian neuropsychologist A. R. Lurija (1973). Based on his lesion patients, Lurija suggested that one region in the brain can be involved in various higher-order cognitive functions. Conversely, he assumed that higher-order cognitive functions are not mediated by only one or two regions but by various regions in the brain. Most importantly, the same higher-order cognitive function may even recruit different regions in different instances depending on the respective psychological and neuronal context. This is what Lurija described as 'dynamic localization'.

Brain and function 7

Localization versus holism: neural overlap of functions

How is the holistic view of brain function considered today? The introduction of functional brain imaging has shifted the pendulum back again toward the localization-based view, with the assignment of specific regions or networks to particular functions like attention, working memory and so on.

These various regions and networks are defined by their involvement in particular tasks or functions requiring stimulus-induced or task-related activity. There is yet another neural network in the brain whose activity seems to be particularly high in the absence of particular stimuli or tasks, the resting state. This network is called the 'default-mode network' (DMN) whose level of neural activity is particularly high in the resting state. These days the DMN is often considered as a specific module for the resting state. As a result, it is tacitly supposed to stand side-by-side with the other networks that are more involved in stimulus-induced or task-related activity.

However, recent imaging studies shed some doubt on the localization theory. The various regions of the DMN – like the anterior and posterior cingulate cortex and the medial prefrontal and parietal cortex – are supposed to subserve psychological and mental activity specifically in the resting state. But the same regions are also recruited during a variety of psychological tasks or functions, including contextual association, navigation and spatial processing, emotion, semantic processing, episodic memory, decision making, execution errors, self-related processing, mind-reading, emotional processing and social interaction.

This sheds some doubt on the functional specificity of the DMN. Conversely, these observations also argue against region- or network-specific localization of the various functions themselves, which seem to recruit more or less the same regions and networks.

This situation with the recruitment of the same regions and network by different functions is not specific to the DMN. The same can be observed in other neural networks, including the bilateral anterior insula, the dorsal anterior cingulate cortex and the thalamus as its core regions (these regions are also subsumed under what is described as the 'salience network'). These regions are active during functions as diverse as interoceptive awareness, empathy, anticipation of emotions, pain, language, anxiety and aversion.

The list of regions and networks that are recruited by different functions goes on. However, the observation of one and the same region and network mediating a variety of different functions sheds some doubt upon the localization-based approach and its attempts to establish a specific one-to-one relationship between regions and functions.

Brain and function 8

Localization versus holism: do we have to revert to holism?

Does this mean that we have to revert to a more holistic view of the brain and its different regions? Based on their data, some neuroscientists, doing either lesion-based studies or functional imaging using electroencephalography, positron emission tomography (PET), or functional magnetic resonance imaging (fMRI) do indeed advocate a more holistic view of brain function. This is further corroborated by neuroanatomy, a subfield that considers single regions as hubs or nodes within the neural network of the whole brain, rather than as modules by themselves.

While the association of a specific region with a specific psychological function must be put into doubt, the data nevertheless show that only a certain set of regions are recruited during various tasks or functions. More specifically, multiple functions seem to recruit from the same set of regions or networks. For instance, the DMN does not only mediate neural activity in the resting state but also shows changes during various social tasks, self-related tasks, autobiographical memory and meditative states. The same applies to the other neural networks that show activity changes during a variety of different tasks, stimuli and conditions. These results suggest that neural activity is globally distributed rather than being localized in particular regions or networks.

On the whole, we can see that throughout its history neuroscience has oscillated between a localizationist and a holistic approach to the brain. This is still ongoing in current neuroscience. While the introduction of functional imaging 20 years ago led to the predominance of a localizationist approach, more recent data seem to suggest a more holistic approach.

What does this tell us about the concepts of localization and holism? We can, at present, not exclude the possibility that this conceptual dichotomy might be more related to us as observers than the independent brain. Whether we assume localization or holism may also depend on the kinds of techniques we employ to investigate the brain, how we analyze the data, and our interpretation of that data. Hence, currently we may not be able to decide whether the conceptual dichotomy between localization and holism is an observer-based dichotomy, or based on the brain itself. The only thing we can do is to continue investigating the brain experimentally, while at the same time being aware and careful of the kinds of concepts we use to describe our data.

View of the brain 1

Intrinsic versus extrinsic view of the brain: history of neuroscience

So far we have discussed how the brain and its various regions and networks are related to the different functions it generates. This leaves open how the brain operates and processes the various kinds of inputs it receives from the body and the environment. Any empirical characterization of the brain as a whole must therefore ask questions about the brain's mode of operation.

What is the brain, and how does it operate? This was the subject of controversial discussion at the beginning of the twentieth century. One view, favored by the British neurologist Sir Charles Sherrington (1857–1952), proposed the brain and the spinal cord to be primarily reflexive.

Reflexive describes the way the brain reacts to stimuli in predefined and automatic ways. This means that the stimuli from the outside of the brain, originating

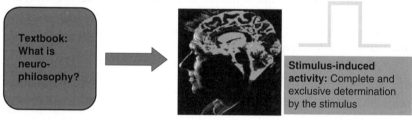

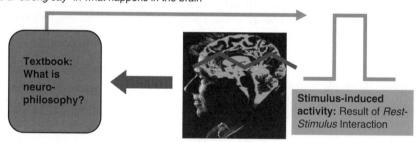

Figure 12.1 Extrinsic and intrinsic views of the brain
This figure illustrates different views of the brain, extrinsic and intrinsic

Upper part: in the extrinsic view of the brain, the neural activity in the brain is related exclusively to the external stimulus, here indicated by the picture on the left. This, in turn, induces what is called stimulus-induced activity, as indicated on the right. That means that the brain itself does not make an active contribution to the neural activity the stimulus generates inside the brain.

Lower part: in the intrinsic view of the brain, the stimulus-induced activity is considered to result from an interaction between the effects of the stimuli in the brain and the intrinsic activity contributed by the brain itself. This means that the brain provides an active contribution to the resulting stimulus-induced activity, namely its own intrinsic activity and how it can interact with the stimulus. Hence, metaphorically speaking, the brain has a say in what happens in the brain as induced by the stimulus.

extrinsically in either the body or environment, are proposed to determine completely and exclusively the subsequent neural activity. The resulting stimulus-induced activity, and more generally any neural activity in the brain can thus be traced back to the extrinsic stimuli. This is what I describe as the 'extrinsic view' of the brain (see upper part of Figure 12.1).

An alternative view, however, was suggested by one of Sherrington's students, Thomas Graham Brown. In contrast to his teacher, he suggested that the brain's activity, that is, in the spinal cord and the brain stem, is not driven primarily by extrinsic stimuli from the outside of the brain. Instead, the spinal cord and the brain stem show spontaneous activity originating intrinsically, or within themselves. Other subsequent neuroscientists like Karl Lashley, Kurt Goldstein and Wolfgang Koehler followed Brown's line of thought and supported the hypothesis of intrinsic activity. This is called the 'intrinsic view' of the brain (see lower part of Figure 12.1).

The assumption of intrinsic activity generated inside the brain itself has major implications for how we conceive stimulus-induced activity. What we as observers describe as stimulus-induced activity, and usually associate with the stimulus itself, must then be regarded as the hybrid result of a specific interaction between the brain's intrinsic activity and the extrinsic stimulus. Stimulus-induced activity and any neural activity in the brain must be traced back to a double input that originates in both the brain's intrinsic activity, and the body and the environment's extrinsic stimuli.

View of the brain 2

Intrinsic versus extrinsic view of the brain: present findings

What does current research tell us about the brain? The dichotomy between intrinsic and extrinsic views of the brain is still as controversial. It has resurfaced in particular in functional brain imaging. Let's start with the extrinsic view.

Many levels of neuroscience, ranging from cellular to regional to behavioral levels of the brain, rely on the experimental application of specific stimuli and tasks to probe neural activity. By comparing different stimuli and tasks, the resulting differences in neural activity are associated with the respective stimuli or tasks. This means that the experimental requirements may predispose and pull us toward an extrinsic view. The extrinsic view has been most predominant in behaviorism. Behaviorism is the characterization of mental states by mere input–output, i.e. stimulus–response relations (see Chapter 6), which finds its continuation in contemporary cognitive neuroscience.

However, the extrinsic view of the brain has been challenged again recently on several grounds. Even in the resting state, that is, in the absence of any kind of particular extrinsic stimuli from either the body or the environment, the brain shows a high degree of metabolic consumption, for instance, about 20 percent of the body's overall energy budget (and oxygen fraction).

Using functional imaging, this high metabolism has been observed especially in a particular set of regions, the default-mode network (DMN), which includes various anterior and posterior cortical midline structures, as well as the bilateral posterior parietal cortex. The high degree of metabolism is indicative of continuously ongoing high levels of neural activity even in the absence of (specific) extrinsic stimuli – that is, in the resting state of the DMN.

However, other regions outside the DMN also show spontaneous neural activity independent of any extrinsic stimuli. This has been, for instance, demonstrated in the auditory and visual cortices, the thalamus, the hippocampus, the olfactory cortex, the cortical midline regions, the prefrontal cortex, the motor cortex, and other subcortical regions like the brain stem and the midbrain. The metabolic and neuronal signs of intrinsic activity are further complemented by behavioral evidence. Such behavioral evidence can, for instance, be observed by what is described as seeking: seeking refers to the continuous search for reward in the environment, even in the absence of extrinsic stimuli. As a result, seeking reflects continuous behavioral activation, e.g. arousal.

View of the brain 3

Intrinsic versus extrinsic view of the brain: interaction

Which view holds – intrinsic or extrinsic? Rather than cashing in one view at the expense of the other, the brain itself may force us to go beyond and to reconcile both views. Any given neural activity in the brain may be proposed to result from the interaction between the brain's intrinsic activity and the extrinsic stimuli from either the body, i.e. interoceptive stimuli, or the environment, i.e. exteroceptive stimuli.

Why? Even in an apparent resting state like sleep, the seemingly intrinsic activity of the brain is nevertheless exposed to continuous extrinsic, i.e. intero- and exteroceptive, input from the body and the environment. For instance, the continuous action of the heart sends interoceptive stimuli to the brain, as do the continuous tactile, auditory, olfactory and gustatory stimuli from the environment during sleep.

Conversely, any extrinsic stimulus first encounters the brain's intrinsic activity before it can be processed at all, and consecutively associated with sensorimotor, affective, cognitive and social functions, as well as their respective neural systems in the brain. Hence, there is no pure intrinsic activity, just as there is no pure extrinsic activity in the brain.

Therefore, in order to understand the brain's neural activity, rather than opposing intrinsic and extrinsic views, we may need to investigate how intrinsic activity and extrinsic stimuli interact with each other. The neural activity we observe in the brain in response to specific extrinsic stimuli from either body or environment may then be regarded as the result of prior modulation of the brain's intrinsic activity by the extrinsic stimulus.

View of the brain 4

Intrinsic activity: purpose of the brain's intrinsic activity

What does the brain do with this huge amount of energy in the resting state? It uses part of this intrinsic energy to process extrinsic stimuli. To do this, it spends only a tiny fraction, around 2–10 percent, of its total energy budget. What happens to the rest? We do not know yet. One may consider the brain's high resting state activity as mere noise in the background of stimulus-induced activity. The resting state activity may then be regarded just as a bad side effect of the stimulus-induced activity, which is the main focus of the brain. But why, then, does the brain waste so much energy and effort for mere noise?

The brain invests the largest part of its energy into its resting state activity. Why is the resting state activity so important for the brain and apparently 'much more' than mere noise? The resting state activity may be the 'tool' by means of which the brain can affect the stimulus and how it is processed in the brain. This means that there must be some kind of interaction between the brain's resting state activity and the stimulus – this is called rest–stimulus interaction.

Rest–stimulus interaction describes the brain's resting state activity as affecting and ingraining itself on the neural activity changes induced by the extrinsic stimuli the brain encounters from the environment. Recent studies demonstrated that this is indeed the case when the resting state activity level in the visual cortex predicts the subsequent stimulus-induced activity and its associated behavioral and mental states.

Despite recent major advances in neuroscience, many issues about the brain's resting state activity remain unresolved these days. What is clear, though, is that we need a view of the brain that is different from the extrinsic and cognitive view as tacitly presupposed in current cognitive neuroscience. Cognitive neuroscience focuses almost exclusively on stimulus-induced or task-related activity. Since the respective stimuli or tasks come from outside the brain, their related neural activities may be described as extrinsic.

This contrasts with the resting state activity that originates from the inside of the brain and may therefore be designated as intrinsic. A shift in focus from stimulus-induced to resting state activity consequently entails a shift from an extrinsic to an intrinsic view of the brain. Interestingly, the quest for an intrinsic view of the brain is not specific to our time, but surfaced in neuroscience at the beginning of the twentieth century (see above).

The implications of the intrinsic view of the brain in understanding the mind remain unclear at this point, and are currently hot research topics in neuroscience. The empirical details of the brain's intrinsic activity and its functional implications will therefore be discussed in the subsequent sections on consciousness and the self.

How does the intrinsic resting state activity of the brain interact with the extrinsic stimuli from the outside world? The relevance of such rest–stimulus interaction is supported by recent findings that show that the level of pre-stimulus resting state activity predicts the neural, phenomenal and behavioral effects of subsequent stimuli. What remains unclear, however, are the exact neuronal features of the resting state itself that make such rest–stimulus interaction possible.

These neuronal features must be intrinsic to the resting state itself, while at the same time they must also be able to create the tendency (i.e. neural predisposition; see Part IV for details) to associate stimulus-induced activity with consciousness and self. Hence, in order better to understand observations during rest–stimulus interaction, we may need to achieve a better understanding of the resting state's intrinsic features. Additionally, we must learn how they predispose rest–stimulus interaction in such a way that the stimulus becomes associated with consciousness and self. In order to do this, we may need to develop an intrinsic–extrinsic interaction model with regard to the brain.

View of the brain 5

Intrinsic activity: Kant's mind and the brain's intrinsic activity

In order to develop an intrinsic–extrinsic interaction model, we may want to venture on a brief excursion into the history of philosophy. René Descartes assumed mental properties intrinsic to the mind to be distinct from the physical features of body and brain. David Hume, who opposed such intrinsic mental properties, disagreed. Instead, Hume advocated an extrinsic view of the mind, believing that mental activity can be traced back entirely to the extrinsic features of stimuli in the world (see Part II for details).

His German successor, Immanuel Kant, combined both intrinsic and extrinsic views of the mind: he claimed that consciousness and self must be considered a hybrid of processes resulting from an interaction between the mind's intrinsic features, and the world's extrinsic stimuli. In order to reveal the nature of such intrinsic–extrinsic interactions, Kant attributed various faculties (i.e. intrinsic features) to the mind, primarily described in *The Critique of Pure Reason* (Kant, 1998).

The mind's intrinsic features included unity of consciousness, self as 'I think', and various templates of spatiotemporal continuity (which were subsumed under the umbrella term 'categories'). According to Kant, the mind uses its intrinsic features to structure and organize the effects of extrinsic stimuli. This, in turn, allows the latter to become associated with consciousness, self and spatiotemporal continuity. Hence, consciousness, self and spatiotemporal continuity are based on the interaction between the mind's intrinsic features and the environment's extrinsic stimuli.

How is Kant's view of the mind and its intrinsic features related to the brain's intrinsic activity? Kant characterized the mind by the process of thinking expressed by 'I think'. In addition, he attributed other intrinsic (i.e. transcendental) features to the mind, like unity, and distinguished them from mere extrinsic (i.e. empirical) stimuli. How can Kant's transcendental view of the mind's intrinsic features inform the neuroscientific investigation of consciousness?

The concept of 'I think' entails that any cognition of extrinsic stimuli must be accompanied by the 'I' (i.e. the self) and its thinking activity as an intrinsic feature of the mind itself. Why is that necessary? This is where neuroscience can shed light on Kant's thought. If the resting state is indeed organized in a self-specific (or self-perspectival; see Part V for details) way, no extrinsic stimulus can 'avoid' the encounter with the resting state, that is, rest–stimulus interaction and its association with the self.

View of the brain 6

Intrinsic activity: Kant's consciousness and the brain's consciousness

How can Kant help inform neuroscientific investigation? Kant deemed 'I think' to be essential for consciousness: we cannot be conscious without the mind's accompanying

'I think'. If 'I think' is indeed related to resting state activity, it may help to decipher the neuronal features of the resting state and its role in consciousness.

Current neuroscientific research focuses mainly on stimulus-induced activity, which is supposed to be sufficient for consciousness, the neural correlate of consciousness (NCC) (see Part IV). This, however, neglects one central feature, 'I think', and, in neuronal terms, resting state activity. Resting state activity itself must contain certain features that are central in constituting consciousness.

In the same way that Kant suggested 'I think' to be necessary for consciousness, we may assume the resting state to be necessary for associating stimulus-induced activity with consciousness. The resting state may then be regarded as a necessary, non-sufficient condition, a neural predisposition of consciousness (NPC) (see Part IV).

What are the features of the resting state that predispose consciousness? Besides 'I think', Kant considered 'unity' (i.e. transcendental unity) to be an intrinsic feature of the mind. Following Kant, one may assume a particular, but currently unknown, unity of neuronal activity in the resting state to predispose consciousness.

Relying on Kant, the British neuroscientist Semir Zeki (2008), assumes such unity to be pre-programmed and central in the neural constitution of visual consciousness. Following Kant, such unity must be described as neurotranscendental, as it must be predisposed by the resting state itself and its specific but unknown spatiotemporal organization. This in turn may, for instance, make the binding and grouping of different stimuli in consciousness, as discussed in the binding problem, possible (see Part IV for details).

What can we learn about the brain from Kant's mind? Future work may want to explore the exact neural mechanisms underlying different forms of rest–stimulus interaction. This, however, is possible only if we achieve a better understanding of the neuronal mechanisms underlying the brain's intrinsic activity, its resting state.

In order to achieve that, we may draw on Kant's insights about the mind's intrinsic features such as 'I think' and unity. This may allow us understand better how the brain's resting state activity is structured and organized. And, most importantly, how that predisposes certain kinds of rest–stimulus interaction and, as Kant might say, consciousness and self.

Critical reflection 1

Why do we need to talk to the brain itself?

How can we further illustrate the brain's intrinsic activity? Let us invoke an imaginary dialogue between a neuroscientist and a brain. 'A brain?' the philosopher may wonder. 'That is not possible because it cannot speak by itself, only a person can speak, not the brain itself.' If you recall from Chapter 8, we discussed a thought experiment about a brain in a vat that receives the same input and sends out the same output as an embodied and embedded brain. Unlike our normal brain, the brain in a vat is connected and wired to a supercomputer that sends out the inputs for the brain and receives its outputs in just the same way as the body and the environment function for our brain.

Why is it important to let the brain itself speak? Because, as discussed in Chapter 11, we cannot access and investigate the brain directly, only indirectly. Such indirectedness introduces observer-related intrusions, where the observer affects his/her own observations. For instance, the observer, that is, the scientist, may use concepts that are more related to the observer and his/her experimental needs than to the brain as it is by itself, independent of his/her observation. We can thus not exclude what I described as observer-related intrusions in our investigation of the brain.

In order to escape the problem of observer-related intrusions, one may wish that one could study and investigate the brain in a more direct, rather than indirect, way. This would mean that our observation of the brain would remain independent of the methodological and technological constraints we are currently facing. Our knowledge about the brain would then no longer be confused by the observer him/herself. Wouldn't that be wonderful?

Let's make it happen. Let's access the brain in a direct way, by talking to it and raising questions that the brain itself has to answer. A talking brain! What remains impossible in our natural world and thus in neuroscience, due to the given indirectedness of our study of mind and brain, we can overcome by performing some purely logical thought experiment in philosophy of brain. That is exactly what we are going to do. Let us thus listen to an imaginary dialogue between a brain (BR) and a neuroscientist (NS).

Critical reflection 2

Concept of intrinsic activity

NS: We recently discovered that you generate intrinsic activity, spontaneously ongoing activity, that changes independent of any extrinsic stimuli.

BR: I am glad you discovered that. You are right: I generate neural activity that is spontaneous and continuously changing. You would probably say that my spontaneous neural activity is dynamic. I am aware that my spontaneous activity is covered and hidden beneath the seemingly dominating stimulus-induced or task-related activity as you seem to call it.

NS: In order to characterize your spontaneous activity and contrast it with the stimulus-induced or task-related activity, we speak of resting state activity. The concept of a resting state signifies the absence of any specific extrinsic stimuli, i.e. stimuli from outside the brain. As, for instance, when a person closes his/her eyes, he/she does not receive any extrinsic stimuli and is therefore supposed to be in a resting state. And then we measure the neural activity during such a resting state using functional imaging like fMRI.

BR: That sounds all well and good. But that definition of resting state seems problematic to me. I continuously receive extrinsic stimuli. The body that is usually attached to me sends me various stimuli from different organs, like from the liver, the stomach, the muscles, etc. – you seem to call these interoceptive stimuli. In addition, I also receive constant stimuli from the environment, i.e. exteroceptive stimuli. Even when the person closes his/her eyes and stops seeing, the other four sensory channels, auditory, tactile, gustatory and olfactory, are still open and provide me with continuous stimulus input.

NS: Do you mean that what we call the resting state is a mixture between your own intrinsic activity and the various extrinsic intero- and exteroceptive stimuli you continuously receive?

BR: Yes, that probably comes close. Your concept of a resting state is paradoxical, though. I am never at rest. If I were, I would be dead and you could not talk to me at all. And the respective person harboring such a brain would be declared 'brain-dead' as you call it.

NS: You thus mean that there is no such thing as a resting state in the brain?

BR: Exactly that. Your concept of the resting state is more driven by and related to you as observer and your experimental needs.

NS: Interesting. What are my experimental needs, though?

BR: The need to distinguish conceptually between different types of neural activity, intrinsic versus extrinsic, so that you can then isolate the former from the latter.

NS: You are right. Such conceptual distinction between intrinsic and extrinsic activity makes the experimental investigation of the former, the intrinsic activity, in terms of the resting state and its various neuronal measures like low frequency fluctuations, energy metabolism and functional connectivity, possible.

BR: All very neat. Good experiments and nice data. But do the data really reflect me, the brain itself, as it is, independent of your observations and its experimental measures?

NS: Are you implying that the concept of a resting state is more of an observer-based concept than a brain-based concept, and as a result is more related to me as an observer than to you as a brain, independent of my observation of you?

BR: Yes. The data you obtain sounds more related to you, your concepts, and their related experimental designs and measures, than to me as an independent brain.

NS: That does not sound too optimistic for me and my approach to unravelling you. What shall I do, then? Can you give me recommendations?

BR: Investigate the structure and organization of my neural activity across both what you call intrinsic resting state and extrinsic stimulus-induced activity. That may enable you to detect the spatial and temporal patterns of my neural activity and its guiding and determining principles and laws.

NS: You mean I should give up investigating and treating the resting state as an isolated entity, and abandon my focus on the activity changes elicited by extrinsic stimuli and their related functions, like sensorimotor, affective, cognitive, and social, functions?

BR: Yes!

Critical reflection 3

Intrinsic activity and consciousness

NS: I am not fully clear why you want me to focus on the structure and organization of your neural activity across the intrinsic resting state and the extrinsic stimulus-induced activity. Why is that so important?

BR: You want to understand what I am and what I do, right?

NS: Correct.

BR: And you want to reveal the purpose of me as a brain?

NS: True, indeed. And I must admit, despite all the impressive progress in our discipline, neuroscience, we haven't yet revealed your characterization as brain, nor your purpose.

BR: Let's compare the situation to the heart.

NS: The heart is nothing but a muscle structure. That defines the heart as heart.
BR: Yes, the muscle structure is an intrinsic design feature of the heart, one that defines the heart as heart.
NS: Why?
BR: Because it is due to the design of the heart as a muscle structure that it can fulfill its purpose, the pumping of blood.
NS: Immediately clear. That is easy. How, though, does that relate to my aim to understand you, the brain?
BR: Very simple. You currently have no ideas about my purpose or my intrinsic design features, i.e. those features that define me as brain, as distinct from other organs like the heart.
NS: Yes, unfortunately, that is true.
BR: But you may, for instance, study and investigate the structure and organization of my neural activity across time, space, different stimuli like intero- and exteroceptive stimuli, and different forms of activity states like resting state and stimulus-induced activity. And this may reveal some clue about my design and the features that define me as brain.
NS: All very fine. Then I know your intrinsic design features. So what? That does not tell me anything about what your purpose is!
BR: Wrong again! By revealing the muscle structure of the heart as its intrinsic design feature, you are able to understand why the heart cannot avoid pumping blood by default. Once you identify the structure and organization of my neural activity as my intrinsic design feature, you will be able to understand why and how I, as brain, predispose and necessitate mental features like consciousness.
NS: You mean the structure and organization of your neural activity across time and space, across intero- and exteroceptive stimuli, and across resting state and stimulus-induced activity is an intrinsic design feature of yours that defines you as brain?
BR: Correct!
NS: And once I reveal the principles and mechanisms of the structure and organization of your neural activity, I will be able to understand how you, as a mere physical organ, can produce mental states like consciousness, self, free will, etc.
BR: Exactly that!
NS: Before I believe that, I need more empirical detail. What you say sounds nice, but it leaves out all the empirical details I need in order to understand it! Therefore, let's journey into the neuroscience of consciousness.
BR: Happy journey!

Take-home message

The previous chapter discussed the nature of explanations and concepts in neuroscience. This left open the discussion of the brain itself, what the brain is as a whole, independent of observation, and how it can be characterized. These issues can be subsumed under the umbrella of philosophy of brain. Philosophy of brain investigates the determination and characterization of the brain in different domains, empirical, epistemic, metaphysical, etc. We characterized the brain in epistemic and metaphysical regard in Chapter 8. This, however, left open its empirical characterization, which was the focus in the present chapter. First, we investigated the various assumptions, including how the brain and its regions and networks are related to the different functions (e.g. sensory, motor, affective, cognitive, etc.) as well as the mental features that are associated with the brain. This led us from localizationist to holistic approaches that relate functions and mental features with either specific regions/networks or the whole brain. In addition, the brain's mode of operation was investigated. This demonstrated either an intrinsically- or extrinsically-determined mode of operation. On the whole, our discussion shows that the empirical characterization of the brain, its relation to mental features as well as its mode of operation, remains currently unclear. Future progress may not only depend on empirical discoveries, but also on conceptual considerations and technological invention, including the development of a philosophy of brain.

Summary

The previous chapter focused on theoretical and methodological problems in neuroscience. This discussion opened the door for an investigation into the characterization of the brain itself, as part of a philosophy of brain. Philosophy of brain aims to characterize the brain as a whole in different domains: empirical, epistemic and metaphysical. The empirical characterization of the brain was the focus in this chapter. One central question is the relationship between the brain's regions and networks, and the various sensory, cognitive and mental functions.

Psychology introduced the concept of modules. The concept of modules refers to special purpose-driven, encapsulated, and domain-specific functional units for processing specific inputs. Therefore modules receive only one specific input and are thus encapsulated. At the same time, modules serve only one particular function in one domain and are thus domain-specific. The visual system and the visual cortex in the brain, as well as other sensory systems, are regarded as typical examples of modules. However, this may be doubted: the sensory cortex shows plenty of connections to other sensory regions as well as to higher-order cognitive regions, which throws some doubt upon the characterization of sensory regions as modules. Further, rather than being organized in modules, different inputs seem to be processed in the same regions, which raises some doubts about encapsulation and domain-specificity.

In addition to sensory functions, cognitive functions have also been characterized by modularity. This touches upon the debate of localization versus holism in past and current neuroscience. Localizationism assumes cognitive and mental functions to be related to specific regions and networks in the brain. Holism suggests that any region, network, and thus the whole brain, contributes to the various cognitive and mental functions. The debate between localizational and holistic approaches to the brain remains undecided and subject to future experimental, technological and conceptual investigation.

In addition to the brain's relation to functions, we may also need to consider its mode of operation. Many neuroscientists in the past and present assume the brain's neural activity to be determined exclusively by extrinsic stimuli. This amounts to an extrinsic view of the brain, which contrasts with the observation of spontaneous or intrinsic activity in the brain independent of and prior to any extrinsic stimuli.

Meanwhile, there has been much recent research into the brain's intrinsic activity, and its exact function and purpose remain currently unclear. Here, one can go back to the German philosopher Immanuel Kant, who suggested the mind's intrinsic structure and organization was central to constituting mental features like consciousness and self. This remains subject to future neuroscientific and neurophilosophical investigation.

Revision notes

- What is the difference between philosophy of neuroscience and philosophy of brain?
- Why could Kant and his account of the mind be fruitful and helpful for neuroscience to better understand the role of the brain's intrinsic activity?
- Define the concept of modules.
- Where does the assumption of modularity originate? In which disciplines?
- What is meant by encapsulation?
- Is modularity empirically plausible when compared to the structure and organization of the brain?
- Explain the distinction between holism and localization.
- What is the difference between intrinsic and extrinsic activity in the brain?
- Why could the brain's intrinsic activity be relevant for the mind, and especially for consciousness?
- Why could Kant's view of the mind be relevant for our understanding of the brain and its structure?
- How do you define the concepts of intrinsic activity and resting state?

Suggested further reading

- Carruthers, P. (2006) *The Architecture of the Mind: Massive Modularity and the Flexibility of Thought* (Oxford/New York: Oxford University Press).
- Fodor, J. A. (1983) *The Modularity of Mind: An Essay on Faculty Psychology* (Cambridge, MA/London: MIT Press).
- Northoff, G. (2012) 'Immanuel Kant's mind and the brain's resting state', *Trends in Cognitive Sciences*, 16(7), 356–9.
- Oosterwijk, S., Lindquist, K. A., Anderson, E., Dautoff, R., Moriguchi, Y. and Barrett, L. F. (2012) 'States of mind: emotions, body feelings, and thoughts share distributed neural networks', *Neuroimage*, September, 62(3), 2110–28. *This is an excellent paper that directly tests the assumption of holism versus localization in the brain with regard to emotions, bodily feelings, and thoughts using functional imaging (fMRI).*
- van Eijsden, P., Hyder, F., Rothman, D. L. and Shulman, R. G. (2009) 'Neurophysiology of functional imaging', *Neuroimage*, May 1, 45(4), 1047–54. *This is an excellent review paper about the discussion of localization versus holism in recent brain imaging.*

Part IV
Neurophilosophy of Consciousness: From Mind to Consciousness

Part I discussed the theoretical framework of the mind in neurophilosophy; Part II outlined the intricacies of the mind–brain relationship; and Part III dealt with the various explanations for the mind and the brain. Now, in Part IV, we will practice neurophilosophy. In a preliminarily way, practicing neurophilosophy means linking theoretical-philosophical questions to the actual neuronal workings of the brain. We have already encountered several questions about the nature and features of the mind as core problems in neurophilosophy (see Part II). The domain in question here is traditionally the metaphysical domain: the domain concerned with existence and reality, and thus also the mind–brain problem (see Part II). Meanwhile, the features of the mind refer to consciousness, self, and intersubjectivity (and others like free will, emotional feelings, etc.). These features of the mind will serve as our starting point for the practice of neurophilosophy.

In this context, we must contextualize questions usually dealt with in the metaphysical and epistemological domains into the context of the empirical domain, the domain of observation in third-person perspective. Such contextualization is necessary in order to fully understand the nature of consciousness and self as the main foci in this and the next part. Such contextualization means that questions usually associated with the metaphysical or epistemological domain are now put into the context of the empirical domain. Contextualization here means that the originally metaphysical or epistemological question is not reduced to an empirical question with the subsequent replacement of their original domain by the empirical domain. This would amount to a brain-reductive neurophilosophy where ultimately there is no real contextualization.

Instead, the aim is to put the originally metaphysical question into the empirical context and to search for its empirical plausibility. Rather than approaching the metaphysical and epistemological issues from a mind-based (as in traditional philosophy) or brain-reductive (as in reductive neurophilosophy and neuroscience) perspective, we thereby pursue a brain-based strategy and thus non-reductive neurophilosophy.

This is apparently visible in the way we will deal with the various arguments that have been developed in philosophy against a physicalistic or materialistic approach to consciousness (see Chapter 13). This is complemented by a discussion of recent theories of the neural correlates of consciousness (NCC) (Chapter 14) and the neural predispositions of consciousness (NPC) (Chapter 15). Finally, the various arguments against the material or physicalistic view of consciousness as raised in Chapter 13 are directly compared with the empirical data and are thus put into the empirical, that is, neuroscientific context of consciousness (see Chapter 16).

More generally, Parts IV and V serve to apply the methodological strategy of non-reductive neurophilosophy to concrete mental features like consciousness and self. This complements the application of the non-reductive neurophilosophical approach in this book: Part II showed how neurophilosophy can address the core problem of philosophy of mind, the mind–brain problem. Rather than discussing the various arguments pro and contra a certain mind–brain solution, we here focused on the definition of the mind and mental features and whether they are in accordance with the empirical data, thus testing for their empirical plausibility.

This was further extended in Part III, when considering the explanation of mental features and the brain itself, that is, as such in a neurophilosophical context. Now this is further extended to the mental features themselves. Instead of focusing on the underlying mind–brain problem and its explanation, we now discuss two concrete mental features, self and consciousness, in a specifically non-reductive neurophilosophical way. Such an approach must be distinguished from discussing the various arguments pro and contra specific philosophical, reductive neurophilosophical, and neuroscientific approaches to the same subject matter, as will become clear in the text.

13
Arguments against the Reduction of Consciousness to the Brain

Overview

Consciousness has long been regarded as the paradigmatic feature of the mind, and its distinction from both the brain and the body. However, this assumption has recently been challenged with the progress in neuroscience over the last 20–40 years. The discovery of the brain's neuronal mechanisms led to discussions about whether consciousness might also be based on the brain and its neural activity. This led philosophers to discuss whether consciousness can or cannot be reduced to the brain and its neural activity. Philosophers suggested various arguments against the possibility of linking, and ultimately reducing, consciousness and its specific phenomenal features to the brain's neuronal features. Consciousness may thus be regarded as the main obstacle to a physicalistic and ultimately neuronal account of the mind. Recent philosophy of mind has therefore argued against the reduction of consciousness to the brain, and against a physical approach to the mind. These arguments are the focus in this chapter.

Objectives

- Understand the phenomenal peculiarities of consciousness
- An explanation of the principal differences between neuronal and phenomenal features
- Insight into the epistemological features of consciousness and its specific type of knowledge
- Major problems and gaps in the explanation of consciousness in terms of the brain and its neural activity
- Distinction between logical and natural conditions and worlds
- Distinction between conceptual-logical and empirical-experimental accounts of consciousness
- Neurophilosophical approach to understanding the hard problem and the explanatory gap problem

Key concepts

Irreducibility of consciousness to the brain, absent qualia, inverted spectrum, what it is like argument, knowledge argument, conceivability argument, hard problem, explanatory gap.

Background 1

Mental states versus physical states

One of the hallmark features of the mind and of our existence is consciousness. You are conscious while reading these lines. You experience the reading of these lines and, I hope, become aware of their content. How is it possible for you to experience something? When you experience something, you have what philosophers describe as 'mental states'.

Mental states are states you experience by yourself. Nobody else can experience your mental states. In other words, mental states are private. Most important, your experiences, and thus your mental states, are specific to you and your person, and its particular perspective. In short, mental states are experienced in the first-person perspective. This is what philosophers call subjectivity.

Let's compare consciousness to the body. Consciousness and its mental states are essentially subjective and private: mental states can be experienced and thus accessed only in the first-person perspective of the respective person, while being inaccessible to any other person and their observation in third-person perspective. This distinguishes mental states, and thus consciousness, from the body and its states. The states of the body, including the brain's neural states, can be accessed by others, and observed and investigated in third-person perspective by any observer. They are not private, but accessible, and can therefore be considered objective rather than subjective.

Why and what makes the difference between mental and bodily states? The states of our body are physical states. Various biochemical processes that can ultimately be traced back to physical processes occur in, for instance, the stomach, when different enzymes digest food. Physical processes also reign in other internal organs like the liver, the pancreas, etc. Even in our brains the observed neuronal states can be traced back to mere physical processes. Bodily states can thus be characterized as physical states that are objective and accessible.

Background 2

Where mental states are located: 'location problem'

Where can we find the mental states and their mental, conscious contents? Bodily states as physical states are merely objective and accessible. This distinguishes them from conscious states, which are subjective and private. Where, then, can we 'locate' the mental states and thus consciousness? When observing the brain and its neuronal, physical states in third-person perspective, all we can see are neuronal states or, in more detail, visualizations of the brain's neuronal states using various technological means like fMRI.

In contrast, when observing the brain and its neuronal states, we do not observe anything that proves or even hints at the existence of mental states and their respective mental contents. The book you experience as mental content in your consciousness cannot be observed in any of our visualizations of the brain's neuronal activity. You may think this is obvious. Why? Because your mental contents are tied to your first-person perspective, while the brain and its neuronal states are observed in third-person perspective. There must thus be an essential epistemological difference, the difference between first- and third-person perspectives.

Does this epistemological difference between first- and third-person perspectives translate into a metaphysical difference of separate existences and realities underlying mind and brain? Philosophers like Descartes assumed a specific substance, a mental substance, to underlie the existence of mental states. This mental substance was called 'mind' and was distinguished from the physical substance of the body (see Chapter 5). The 'mind' was supposed to be everything the body wasn't. It was assumed that the mind was subjective rather than objective, and was private rather than accessible. Most important, the mind was characterized by mental states rather than physical ones.

Since consciousness cannot be observed in the brain and the body, it was originally associated with the mind as distinct from body and brain. This led to what I discussed as mental approaches to the mind–brain problem in Part II. According to mentalist approaches, consciousness cannot be reduced to the brain and its physical, neuronal features. We cannot locate consciousness anywhere in the brain. We therefore assume that consciousness must be associated with some mental features as distinct from the brain's physical features. This leads us ultimately to assume the existence and reality of a mind. Our assumption of a mind may thus have its roots in our inability to 'locate' mental features in the brain and body. One may thus want to speak of a *'problem of the location of mental states', the 'location problem'*.

The location problem consists in the question of where the mental features are situated. Since they are mental, these features do not seem to be situated in the body that shows only physical features. Moreover, since mental features also cannot be found in the brain, they apparently must also be distinguished from neuronal features. This raises the question of the location of mental features. That, in turn, leads to the assumption of a mind and subsequently to the mind–brain problem. Hence, the location problem may be considered a necessary presupposition of the possible mind–brain problem: if there were no such location problem, the question of the mind–brain problem could not even be raised.

Background 3

Is the 'location problem' a neurophilosophical problem?

Both the philosopher and the neuroscientist, as well as the reductive neurophilosopher, may now be confused. Why? Let us start with the philosopher. The philosopher may want to argue that the location problem is not a problem at all. The real problem is in the question of the relationship between mind and brain, the mind–brain problem. This, in turn, will solve the location problem in that it will tell us that the mental features are located in either the mind or the brain. Hence, if it makes sense at all, the location problem must be considered secondary to the mind–brain problem rather than being a presupposition.

This is obviously different in the non-reductive neurophilosophical approach. Here, the mental features themselves rather than the assumption of a mind and its relationship to the brain (as in philosophy) are the starting point. This raises the question of the localization of the mental features in the brain and body. This is the location problem. If the location of the mental features in the brain and body is denied, one has to assume some kind of mind.

This in turn raises the question of the mind's relationship to the brain and body. The mind–brain problem is here thus secondary to and a consequence of the location problem rather than the latter being a consequence of the former as in philosophy. Accordingly, what looks like a confusion from a philosophical perspective, the location problem, turns out to be an essential presupposition or necessary condition of the possible mind–brain problem.

How would the reductive neurophilosopher see the location problem? He/she would not consider it a problem at all. Instead, it is his/her basic assumption that mental features are located in the brain. Hence, he/she does not see that as problem, but takes it instead as given. And if something is evident, namely that mental features are located in the brain, there is no problem. Hence, the reductive neurophilosopher would see the location problem as not being a problem in itself. The same applies to the neuroscientist, who obviously locates the mental features in the neuronal features of the brain.

In sum, while being considered superfluous or confused in the perspectives of the philosophers, the reductive neurophilosopher, and the neuroscientist, the location problem must be considered a valid problem in the context of a non-reductive neurophilosophical framework. There it provides the basic assumption and a necessary presupposition for raising the question of the relationship between mind and brain, the mind–brain problem: if the mental features cannot be located in the brain and body, one must assume their location in the mind, which, in turn, raises the mind–brain problem.

Background 4

'Location problem': is consciousness in the brain?

However, recent progress in neuroscience sheds some doubt on the location and association of consciousness with a mind and its mental features as distinguished from the physical features of the brain and the body. Certain conscious states in perception, for instance, are related to particular neuronal mechanisms (see following chapters for details). And abnormal alterations in our consciousness, and even the loss of consciousness in neurological and psychiatric disorders, seem to accompany abnormal changes in the brain (see following chapters). All this suggests that consciousness is associated with, and can be 'located' in, the brain, rather than in some kind of mind as distinct from the brain.

How can we further illustrate the 'location problem'? Let's introduce an imaginary dialogue between a philosopher and a neuroscientist.
The philosopher may ask the neuroscientist: 'Where is consciousness?'
'In the brain, of course,' the neuroscientist rushes to answer.
'But', the philosopher argues, 'even the most colorful brain scans do not demonstrate any experience, let alone its contents. All they reveal is some color-coded degrees of neuronal activity.'
'And that color-coded neuronal activity is exactly what you describe as consciousness,' the neuroscientist explains.

'No, that is impossible,' answers the philosopher.
'Why?' asks the neuroscientist.
'Because you will never be able to find any mental contents and states in the neuronal states of your brain. Your colorful pictures of the brain are colorful pictures, nothing else. At best they correspond to neuronal activity in the brain. But mental states? No, they cannot be "located" and observed in the brain and its neuronal activity.'
'You thus assume mental states and hence consciousness to be principally different from the brain, correct? If so you must provide some arguments for that.'

Background 5

'Location problem': different types of argument against the reduction of consciousness

How can the philosopher further substantiate his/her point? The philosopher could argue against the reduction of consciousness and its phenomenal features to the brain. These arguments should point out some reasons why consciousness and its phenomenal features cannot in principle be reduced to the brain and its neuronal, physical states. Rather than referring to some specific empirical, neuronal features of the brain itself, these arguments consider the various features of consciousness itself and assume them to be principally different from the ones of the brain.

One may want to divide these arguments into four different types: first, there are those that argue for the irreducibility of the different phenomenal features of consciousness to the brain's neuronal (and ultimately the body's) physical features. Second, there are arguments for knowledge about consciousness as being principally different from knowledge about the brain, body and behavior. Third, other arguments focus on the explanation of consciousness – what kinds of explanatory concepts and relations are needed to properly explain consciousness? And what would count as a proper explanation of consciousness?

Fourth, there are arguments that invoke zombies. Zombies describe imaginary creatures with the same neuronal and behavioral states as humans while showing no consciousness at all. The zombies thus show the same brain and body as humans, but no consciousness. What does the purely logical scenario of zombies tell us about consciousness in our natural world? This is the kind of question that is discussed in these arguments. In the following, we will provide an overview of the different types of arguments.

Arguments about the phenomenal features of consciousness 1

Absent qualia argument

The central feature of consciousness is supposed to consist in the 'what it is like' that describes a 'phenomenal-qualitative feel' (see next chapter for details). This 'phenomenal-qualitative feel' is often described by the term qualia, signifying a hallmark feature of consciousness. The typical example is the experience of colors. You experience a redness associated with the color red of the book in front of you. It is this redness that signifies the phenomenal-qualitative feel and thus qualia.

Table 13.1 Philosophical arguments against the reduction of consciousness to a physical–material basis

	Arguments about phenomenal features of consciousness	Arguments about the knowledge of consciousness	Arguments about the logical possibility of consciousness	Arguments about the explanation of consciousness
Absent qualia	Absence of 'what it is like': Relationship between phenomenal and neuronal features			
Inverted qualia	Inversion of 'what it is like': Relationship between phenomenal features and content			
'What it is like' argument		Point of view: Difference between subjective and objective knowledge		
Knowledge argument		Superscientist Mary: Difference between first- and third-person knowledge		
Conceivability argument			Philosophical zombies showing no consciousness: Difference between logically possible worlds and the actual natural world	
Hard problem				Difference and bridging between possible consciousness and nonconsciousness
Explanatory gap problem				Empirical, epistemological, ontological, and conceptual gaps in our explanations of the brain ('how it works') and consciousness ('what it is like')

Now let's imagine the following: you see the red book, but no longer experience the specific phenomenal-qualitative feel of redness. There is no 'what it is like' and thus no phenomenal-qualitative experience that accompanies you seeing the book. In this case, you suffer from what the philosophers describe as 'absent qualia'. Most important, while your qualia remain absent, everything else remains the same, including the wiring of the brain, your behavior and the book you see. The only thing missing is the phenomenal-qualitative experience of the 'what it is like'.

How are absent qualia possible? The only way for them to be possible is that they must be different from the brain, its neuronal processes and wiring, as well as from your behavior. Given that everything is identical except the presence versus absence of qualia, the latter cannot be reduced to either your brain, behavior, or to some other computational processes. This is possible only if qualia and thus consciousness are not reducible to the brain, behavior, or computational processes. Instead, qualia must be related to some kind of underlying mechanism that is different and separate from all three – the brain, behavior and computational processes.

What might the neuroscientist reply? Let's think about another imaginary dialogue.

Neuroscientist: Do you really infer from the thought experiment about absent qualia that qualia must be different from the brain, behavior and computational processes?

Philosopher: Yes, otherwise the possibility of an absence of qualia would not be given in the case of an otherwise identical brain with the same neuronal, behavioral, and computational states. Only if the qualia are different from neuronal, behavioral, and computational states, can they also remain absent without changing these states.

Neuroscientist: But this is a purely logical argument. In reality – what you call the natural world – qualia cannot remain absent while the brain, behavior, and computational processes remain identical. Hence, what you call absent qualia is a logical artifact, a pure thought experiment in the purely logical world that bears no relation to the actual reality in the natural world.

Philosopher: That is wrong. I can, indeed, infer from a logical thought experiment in the logical world that qualia and the brain must be principally different. If absent qualia were to be impossible, I would infer that qualia and the brain must be related to each other and thus that qualia may be 'localized' in the brain. This is not, however, what this thought experiment tells me. Rather, it tells me the opposite: that qualia can remain absent in the presence of the same neuronal features of the brain. If so, qualia cannot be 'localized' in the brain and its neuronal features.

Arguments about the phenomenal features of consciousness 2

Refutation of the absent qualia argument

The absent qualia argument is often considered as an argument against the possible reduction of consciousness to the brain in particular, and a physical mind–brain approach in general. Let us repeat what the philosopher in our imaginary dialogue has already put forward. If the qualia can remain absent while the brain and its physical, neuronal features remain the same, the former cannot be reduced to the latter. Hence, qualia as a phenomenal hallmark feature of consciousness must be different from the brain and can therefore not be 'located' in the brain and its neuronal states.

The inability to 'locate' qualia in the brain implies the need for a mental, rather than physical, approach to consciousness and the mind–brain problem. The absent qualia argument must consequently be considered a threat to the neuroscientist's attempt to associate consciousness with the brain, let alone 'locate' it in the brain's neuronal states. The argument also threatens those philosophers who argue in favor of a physical approach to consciousness and the mind.

How can one counter the absent qualia argument? This philosophical discussion has focused on the meaning of the two concepts 'absent' and 'qualia'. Let's start with the latter, the concept of qualia. The proponents of a physicalistic approach, like the American philosophers Daniel Dennett and Patricia Churchland, argue that there are no private and subjective features in our mental states; instead, consciousness is accessible and objective, and therefore accessible to third-person perspective-based observation. In this model, there are no qualia. And if there are no such qualia, they cannot be absent, which in turn makes the absent qualia argument futile.

In addition to the concept of qualia, one may also question the concept of 'absent'. Do the qualia truly remain absent? One way to answer this question is to refer to absent knowledge rather than absent qualia: qualia do not truly remain absent, but are rather not properly recognizable. If so, what the philosopher describes as 'absent qualia', amounts to nothing but an absence of knowledge, rather than a true absence of qualia.

The absent qualia argument may then not be about the absence of the qualia themselves, but rather an argument about an inability to recognize and know qualia. The presence of qualia is thus compatible with the absence of knowledge about them. In short, what is absent are not qualia, but the knowledge of qualia. This undermines the absent qualia argument, making it invalid in any support of a mental rather than physical approach to consciousness and mind.

Arguments about the phenomenal features of consciousness 3

Inverted spectrum argument

The absent qualia argument argues in favor of the absence of qualia in the presence of the same brain states. This means that both experience and the respective mental content, like the book in front of you, remain absent. Another argument, the 'inverted spectrum argument', takes this distinction between experience and mental content as its starting point.

The 'inverted spectrum argument' presupposes an analogous thought experiment. Suppose that the neuronal states of the brain, your behavior and the computational processes are the same in two situations or between two persons. The main difference consists in that one person experiences blue, i.e. 'blueness', when seeing red, and experiencing 'redness' when seeing blue. Meanwhile, the second person experiences 'blueness' when seeing blue and 'redness' when seeing red.

Unlike the absent qualia argument, this argument does not ask for the absence but the presence of qualia, and thus of both experience and content. Instead, it rather raises the possibility of an inversion of the content of qualia with regard to the experience of that very content. In other words, the 'what it is like' is associated with different contents: the content blue is now associated with the experience of 'redness' rather than 'blueness'. Meanwhile, the content red comes with the experience of 'blueness' rather than 'redness'.

How can the 'inverted spectrum argument' be distinguished from the 'absent qualia argument'? The 'absent qualia argument' is about the relationship between phenomenal and neuronal/behavioral/computational features. Even if the latter are identical, the former, including both experience and content, can nevertheless remain absent.

This is different in the inverted spectrum argument. This argument presupposes the distinction between content and experience as well as their possible dissociation. As a result, the mental or phenomenal contents themselves are associated with certain neuronal processes in the brain. If the phenomenal contents themselves, and the phenomenal experience of them, can dissociate from each other – for instance, if the content blue accompanies the experience of 'redness' – the phenomenal experience cannot be associated with the brain and its neuronal states.

Let's go into more detail. The proponents of a mental approach argue that qualia cannot be reduced to the brain. If qualia, i.e. the experience, can be inverted in the presence of the same phenomenal content and its underlying neuronal states, the former cannot be reduced to the latter. More specifically, if the neuronal states of the brain process the content red while you experience 'blueness' rather than 'redness', the latter must be different from the former. This means that your experience of 'blueness' cannot be reduced to and 'localized' in the brain. There must be some special, non-physical, mental feature about phenomenal experience and thus qualia (the blueness) that make it distinct from the brain's physical features that account for the content of qualia (the redness).

Arguments about the phenomenal features of consciousness 4

Converse spectrum inversion

Are inverted qualia empirically plausible? This means, can they occur in the natural world? The proponent of a physical approach to consciousness denies this. According to him, inverted qualia cannot occur in the natural world and have, for instance, never been observed in neurological or psychiatric patients.

These patients show alterations in their brain that accompany changes in both experience and the content of their mental states. In contrast, cases with dissociation between content and experience have not as yet been observed. The inverted qualia argument is consequently not empirically plausible. In short, in accordance with the empirical data, the qualia argument must therefore be considered a purely logical argument that applies only to the logical world, but not to our actual world, the natural world.

Alternatively, the proponent of a physical approach could admit that inverted qualia are possible. In that case he/she may want to distinguish between content and context. The qualia and their inversion may be related to the context, as distinguished from the content with which they are associated. Let's imagine you are in a library and are looking for a red book. The red color of the book is associated with the book itself as the content of your consciousness.

How about your experience of the book's red color as its content? Imagine now that your red book is located on the shelves among blue books. Now you may experience the red book as blue (blueness) despite the fact that it is red. The context of the book and its content may thus determine your experience, and deviate and dissociate it from the content. Hence, content and context may dissociate from each other. The inverted qualia as an argument about the dissociation between experience and content may then be reinterpreted as an argument about the relation between context and content.

How does the assumption of this content–context relationship accompany the physical approach? The proponent of a physical approach could argue that content and context are mediated by two different neuronal systems in the brain, and that these two neuronal systems may be out of sync when there is an instance of inverted qualia as manifested in the dissociation between content and context. Hence, inverted qualia are here re-interpreted as a special form of the content–context relationship and their respective underlying neuronal systems.

If the assumption of inverted qualia as a content–context relationship holds true, one would also expect the reverse scenario to be possible. Not only can the context, and thus qualia, dissociate from the content, as in the inverted spectrum argument, but the content itself should then also be dissociable from the context. In this case, the experience of 'redness' should remain while the content, the color red, should change into, for instance, a blue content. This scenario has been discussed as the 'converse spectrum inversion' by the American philosopher Ned Block.

Neurophilosophical discussion 1

Neuronal–mental versus neuronal–neuronal dissociation

What does the reinterpretation of qualia as a content–context relationship tell us about the relationship between qualia and the brain in general? The proponents of a mental approach to consciousness are eager to claim that qualia are extrinsic to the brain and its neuronal states. The term 'extrinsic' signifies here that nothing about the brain and its neuronal features implies mental features like experience, and thus qualia, in a necessary and unavoidable way.

If, however, the brain and qualia are not intrinsically linked to each other in a necessary and unavoidable way, both could dissociate from each other. This in turn makes it necessary to assume some non-physical, mental feature to account for qualia, and thus experience. This is suggested by philosophers like McGinn and others (see Chapters 6 and 7), who opt for a mental rather than physical approach to consciousness.

The response from the mental proponents is undermined by the distinction between content and context. The mental proponents implicitly assume that contents and experience are different and can dissociate. From this they infer that only content, but not experience, is associated with the brain. This assumption is undermined by the distinction between content and context, and their association with different neuronal mechanisms. Experience is in this case linked to the context. The context itself is in turn closely related to a specific neuronal mechanism that is different from the one that processes the content. Content and context are consequently related to different neuronal mechanisms.

Now imagine some changes. For instance, the neuronal mechanisms underlying the processing of context may be altered, while the one related to the contents is preserved. In that case, context and content dissociate from each other. This means that the experience and content also dissociate from each other, which makes the scenario of inverted qualia empirically plausible. The mental proponent who bases his/her opposition to a physical approach on the empirical implausibility of inverted qualia may then be rebuked.

The dissociation between qualia and neuronal states is, in this model, no longer a mental–neuronal dissociation where – as presupposed by the mental proponents – a separate mental feature dissociates from the brain's physical features. Instead, it is a dissociation between two different neuronal systems – one that underlies context, and another that processes content. Hence, the alleged mental–physical dissociation is replaced by neuronal–neuronal dissociation.

Neurophilosophical discussion 2

Empirical plausibility of qualia as context–content relationship

Is the assumption of two different neuronal mechanisms underlying content and context empirically plausible? The mental proponents only associate content, and not experience, with the brain. Why? The mental proponents tend to presuppose the brain as an organ that is isolated from its context or the respective environment. This implies that context-dependence is not considered a feature of the brain and its neuronal states. Instead, context-dependence is associated with mental features and thus with the mind rather than with the brain and its neuronal features. In other words, the mental proponents presuppose the brain to be 'isolated' from context, rather than embedded in its respective environmental context.

The assumption of different neuronal mechanisms underlying the processing of content and context now puts the brain back into its context. This presupposes an 'embedded brain', rather than an 'isolated brain'. The brain as an 'embedded brain' can be defined by its intrinsic relation to both the body and the environment.

What are the neuronal mechanisms by which the brain is necessarily and continuously linked to its context, the body and the environment? Currently, we do not really know. What we do know is that if experience, and thus the 'what it is like' qualia, are indeed identified as context, the brain's intrinsic relation to the body and the environment may be central for the generation of qualia in particular and consciousness in general.

What does this imply for the inverted spectrum argument? This argument might be reinterpreted as an argument about the difference between isolated and embedded brains. Contents and experience may be inverted and thus dissociated from each other when the brain is embedded because of the different neuronal mechanisms underlying the neuronal processing of content and context. Meanwhile, content and context may not dissociate from each other in the case of an isolated brain, because then the neuronal differentiation between content and context is eliminated.

Since there are strong empirical hints that human brains are embedded, rather than isolated, the inverted spectrum argument must be considered empirically and neuronally plausible. The possibility of inverted spectra consequently does not necessitate the assumption of a special mental feature outside the brain to account for experience. Any such inference of a mental feature may then be deemed empirically and neuronally implausible while, admittedly, still being logically consistent. This would mean that the inverted spectrum argument is not in fact concerned with our actual, natural world, but rather with a logically imagined world.

Arguments about the knowledge of consciousness 1

'What it is like' argument

The American philosopher Thomas Nagel introduced the 'what it is like' argument in his famous article 'What is it like to be a bat?'. The bat is obviously very different from humans; it sleeps during the day and is active at night. Most important, its sensorium is completely different than that of humans – it is attuned to ultrasonic frequencies, which we, as humans, remain unable to perceive. Nagel argues that even if we knew everything about the bat's neurophysiology, we would still not know anything about how the bat experiences its environment in terms of its ultrasonic waves. Hence, the merely objective knowledge about the bat's brain does not tell us anything about the subjective knowledge of the bat's experience and its consciousness.

Why? Because we, as humans, remain unable to assume the point of view of a bat. Since the point of view is very much dependent upon the biophysical features of the organism (for instance, the physical properties of its sensorium), different biophysical features may imply different possible points of view.

If another organism has completely different biophysical features, it will have a different point of view in comparison to humans. Most important, because of the biophysical differences, we as humans will remain unable to comprehend and ultimately experience the bat's point of view. Any knowledge of the bat's brain does not provide us with any knowledge of the bat's point of view and its experience.

The 'what it is like' argument is essentially centered on point of view and its phenomenal-qualitative manifestation – the 'what it is like' – as hallmark features of consciousness. In order to account for the 'what it is like' of consciousness, we need a specific type of knowledge that is different from the knowledge that accounts for the brain and its neuronal states: the 'how it works'. The brain's 'how it works' describes the neuronal-quantitative knowledge (of the brain) that is objective, rather than subjective, and accessible rather than private.

This is different in the case of 'what it is like' knowledge. That knowledge is subjective rather than objective, and private rather than accessible, primarily because of its link to experience in first-person perspective. Because both types of knowledge – the 'how it works' versus the 'what it is like' – are different, the 'how it works' knowledge cannot account for the 'what it is like' knowledge. However, if their respective knowledge differs, the 'what it is like' must also be principally different from the 'how it works' knowledge. This ultimately means that consciousness cannot possibly be reduced to the brain.

Arguments about the knowledge of consciousness 2

Refutation of the 'what it is like' argument

How can the neuroscientist and neurophilosopher escape the 'what it is like' argument and its assumption of the irreducibility of consciousness to the brain? He/she might say that the difference between both types of knowledge concerns only a difference in the access to, and/or description of, one and the same underlying knowledge. In short, perhaps both the 'what it is like' and the 'how it works' knowledge may refer to one and the same underlying feature. Both 'what it is like' and 'how it works' knowledge access the same feature in different ways: subjective or objective. Or it is possible that 'what it is like' and 'how it works' knowledge describe that feature in different ways using distinct concepts like phenomenal and neuronal concepts.

Why is the 'what it is like' argument relevant in the current discussion? The 'what it is like argument' is of extreme relevance in the current discussion about mental versus physical approaches to consciousness and the mind. First, it describes the 'what it is like' as a phenomenal hallmark feature of consciousness. This implies what Thomas Nagel calls a 'point of view'. What is a point of view? Why is a point of view relevant for consciousness and its 'what it is like'? The 'what it is like' argument presupposes the point of view to be essential for consciousness, implying that without a point of view there would be no consciousness. This will be explained in more detail in the next chapter.

Second, the 'what it is like' argument provides an objection to a physical approach to the mind. Since the 'what it is like' and its point of view cannot be accounted for by the neuronal, physical workings of the brain, consciousness cannot be reduced to the brain and its neuronal states. Hence, the physical approach to consciousness and the mind fails. The 'what it is like' argument is thus a major obstacle to a physical and ultimately neuroscientific approach to consciousness.

How can the physicalist and the neuroscientist deal with the 'what it is like' argument? One possibility is that they can deny the 'what it is like' and consider point of view as a mere illusion. This is the position taken by American philosophers Daniel Dennett and Patricia Churchland, who aim to clear the way for a physical and thus neuroscientific approach to consciousness. According to them, the 'what it is like' simply does not exist. This implies that there is nothing like point of view. What appears to be a special feature is nothing but a misguided illusion.

What, however, if denial of qualia including a point of view presupposes by itself a point of view? One would then argue that without a point of view any kind of denial of the point of view would remain impossible. This is the way some recent philosophers like Bennett and Hacker argue against such a reductive approach. More generally, this means that even the denial of consciousness is possible only on the basis of consciousness. Following these arguments, there is no way of escaping a point of view, and more generally consciousness. This means that we may need to take the point of view and consciousness seriously. Simple denial and their outing as mere illusions is not possible. How can we do that? We will discuss this in detail in Chapter 15.

Arguments about the knowledge of consciousness 3

Knowledge argument

The 'what it is like' argument is about the knowledge of a point of view and the associated experience, the 'what it is like'. The Australian philosopher Frank Jackson considers this problem within the context of the human species and is well known for his thought experiment concerning an imaginary neuroscientist called Mary.

In his imaginary thought experiment, Mary is a brilliant neuroscientist who knows everything about color perception. More specifically, she has complete knowledge about the neurophysiological processes underlying our color perception. And she also knows everything about the physics of light and its color spectrum. In other words, Mary is the kind of neuroscientist every neuroscientist would like to be, especially those researching consciousness.

But Mary is even more special, Jackson maintains in his imaginary thought experiment, for she has been brought up in an entirely black and white environment. She has never seen or experienced any colors apart from black and white. Even though she knows everything about the neurophysiological processes underlying the different colors like red, green, yellow, etc., she has no knowledge about the experience of these colors herself. In other words, she has no knowledge about the 'what it is like' of colors, but she has complete knowledge about the brain's 'how it works'.

Let's stage a brief imaginary debate between Frank Jackson and a current neuroscientist focusing on consciousness.

'Doesn't she lack something?' Frank Jackson asks.
'Yes, she does, she lacks knowledge of the phenomenal-qualitative features of the consciousness of colors,' the neuroscientist may respond.
'Her knowledge is thus incomplete,' answers Jackson.
'But that does not mean that consciousness and its "what it is like" cannot be reduced to the "how it works" of the brain!' the neuroscientist is quick to remark.
'It does,' Jackson replies. 'Why? Because Mary, as such a brilliant neuroscientist, should not lack any knowledge about any color; she is supposed to have complete knowledge about the brain and "how it works". If the "what it is like", and thus consciousness altogether, can be reduced to the brain and its neuronal features, Mary should not have any deficits in her knowledge given her complete account of the brain's "how it works". The fact that Mary lacks knowledge thus signifies that the "what it is like" of consciousness is different from the "how it works" of the brain.'

Like the 'what it is like' argument, the 'knowledge argument' is about knowledge, the kind or type of knowledge we need in order to account for consciousness and its hallmark feature, the 'what it is like'. This type of knowledge is further specified by the case of Mary. Mary has complete knowledge about the brain's 'how it works', which is possible because of her brilliant neuroscientific observations in third-person perspective (TPP). However, despite her complete knowledge about the brain's 'how it works' in TPP, she lacks the knowledge about the 'what it is like' of colors in her first-person perspective (FPP).

Taken together, consciousness and its 'what it is like' require first-person knowledge. Such knowledge must be distinguished from the third-person knowledge that is required to know the 'how it works' of the brain. Since both types of knowledge are different, they must refer to different properties and features that cannot be reduced to each other. The knowledge argument is thus taken as an argument against the reduction of consciousness to the brain and its physical features. Hence, any neuroscientist who aims to explain consciousness in terms of the brain's neuronal processes is confronted by the knowledge argument and its claim for the non-physical basis of consciousness.

Arguments about the knowledge of consciousness 4

Refutation of the knowledge argument

How can the neuroscientist and neurophilosopher escape Jackson's argument? Let's imagine that Mary now encounters, for the first time in her life, a colorful environment. Does she gain new knowledge in addition to all her neurobiological knowledge about the brain's 'how it works' underlying color? Jackson would argue, 'Yes, Mary does gain new knowledge, first-person knowledge.' If so, one would also assume that novel neuronal mechanisms and systems will be activated and recruited in her brain. This means that new knowledge, and thus epistemic novelty, goes hand in hand with new neuronal mechanisms, i.e. neuronal novelty.

Others would agree with Jackson that Mary does indeed gain new knowledge, but that it does not concern anything novel, e.g. new or different features. If so, one would expect that epistemic novelty is not accompanied by neuronal novelty, meaning that no principally novel neuronal mechanisms and systems are recruited. Her knowledge is only about a novel 'knowing-how' while the 'knowing-that' remains the same, e.g. the referent, the 'that' of her knowledge, has not changed. One may also say that she acquires the same knowledge in a new mode that was not available to her previously. Meanwhile, the new knowledge, e.g. the new mode, still refers to the same referent as her previous knowledge.

One could also argue that both types of knowledge, first- and third-person knowledge, differ only in their manner of representation: 'third-person knowledge requires verbal-linguistic representation, while first-person knowledge is associated with a non-verbal, phenomenal representation mode'. Knowledge about the same referent is thus represented in different ways, verbally and non-verbally. The epistemic novelty is characterized here as mere representational novelty, while no neuronal novelty is assumed.

Neurophilosophical discussion 3

Epistemic–metaphysical inference

Rather than discussing the concept of knowledge by itself, and what is novel about it, one could also shed some light on an implicit presupposition in the knowledge argument. The knowledge argument (as well as the 'what it is like' argument) assumes that one can infer from different kinds of knowledge – first- versus third-person knowledge – that differences in existence and reality, mental versus physical, exist and are real.

This is an inference from the epistemological domain, the domain of knowledge, to the metaphysical domain, the domain of existence and reality. The question now is whether such an inference is justified: do different types of knowledge imply different kinds of existences and realities? Or are knowledge and existence/reality two different entities that should not be linked and thus confused with each other?

How do philosophers evaluate such an inference? The philosopher may consider such an epistemic–metaphysical inference on the basis of logical criteria and thus test for its logical plausibility: do the epistemic assumptions and their respective concepts imply anything about the underlying existences and realities? If so, the epistemic–metaphysical inference is justified and thus logically plausible; if not, it must be considered as fallacious and logically implausible.

What about the neurophilosopher? Since he/she is interested in focusing on the brain, he/she may approach the epistemic–metaphysical inference in two ways. First, he/she can investigate whether this epistemic–metaphysical inference is empirically plausible. This means that he/she must investigate the neuronal mechanisms underlying epistemic assumptions and thus our knowledge in general.

If the recruitment of those neuronal mechanisms implies the existence of some real and existing object, the epistemic–metaphysical inference may be empirically plausible, or in accordance with the neuronal mechanisms. If not, it may be discarded as empirically implausible, while it can still be tested for its logical plausibility.

In addition, the neurophilosopher may want to use a second strategy. The first strategy presupposed the consideration of the brain in exclusively the empirical domain. However, as discussed in Part II, one may consider the concept of the brain also in the epistemological and metaphysical domains. If so, one could discuss the brain's intrinsic features, especially its epistemic features, which are those that define the brain as brain (in the epistemic domain). If these intrinsic epistemic features of the brain refer to and imply anything about existence and reality, one may diagnose the epistemic–metaphysical inference to be based on the brain and its intrinsic features. The epistemic–metaphysical inference is thus a consequence of the way our brains work and may therefore be considered brain-based and part of the natural world.

If, in contrast, the intrinsic epistemic features of the brain do not refer to anything about existence and reality, we are confronted with a different scenario. The epistemic–metaphysical inference cannot then be regarded as a consequence of our brains' workings, and are thus not brain-based and part of our natural world. Instead, the epistemic–metaphysical inference is then mind-based and only part of the logical world (rather than the natural world).

The inference would consequently be valid within the context of philosophy, whereas it would remain irrelevant to neurophilosophy. Why? Both reductive and non-reductive neurophilosophy focus on the natural rather than the logical world. However, unlike reductive neurophilosophy, the non-reductive approach will consider the borders or transitions from the logical to the natural world. How can we do that? For that, yet another type of argument, about the logical possibility of consciousness, is relevant.

Arguments about the logical possibility of consciousness 1

Zombies and the conceivability argument

The 'conceivability argument' is about zombies. What are zombies? As noted on page 357, zombies are imaginary creatures that show the same physical, neuronal and behavioral states as humans, except for consciousness. Zombies are supposed to show no subjective experience, no mental life, and hence none of the phenomenal features of consciousness. Not only do qualia remain absent in the life of zombies, but also self-perspective, unity of self, intentional organization, and inner time and space consciousness are absent.

Despite the absence of any trace of consciousness, the philosophical zombie nevertheless shows exactly the same neuronal states, the same computational processes and the same behavior. Therefore, the only feature differing between zombies and humans is the absence of consciousness. We can thus conceive of zombies in our imagination or in thought experiments. They are logically possible as part of a logical world. This is described by what philosophers call the 'conceivability of zombies'.

Is this philosophical zombie possible and conceivable? The proponents of the conceivability argument, like Thomas Nagel and David Chalmers, say yes. Why? Because we can well imagine in a purely logical way that the same brain and wiring might exist without consciousness. This is a purely logical thought experiment and applies to the world of logical possibilities, clearly distinct from our natural world (see below). Even if they are naturally impossible, zombies may at least be logically possible and thus conceivable in a purely logical world.

What does the conceivability of a zombie tell us about the relationship between brain and consciousness? If we can conceive the absence of consciousness in the presence of the same brain, consciousness cannot be reduced to the brain. Only if consciousness is different from the brain can it remain absent in the presence of the same brain. Hence, the conceivability argument provides support for the irreducibility of consciousness to the brain, and thus a non-physical approach to consciousness and the mind in general.

Let us focus in more detail on the relationship between brain and consciousness. One may imagine that the brain is based on intrinsic features, which define it as brain, but do not imply anything about consciousness. In this case, brain and consciousness may indeed be independent of each other and thus dissociated, as is suggested in the case of the philosophical zombie.

However, another scenario would look different. In this case, the definition of the brain and thus its intrinsic features (those that define the brain as brain) do indeed imply something about consciousness. Consciousness is then necessarily linked to the brain and therefore unavoidably accompanies any brain and its neural activity by default.

What does this mean for the conceivability of zombies? The philosophical conceivability of the zombie depends on logical plausibility. This is not the same thing as the neurophilosophical conceivability of the zombie, which would depend on the definition of the brain and its intrinsic features – in short, whether the definition of the brain and its intrinsic features that define the brain as brain imply anything about consciousness. Hence, the conceivability of a neurophilosophical zombie depends on the definition of the brain.

Arguments about the logical possibility of consciousness 2

Natural and logical worlds

Is the conceivability argument relevant for us? To answer this question requires the shedding of light on the presuppositions that support the conceivability argument, which presupposes the distinction between logical and natural worlds. The natural world is the one we live in, the actual world of us as humans and other species, and is governed by specific physical and biological laws. The natural world is to be distinguished from the logical world, which is a world that does not actually exist. Zombies are only possible as yielded in our thought experiments; they are not really part of our actual world.

The logically possible world(s) are not governed by the physical and biological laws of the natural world, but by the laws of logic, to which our thought experiments must adhere. There are thus no biological and physical constraints to the logical world, only logical constraints. Even if physically and biologically impossible, the logical world may nevertheless be logically possible as long as it is in agreement with the logical laws. Accordingly, the conceivability argument as a purely logical argument about zombies exploits this distinction between biological/physical laws of the actual natural world, and the logical laws of the logically possible worlds.

You may now be confused. Today's neuroscientists tell us that the brain is necessary for consciousness to occur. Even if they do not yet know the exact neuronal mechanisms, it is clear that, without the brain, consciousness remains impossible. How, then, can philosophers argue for the conceivability of zombies? Let's envision the following imaginary dialogue between a philosopher and a neuroscientist.

The philosopher may want to make his point: 'Zombies as human-like creatures without consciousness are possible. I can imagine that there are brains with the same neuronal states as ours that do not yield consciousness.'

'Wait,' the neuroscientist may want to counter, 'that is not in accordance with what the empirical data shows. Lesions in the brain accompany changes of consciousness or, in the vegetative state, even the absence of consciousness.'

'That is all well and good,' the philosopher may respond. 'This concerns only the natural world, the world we live in. The conceivability argument, in contrast, refers to a logically possible world, which is beyond and wider than the natural world.'

'There is only one world – the one we live in,' the neuroscientist stumbles.

'Yes, one actual world, the natural world. But that does not mean that we cannot imagine other possible worlds in our thought experiments. And these possible worlds may go beyond our actual natural world. For instance, in a logically possible world, consciousness may no longer necessarily be tied to a particular brain state. If we can at least conceive that in a purely logically possible world, consciousness cannot be reduced to the brain.'

'I see you distinguish between natural and logical worlds,' the neuroscientist replies. 'But how do you know that the assumptions you make for the logically possible worlds – like the conceivability of a zombie – also apply to our actual natural world? Only if you are able show the link between logical and natural worlds can the conceivability argument become relevant for us as humans. Otherwise, in the absence of such an important link between logical and natural worlds, your conceivability argument remains a purely logical thought experiment that has no relevance for us and our natural world.'

Neurophilosophical discussion 4

Conceivability argument in reductive neurophilosophy: irrelevance of zombies

The neuroscientist is right. In order for the conceivability argument to be relevant in our human world, the step from the logically possible world to the actual natural world must be demonstrated. This is indeed one major focus in the work of David Chalmers, who aims to show the transition from the logical to the natural world. If this can be done, only then will the conceivability argument and its zombies be relevant in our natural world.

If this link is proven it will support the assumption of the irreducibility of consciousness to the brain and disprove the possibility of physical approaches to consciousness. If, in contrast, no such link between natural and logical worlds can be demonstrated, the conceivability of zombies in the logical world may accompany the reducibility of consciousness to the brain, and the assumption of a physical approach to consciousness in the natural world.

What does this mean for philosophy? If the philosopher can show the link between his/her purely logical claim of the conceivability of zombies and the natural world, he/she will be able to extend his/her claim to our actual brain and consciousness. The conceivability argument would then be not only relevant philosophically, that is, in a purely logical world, but also neurophilosophically, that is, in our natural world.

How about the relevance of the conceivability argument and zombies for neurophilosophy? Let us start with reductive neurophilosophy. As detailed in Chapter 4, the reductive neurophilosopher focuses only on the empirical domain that by default implies the natural world. In contrast, the logical world is excluded from the empirical domain, which is based on observation rather than argument as used to support purely logical claims. In short, reductive neurophilosophy is exclusively about the natural rather than the logical world.

How does this stand with regard to the conceivability argument? The conceivability argument and zombies are considered mere logical possibilities that as such presuppose the logical world and, even more important, would remain impossible without such presupposition. If that very same logical world is now excluded from any consideration, as in reductive neurophilosophy, the conceivability argument and zombies become simply impossible. In the same way that neither chair nor table can stand on a floor if there is no floor at all but only a big dark hole, the conceivability argument and zombies become impossible if there is no longer any logical world beside the natural world. In other words, the conceivability argument and zombies are simply irrelevant at best and confused at worst for the reductive neurophilosopher (and the neuroscientist who relies on the same presupposition, the exclusive focus on the natural world).

Neurophilosophical discussion 5

Conceivability argument in non-reductive neurophilosophy: transition of logical zombies from logical to natural worlds

How about the non-reductive neurophilosopher? In contrast to his/her reductive colleague, the non-reductive neurophilosopher takes the distinction between logical and natural worlds seriously. This is reflected in his/her aim to achieve both logical and empirical plausibility. Logical plausibility is about the accordance of the empirical data and facts from the natural world of neuroscience with the logical claims and arguments in the logical world of philosophy. Conversely, empirical plausibility concerns the correspondence of the purely logical claims and arguments in the logical world of philosophy with the empirical data and facts in the natural world of neuroscience (see Chapter 4 for details of both empirical and logical plausibility).

Both empirical and logical plausibility thus directly compare and match empirical data/facts and logical arguments. This is possible only by presupposing the distinction between natural and logical worlds. More specifically, empirical and logical plausibility are possible only if one presupposes some kind of border between natural and logical worlds which, while marking a distinction, can also be crossed and passed at the same time. In addition to the border, there must thus be some border stations which, given proper documents like a passport or, in our case, empirical and logical plausibility, can be passed in both directions.

How can we put this in more concrete terms? David Chalmers' (see below) attempt to provide the transition from the logical world of the conceivability argument to the natural world of our brain can be considered as an attempt to lend empirical plausibility to his otherwise purely logical claim. Metaphorically speaking, he aims to cross the border from the logical to the natural world and uses the passport of empirical plausibility to pass the border guard. The direction is here from logical to natural worlds, and thus from philosophy to neuroscience, so that the purely logical zombies are taken from their logical world to our natural world.

Neurophilosophical discussion 6

Conceivability argument in non-reductive neurophilosophy: plausibility of empirical zombies

How can we support Chalmers? The non-reductive neurophilosopher may want to investigate the brain and its relation to consciousness. Given Chalmers' claim that consciousness and brain can dissociate, as, for instance, in zombies, we may want to search for empirical cases where, for instance, the same lesion in the brain is in one case accompanied by loss of consciousness and in another by preserved consciousness. Technically put, we search for instances that demonstrate dissociation between neural and phenomenal features, and thus between brain and consciousness. In other words we search for nothing less than empirical zombies.

Empirical zombies must show that exactly the same kind of brain function or lesion can be linked to the presence of consciousness or to the absence of consciousness. That would show that the brain and its neural function remain irrelevant to the presence or absence of consciousness. Can we really demonstrate such empirical zombies? In order to demonstrate that, we ultimately need to investigate cases without a brain that nevertheless show consciousness. That, however, remains impossible empirically, since the absence of the brain implies the presence of death. Hence, we encounter a very basic empirical limitation in our search for empirical zombies.

Is the absence of the brain linked to the absence of consciousness? We currently do not know. We equate death with the absence of consciousness. However, we cannot exclude that the mechanisms underlying death and the absence of consciousness are different and thus dissociate from each other. If so, even brain death would not exclude the natural possibility of empirical zombies. Empirical zombies may still be empirically plausible.

This changes if one defines the brain by some kind of intrinsic feature, and link that, as a neural predisposition, with consciousness. The absence of the brain and its particular intrinsic features would then entail the absence of consciousness. In that case, empirical zombies would remain impossible since changes in the brain, that is, in the intrinsic feature as neural predisposition, entail changes in the presence or absence of consciousness.

Empirical zombies would consequently remain impossible and would thus be empirically implausible. The conceivability argument and zombies would then be plausible only within the logical world, that is, logically plausible, but not in the natural world, that is, empirically plausible. They are consequently only philosophically but not neurophilosophically relevant. If so, we can only speak of logical zombies in a meaningful sense, whereas the term empirical zombies would then be not only irrelevant but simply false.

Arguments about the explanation of consciousness 1

Easy problems

After Frank Jackson, who introduced us to Mary the brilliant neuroscientist and the 'knowledge argument', we now encounter yet another philosopher from Australia. David Chalmers studied in the United States before returning to his native Australia. His philosophical home turf is consciousness and, more specifically, what he coined as the 'hard problem' and the 'easy problems'.

Let's start with the 'easy problems', which are not only assumed to be solved easily, but are also much easier to describe than the 'hard problem'. The 'easy problems' concern the relationship between the various cognitive functions and consciousness. The cognitive functions include the ability to discriminate, categorize and react to environmental stimuli, the integration of different information, the reportability of mental states, to access one's own internal and mental states, the focus of attention and the deliberate control of behavior. These may be associated with more specific cognitive functions like attention, working memory, executive functions, etc., which in turn are related to specific neuronal mechanisms and are investigated in cognitive psychology and cognitive neuroscience (see Chapter 3).

The easy problems may therefore ultimately be addressed in cognitive neuroscience and its search for the neuronal underpinnings of cognitive functions and how they relate to consciousness. The main experimental target, here, is how the

Arguments about the explanation of consciousness 2

Hard problem

Those were the easy problems. It becomes much harder when we speak of the 'hard problem'. Rather than targeting the cognitive functions associated with consciousness as the easy problems, the hard problem focuses on how the phenomenal features characterize consciousness – the phenomenal-qualitative and subjective aspects of consciousness, the 'what it is like' (see above). How are these phenomenal features of consciousness linked to the brain and its neuronal features? How and why do the brain and its neuronal features yield phenomenal features and consciousness at all? The brain could instead also yield no consciousness. The guiding question of the hard problem is thus why and how is there consciousness rather than nonconsciousness?

Let's be clear. When formulating the hard problem, Chalmers is not so much interested in the distinction between unconscious and conscious; rather, he focuses on why there are any conscious states at all, as distinguished from nonconscious states. Even unconscious states have at least the potential or possibility of becoming conscious. Both consciousness and unconscious must be distinguished from those states that do not have the potential or possibility of becoming (phenomenally) conscious. These states that do not even have the potential or the possibility of ever becoming conscious may be designated as nonconscious states.

How are these conceptual distinctions related to the hard problem? Chalmers is interested in how the seemingly nonconscious states of the brain can contribute to the generation of states that, in principle, can become conscious; that is, conscious or unconscious. The hard problem is therefore about the distinction between consciousness/unconsciousness and nonconsciousness. The hard problem is not about the distinction between consciousness and unconsciousness, as that is the primary focus of the easy problems and the neuroscientist (see below, as well as Chapters 14 and 15).

Arguments about the explanation of consciousness 3

Easy and hard problems and the brain

Let us reformulate the 'easy problems' and the 'hard problem' within the context of the brain. The 'easy problem' raises the following questions: 'What are the neuronal mechanisms underlying the cognitive functions associated with consciousness?' and 'What are the neuronal mechanisms that allow for unconscious and conscious states to be differentiated?' These problems are easy because they rely on cognitive neuroscience and its experimental investigation of the neuronal mechanisms underlying cognitive functions and the distinction between conscious and unconscious.

This must be distinguished from the question guiding the 'hard problem': 'What and how do these features in the brain make it possible to associate, in principle, the brain's neuronal states with phenomenal states – phenomenal consciousness, rather than non-phenomenal states?' The 'hard problem' thus asks for the what and how of those neuronal mechanisms that allow and make possible the assignment of phenomenal states, including their 'what it is like', to the brain's neuronal states.

Rephrased in this way, the 'hard problem' ultimately raises the question: Why and how are the brain's neuronal states associated at all with experience and consciousness, rather than not? Since our mind can be characterized by consciousness as its hallmark feature, the hard problem is about the mind in general. If we remain unable to address the hard problem, we will therefore not be able to understand the mind fully. There is thus much at stake when it comes to the hard problem.

Arguments about the explanation of consciousness 4

Explanatory gap problem

The easy and hard problems are about our explanation of consciousness; in short, how we can explain the relationship of cognitive functions and phenomenal features to consciousness. We can explain consciousness in, for instance, purely phenomenal and thus mental terms by assuming specific mental processes and functions. We can then remain purely within the realm of the mental without requiring any reference to the brain and its neuronal states. The same may be possible on the level of the brain, which – analogously – can be explained in purely neuronal terms without any reference to the phenomenal and mental realm.

However, when taken together, neuronal and phenomenal explanations cannot be linked. There remains a gap between neuronal and phenomenal explanations. Why is there this gap, and how can we define it? This is what the philosopher Joseph Levine described as the 'explanatory gap problem'.

What is the explanatory gap problem about? Levine argues that there is a gap in our explanation: on the one hand, we can explain the physical and neuronal states of the brain quite well, the 'how it works' (see above). However, this leaves open the explanation of the phenomenal states, and thus the 'what it is like' of consciousness. In particular, we cannot explain how the 'how it works' of the brain is related to the 'what it is like' of consciousness, and thus how the former explains the latter. There is thus a gap in our explanations, which Levine describes as the 'explanatory gap'.

How does the explanatory gap problem relate to the hard problem? Like the 'hard problem', the 'explanatory gap problem' relies on the distinction between consciousness and nonconsciousness. Our explanations of the brain and its nonconscious neuronal states leave open a gap with regard to the explanation of the phenomenal features of consciousness. The phenomenal explanations about consciousness do not imply anything about the neuronal features of the brain, nor do the neuronal explanations of the brain imply anything about the phenomenal features of consciousness. Very much like the hard problem, the explanatory gap problem presupposes and taps into the difference between consciousness and nonconsciousness. More specifically, the explanatory gap problem argues that there is a gap between our explanations of the brain and our explanations of experience/consciousness.

Arguments about the explanation of consciousness 5

Explanatory gap problem as epistemological or metaphysical claim

What is the difference between the hard problem and the explanatory gap problem? While both focus on the difference or gap between consciousness and nonconsciousness, phenomenal and neuronal, they presuppose distinct aspects. The hard problem concerns the gap itself, the gap between neuronal and phenomenal states and thus between nonconsciousness and consciousness, while the explanatory gap problem targets our explanations and, more specifically, the relationship between our explanations of the two sides of the gap – phenomenal and neuronal explanations.

Let us reformulate the difference between the two problems. The hard problem asks: why are there phenomenal states at all, rather than non-phenomenal states? The hard problem is thus about the gap itself and why there is such a gap between consciousness and nonconsciousness. This is different in the explanatory gap problem, which is not so much concerned with the gap itself but rather with our explanations of the gap. The explanatory gap problem thus raises the questions: why is there a gap between our explanations of neuronal and phenomenal features? And how can we bridge this explanatory gap in our explanations of brain and consciousness?

The explanatory gap may be considered primarily within the epistemological domain, the domain of knowledge. The knowledge about the brain's 'how it works' may differ in principle from the knowledge about the 'what it is like' of consciousness. Both types of knowledge may require different abilities, which implies an epistemological gap. Even if the processes and mechanisms underlying the brain's 'how it works' and the 'what it is like' of consciousness are identical, the epistemic abilities and thus our knowledge of them may differ, as manifested in neuronal and phenomenal explanations. One may consequently want to speak of an 'epistemological explanatory gap'.

In addition, one may put the explanatory gap in an ontological (or metaphysical) context. One may raise the question: why there is an empirical and/or epistemological explanatory gap to begin with? Since metaphysics is the philosophical discipline dealing with questions of existence and reality, one may speak here of a 'metaphysical explanatory gap'. The 'metaphysical explanatory gap' raises the question of why and how there is an explanatory gap at all; it is thus about the existence and reality of the explanatory gap and is therefore metaphysical.

The 'metaphysical explanatory gap' needs to be distinguished from the explanatory gap in the epistemological context that is rather about a gap in our explanations and thus in our knowledge. The explanatory gap in the epistemological context; that is, the 'epistemological explanatory gap' is supposed to be manifested in our knowledge. Whether such a gap in our knowledge, the epistemological explanatory gap, corresponds also to a gap in existence and reality, the metaphysical explanatory gap, is a second and more or less independent question. Accordingly, depending on the presupposed domain, whether epistemological or metaphysical, we need to distinguish epistemological and metaphysical explanatory gaps from each other.

I focus here predominantly on the epistemological explanatory gap. Below I will consider the explanatory gap as a primarily epistemological argument that argues for different types of knowledge: phenomenal and neuronal. I will refrain, however, from regarding the explanatory gap argument as a metaphysical argument, which has been addressed in detail in Part II about the mind–brain problem.

Neurophilosophical discussion 7

Hard problem and the brain's intrinsic features

Why is the hard problem central for the current mind–brain discussion, and especially for the neuroscientist? The hard problem can be considered a central criterion for a neuroscientific approach to consciousness that must be solved in order for it to be regarded as successful. In other words, a neuroscientific solution to the question of consciousness must include an answer to the hard problems. The neuroscientist must thus show why the brain, on the basis of what kinds of features, yields consciousness rather than nonconsciousness. If the neuroscientist leaves out the answer to this question, the neuronal explanation of consciousness will be considered incomplete.

How can the neuroscientist find an answer to the hard problem? Currently, neuroscientists seem to focus more on the easy problems than the hard problem. They search for the cognitive functions underlying and associated with consciousness, and how they allow for the distinction between conscious and unconscious states. The guiding question here is: what are the cognitive functions and their underlying neuronal mechanisms that allow for the distinction between conscious and unconscious states? This reflects the design and set-up of the easy problems. In contrast, the current neuroscientific focus is not so much on the distinction between phenomenal and non-phenomenal states, and hence between consciousness and nonconsciousness, as outlined in the hard problem. For that, the neurophilosopher may need to step in. How can the neurophilosopher shift the neuroscientist's focus from the easy problems to the hard problem? You'll recall from above that in order to link the brain to consciousness, we need to find some kind of feature in the brain itself that makes the constitution of consciousness, as distinct from nonconsciousness, necessary and unavoidable.

What does such a feature in the brain – one that makes consciousness as distinct from nonconsciousness possible – look like? If such a feature of the brain makes the constitution of consciousness, as distinct from nonconsciousness, necessary, and thus unavoidable, it may be an intrinsic feature that defines the brain as brain. The brain constitutes consciousness rather than nonconsciousness. Thus, there must be some feature intrinsic to the brain that defines the brain as brain and thereby makes the possible constitution of consciousness necessary and unavoidable.

The guiding question of neurophilosophy may thus be: what are the brain's intrinsic features that define the brain as brain and thereby make the constitution of consciousness unavoidable? We unfortunately do not know what the brain's intrinsic features are currently. We discussed the brain's intrinsic activity, resting state activity, as distinguished from its extrinsic activity, stimulus-induced activity (see Chapter 12). However, nothing in the brain's intrinsic activity entailed and implied the constitution of consciousness. We must continue to search for further features in the brain's intrinsic activities that make the constitution of consciousness necessary.

Metaphorical comparison

Intrinsic features in heart and brain

Let's compare the situation to the heart. We know the intrinsic feature that defines the heart as heart. The heart can be defined as heart by virtue of its pumping function – pumping blood throughout the whole body. This means that the heart cannot avoid pumping blood because otherwise it would no longer be defined as heart. How does the heart realize this pumping function? By means of its structure and organization as a muscle, which is thus an intrinsic feature of the heart. The same might hold true in the case of the brain. The neurophilosopher may want to assume some kind of intrinsic feature in the brain that defines the brain as brain. Analogous to the muscle organization that makes the heart's pumping function possible, even necessary and unavoidable, the yet unknown intrinsic feature in the brain may also make the constitution of consciousness, rather than nonconsciousness, necessary and unavoidable. What could such an intrinsic feature look like? One may tentatively assume it to be related to the intrinsic activity in the brain that possesses a certain yet unknown structure and organization.

This intrinsic feature may make it necessary to constitute consciousness rather than nonconsciousness. Accordingly, the intrinsic feature of the brain may then provide the answer to the hard problem: why there is consciousness rather than nonconsciousness. This is analogous to the case of the heart, where the heart's muscle organization provides the answer to the question of why there is pumping rather than non-pumping.

Neurophilosophical discussion 8

Lack in our current knowledge

The explanatory gap problem is about the gap or difference between our neuronal explanations and phenomenal explanations. Even if both neuronal and phenomenal explanations refer to one and the same feature – as in consciousness – there remains nevertheless a gap between the explanations. Why? Because the phenomenal explanations do not imply anything about the neuronal explanations, nor do the neuronal explanations imply anything about consciousness.

The neurophilosopher may now take a twofold approach. First, he/she may want to search for those features that define the brain as brain, the intrinsic features. This may be followed by an investigation of whether the brain's intrinsic features, and our purely neuronal explanations of them, imply anything for consciousness and phenomenal explanations, which might allow us to, at least methodologically, bridge the gap and search for the 'right' kinds of neuronal explanations that refer to and imply consciousness and our phenomenal explanations.

In addition, the neurophilosopher may also look for the necessary neuroepistemic conditions that first and foremost make the explanatory gap problem possible at all. Maybe our brain is designed in such way that we cannot avoid the gap in our explanations, and thus the explanatory gap; the explanatory gap may then by itself be necessary and unavoidable (at least in an empirical sense).

Let us start with the first point: how to bridge the explanatory gap in refining our neuronal explanations. For that we may want to compare our situation again with that of the heart. What would such an explanatory gap look like, in the case of the heart? Imagine that we knew everything about the heart and its biochemical and electrical processes. The only thing we don't know is that it is a muscle structure and that muscles have the ability to pump.

In that case we would be confronted with two different explanations: the physiological explanation about the heart itself, and the engineering explanations about its pumping functions. And because of the lack of our knowledge about the heart itself and its intrinsic features – its muscle structure – we remain unable to close the gap between the physiological and engineering explanations. There is thus an explanatory gap in our explanations of the heart.

This is the situation we currently face in the case of the brain and consciousness. We have plenty of physiological, neuronal explanations of the brain, and in addition to these we have phenomenal explanations about consciousness as the equivalent to the heart's engineering explanations about its pumping function. However, there is no linkage or connection between the neuronal and phenomenal explanations. Why? Nothing in our current neuronal explanations refers to and implies anything about consciousness. Hence, we currently lack the knowledge of the brain's equivalent to the heart's muscle structure.

Neurophilosophical discussion 9

Explanatory gap as brain-based problem

What is the therapeutic advice for the neuroscientist by his/her psychiatrist, the neurophilosopher? He/she, the neuroscientist, will extend his/her neuronal explanations by searching for those neuronal features in the brain, the intrinsic features that imply and refer to the phenomenal features of consciousness. Analogous to the heart's organization as muscle, he/she should search for some features concerning the general structure and organization of the brain's neural activity, rather than some of its physiological details. If he/she does so, the neuroscientist's neuronal explanations will, one hopes imply some reference to consciousness.

In addition to the kinds of intrinsic features that may bridge the explanatory gap between neuronal and phenomenal explanations, the neurophilosopher may also want to search for necessary neuronal conditions that enable or predispose us to the gap in our explanations. For that, the neurophilosopher may want to investigate the neuronal and neuroepistemic mechanisms underlying our ability to make both neuronal and phenomenal explanations.

There may not be much neuronal overlap between the respectively neuroepistemic mechanisms underlying neuronal and phenomenal explanations. If so, the brain itself and its specific neuronal design may make the occurrence of an explanatory gap in our explanation of consciousness necessary and thus unavoidable. In other words, the explanatory gap problem may then be regarded as a brain-based problem (see Part III for brain-based concepts).

Take-home message

So far we have discussed the concept of neurophilosophy (Part I), the mind–brain problem (Part II), and philosophical problems in psychology and neuroscience

(Part III). Part IV addresses concrete mental features like consciousness that characterize the mind. Methodologically, the investigation of consciousness may be 'located' at the interface between conceptual and empirical issues and, more generally, between metaphysical, epistemological and empirical domains. This chapter focuses on theoretical arguments provided by the philosophers who are against a neuroscientific, physicalistic explanation of consciousness. These problems include the absent qualia and the inverted spectrum problem, the 'what it is like' and the knowledge problem, the conceivability problem, and the hard problem and the easy problems. These arguments against a neuroscientific approach to consciousness provide useful ideas and criteria that need to be addressed in future neuroscientific and neurophilosophical explanations of consciousness. It is also clear that these arguments may have different explanatory powers and validity, depending on the respective context: whether these arguments are discussed within the purely logical context of philosophy, or whether they are framed within the natural context of the actual world, as presupposed in neurophilosophy and neuroscience.

Summary

This chapter focuses on various arguments that are put forward from philosophy to discuss the obstacles neuroscience must overcome to explain the phenomenal features of consciousness by the neuronal features of the brain. More generally, these arguments are arguments against a neuroscientific and thus physical approach to the mind and the mind–brain problem. The first set of arguments is about the phenomenal features of consciousness. There is the absent qualia argument that argues for the possible absence of qualia. The inverted spectrum argument claims that the same contents can be associated with different experiences, e.g. phenomenal-qualitative states, as signified by 'what it is like'. Other arguments are more about our knowledge of consciousness and its phenomenal features. These include the 'what it is like argument' and the 'knowledge argument'. Both arguments claim that consciousness and its phenomenal features require a specific and different type of knowledge when compared to knowledge about the brain and its neuronal features. The main difference consists in that knowledge about consciousness is based on the first-person perspective while knowledge about the brain is third-person-perspective based. Another line of argument pertains to the logical possibility and conceivability of zombies within the logical world, and the implication of this for the occurrence of consciousness within the natural world. This concerns mainly the conceivability argument that infers from the logical imaginability of zombies the irreducibility of consciousness to the brain. Finally, arguments targeting the explanation of consciousness are discussed. These include the hard problem and the explanatory gap problem. The 'hard problem' raises the question: Why is there consciousness at all, rather than nonconsciousness? This presupposes the distinction between consciousness and nonconsciousness that will be addressed in further detail in the next chapter. The other central argument here is the 'explanatory gap problem' that describes a gap between our neuronal explanations of the brain and the phenomenal explanations of consciousness. Both the hard problem and the explanatory gap problem can be regarded as set standards or criteria for future neuroscientific explanations of consciousness. Only if both arguments, the hard problem and the explanatory gap problem, can be answered satisfactorily can consciousness be explained completely in neuroscientific terms, implying ultimately a purely physical approach to the mind. Finally, the differences between philosophical and neurophilosophical approaches are pointed out, especially with regard to the hard problem and the explanatory gap problem.

Revision notes

- Why is the possible reduction of consciousness to the brain so important? Why can consciousness no longer simply be associated with a mind?
- Describe the different types of arguments against the irreducibility of consciousness to the brain?
- What is the absent qualia argument? How does it work, what does it argue for, and what are its presuppositions?
- How does the inverted spectrum argument differ from the absent qualia argument? What are the implicit presuppositions of the inverted spectrum argument?
- What types of knowledge can be distinguished within the context of consciousness and the brain?
- What do the 'knowledge argument' and the 'what it is like' argument argue for? How can they be distinguished from each other?
- What are zombies, and why are they relevant?
- How are logical and natural conditions and worlds defined? How can you infer from logical to natural conditions, as implied by the conceivability argument?
- What is the 'hard problem' and how is it to be distinguished from the 'easy problems'?
- Describe what the explanatory gap consists of. Are there different types of explanatory gaps?

Suggested further reading

- Chalmers, D. (1998) 'The problems of consciousness', in H. Jasper et al. (eds), *Consciousness at the Frontiers of Neuroscience: Advances in Neurology*, 77, 7–19.
- Jackson, F. (1982) 'Epiphenomenal qualia', *Philosophical Quarterly*, 32, 127–36.
- Levine, J. (1983) 'Materialism and qualia: The explanatory gap', *Pacific Philosophical Quarterly*, 64, 354–61.
- Nagel, T. (1974) 'What is it like to be a bat?', *The Philosophical Review*, 83(4), 435–50.
- Searle, J. (2004) *Mind: A Brief Introduction* (Oxford/New York: Oxford University Press).

14
Neural Correlates of Consciousness (NCC)

Overview

The previous chapter focused on various philosophical arguments against reducing consciousness and its phenomenal features to the brain and its neuronal features. This left open the neuroscientific mechanisms underlying consciousness. This is the focus of the present chapter. We will discuss various suggestions and hypotheses for neuronal mechanisms that may be central in yielding consciousness. In short, the main focus is on the neuroscience of consciousness. This will be complemented by a discussion of the empirical shortcomings of the present neuroscientific approaches. Finally, we will compare the neuroscientific suggestions about consciousness with the philosophical arguments against the reduction of consciousness to the brain, as discussed in Chapter 13.

Objectives

- Learn about the neuroscience of consciousness
- Understand the different neuronal mechanisms that possibly underlie consciousness
- Learn more about the neuronal mechanisms and how they can be investigated experimentally
- Be able to distinguish empirical and conceptual issues, while at the same time linking them in a non-reducible way
- Make a comparison between neuroscientific mechanisms and conceptual arguments about consciousness

Key concepts

Contents of consciousness, neural correlates of consciousness (NCC), re-entrant circuits, neuronal synchronization, global workspace, cognitive functions, neuronal and experimental specificity.

Background 1

Phenomenal features of consciousness as starting point in philosophy

In the previous chapters, we touched upon the phenomenal features of consciousness. These included 'what it is like' as a core feature that was specified as a 'point of view' with phenomenal-qualitative features. These are also known as qualia. In addition, consciousness can be characterized by unity, spatial and temporal continuity, and self-perspectival and intentional organization. These features will be discussed in greater detail in Chapter 16. They are features of our subjective experience and are therefore called 'phenomenal features'.

These phenomenal features must be distinguished from the neuronal features of the brain. The brain's neuronal features can be observed in third-person perspective. This means that they are regarded as objective features. In contrast, the phenomenal features can only be experienced in first-person perspective, so they are considered subjective features. Most important, phenomenal features are associated with consciousness, implying that they cannot be found in the brain and its neuronal states. It also means that, conversely, the brain's neuronal features cannot be located in subjective experience or in consciousness.

How, though, can we investigate consciousness in neuroscience, if it cannot be observed in the brain and its neuronal states? Neuroscience presupposes a third-person perspective and the objective investigation of the brain. Therefore, its starting point cannot be the phenomenal features of consciousness, as these are subjective and tied to the first-person perspective.

Background 2

Neural correlates of the contents of consciousness as a starting point in neuroscience

What are the starting points for the neuroscientific investigation of consciousness? One is the content of consciousness. For example, we are conscious of a book and its red color. The book is thus the content of consciousness, also known as the phenomenal content. Our consciousness is always 'consciousness about contents'. These can include events, people, or objects in the environment. Alternatively, the contents of consciousness can consist in our own thoughts or some imaginary scene about people, objects, or events in our dreams when we sleep. Hence, phenomenal contents must be considered a central dimension of consciousness.

These contents can now be taken as the starting point for neuroscientific investigation. Neuroscience can, for instance, investigate the neuronal differences between those contents that are conscious and those that remain unconscious. How can we illustrate this further? The unconscious content may still exert an impact on our behavior, as in, for instance, when you avoid snakes because you were poisoned and almost died from a snake bite as a young child. When avoiding snakes as an adult, you may not be aware that the event in your past makes you avoid the path through the forest. However, your unconscious may become conscious, as you realize that it is because of snakes that you avoid certain activities like camping, hiking and walking in the woods.

This example illustrates that the same content – like a particular event, person, or object – can be presented both in a conscious mode with subjective experience, and in an unconscious mode. The neuronal difference between the conscious and the unconscious modes must then be related to consciousness. This is denoted by the term 'neural correlates of consciousness' (NCC).

Biography 1

Francis Crick

The neuronal difference between conscious and unconscious contents must signal the neuronal conditions that are sufficient for consciousness to occur. Now is an appropriate time to introduce Francis Crick (and Christof Koch). Francis Crick became famous when he received the Nobel Prize, together with James Watson, for discovering DNA as the genetic code. After making this discovery, he then worked on the neural code of consciousness. Together with his collaborator, Christof Koch, he was searching for the neuronal mechanisms underlying consciousness.

Crick and Koch introduced the term neural correlates of consciousness. Technically speaking, NCC describes the search for the minimum neuronal conditions that are jointly sufficient for any specific content to be associated with consciousness. In other words, NCC refers to those neuronal mechanisms that account for the difference between unconscious and conscious contents. What are the neuronal mechanisms that allow for particular contents to be associated with consciousness? Several neuronal mechanisms have been discussed as possible candidate mechanisms for the NCC. The main ones are presented below.

Neural correlates of consciousness (NCC) 1

Gamma oscillations and neuronal synchronization

Crick and Koch strongly favor neural synchronization as a candidate mechanism for consciousness. Neural synchronization describes the temporal coordination and integration of neural activity changes across different brain regions. Neural activity is not static. Instead it is dynamic and fluctuates in different rhythms. These rhythms or fluctuations can occur rather slowly, in the low frequency range (1–4 Hz), or even lower in the range of 0.001 to 0.1 Hz, as in the resting state of the brain (see Chapter 12 for details). The rhythms can also be fast, as in, for instance, the range of 30–40 Hz, the gamma range. This range can be observed especially during stimulus-induced activity.

Such neuronal synchronization allows the neural activity of different regions to be coordinated, which in turn makes it possible to tie and link the stimuli they process. In other words, the stimuli processed in the different regions are linked together via the neuronal synchronization of their underlying regions. The neurons thus form a 'neural coalition', resulting in what is described as 'binding by synchronization'.

'Binding by synchronization', under observation, seems to be associated especially with gamma synchronization in the range of 30–40 Hz. This gamma synchronization was initially observed in the visual cortex 20–30 years ago by the German neuroscientist Wolf Singer. In the visual cortex, it seems to bind together the different visual features like motion, color, shape, etc. of visual stimuli. Since then, the gamma oscillations have been observed in other regions and functions – ranging from other sensory cortices and the prefrontal cortex to include sensory, motor and cognitive functions.

How is this binding by synchronization related to consciousness? Consciousness is about contents. Rather than experiencing a single stimulus in isolation, we experience different stimuli together in one content. Since the binding of different stimuli is essential for constituting the contents of consciousness, Crick and

Koch assume the often-observed gamma synchronization as a neural correlate of consciousness. Without gamma synchronization, consciousness remains impossible. This implies that gamma synchronization must be considered a sufficient condition of consciousness, also known as a neural correlate of consciousness.

Biography 2

Gerald Edelman

Gerald Edelman is an American neuroscientist who, like Francis Crick, was awarded a Nobel Prize for his work in immunology, and for his discoveries concerning the mechanisms operating in the immune system. The immune system is responsible for producing biochemical substances that protect and defend the body from foreign and potentially dangerous substances. After he discovered the immunological code, Edelman, much like Crick, also turned to the brain in the hope of revealing the neural code of consciousness.

Edelman (2003, 2005) considers cyclic processing, and thus circularity, within the brain's neural organization as central for constituting consciousness. Cyclic processing describes the re-entrance of neural activity in the same region after looping and circulating through other regions in so-called re-entrant (or feedback) circuits.

This is the case in the primary visual cortex (V1): the initial neural activity in V1 is transferred to higher visual regions, such as the inferotemporal cortex (IT), in feed-forward connections. From there, it is conveyed to the thalamus, a subcortical region, which relays the information back to V1 and the other cortical regions. This implies thalamo-cortical re-entrant connections. Consciousness is assumed to be constituted on the basis of such feedback or re-entrant connections that allow for cyclic processing.

Neural correlates of consciousness (NCC) 2

Re-entrant processing and information integration

What is the purpose of the feedback or re-entrant circuits? Re-entrant circuits integrate information. This led Guilio Tononi, an Italian researcher, who now lives and works in the USA, to emphasize the integration of information as a central neuronal mechanism in yielding consciousness. He developed what he calls 'Integrated Information Theory' (IIT). The IIT postulates that the degree of information that is linked and integrated is central for consciousness.

The integration of information can, for instance, be achieved by establishing functional connectivity between different regions. Functional connectivity describes the temporal coordination between different regions' neural activities. Increased functional connectivity between different regions may allow us to link together the different regions' information, and can therefore be regarded as a measure of increased information integration. If, in contrast, functional connectivity between the different regions is low and disrupted, their respective information can no longer be integrated. Tononi assumes that low information integration makes consciousness impossible.

The re-entrant circuits, and especially the functional connectivity between thalamus and cortex – as postulated by Tononi's former supervisor, Edelman, seem to play a central role in integrating information. If they are disrupted, information integration is low, and it is assumed that consciousness can no longer be generated.

How does Tononi support his assumption? He investigated various disorders of consciousness, including the vegetative state, sleep (especially NREM sleep) and anesthesia, all of which can be characterized by reduced or absent consciousness (see Chapter 16 for further details). Despite their differences, he found, in all three examples, reduced functional connectivity throughout the brain, and in particular in the thalamo-cortical re-entrant connections. All three cases of decreased or nonconsciousness suffered from decreased information integration, which Tononi assumes to be central in their decrease in consciousness.

Why are the thalamo-cortical re-entrant connections so central to consciousness? These re-entrant connections process all kinds of stimuli independent of their origin and their respectively associated contents. What, then, do these thalamo-cortical re-entrant connections do? According to Tononi, they allow the brain to generate particular qualia, which can be associated with the various contents, and become conscious. If, in contrast, the functional connectivity, and thus the thalamo-cortical re-entrant processing, is low or completely disrupted, and the contents can no longer be associated with qualia. Consequently, the addition of the specific quality, the qualia, remains impossible. This makes it impossible for the contents to achieve consciousness.

Neural correlates of consciousness (NCC) 3

Global workspace

Another suggestion for a neural correlate of consciousness comes from other neuroscientists like Bernard Baars and Stanislas Dehaene. They assume a global distribution of neural activity across many brain regions, in a so-called global workspace, to be central for yielding consciousness. The information and its contents processed in the brain must be globally distributed across the whole brain in order for them to become associated with consciousness.

If information is only processed locally, within particular regions, but not globally throughout the whole brain, according to Baars and Dehaene, it cannot be associated with consciousness. The main distinction between unconsciousness and consciousness is then supposed to be manifested in the difference between the local and the global distribution of neural activity. Hence, the global distribution of neural activity is considered a sufficient condition, and thus neural correlate, of consciousness.

Dehaene and Jean-Pierre Changeux propose what they describe as a 'global neuronal workspace theory'. They assume that the prefrontal–parietal network has to be recruited by the stimulus, which then allows the linkage between the different networks, and consequently the global processing of the stimulus.

They also assume that the brain's intrinsic activity plays a central role in consciousness. The timing of the stimulus relative to the ongoing spontaneous phase fluctuations is central here. Why? Depending on the timing, the stimulus may or may not lead to the recruitment of the fronto-parietal neurons and network, which is, in turn, central in allowing for consciousness.

If, for example, the spontaneous firing activity in the fronto-parietal network is too strong and continuous, it can block, and thus prevent, its ignition by the external stimulus. Since Dehaene and Changeux assume the fronto-parietal network to be a global neuronal workspace that is necessary for consciousness, the stimulus may consequently be 'denied' conscious access, and thus remain unconscious or pre-conscious (see below for conceptual details).

Neurophilosophical discussion 1

Specificity of the NCC

The above suggestions for the NCC are hypotheses about the neuronal mechanisms underlying conscious contents. This raises the question of whether the suggested neuronal mechanisms are specific for that particular content. Or whether the same neuronal mechanisms can be associated with other contents too.

The NCC must hypothesize a specific neuronal mechanism like re-entrant circuits or synchronization. Most important, the suggested neuronal mechanism must be related specifically to consciousness, rather than other cognitive (and non-cognitive) functions, otherwise the candidate neuronal mechanisms cannot be regarded as specific for consciousness. Hence, the NCC must show what I call 'neuronal specificity'.

Second, the NCC must be able to give concrete suggestions for the experimental testing of the neuronal mechanisms in question. The NCC must also specify the kind of methodological approaches required for such experimental testing. This includes a valid and reliable test to measure consciousness. In the case of the NCC, this is most often the content of consciousness, which is a rather problematic dimension to use as a measure for consciousness.

One would prefer to have a more valid – i.e. a measure of what they are supposed to measure – and reliable – i.e. stable measures across different experimental settings and investigators – measure for consciousness, a so-called 'consciousness-meter'. Only if we have valid and reliable measures that are specific for consciousness are we able to specify our experimental designs with regard to consciousness. In other words, the NCC needs to show what can be described as 'experimental specificity'. One would like to have different experimental measures for the different phenomenal features of consciousness like qualia, unity, etc. (see below and Chapter 16), which would be phenomenally specific.

Neurophilosophical discussion 2

Neuronal and experimental specificity of 're-entrant connections'

Let's start with the neuronal specificity of the various suggestions for the NCC. The first suggestion was the crucial role of feedback or re-entrant circuits – like the thalamo-cortical connection – in generating consciousness. This raises the question: what is the role of the feed-forward connections, the connections that lead from input regions like the visual cortex to higher regions like the inferotemporal cortex (IT)?

Are feedback or re-entrant circuits critical? Can they be regarded as sufficient for consciousness? Koch and Tononi doubt that they can. There is, at present, no sufficient experimental support demonstrating that without feedback or re-entrant circuits consciousness remains impossible. Even in the absence of feedback or re-entrant circuits, consciousness may already be possible.

Koch and Tononi assume that the feedforward connections from V1 to IT may be sufficient in themselves to induce consciousness. This means that the feedback or re-entrant connections may not be specific for consciousness. Instead, they may be specific for some as yet unknown function that remains independent of (though eventually based on) consciousness. If so, the re-entrant connection must be regarded as unspecific for consciousness and should be distinguished from other cognitive functions. This fact would make re-entrant connections neuronally unspecific.

Are the feed-forward connections from V1 to IT specific for consciousness? This can be assumed if they are not involved in unconscious processing and states. However, this is rather unlikely, since even during unconscious states the visual information is conveyed from V1 to IT. Hence, even the feed-forward connections may not be specific for consciousness. If this is the case, the V1–IT connections may not be neuronally specific for consciousness.

In addition to neuronal specificity, one may raise questions about experimental specificity. We currently lack neuronal measures of re-entrant connections that clearly distinguish them from feedforward connections. To distinguish between them we would need to have separate neuronal measures for both feed-forward and re-entrant connections that illustrate how they remain independent of each other. While this is an experimental-methodological requirement, it may be difficult to achieve neurophysiologically. This is because the activation of the re-entrant connections may presuppose prior input from the feed-forward connections, which puts their experimental independence into doubt. In short, there is experimental unspecificity.

Furthermore, we also need to account for experimental measures that allow access to specifically phenomenal consciousness, as distinguished from access to consciousness (and the reverse). What is phenomenal consciousness? Phenomenal consciousness, as described in the previous chapter, is characterized by 'what it is like'. In contrast, access to consciousness is characterized by detection, reporting and access; this may be related more closely to cognitive functions like attention, working memory, etc. (see below). Hence, future investigation will need to target specific experimental measures for phenomenal consciousness, as distinguished from access consciousness and cognitive functions.

Neurophilosophical discussion 3

Specificity of the global workspace and neuronal synchronization

What about the global workspace theory? This theory assumes a widespread globalization of neural activity throughout the whole brain as essential for inducing consciousness. If, in contrast, there is no such extension, consciousness cannot be induced. This means that the respective content would remain unconscious. Hence, the global distribution of neural activity, as distinguished from local or regional localization of neural activity, is considered specific for consciousness. Is the degree of globalization of neural activity thus a specific neuronal measure of consciousness?

Is globalization really specific for phenomenal consciousness and 'what it is like'? Or is it specific to the integration and accessibility of information, as associated with access consciousness? In order to access the contents of consciousness for subsequent reporting, detection and awareness, the content must be linked to the respective cognitive functions. This may indeed be made possible by the globalization of neuronal activity.

What does this imply for the neuronal specificity of consciousness? Since the cognitive functions are associated with access consciousness, the globalization of neural activity may be specific for accessing consciousness and the integration of cognitive information. This leaves open whether the globalized neural activity is really necessary and sufficient for phenomenal consciousness and 'what it is like' to occur.

What about synchronization as another suggestion of NCC? Neuronal synchronization is assumed to be critical for both the binding of different stimulus features and consciousness. However, as Koch and Tononi themselves point out, it is not clear whether such neuronal synchronization is really critical for consciousness to occur. Neuronal synchronization may occur in the brain, independent of consciousness. As a result, it allows different stimuli with 'integration remaining', though independent of consciousness altogether, to bind and integrate.

Consequently, one may need to distinguish between binding and consciousness: neuronal synchronization may be critical for the binding of different stimuli into one content or percept. This, though, does not mean that neuronal synchronization is also critical for the association of that content with consciousness. Hence neuronal synchronization may be specific for the linkage and binding of different stimuli into one content, while it may lack neuronal specificity for consciousness.

Neurophilosophical discussion 4

'Brain without oscillations' and the brain's intrinsic features

Another question is the consideration of low frequency fluctuations. Neuronal synchronization focuses predominantly on higher frequency fluctuation in the gamma range (30–80 Hz). This raises questions about the role of lower frequency fluctuations, lower than 30 Hz during the synchronization of higher ones.

What is the role of the low frequency oscillations in the range 1–4 Hz and lower (0.001–0.1 Hz) during the synchronization of higher frequency oscillations? The data suggest that the timing and power of higher frequency synchronizations (like gamma with 30–40 Hz) are dependent on the lower frequency fluctuations (like delta with 1–4 Hz).

What may be neuronally specific for the binding of stimuli into contents, is no longer the gamma synchronization, but rather its temporal relationship to low frequency fluctuations. How the high–low frequency relationship relates to the occurrence of consciousness remains unclear at this point. The neuronal specificity of the high–low frequency relationship with regard to consciousness is yet to be explored.

How about experimental specificity? To demonstrate experimentally the critical role of synchronization in the gamma range for consciousness, we would need to investigate a brain without synchronization. This leads us to the Hungarian-American scientist Görgy Buszáki and his impressive book, *Rhythms of the Brain*

(Buszáki, 2006). While Buszáki remains unsure whether he embedded his empirical data into a novel theory, he is certain that the brain's neuronal activity is embedded in continuous fluctuations or oscillations. He considers it an intrinsic feature of the brain that defines the brain as brain. A brain without synchronization, and thus a 'brain without oscillations', remains an impossibility and cannot even be imagined. The problem remains, however, that we would need such a 'brain without oscillations' to demonstrate experimentally the causal role of neuronal oscillations and synchronization for consciousness. There are thus principal constraints and limits to the degree of experimental specificity we can achieve.

One may, however, approach this issue in an indirect way. For example, certain states with impaired consciousness, including vegetative states, patients under anesthesia, and particular stages during sleep (i.e. the non-REM stages where we do not dream), show predominantly low frequency oscillations (1–4 Hz and lower). At the same time, higher frequency oscillations are weaker, if not absent, in these states. This hints indirectly at the possible relevance of the relationship between high and low frequency oscillations for consciousness. These data may then serve to investigate the causal relationship between low and high frequency oscillations and the relevance of that relationship for consciousness. Before we pursue this strategy in the next chapter, we need to discuss briefly some of the other suggestions for a possible candidate for the NCC.

Neural correlates of consciousness (NCC) 4

Cognitive functions: working memory and attention

The NCC candidates discussed so far emphasized the need for integration and coordination throughout the whole brain. As a result, they may be considered global approaches to consciousness that do not emphasize specific functions as being central. In contrast, other neuroscientific authors assume specific cognitive functions – like working memory, attention, or executive functions – to be central.

Let us start with working memory. Working memory describes the online manipulation and maintenance of stimuli in short-term memory. Access to short-term memory may be essential for the stimulus and its associated content to become conscious. Why? Because we need some time to process the stimuli in such a way that they can be associated with consciousness. Since working memory has been associated with neural activity in the prefrontal cortex, this region has been considered central in yielding consciousness.

Another cognitive function that is highly debated as a possible candidate for consciousness is attention. Attention allows us to focus on particular stimuli and contents. For instance, while reading this book, I focus on the book itself, as distinguished from the table and the flowers standing in a vase on the table. I also do not focus on the birds singing outside, but rather on the lines written in the book. In short, my attention is fully on the book and its contents.

This is an example of visual attention, where we focus on a particular visual object, in this case the book. Neurobiologically, attention is associated with a specific network, the attention network. This comprises mainly the lateral prefrontal cortex and the lateral parietal cortex, as well as the temporal cortex. Most important, these regions have direct access to the sensory cortex, including the visual cortex, whose activity they can modulate. Modulation from higher cortical to lower sensory regions is called top-down modulation.

Neural correlates of consciousness (NCC) 5

Cognitive functions: attention and representation

How is attention related to consciousness? Once we focus on a particular content, we become conscious of it. Hence, attention may be indispensable to consciousness and may therefore be regarded as a sufficient condition of consciousness. Top-down modulation from the prefrontal and parietal cortex to the sensory cortex has therefore been suggested as a neural correlate of consciousness.

However, recent investigations have shed some doubt on the central role of attention in consciousness. Observation tells us that consciousness can occur independently of attention. For instance, even if I am not paying attention to the book, I nevertheless can experience the book and its redness. In other words, I can be conscious of the book without paying attention to it. Neurobiologically, this goes along with the consciousness that can occur without the involvement of the lateral prefrontal–parietal cortical network and top-down modulation. If, however, consciousness and attention can dissociate from each other on both levels – psychologically and neurobiologically – attention and its subserving neural networks may no longer be regarded as a candidate mechanism for the neural correlate of consciousness.

The central role of cognitive functions is also emphasized by philosophers who associate consciousness with special forms of representation. The original stimulus is processed in the sensory cortex and therefore represented in the respective neural activity. Then it is further processed in other regions, like the prefrontal cortex, that allow for the original representation of the stimulus to be represented in association with cognitive functions. This is a type of re-representation – a meta-representation or higher-order representation – of the stimulus in the neural activity.

The philosophers have developed different forms of representation that they associate with consciousness. The original representation of the stimulus in the initial sensory processing may not be sufficient for the stimulus to become conscious. In order for it to be associated with consciousness, it must be represented in a special way: via re-representation or meta- or higher-order representation. This in turn will allow for consciousness. These theories have been described as representational theories of consciousness. Since they are predominantly conceptual, rather than neurobiological, I will not discuss them in further detail here.

Neural correlates of consciousness (NCC) 6

Cognitive functions in vegetative states

Are cognitive functions specific to consciousness? This is the right moment to introduce Adrian Owen. Adrian Owen investigated unresponsive patients in vegetative states (VS) that did not show any reaction to the environment. Moreover, they were considered nonconscious (see Chapter 16 for more details). This type of VS can occur in the wake of trauma and brain lesions, as in, for instance, motorcyclists who suffer an accident.

What did Adrian Owen do with these patients? He put them into an fMRI scanner and gave them specific instructions on a screen. These instructions included to imagine playing tennis, or to imagine moving through the various rooms of the patients' own house. When imagining these tasks, healthy subjects show activity in those regions that are involved in the actual execution of playing tennis – the sensorimotor cortex – and in house navigation – the parietal cortex and the hippocampus.

What about the vegetative patients? Despite being apparently nonconscious and non-responsive, the vegetative patients could nevertheless activate the same regions as healthy subjects when instructed to perform these tasks (via the screen). This has been replicated in other studies. In some cases, the patients are even able to indicate answers to specific questions by associating 'Yes' with tennis playing and 'No' with 'house navigation'.

What do these results tell us about consciousness? First and foremost, these data tells us that some patients with VS may still have cognitive abilities. Does this imply that they are still conscious? There is much debate today about whether VS patients have some preserved islands of consciousness that allow them to perform these imaginary tasks and consequently to activate the respective regions in the brain.

Neurophilosophical discussion 5

Relationship between cognitive functions and consciousness

These findings raise the question of the relationship between cognitive functions and consciousness. Does the ability to perform cognitive functions, including using imagination (and others, like attention, working memory, etc.; see previous chapter), imply the presence of consciousness? If so, one would assume that a close relationship between cognitive functions and consciousness exists. Or, if not, cognitive functions and consciousness may remain, at least to some degree, independent of each other.

Since we currently do not fully understand the relationship between cognitive functions and consciousness, we may sketch different possible scenarios. Consciousness may be necessary for cognitive functions to occur, though it also may require additional factors in order to be constituted. If so, consciousness may be considered more basic and prior to cognitive functions. In short, cognitive functions might actually build on the existence of consciousness. In this model, consciousness might not be regarded as a cognitive function, while it may be considered an affective or sensorimotor function. Sensorimotor functions are often considered more basic than cognitive functions.

Alternatively, one could also imagine the reverse scenario, in which cognitive functions are more basic than the functions necessary for consciousness to occur. This is presupposed in the higher-order theories of consciousness, where consciousness is considered to be the result of the representation of the contents as contents. Because cognitive functions like working memory and attention are required for the re-representation of contents, consciousness might be the result of the cognitive functions. In this case, consciousness itself might be regarded as a cognitive function – a higher-order one.

Finally, one could also imagine consciousness and cognitive functions occurring independently of each other. In this case, consciousness and cognitive functions could be mediated by different regions that are not related to each other. Consciousness would then occur without cognitive functions and vice versa. This model presupposes that consciousness is no longer considered a cognitive function. This leaves open whether consciousness is related to an affective or a sensorimotor function, or some other special function.

Why is all this relevant for VS patients and the clinical context? It is relevant precisely because it leads to different diagnoses. For instance, the occurrence of cognitive imagery is compatible with the absence of consciousness. VS patients may be able to perform these imaginary tasks, but are nevertheless diagnosed as unconscious. The same holds true in the second scenario, where consciousness is considered to be the result of cognitive functions. In this case, the occurrence of cognitive functions does not imply the occurrence of consciousness.

What about the first scenario, where cognitive functions are based on consciousness? In this case, cognitive functions, like imagination, cannot occur without the prior occurrence of consciousness. This means that in order to be able to perform cognitive imagery, the VS patients must be conscious. In short, there can be no cognitive functions without consciousness. This is what the current clinical debate about VS and consciousness seems implicitly to presuppose.

Neural correlates of consciousness (NCC) 7

Affective functions and consciousness

Other functions besides cognitive functions have also been associated with consciousness. Here, one central function is emotions, or affective functions. This leads invariably to a discussion of Jaak Panksepp (1943–). Panksepp was one of the founders of what is described as 'affective neuroscience'. He was born in Estonia and now lives in Oregon, USA. Among his animal-based research on subcortical regions and emotions, he is also well known for finding the neural correlates of laughter in rats.

Leaving the laughing rats aside, what does Panksepp think about consciousness? He assumes affective functions to be central in yielding consciousness. More specifically, emotions are always already associated with an experience or a feeling. For example, while reading this book you may experience the feeling of frustration about these neurobiological details that do not explain consciousness at all.

Emotional feeling accompanies any experience, and thus consciousness. Hence emotional feelings may open the door to the neurobiological correlates of consciousness. Based on his extensive experiments with animals, Panksepp now posits the subcortical regions as the prime sites for the different emotional feelings like pleasure, anger, etc.

Because the subcortical regions receive inputs from both the environment and the body, they must integrate these two. This integration, it is presumed, is manifest in emotional feeling, and thus consciousness. Since the subcortical regions are more basic and developed earlier in evolution than cortical regions, such affective consciousness must be considered as the core nucleus, and thus the most basic and primary manifestation of consciousness in general.

Neural correlates of consciousness (NCC) 8

Sensorimotor functions and consciousness

In addition to affective functions, sensorimotor functions have also been associated with consciousness. Sensory functions like visual and auditory functions allow for the respective visual and auditory stimuli to be processed in the brain. This occurs in the different sensory cortices and their five different modalities: visual, auditory, gustatory, tactile, and olfactory. Motor functions are processed in the motor and the premotor cortex, allowing for the generation of movements and action. Most important, the sensory and motor cortex have been shown to be closely coordinated with each other. This means that their activities adjust to each other.

Why are sensorimotor functions assumed to be related to consciousness? Sensorimotor functions are less complex and more fundamental, that is lower-order, than cognitive functions, which are more complex and thus higher-order. How can such lower-order functions be essential for consciousness, which is usually regarded as a higher-order function? The sensorimotor functions target the body. Thanks to the sensorimotor functions in the brain, we can experience our body subjectively as a lived body, which is to be distinguished from the merely physical body – also known as the objective body (see Chapter 1 for details). In short, we are conscious of our body.

By experiencing our body as a lived body – as subjective rather than objective – we are able to experience the environment in our consciousness. The conscious experience of our body as a lived body provides access to consciousness in general. If this is the case, the functions underlying the consciousness of our body, the sensorimotor functions, must be considered central in constituting and allowing for consciousness in general. Hence, sensorimotor functions and their underlying neuronal substrates must be considered possible candidate mechanisms for the neural correlates of consciousness.

The sensorimotor functions and their role in consciousness are emphasized by the proponents of an embodied approach to mind and consciousness (see Chapters 1 and 7). In short, the embodied body emphasizes the need to consider the body as having a potentially central role in constituting consciousness.

Without a body and its sensorimotor functions, consciousness remains impossible. The body provides the point of view from which the body, the environment and ourselves are experienced in our consciousness. Without this point of view, consciousness of both the body and the environment (and thus consciousness in general) would be impossible. Hence, consciousness is necessarily embodied, and such embodiment is realized by the sensorimotor functions in the brain that control the body.

Neurophilosophical discussion 6

Consciousness as higher- or lower-order function

The shift from cognitive functions to affective and sensorimotor functions implies a major difference in the characterization of consciousness. Cognitive functions are considered higher-order functions based on lower-order functions like sensory and affective functions. If consciousness is associated with cognitive functions, then consciousness itself must be regarded as a higher-order function, if not the pinnacle of all higher-order functions.

As higher-order functions, cognitive functions are assumed to allow for the integration of different stimuli into one coherent and coordinated content. This coordination and integration is assumed to be central for consciousness, as is reflected in the initially discussed global NCC candidates, like global workspace, re-entrant processing, neuronal synchronization, and as in the cognitive theories.

These neuroscientific hypotheses emphasizing global integration and coordination as well as higher-order cognitive functions are in accordance with the predominant philosophical presupposition of consciousness (see Part II). Philosophically, consciousness has been regarded as the highest hallmark feature of the mind, as distinguished from the body. The mind is the highest and most complex, whereas the body is considered purely mechanical and therefore lower order. Hence, the philosophical assumption of mind and consciousness as higher-order functions is mirrored nicely in the current neuroscientific candidate mechanisms of the NCC (see left part of Figure 14.1).

How does this relate to affective and sensorimotor functions as correlates of consciousness? The focus on affective and sensorimotor functions departs from both the global and cognitive NCC candidate mechanisms in that they no longer emphasize the need for higher-order mechanisms like coordination, integration, or cognitive functions. Instead, they focus more on lower-order functions, including affective and sensorimotor functions. Affective and sensorimotor functions are considered more basic in comparison to cognitive functions, making them lower-order, rather than higher-order.

Thus the shift from cognitive to sensorimotor and affective functions implies a shift from higher- to lower-order functions in the characterization of consciousness. In this case, consciousness is no longer considered the pinnacle of all functions, the highest point of integration and coordination. Instead, consciousness is considered a more basic function, no longer dependent upon cognitive functions. In short, consciousness is regarded as a lower-order function.

Metaphorical comparison

Floors, tables and flowers

This is a major shift from the philosophical tradition that considered consciousness, and ultimately the mind, as the highest achievement of integration and coordination. When taking this shift to lower-order functions seriously, consciousness is suddenly considered to be a rather basic function, rather than the highest achievement. Compare this to a table with a big flower bouquet standing on it. Philosophers like Descartes (see Chapter 5) considered the legs of the table to be the body, while its tabletop may be compared to the brain. This is then somehow linked to the mind, the big and shiny flower bouquet.

This scenario is now reversed when characterizing consciousness by sensorimotor or affective functions. When associating consciousness with affective functions, consciousness may be compared to the tabletop in our scenario (with the subcortical regions in midbrain and brain stem as the equivalent in the brain). One may, though, also link consciousness with sensorimotor functions and the body itself. In that case, consciousness can be compared to the legs of the table upon which all

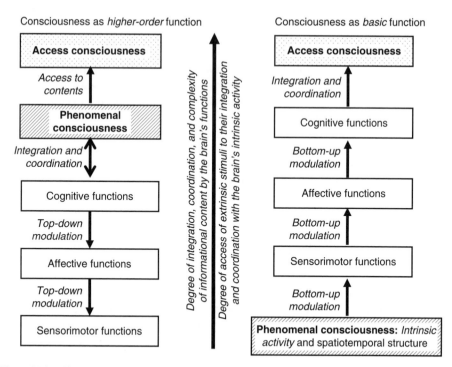

Figure 14.1 Characterization of consciousness in relation to other functions
The figure illustrates schematically the relationship of consciousness to other functions, including cognitive, affective and sensorimotor.

Left part: The figure indicates that consciousness can be considered as a higher-order function (hatched box) that is the output or outcome of the various other functions which are a necessary condition for consciousness. In other words, before consciousness occurs, the other functions must occur that can operate prior to and independently of consciousness. Consciousness can then be regarded as a matter of integration or complexity, as indicated on the right.

Right part: The figure illustrates a different view of consciousness. Now the most basic form of consciousness, phenomenal consciousness (hatched box), is considered to be the very basis or necessary condition of all other functions. This, unlike the other functions, is made possible by the brain's intrinsic activity that provides the background for any stimulus-induced activity and the other associated functions. Rather than consciousness per se, only a specific cognitive form of consciousness, access consciousness, can be considered the output (dotted box at top). Consciousness that is phenomenal consciousness can no longer be reduced to mere integration and complexity, but rather by the degree of access to integration and complexity as provided by the brain's intrinsic activity.

other functions depend. What, then, is the shiny and colorful flower bouquet? This represents the cognitive functions, the higher-order functions that are associated with the cortex of the brain.

The critical spirit may now remark that we forgot something in our scenario: namely the floor upon which the table and its accoutrements stand. In order for the table and its legs to stand solidly, the floor must be level. If it is not level, the table itself, let alone the flower bouquet, will fall over. The floor is thus even more basic than the legs (the body's sensorimotor functions), the tabletop (the brain's subcortical regions and their affective functions) and the flower bouquet (the brain's cortical regions and their cognitive functions).

Neurophilosophical discussion 7

Consciousness as basic function

How can we account for the floor in neuroscientific terms? Can the floor correspond to consciousness itself? That would be in accordance with the fact that all functions, sensorimotor, affective and cognitive, can go along with consciousness. On the other hand, we are not always conscious, and most of the sensorimotor, affective and cognitive processing seems to remain unconscious. Is the floor the unconscious?

How does the floor make the association of consciousness to the different functions possible? We will see in Chapter 15 that this leads us back to the brain itself, its intrinsic activity independent of any extrinsic stimuli and the respectively associated functions, sensorimotor, affective and cognitive. Hence, as we will see, the brain itself may provide the floor upon which the table, its legs and the flower bouquet stand.

If so, consciousness must not only be considered a lower-order function, but also the most basic function underlying both lower-order sensorimotor/affective and higher-order cognitive functions. Consciousness as a basic function may then concern phenomenal consciousness as characterized by phenomenal features like 'what it is like'. Meanwhile access to consciousness, the ability to detect, report, discriminate, etc. (see above as well as Chapter 16 for details about the distinction between phenomenal and access consciousness), may then be considered a higher-order function (see right part of Figure 14.1).

In short, phenomenal consciousness might be considered the invisible floor, while access to consciousness may correspond to the colorful and shiny flower bouquet. Before investigating the floor in further detail in the next chapter, we want to put the various NCC candidates briefly into the context of the various philosophical arguments against the reduction of consciousness to the brain (as discussed in Chapter 13).

Neurophilosophical discussion 8

NCC and philosophical arguments

How do the suggestions discussed above for the neural correlates of consciousness (NCC) relate to the various philosophical arguments against reducing consciousness to the brain? These philosophical arguments, as discussed in the previous chapter, raised problems that obstruct the arguments in favor of reducing consciousness to the brain and its neuronal features. The problems raised pointed out that consciousness requires specific features known as phenomenal features, which are to be distinguished from the brain's neuronal features. This was discussed in the absent qualia and the inverted spectrum arguments.

Moreover, consciousness required a specific type of knowledge known as first-person knowledge, which is to be distinguished from the third-person knowledge of the brain. This was discussed in the 'what it is like argument' and the 'knowledge argument'. Finally, our current explanation of consciousness suffers from a gap, known as the explanatory gap, between the brain's neuronal features and the phenomenal features of consciousness. To bridge that explanatory gap we need to solve what is described as the 'hard problem', which can be described as: 'why and how there is consciousness at all rather than nonconsciousness'.

One may now be keen to know whether and how the suggested neural correlates of consciousness (NCC) provide answers to these various problems. To help explain this, I have invoked an imaginary dialogue between a neuroscientist (NS) and a philosopher (PH).

Neurophilosophical discussion 9

NCC and absent and inverted qualia

NS: Finally, we come to the empirical side of business and to the concrete data about the neural mechanisms underlying consciousness. Doesn't this make the various arguments you discussed superfluous?

PH: All you showed me were neuronal mechanisms; for instance, neuronal synchronization or global synchronization. These are indeed very interesting when trying to understand how the brain functions. I agree with that.

NS: They not only show us how the brain functions, but also how and why the brain yields consciousness.

PH: I am not so sure about that. First, you did not really show that neuronal mechanisms like globalization, re-entrant circuits and synchronization yield consciousness. They only yield globalized, recurring and synchronized neuronal activity. But do they yield consciousness? That you did not demonstrate adequately.

NS: Are you saying that the suggested NCC are compatible with the absence of qualia in the presence of the same brain and neuronal mechanisms, as raised in the absent qualia argument?

PH: Yes, exactly that. And some of the remarks about the lack of neuronal and experimental specificity address exactly this point. All the neuronal mechanisms you suggested can also occur in the absence of consciousness, as, for instance, in an unconscious or even nonconscious state. They are thus very compatible with the absent qualia argument.

NS: Hard stuff to digest. You say that the NCC do not provide any solution or answer to the absent qualia argument. Hence, to reveal the neuronal mechanisms underlying the processing of conscious contents is not sufficient.

PH: Yes. One of the problems the current NCC neglects is that the processing of contents themselves does not reveal or imply anything about consciousness. That is even more pronounced by the inverted spectrum argument. Here, the same content may be associated with different experiences, qualia and thus 'what it is like'. This implies that the experiences or qualia, 'what it is like', cannot be traced back to the contents themselves and their underlying neuronal mechanisms.

NS: You suggest an additional neuronal factor? One that operates on top of the neuronal processing of the contents and thereby adds consciousness to them?

PH: Yes, there must be at least an additional factor, neuronal or non-neuronal, to explain consciousness. Most important, this extra ingredient cannot be traced to the contents and thus to the neural processing of the related stimuli in the brain.

NS: Doesn't the globalization, synchronization, or re-entrance of neuronal activity as described above provide this extra ingredient?

PH: No, because these suggestions for the NCC only provide mechanisms for the neural processing of stimuli, and thus of contents. We need another completely different ingredient – independent of contents altogether – in order to account for how 'what it is like' is associated with the processing of stimuli and their respective contents.

NS: Sounds mysterious to me. Where does this extra ingredient come from? From the brain itself and its neuronal mechanisms?

PH: As long as you assume that this extra ingredient is associated with the brain itself and its neuronal states, it is not at all mysterious. It is a neuronal ingredient of which we are simply unaware. It might, however, be detected in the future. But if this extra ingredient is non-neuronal and thus non-physical – mental or otherwise – it may be regarded as mysterious.

NS: You are right. Maybe there is more to the brain than the mere processing of stimuli and content, the stimulus-induced activity – as in, for instance, its intrinsic activity. This could provide the kind of neuronal specificity that is lacking so far in the NCC. I agree that would not be mysterious at all.

Neurophilosophical discussion 10

NCC and 'hard problem' and 'explanatory gap'

NS: Considering what you have said so far, I assume that you deny that the NCC provides an answer to the 'hard problem': why there is consciousness rather than nonconsciousness.

PH: You are right. At best the various suggestions for the NCC provide an answer to the easy problems, the association of consciousness with cognitive functions.

NS: You need to explain that in more detail.

PH: The easy problems target the differentiation between phenomenal and access consciousness: how can we access the contents of consciousness such that we are able to detect, recognize, and become aware of them as contents of our consciousness? The suggestions for the NCC address exactly that. They describe additional steps in the neuronal processing of contents like re-entrance, synchronization and globalization.

NS: You mean that re-entrance, synchronization and globalization process the content further to recruit cognitive functions that in turn allow access to, and recognition of, content as content.

PH: Yes, exactly that. The NCC may thus, at best, provide answers to what David Chalmers described as the 'easy problems', the questions that ask about the relationship between cognitive functions and consciousness.

NS: This does not answer the 'hard problem'. As I understand it, access consciousness is only possible on the basis of phenomenal consciousness. Nothing can enter or access consciousness if it is not already phenomenally conscious. Hence, experience itself – and thus phenomenal consciousness – must precede and be more basic than the accessing of contents.

PH: True, phenomenal consciousness is prior to and thus more basic than access consciousness. Why, though, is that of relevance to you?

NS: It would mean that the neuronal mechanisms underlying phenomenal consciousness must occur prior to the ones underlying access consciousness.

PH: Interesting thought. That goes well with the suggestion of an 'extra ingredient' that cannot come from the neural processing of the stimuli and the related contents themselves. Philosophers in the past, like Descartes, assumed that it must come from a mind that may somehow be attached to the brain. Some current philosophers like Colin McGinn and Thomas Nagel 'locate' such an 'extra ingredient' in the brain itself, and they speak of 'proto-mental properties'.

NS: That sounds rather mysterious to me. And if I remember correctly, the previous chapters about the mind–brain problem in Part II, their positions are called 'mysterianism'.

PH: Correct. Whatever one may think about these positions, they tell us at least one thing: that is, it would be worth taking a second look at the brain itself and its intrinsic features.

NS: You mean we need to raise the following question: what does the brain itself and its intrinsic features provide and add to the neuronal processing of extrinsic stimuli and their contents?

PH: Yes. That may make the difference between consciousness and nonconsciousness. The brain itself and some as yet unknown intrinsic neuronal feature may reveal why there is consciousness rather than nonconsciousness. In other words, the brain's intrinsic features may account for the fact that usually we have no other choice than to be conscious, rather than remaining nonconscious.

Take-home message

Chapter 13 focused on various arguments from philosophy against reducing consciousness and its phenomenal features to the brain and its neuronal features. This left open for discussion the neuronal mechanisms underlying consciousness. These are the focus in this chapter. Various suggestions for neuronal mechanisms underlying consciousness were discussed; these included synchronization, global workspace and re-entrant connections. However, all these suggestions for a neural correlate of consciousness (NCC) suffered from unspecificity in neuronal, phenomenal and experimental terms. Because of this unspecificity, the current suggestions for NCC remain unable to account for the various problems raised by the philosophical arguments against reducing consciousness to the brain. Hence, we may need to search for neuronal features in the brain other than the current NCC to get a better understanding of the neuronal mechanisms underlying consciousness.

Summary

The previous chapter focused on various arguments in philosophy against reducing consciousness to the brain. This left open for discussion the empirical and neuroscientific mechanisms. These were the focus in the present chapter. The current neuroscientific investigation of consciousness focuses on the neuronal mechanisms that allow for the distinction between conscious and unconscious contents. For that, the concept of the neural correlates of consciousness (NCC) was established. It describes the sufficient neural conditions of consciousness. Various suggestions for the NCC have been made recently. The main ones concern the re-entrance of neuronal activity, where the neural activity related to particular stimuli and contents enters in one region, circulates and loops through other regions, and finally comes back to the original region. The subcortical thalamus and its connections to the cortex seem to be central here. Another suggestion for the NCC assumes that the degree of spread and extension of neural activity to other regions like the prefrontal and parietal cortex may be central for yielding consciousness. By providing such a global workspace, contents can become conscious. The globalization of neuronal activity across the brain may be accompanied by the increased synchronization of the neural activities within and between different regions. This neuronal synchronization may be central in binding together different stimuli into one content, and in bringing the content into consciousness. Other NCC suggestions assume that particular cognitive functions – like attention, working memory, etc., and their underlying neuronal mechanisms – are central in yielding consciousness. After presenting these suggestions for the NCC, their problems were discussed by focusing on their lack of neuronal and experimental specificity with regard to consciousness. The chapter concluded with a discussion of whether and how the NCC can provide an answer to the various philosophical arguments against the reduction of consciousness to the brain and its neuronal features.

Revision notes

- What does the abbreviation NCC stand for and what does it mean? What aspect of consciousness is the main focus of the NCC?
- What are the different neuronal mechanisms suggested as the NCC that allow consciousness and unconsciousness to be distinguished?
- Why do some of the NCC lack specificity? What different forms of specificity do you know?
- Can the NCC provide a positive answer to the various philosophical arguments for the irreducibility of consciousness to the brain? If yes, why and how? If no, why not?
- What is needed from neuroscientific research to overcome the philosophical objections to a neuroscientific theory of consciousness?

Suggested further reading

- Crick, F. and Koch, C. (2003) 'A framework for consciousness', *Nature Neuroscience*, 6(2), 119–26.
- Dehaene, S. and Changeux, J. P. (2011) 'Experimental and theoretical approaches to conscious processing', *Neuron*, 70(2), 200–27.
- Edelman, G. M. (2003) 'Naturalizing consciousness: A theoretical framework', *Proceedings of the National Academy of Sciences of the United States of America*, 100(9), 5520–4. doi:10.1073/pnas.0931349100.
- Northoff, G. (2013b) *Unlocking the Brain. Vol. II: Consciousness* (Oxford/New York: Oxford University Press).
- Tononi, G. and Koch, C. (2008) 'The neural correlates of consciousness: An update', *Annals of the New York Academy of Sciences*, 1124, 239–61. doi:10.1196/annals.1440.004.

15
Neural Predispositions of Consciousness (NPC)

Overview

The previous chapter discussed the neural correlates of consciousness (NCC). Various suggestions for the neuronal mechanisms that underlie conscious contents, as distinguished from unconscious contents, were put forward. The present chapter delves deeper into the neuroscience of consciousness by focusing on the neuronal mechanisms underlying the level or state of consciousness. To do this, various disorders of consciousness are discussed. Empirically, this leads away from the neuronal activity underlying contents, the extrinsic stimulus-induced activity, to the brain's intrinsic activity. The brain's intrinsic activity is thus the focus of this chapter. More specifically, the potential relevance of the intrinsic activity for consciousness as a neural predisposition of consciousness (NPC) is discussed, alongside its implications for philosophical problems like the hard problem.

Objectives

- Distinguish between contents and level/state of consciousness
- Learn about disorders of consciousness like the vegetative state
- Understand the distinction between stimulus-induced activity and intrinsic activity
- Become acquainted with the features of the brain's intrinsic activity
- Understand the difference between neural correlates and neural predispositions of consciousness, as well as the difference between actual and possible consciousness
- Link the neural predispositions of consciousness to the philosophical arguments against the linkage between brain and consciousness

Key concepts

Level/state of consciousness, content- and level-based NCC, neural predisposition of consciousness (NPC), actual versus possible consciousness, global metabolism and energy demand, low frequency fluctuations, 'slow wave hypothesis', form of consciousness.

Background 1

Disorders of consciousness

Almost all suggestions for the NCC discussed above have focused on particular contents of consciousness: how can some particular contents enter consciousness while others do not? In addition to the contents of consciousness, we also need to distinguish between different levels of consciousness. The definition of a level or state of consciousness concerns the degree of arousal and wakefulness, which is, for instance, reduced in what clinical neurologists, the specialists in disorders of the brain, describe as 'disorders of consciousness'. Most common among the disorders of consciousness is the vegetative state.

The vegetative state is a state where patients suffering from a severe trauma to the brain, as in a motorbike accident, for example, do not demonstrate any signs of behavior that indicate consciousness in any form. They only show physiological and vegetative functions, but no signs of any mental states or consciousness. This means that the patient seems to remain unresponsive. Some patients, however, do demonstrate some small signs of possible consciousness. These include certain reactions to specific perceptual stimuli. In those cases, doctors might call this a minimally conscious state, rather than a vegetative state. Finally, patients in a coma no longer even open their eyes.

The exact causes of the vegetative state remain unclear. Damage to the brain is clearly essential, but it is not a circumscribed lesion in one particular region of the brain that leads to a vegetative state. Recent imaging studies show changes in the functional connectivity between different regions, as in, for instance, the thalamo-cortical connectivity that is presumed to be essential for consciousness. Whether decreased thalamo-cortical connectivity causes loss of consciousness in the vegetative state remains to be fully investigated.

In addition, functional connectivity between other regions, like the regions in the middle of the brain – the midline regions – which show high intrinsic activity, seem altered in these patients. How this contributes to the loss of consciousness is, however, not yet clear.

Anesthesia is another example of a disorder of consciousness. Anesthesia is induced when a person undergoes surgery to reduce pain and eliminate the memory of the surgery. Specifically, patients under anesthesia demonstrate a decrease in global metabolism. Despite the successful use of anesthetics in surgery for many decades, exactly what happens to the patient's neuronal mechanisms under anesthesia is still unclear. Hence, anesthesia cannot provide an answer to the question of the neural correlates of consciousness at this point in time.

Finally, sleep is another example of a state where consciousness is altered. For example, when you have nightmares, you make rapid movements with your eyes, called rapid eye movements (REM sleep). In dreams, your consciousness is altered with bizarre contents that show strange spatial and temporal structures. For instance, you may experience that this book as hanging in the sky, surrounded by dark clouds, so that you are forced to climb up a ladder to read it. This type of abnormal space in your dreams may be accompanied by an abnormal sense of time. For example, you may experience yourself as a far younger or far older individual.

What happens when dreams cease during sleep? When you are not dreaming, you 'lose' your consciousness completely and do not make any rapid eye movements. This is called non-REM sleep (NREM). What are the neural mechanisms underlying non-REM sleep? Non-REM sleep shows a significant decrease in thalamo-cortical functional connectivity. There is also considerably less extension of neuronal activity throughout the cortex in non-REM sleep compared to REM sleep. This means that REM and non-REM sleep – and thus perhaps even consciousness and nonconsciousness – seem distinguishable by means of the degree of available thalamo-cortical connections and the global extension of neural activity. How and why those connections yield consciousness, however, remains unclear.

Background 2

Content-based NCC versus level-based NCC

What do these disorders of consciousness tell us about consciousness in general? Despite their different origins, these disorders share a reduced level of consciousness. Hence, they tell us that consciousness may include different dimensions, contents and levels (or states). The current literature thus distinguishes between levels (or states) and contents of consciousness.

If one follows the logic behind distinguishing between the level and content of consciousness, one may then also distinguish between corresponding neuronal mechanisms. More specifically, the neuronal mechanisms underlying the contents of consciousness may differ from those underlying the level or state of consciousness. If this is the case, it would be important to distinguish the neural correlates of the contents of consciousness from the neural correlates of the level of consciousness.

Let's go into more detail. The neural correlates of the level of consciousness, also known as the level-based NCC, concern the sufficient neural conditions of the level of consciousness. As such, the level-based NCC must be distinguished from those sufficient neural conditions that underlie the processing of conscious contents, also known as the content-based NCC. In short, we need to differentiate between the content-based and the level-based NCC. In the previous chapter, we focused on the neuronal mechanisms underlying the contents of consciousness. We demonstrated how neuronal mechanisms, like re-entrant processing, global workspace and neuronal synchronization, allow for the neural processing of contents and their association with consciousness. We then focused on content-based NCC. Now we will shift our attention from content-based NCC to level-based NCC and to discuss several possible candidate neural mechanisms.

Level-based NCC 1

Intrinsic activity and low frequency fluctuations

The suggestions for the content-based NCC (see the previous chapter) focused mainly on neuronal activity related to stimuli, the so-called stimulus-induced activity. The reason for this is that the focus was on contents that are based on extrinsic stimulus input to the brain, and how they can be associated with consciousness. Since the stimuli originate in either the body or the environment, stimulus-induced activity must be characterized as extrinsic activity.

In addition to the extrinsic stimulus-induced activity, we can also observe some activity in the brain that does not originate from the body or the environment. Instead, this neural activity seems to be generated from within the brain itself and may consequently be described as intrinsic activity. *The term intrinsic activity describes spontaneous activity generated inside the brain itself (see also Chapter 12). This activity is independent of any extrinsic stimuli and is thus purely intrinsic. All regions in the brain show such high spontaneous activity, e.g. intrinsic activity.*

However, some regions show particularly high spontaneous activity. These regions include the regions in the middle of the brain, the midline regions and the bilateral parietal cortex. Because these regions seem to show high activity and high metabolism, and are closely connected to each other, the neurologist Marcus Raichle coined a phrase to describe the regions as the Default-Mode Network (DMN). Since its introduction in 2001, the DMN has received considerable attention in neuroscience. Thus far, many imaging studies support the existence of the DMN as a special network that functions alongside others in the brain.

Since the observation of spontaneous activity implies the absence of extrinsic stimuli, the term intrinsic activity is often used interchangeably with 'resting state activity' (see also Logothetis *et al.* (2009) for a discussion on the concept of the resting state).

Most recently, the brain's intrinsic activity and the DMN have also been regarded as possible candidate mechanisms for consciousness. Here, we need to introduce a typical feature of the brain's intrinsic activity, namely that its level fluctuates rhythmically in either a slower or a faster way. This leads to what is described as fluctuations in different frequency ranges.

Intrinsic activity can be characterized by strong and predominantly slow fluctuations in the low frequency range of 0.001 to 0.1 Hz. The predominance of such slow wave fluctuations distinguishes the intrinsic activity from stimulus-induced activity or extrinsic activity. In stimulus-induced activity the balance shifts towards higher frequency ranges of 30–40 Hz – that is, the gamma fluctuations and synchronization as discussed in the previous chapter.

Level-based NCC 2

Low frequency fluctuations and consciousness

How are these slow wave fluctuations related to consciousness? To help answer this question, we should introduce the Asian-American neuroscientist, Be He. Be He works with another neuroscientist, Marcus Raichle, and is very interested in the DMN and its spontaneous activity, as well as in consciousness. The two scientists wrote a paper about intrinsic activity and consciousness (He and Raichle, 2009).

What does He suggest about consciousness and intrinsic activity? She suggests that the resting state's slow wave fluctuations in the frequency ranges between 0.001–4 Hz are central in yielding consciousness (He *et al.*, 2008; He and Raichle, 2009; Raichle, 2009). Because of the long durations of their ongoing cycles, also known as phase durations, the slow wave fluctuations may be particularly suited to integrating different types of information. This information integration may then allow for the respective contents to become associated with consciousness. Because it is based on low-frequency fluctuations, this hypothesis is called the 'slow wave hypothesis'.

The assumption of information integration is supported by the supposed anatomical origin of the slow wave fluctuations: they are generated in cortical layers I and II, where the afferences – the inputs from different regions and cells from many different cortical layers and regions – converge with each other. This predisposes the slow wave fluctuations to integrate the information from these afferences. This integration meshes nicely with the assumption of consciousness being based on information integration (see the previous chapter): the more information is integrated, the more likely it will become conscious.

Moreover, the 'slow wave hypothesis' can be regarded as complementing the hypothesis on neuronal synchronization discussed in the previous chapter: low frequency fluctuations (0.001–4 Hz) are observed mainly in the resting state. Meanwhile, neuronal synchronization targets predominantly higher frequency fluctuation in the gamma range (30–40 Hz) during stimulus-induced activity. What remains unclear, however, is how low- and high-frequency fluctuations must interact in order to yield consciousness.

What does the 'slow wave hypothesis' imply for the NCC? As it stands, the 'slow wave hypothesis' is still focused on contents and their association with consciousness. Due to their long time windows, the low-frequency fluctuations are suitable candidates for integrating different stimuli into one coherent content. Here, one must again raise the issue of neuronal specificity: the low-frequency fluctuations may be specific for integrating stimuli into contents (i.e. binding). It remains open as to whether they are really specific for associating contents with consciousness.

Is the 'slow wave hypothesis' a 'content- or level-based NCC'? Because of its focus on content, the 'slow wave hypothesis' seems to be more a content-based NCC, rather than a level-based NCC. However, due to the continuous presence of the slow waves even in the resting state, they may also impact the level of consciousness. For example, patients in VS show reduced slow frequency fluctuations in their resting state activity. However, it remains unclear at this point whether the slow wave reduction causes the loss of consciousness in VS. If yes, the 'slow wave hypothesis' may be considered a possible candidate for a level-based NCC.

Level-based NCC 3

Metabolism and energy demand of the brain

Another suggestion for the relevance of the brain's resting state activity in consciousness comes from the American biophysicist, Robert Shulman (1924–). Shulman is a physicist who was instrumental in introducing the technology and physics necessary to image and visualize the brain's function in, for example, fMRI. Robert Shulman is still an active member of the brain imaging center at Yale University in New Haven, Connecticut.

Having introduced the physical and technological basics of imaging techniques to the neuroscientific community, Shulman now aims to convey the importance to neuroscientists of the brain's metabolism for consciousness. Based on his experimental observations (see the recent excellent book – Shulman, 2012), Shulman assumes the resting state's baseline metabolism, and thus the brain's energy demand, as a necessary condition of consciousness. The brain shows spontaneous neural activity, and this activity is possible only on the basis of energy. Where does the energy come from? The brain receives metabolic input from glucose (sugar) especially, which provides (via some chemical processes) energy to the brain. Without metabolism, no energy could be provided to the brain, and without energy, there can be no neural activity.

Moreover, the brain seems to invest its energy predominantly in its intrinsic activity. Based on his own investigations, about 80–85 percent of the total glucose, the main energy provider, is used to maintain and sustain high neuronal activity, even during the absence of specific stimulation in the resting state. Meanwhile, the remaining 15–20% of the total glucose and energy are used for increments in energy metabolism during stimulus-induced activity. Hence, the brain seems to invest much more of its energy into its intrinsic activity, rather than extrinsic stimulus-induced activity. In other words, the brain's own intrinsic activity seems to be more important than the stimulus-induced activity.

Level-based NCC 4

Energy and consciousness

Why does the brain invest so much of its energy in its intrinsic activity? It suggests that the intrinsic activity must bear a special significance for the brain's functioning. This is where consciousness comes in. Shulman's observations in patients under anesthesia go along with strong reductions in energy metabolism. The global metabolism across the whole brain and all of its regions is reduced by 40–50 percent during anesthesia. Analogous reductions in global metabolism can be observed in the vegetative state as well. This has been shown in investigations using positron emission tomography (PET) that measures the glucose metabolism in the whole brain.

These findings led Shulman to posit the baseline level of energy and metabolism during the brain's intrinsic activity as being central for consciousness. According to him, a certain level of baseline metabolism and energetic activity is necessary to develop consciousness.

If, in contrast, the levels of baseline metabolism and energy supply are too low, one glides into a nonconscious state, as is the case in patients under anesthesia. In short: without metabolism, there is no energy. Without energy, there is no neural activity. Without neural activity, consciousness is impossible. Hence, the brain's baseline metabolism – its degree of energy in the resting state – must be central for consciousness to occur.

Shulman suggests that metabolism and energy are central in yielding consciousness, and more specifically the level of consciousness. As a result, his suggestion is a level-based NCC. One may, however, question the specificity. One could argue that any kind of cognitive function independent of consciousness requires metabolism and energy.

For example, attention, working memory, or executive functions may require plenty of energy. Cognitive functions, as well as affective and sensorimotor functions, may require energy even though they may remain independent of consciousness. This means that there is no necessary link from the brain's intrinsic activity to consciousness. In short, energy and metabolism do not imply anything about consciousness. Hence, Shulman's suggestion that baseline metabolism serves as a candidate for level-based NCC may be too unspecific.

However, his suggestion points to the central role played by the brain's intrinsic activity in consciousness. Intrinsic activity cannot only be characterized by neural activity, but also by a demand for energy and a high metabolism rate. And since this high metabolism rate appears to be reduced in disorders of consciousness, like the vegetative state and in patients under anesthesia, it may be considered central for consciousness.

Why, though, is the high metabolism central for the intrinsic activity and its possible involvement in generating consciousness? That we do not yet know. One may assume that the metabolism and energy make possible the constitution of certain as yet unknown features in the intrinsic activity. These features, as based on the brain's intrinsic activity, may be central for consciousness to occur. However, at this point in time, we have not yet discovered these features nor figured out how they are related to consciousness.

Neurophilosophical discussion 1

Neural correlates versus neural prerequisites

The NCC describes the sufficient neural conditions of both the content and level of actual consciousness, concerning both its level and contents. One can therefore speak of content- and level-based NCC. Left for discussion, here, are questions about the necessary neural conditions of consciousness.

Recent authors, like the German-born and US-based scientist Christof Koch and others, speak of 'enabling conditions' (Koch, 2004; Dehaene et al., 2006; van Eijsden et al., 2009) or 'neural prerequisites' (de Graaf et al., 2011) of consciousness. 'Enabling conditions' or 'neural prerequisites' are those neuronal mechanisms that are necessary to actually yield consciousness. Though they remain unable to generate consciousness by themselves, independent of additional neural conditions, the 'enabling conditions' or 'neural prerequisites' are necessary, but non-sufficient, for the occurrence of actual consciousness.

What would these enabling neural conditions look like? Let's recall that the NCC are the sufficient neural conditions of consciousness. In the previous chapter, we discussed several candidate mechanisms for the content-based NCC. These included neuronal synchronization, re-entrant connections and global workspace. These neuronal mechanisms were considered sufficient for constituting the consciousness of contents. Because of their focus on contents, these suggestions concerned mainly stimulus-induced activity and extrinsic activity. The sufficient neural conditions of consciousness, the NCC, in this model, may be related more highly to stimulus-induced activity and thus to extrinsic activity.

This is different in the case of the necessary neural conditions. Unlike the sufficient conditions, the necessary conditions may be related more to the brain's intrinsic activity. Instead of extrinsic activity, here the focus is on the brain's intrinsic activity. Slow wave fluctuations and baseline metabolism are regarded as neural conditions that may be central for the level of consciousness.

Are slow wave fluctuations and baseline metabolism necessary enabling conditions or sufficient for consciousness? Shulman himself argues that the baseline metabolism is not a sufficient condition by itself for consciousness to occur. Instead, it is only a necessary or enabling condition of consciousness. Hence, baseline metabolism may be regarded as an enabling condition or neural prerequisite, rather than a neural correlate.

Another candidate for the enabling condition or neural prerequisite may be neural activity in the brain stem and/or the midbrain. Lesions in the midbrain and/or the brain stem may lead to loss of consciousness. However, neural activity in the midbrain/brain stem alone may not be sufficient to yield consciousness. For consciousness to occur, neural activity in additional regions like the cortical regions may be required (which has also been heavily debated and remains unclear at this point).

What does this short discussion of the enabling discussion tell us with regard to the neural mechanisms of consciousness? It tells us that we may need to consider different kinds of neuronal mechanisms with different roles in constituting consciousness. More specifically, we need not only to distinguish between content and levels of consciousness but also between necessary non-sufficient and non-necessary sufficient neural conditions of consciousness. In short, we may need to distinguish between neural prerequisites and neural correlates of consciousness.

Neurophilosophical discussion 2

Neural predispositions and the hard problem

In Chapter 13, we discussed the hard problem as a central dilemma that needs to be addressed if consciousness is to be explained. Briefly, the hard problem describes the following question: why is there consciousness rather than nonconsciousness? The hard problem thus raises questions about those neural conditions that make possible the distinction between consciousness and nonconsciousness.

Can the hard problem be addressed by the candidates for the neural prerequisites and neural correlates? Let's consider what exactly the neural prerequisites and neural correlates target. They target those neuronal mechanisms that are necessary and sufficient for the constitution of conscious states. The central question here is: what kinds of neuronal mechanisms are necessary and sufficient to convert (or transform) an unconscious state into a conscious one?

The main distinction in the search for the neural prerequisites and neural correlates is thus the distinction between consciousness and unconsciousness. This, however, does not correspond to the distinction targeted in the hard problem – the distinction between consciousness and nonconsciousness. More specifically, the concept of consciousness, as presupposed in the hard problem, includes both consciousness and unconsciousness, since both have at least the potential to become conscious. This is not the case in nonconsciousness, whose contents cannot become conscious at all. Hence the hard problem concerns what conceptually may be described as 'possible consciousness' as distinguished from nonconsciousness or impossible consciousness.

The concept of 'possible consciousness' must be distinguished from the concept of actual consciousness as presupposed in the concepts of neural prerequisites and neural correlates. Both neural prerequisites and neural correlates imply actual consciousness as distinguished from unconsciousness, rather than possible consciousness as distinguished from nonconsciousness. This implies that the various neural candidates for neural correlates and neural prerequisites may not be able properly to address and solve the hard problem on their own.

What else do we need to address the hard problem? The concepts of neural perquisites and neural correlates describe the necessary and sufficient conditions of actual consciousness as distinguished from unconsciousness.

In order to address the hard problem properly, we need to reveal those neural conditions that predispose the brain to constitute possible consciousness – actual consciousness and unconsciousness, rather than nonconsciousness. One may consequently speak of what can be described as the neural predispositions of consciousness (NPC). *The neural predispositions of consciousness (NPC) concern those neural conditions that predispose the brain to constitute possible consciousness, e.g. actual consciousness and unconsciousness, rather than nonconsciousness.*

What are the possible candidate neural mechanisms for the NPC? One possible candidate mechanism could be the brain's intrinsic activity. However, as described above in the case of its energy and metabolism, the brain's intrinsic activity does not imply anything about consciousness. Put into more technical terms, the presence of the intrinsic activity by itself does not necessitate consciousness.

To link the brain's intrinsic activity more closely to consciousness, and to explain why intrinsic activity makes consciousness possible, we have to describe it in greater detail. The remaining sections of this chapter will focus precisely on this. Before moving on to this, however, let us illustrate briefly the current situation by an analogous comparison with diabetes.

Metaphorical comparison 1

Insulin and resting state activity

In order to reveal the possible role of the brain's intrinsic activity in consciousness, I first want to compare the current situation in neuroscience to the relationship between insulin and diabetes. Diabetes mellitus is a disorder where one suffers from high blood sugar levels. This leads to various symptoms throughout the body, including gangrene and blindness. Diabetes is caused by a decreased or depleted production of insulin in the pancreas. Let's imagine the following scenario.

Imagine that all we know about diabetes is that in patients presenting its symptoms, there exists an abnormally low level of insulin. Meanwhile, we do not know why low levels of insulin are accompanied by the various symptoms typical of diabetes. We know that the pancreas produces insulin, but we do not know how insulin is connected to the different symptoms across the various body organs (eyes, legs, etc.). We know that insulin is somehow connected to diabetes, since our recent scientific data show some kind of correlation between the level of insulin and the degree of symptoms. But we do not know anything about the underlying mechanisms and processes that make such a correlation possible.

Physiologically, we do not know that insulin controls and modulates the level of glucose. Knowledge about glucose is central to understanding why and how the various kinds of symptoms are yielded in diabetes. Glucose by itself may even be known. What remains unclear though is how glucose is connected to both insulin and the diabetic symptoms.

This is the situation we are currently facing in the neuroscience of consciousness. We know something about the brain's resting state that may be considered analogous to our knowledge of insulin in our imaginary thought experiment on diabetes. And at best we also know that there seems to be some kind of correlation between the level of the brain's neuronal activity and the degree of consciousness, especially as suggested by the results from Shulman and the observations made on patients under anesthesia and in a vegetative state (see above).

In contrast, the underlying neuronal mechanisms and processes that yield this correlation remain unclear. We do not know how the neuronal activity yields the phenomenal-qualitative feel, the point of view, or the subjective nature of consciousness and its other phenomenal features. What we are missing are thus the neuronal mechanisms and processes that allow the brain's neuronal activity to yield consciousness and its various phenomenal features.

This is very much analogous to our imaginary scenario, where we remain unable to causally connect insulin to glucose and the various symptoms of diabetes: we are missing the link between insulin and the various diabetic symptoms in our imaginary scenario. Analogously, we currently remain unable to account for the link between the brain's intrinsic activity and the various phenomenal features of consciousness.

Metaphorical comparison 2

Diabetes and consciousness

One may now want to argue that the analogy between our current knowledge of consciousness and the imaginary diabetes scenario does not hold true. We do have plenty of knowledge of stimulus-induced activity, which may correspond to our knowledge about glucose. Meanwhile, our knowledge about resting state activity corresponds to insulin.

However the analogy is not affected by that argument. Why? In our 'diabetic thought experiment' we lack the knowledge that the level of glucose is controlled and predisposed by insulin. Analogously, we currently lack the knowledge about the relationship between resting state and stimulus-induced activity. More specifically, we do not know how the resting state activity controls, and thus predisposes, subsequent stimulus-induced activity.

How is all this related to consciousness? Consciousness is usually associated with extrinsic stimuli and consecutively with stimulus-induced activity in the brain. Since currently we do not know how resting state and stimulus-induced activity are related to each other, we also remain unable to link the brain and its intrinsic activity directly to consciousness. Once we were able to connect insulin to glucose, we could understand the physiological mechanisms underlying the manifestation of the different diabetic symptoms in the various organs across the whole body. We then understood how the level of insulin predisposes the various symptoms via its impact on glucose.

The analogous scenario may now be assumed in consciousness. We aim to link resting state activity and stimulus-induced activity. This may enable us to see how the brain's intrinsic activity predisposes the neuronal mechanisms that allow for the constitution of the phenomenal-qualitative and subjective features of consciousness during stimulus-induced activity. Analogous to insulin, the resting state activity itself may then be supposed to provide the very basis – or neural predisposition (see below) – for what the American philosopher Thomas Nagel described as point of view.

The analogy goes even further. Too little insulin will affect the glucose level and hence our general level of alertness. We may lose consciousness and glide into a diabetic coma, as it is called. The same seems to hold true of the brain's resting state activity: if the resting state activity level is too low, we lose consciousness and end up in a vegetative state or, even worse, in a coma. In short, both too little insulin and not enough activity in resting state lead to a coma.

Metaphorical comparison 3

Glucose and stimulus-induced activity

What does the figurative comparison with our fictive 'diabetic thought experiment' tell us? The discovery of insulin's intrinsic link to glucose revealed insulin's role in both controlling glucose levels and other various body functions (as visible in the symptoms of diabetes). Analogously, we may need to decipher the brain's intrinsic link between resting state activity and stimulus-induced activity, and how their associated phenomenal features manifest in consciousness.

More specifically, we need to understand how the brain's resting state activity impacts and predisposes the subsequent stimulus-induced activity. This will shed light on the neuronal mechanisms underlying stimulus-induced activity and, most important, on how the respective stimulus can be associated with the phenomenal features of consciousness.

How can we investigate consciousness? In order to reveal the neuronal mechanisms underlying consciousness itself, we must better understand stimulus-induced activity. For that, we need to go back to the resting state that predisposes stimulus-induced activity. In the case of diabetes, we have to make a detour via insulin and its intrinsic features to better understand the mechanisms of glucose itself, and how and why it yields diabetes and its various diabetic symptoms. Analogously, we need to make the detour via the brain's intrinsic activity and its resting state in order to understand how and why stimulus-induced activity can yield consciousness and its various phenomenal features.

Current research themes 1

Spatial characterization of the brain's intrinsic activity

How can we describe the brain's resting state activity in further detail? This is the subject of current and ongoing research. High resting state activity has been observed not only on the cellular or microscopic level of the brain, but also on a macroscopic level – the level of macroscopically identifiable regions.

Early studies on humans using PET identified high oxygen and glucose consumption during the resting state in a particular set of regions. These include the anterior and posterior cortical midline regions like the ventromedial prefrontal cortex (VMPFC), the dorsomedial prefrontal cortex (DMPFC), the different parts (sub-, pre-, and supragenual) of the anterior cingulate cortex (ACC), the posterior cingulate cortex (PCC) and the precuneus, as well as other regions like the lateral parietal cortex and the hippocampus.

These regions have consequently been subsumed under the concept of the default-mode network (DMN) that includes what has been called the cortical midline structures (CMS) and the bilateral parietal cortex. The DMN seems quite special when compared to the other regions of the brain. This is because it shows a higher degree of metabolism and neural activity in the resting state when compared to the other regions.

Analogous observations were made using fMRI. During the presentation of external stimuli like emotional or cognitive tasks, these regions demonstrate predominantly negative signal changes. These negative signal changes are called deactivation or negative BOLD response (NBR); their exact underlying physiological mechanisms remain unclear, though, at this point in time. The regions showing such deactivation or NBR during stimulation must be distinguished from those that show activation or positive BOLD responses (PBR) in fMRI.

This has led to the distinction between task-positive and task-negative regions (see below for details). Task-positive regions include regions like the lateral prefrontal cortex. These show positive signals (PBR) during tasks. Meanwhile, task-negative regions are those that show negative signals (NBR) during tasks or stimuli in the regions of the DMN. The regions within the DMN are connected positively to each other, as are the regions within the task-positive network. In contrast, task-positive and task-negative regions are correlated negatively to each other. This means that increases in activity and connectivity in one network accompany analogous decreases in the other.

On the whole, current research shows a certain spatial pattern of neural activity across different regions and networks in the resting state. Different networks can be distinguished from each other. Moreover, the configuration of these networks and their inclusion and exclusion of regions may change over time. However, the exact spatial patterns and the principles underlying their continuous dynamic changes currently remain unclear. It is clear that the brain's intrinsic activity is characterized by a specific, yet-to-be-detailed spatial structure or organization.

Current research themes 2

Temporal characterization of the brain's intrinsic activity

We have already discussed that the level of intrinsic activity fluctuates over time. These spontaneous fluctuations occur in different rhythms and thus at different frequencies. Using electrophysiological recordings such as EEG, Llinás (2002) has observed that the intrinsic brain activity exhibits auto-rhythmic electrical oscillations (or synchronizations) across different brain regions. Examples of these regions include the thalamic nuclei and cortical regions. Thus the brain's intrinsic activity can be characterized by rhythmic fluctuations.

Spontaneous signal fluctuations in the low frequency ranges of the BOLD signal can be observed using fMRI (rather than EEG). The spontaneous BOLD fluctuations are to be found in lower frequency ranges including the delta band (1–4 Hz), up-and-down states (0.8 Hz) and infra-slow fluctuations (ISFs) (0.01–0.1 Hz). All three, delta, up-and-down states, and ISFs, are often subsumed under the concept of slow cortical potentials (SCP). This is because they can be measured using an EEG (with a special amplifier).

These SCP seem to be related to the spontaneous BOLD fluctuations; both represent fluctuations in cortical excitability across time. This, in turn, may affect the spiking activity of neurons at the cellular level, as well as activity changes at the regional level. However, it remains unclear how spontaneous cellular activity at the level of neurons translates into spontaneous fluctuations in (and across) regions during the resting state.

In addition to the low frequency fluctuations, we have already encountered higher frequency fluctuations in the gamma range between 30 and 40 Hz. These higher frequency fluctuations of neural activity are also present in the resting state, signifying intrinsic activity. They are, however, rather weak in comparison to low frequency fluctuations. However, the higher frequency fluctuations, and especially the gamma rhythms, become stronger during stimulus-induced activity. Hence, the extrinsic stimuli shift the balance from the lower frequency to the higher frequency fluctuations.

How are low- and high-frequency fluctuations related to each other? Each frequency fluctuation can be characterized by its amplitude (its power) and the timing of its onset (its phase onset). Recent investigations have demonstrated that the phase onset – and thus the timing of the low frequency fluctuations – predicts and determines the degree of power of the high frequency fluctuations. For instance, the phase onset of low frequency fluctuations like delta (1–4 Hz) predict the degree of power of the higher gamma frequency fluctuations. More technically, the low frequency fluctuations align the higher frequency fluctuations' power to their phase onsets. This is called phase–power coupling.

Another finding is that the phase onset of the high frequency fluctuations may be linked to the phase onset of the lower frequency fluctuations. For example, two higher frequency fluctuations may occur within longer phase durations, the low frequency fluctuations. More technically, the phases of low- and high-frequency fluctuations become aligned with each other. This is called phase–phase coupling.

Phase–phase coupling suggests that low- and high-frequency fluctuations in neural activity are aligned and thus related to each other. By showing phase–phase and phase–power coupling, low- and high-frequency fluctuations become structured and organized in a certain way, constituting what is called temporal structure or the organization of the intrinsic activity. As in the case of the spatial structure, the temporal structure is highly dynamic and changes continuously, even in the resting state.

Current research themes 3

Interaction between resting state and stimuli

We currently do not know the exact mechanisms and principles by which the brain constitutes the spatial and temporal structure during intrinsic activity. Moreover, we do not know how the spatial and temporal structure of the intrinsic activity is modified by extrinsic stimuli. What we do know from recent experiments is that the intrinsic activity may be relevant in constituting extrinsic stimulus-induced activity. This can be described as the interaction between resting state and stimulus, also known as the rest–stimulus interaction.

Recent studies do indeed lend empirical support to such rest–stimulus interaction. The Dutch researcher N. J. Maandag (Maandag et al., 2007) manipulated the resting state activity level in rats with anesthetic drugs and investigated their neural activity changes. The different resting state levels led to different neuronal activity patterns in the cortex during subsequent movement: some regions were active only during a specific resting state activity level, while others were recruited only in the other resting state level.

Analogous results have been observed in humans. Here, the degree of pre-stimulus resting state activity predicts the degree of subsequent stimulus-induced activity. The higher the preceding level of the resting state activity, the higher the level of the subsequent stimulus-induced activity. This has been demonstrated in both the visual and the auditory cortex.

In addition, the preceding resting state activity level in other regions, including the lateral prefrontal cortex and the midline regions, may also predict the subsequent stimulus-induced activity level in the visual and auditory cortex. The interaction between resting state activity and stimulus thus occurs not only within the same region (intra-regionally), but also across different regions (trans-regionally).

Current research themes 4

Non-linear interaction between resting state and stimuli

How exactly do resting states and stimuli interact with each other? The stimulus-induced activity, as related to the extrinsic stimulus, is not simply added or superimposed on the ongoing intrinsic activity of the brain. This would imply that the latter, the intrinsic activity, remains more or less unaffected. Instead, the results suggest a more intimate and closer interaction between resting state and stimulus.

Resting state and stimulus seem to interact in such a way that the resulting degree of stimulus-induced activity is either higher or lower than simply their addition. The stimulus must thus trigger certain currently unknown features in the resting state that allow it to interact with the stimulus in a non-additive and thus non-linear way.

How could non-linear rest–stimulus interaction be mediated biochemically? Such rest–stimulus interaction may be strongly dependent upon certain biochemical substances – transmitters like GABA, glutamate or dopamine. Studies performed on humans have demonstrated that the degree of stimulus-induced activity in the perigenual anterior cingulate and the visual cortex is dependent upon the resting state concentration of GABA in the same regions. The higher the concentration of GABA in the resting state, the higher the degree of non-linearity in the rest–stimulus interaction, and the higher the subsequently resulting stimulus-induced activity. Despite these results, this remains a tentative hypothesis that needs to be investigated in further detail.

Besides GABA, the resting state concentration of other transmitters like glutamate, serotonin, or dopamine, also predict the degree of stimulus-induced (or task-related) activity during, for instance, reward, cognitive conflict, or empathy.

In addition to rest–stimulus interaction, there is also reverse traffic that is stimulus–rest interaction. In that case the stimulus-induced activity affects and modulates the subsequent resting state activity. Lewis et al. (2009) demonstrated that visuospatial learning did not only lead to activation changes in the visual cortex but also to functional connectivity changes between the visual cortex and the DMN during the following resting state period (see Northoff *et al.*, 2010a, 2010b for review).

Similar observations were made in studies on working memory and self-relatedness. Higher degrees of working memory and self-relatedness lead to stronger connectivity and activity changes within the DMN during subsequent resting state periods when compared to lower degrees of working memory and self-relatedness. These results suggest that there may be bilateral traffic between resting state and stimulus-induced activity, and thus between intrinsic and extrinsic activity. Whether and how such bilateral traffic is important in yielding consciousness remains unclear at this point.

Ideas for future research 1

Intrinsic activity and consciousness

What does the brain do in the resting state? There is nothing but a bunch of neurons in our brains. But these neurons themselves seem to use a lot of energy. This suggests that our neurons are always active – they are never at rest; even if we feel at rest, our brains are never completely inactive.

Unlike a car's engine that we can simply turn off when we no longer want to drive, the brain is continuously active. This means that it seems continuously to constitute some kind of temporal and spatial structure. The brain is an engine that never stops working and, unlike the car, keeps us continuously in motion. This can be observed, for example, even in sleep when we are dreaming. Why does the brain seem to continuously and dynamically constitute some kind of temporal and spatial structure? We currently do not know the exact features of the brain's spatial and temporal structure or organization. Nor do we know why and how the brain's intrinsic activity developed such a spatial and temporal structure.

How is rest–stimulus interaction related to consciousness? The research group around the German-French neuroscientist and neurologist Andreas Kleinschmidt investigated what is described as 'bi-stable perception'. One may perceive the same picture as either a vase or an old woman. This is called bi-stable perception. It describes the fluctuation between two different percepts with regard to one and the same stimulus.

How is bi-stable perception possible? Kleinschmidt and his colleagues observed that the degree of preceding resting state activity in those regions processing the vase and the women predicts whether one perceives the stimulus – in this case, a picture – as either a vase or an old woman. Hence, the resting state activity itself seems to be central in selecting the kinds of contents for subsequent consciousness.

These studies shed light on why certain contents and not others are associated with consciousness. The studies are thus concerned with what we described above as the content-NCC; this implies a distinction between consciousness and unconsciousness. In contrast, the studies do not tell us anything about the distinction between consciousness and nonconsciousness: that is, the hard problem. In order to address the latter distinction, we need to account for the temporal and spatial structure of the brain's intrinsic activity.

Ideas for future research 2

Spatiotemporal structure and consciousness

By constituting a temporal and spatial structure, the brain's intrinsic activity has a tool or means that it imposes upon all extrinsic stimuli. As such, the brain's intrinsic activity may provide a grid, template, or schemata along whose lines all subsequent forms of neural activity, intrinsic and extrinsic activity, is organized and structured.

That the brain's intrinsic activity seems to provide an organizational template has already been described by the American psychologist and neuroscientist Karl Lashley (1890–1958). He says, 'A second point of major importance is that the nervous system is not a neutral medium on which learning imposes any form of organization whatsoever. On the contrary, it has definite predilections for certain forms of organization and imposes these upon the sensory impulses that reach it. In its functional organization, the nervous system seems to consist of schemata or basic patterns within which new stimuli are fitted' (Lashley, 1949: 35).

How are these schemata of intrinsic activity related to consciousness? We currently do not know. We know that there is a certain temporal and spatial structure in the brain's intrinsic activity, but we do not know how it is related to consciousness. More specifically, we do not know whether it creates the brain's tendency to constitute consciousness rather than nonconsciousness. In other words, we currently do not know whether the brain's intrinsic activity correspond to what we describe as neural predispositions of consciousness (NPC) or not.

Despite our lack of knowledge, we may nevertheless formulate at least some kind of criteria for how the brain's intrinsic activity must be spatially and temporally structured. One of the central phenomenal features of consciousness are spatial and temporal continuity. Spatial and temporal continuity describe how the contents of consciousness are always already embedded in a spatiotemporal grid that provides temporal flow and spatial integration in our experience. Instead of being segregated in time and space, the different contents are spatially and temporally linked in our consciousness.

Despite their occurrence at different discrete points in physical time, we nevertheless experience a temporal continuum, a transition, between the different contents in our consciousness. This temporal continuum in consciousness does not seem to correspond one-to-one to the objective time and its discrete points in time as we observe them in third-person perspective. Instead, we experience a continuum between different discrete points in our consciousness and thus what is described phenomenally as 'dynamic flow' (James) or 'phenomenal time' (Husserl).

While there has been much debate about time and consciousness, there has been less discussion about the experience of space in consciousness. Analogously to time, one may want to make a similar distinction. The contents in consciousness are not experienced at their different discrete points in physical space. Instead, they are embedded and integrated into a spatial continuum with multiple transitions between the different discrete points in physical space. As in the case of time, the contents are woven into a spatial grid or template that emphasizes continuity and transition over discontinuity and segregation.

How is a spatiotemporal grid and its spatial and temporal continuity constituted? One may now assume the brain's intrinsic activity to be central in constituting the spatiotemporal grid. If so, one would assume that the neuronal features underlying the spatial and temporal structure of the brain's intrinsic activity discussed above would account for the spatial and temporal features as we experience them in consciousness. One would then expect the temporal distances experienced between different contents in consciousness to correspond to the temporal distances constituted in the temporal structure of the brain's intrinsic activity.

Neurophilosophical discussion 3

NPC and the hard problem in non-reductive neurophilosophy

How is the brain's intrinsic activity and its suggested spatiotemporal structure related to the hard problem and the explanatory gap problem? Recall that the hard problem raises the question: why is there consciousness rather than nonconsciousness? The answer is clear. Because the brain shows intrinsic activity that, because of its presumed spatiotemporal structure, predisposes possible consciousness. The predisposition for possible consciousness can be transformed into the manifestation of actual consciousness in the 'right' context – the 'right' extrinsic stimulus and its 'right' interaction with the intrinsic activity.

While possible consciousness is supposed to be mediated by the neural predispositions, the intrinsic activity and its spatiotemporal structure, actual consciousness, may be related to neural prerequisites and correlates. Neural prerequisites concern the necessary or enabling conditions for consciousness, while neural correlates refer to the sufficient conditions of actual consciousness. All three neural predispositions, neural prerequisites and neural correlates, may be related to different neuronal mechanisms which have to act in conjunction in order to constitute consciousness rather than nonconsciousness.

In short, I assume the conjunction of neural predisposition, prerequisites and correlates as an empirical answer to the hard problem. Only if all three go together will we be able to explain consciousness fully. The neural predispositions will account for the necessary conditions of possible consciousness; the neural prerequisites for the necessary conditions of actual consciousness; and the neural correlates will reveal the sufficient conditions of actual consciousness.

Such an approach presupposes non-reductive neurophilosophy. It is neurophilosophical in that it puts a mental feature, that is, consciousness, and an associated philosophical problem, the hard problem, into the context of the brain and neuroscience. Moreover, it is non-reductive because it does not limit the neural conditions only to the actual manifestation of consciousness but also includes those for its possible manifestation. This means that the originally philosophical distinction between possibility and actuality is here drawn and transferred from the logical world of philosophy to the natural world of neurophilosophy and neuroscience.

Neurophilosophical discussion 4

NPC and the hard problem in neuroscience and reductive neurophilosophy

How can such a non-reductive neurophilosophical approach to consciousness be distinguished from philosophical, reductive neurophilosophical and neuroscientific

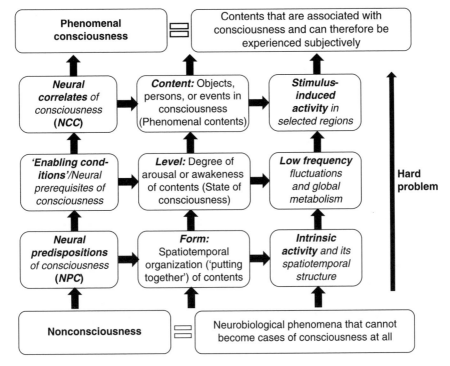

Figure 15.1 Hard problem and neural predispositions, prerequisites and correlates of consciousness

The figure shows the three main dimensions of consciousness (left middle row): consciousness (left row); the underlying neuronal mechanisms (right middle row); and alterations in corresponding disorders (right row).

Content refers to the objects and events in consciousness, the phenomenal contents as the philosophers say. The contents are the main focus in the various neuroscientific suggestions for the neural correlates of consciousness (NCC). They imply stimulus-induced activity and are altered in patients with selective brain lesions.

Level refers to the different degrees of arousal and awakeness, and thus to the state of consciousness. The level or state of consciousness is related to global metabolism and energy supply, which are found to be impaired and highly reduced in disorders of consciousness like vegetative state and coma. Moreover, neural activity in the brain stem and midbrain is supposed to play an essential role in maintaining arousal. This reflects what is described as the 'enabling conditions' or 'neural prerequisites' of consciousness.

Form describes the spatiotemporal organization and structuring ('putting together') of contents in consciousness. As such they make for the neural predispositions of consciousness (NPC) and are related to the resting state activity and its spatial and temporal structure. This seems to be abnormal in, for instance, psychiatric disorders like depression or schizophrenia.

In the upper and lower parts of the figure the concepts of phenomenal consciousness and nonconsciousness are signified. The upwards arrows indicate that each level is based on the respective previous one. Hence the move from neural predisposition may allow us to bridge the gap between nonconsciousness and consciousness.

approaches? Let's start with neuroscience. The neuroscientific approach considers only actual consciousness, that is, what is and actually occurs and can be observed in the natural world. This limits the neuroscientist to the necessary and sufficient conditions of actual consciousness that are well reflected in the current concepts of 'enabling conditions', 'neural prerequisites' and neural correlates.

The concept of neural predisposition may, in contrast, sound strange to the neuroscientist. Why? Since he/she focuses exclusively on the actual occurrence of consciousness and its neural conditions, he/she may consider the search for neural conditions of possible (rather than actual) consciousness as strange, if not superfluous and confused. If at all, he/she may claim, that this is the domain of philosophy rather than neuroscience.

How about the reductive neurophilosopher? Given his/her background in philosophy, he/she may be sensitive to the distinction between possible and actual consciousness. However, he/she will most likely associate that distinction with the logical rather than the natural world. The distinction between possible and actual worlds may be relevant from a purely logical point of view: logical worlds are only possible, that is, logically possible (as claimed by logical arguments) but not actually realized and thus a mere fictive possibility for the reductive neurophilosopher. In contrast, the natural world is actual and real, a real actuality, since otherwise we would not be able to observe it.

What does this imply for the stance of the reductive neurophilosopher on neural predispositions? The reductive neurophilosopher focuses exclusively on the natural world while declaring the logical world to be irrelevant. Such proclaimed irrelevance of the logical world entails the irrelevance of those conditions underlying the possibility of the logical world. Consequently, the search for neural predispositions as the necessary conditions of possible consciousness may be considered a confusion between natural and logical worlds, and thus between real actuality and fictive possibility.

How does this affect the hard problem? If at all, the hard problem is here regarded as merely an empirical problem. Once the neural prerequisites and the neural correlates are found, the hard problem will be solved. The hard problem thus becomes a mere empirical problem that can be solved by neuroscience alone so that there is no need for philosophy any more.

Neurophilosophical discussion 5

NPC and the hard problem in philosophy

How about the philosopher? In contrast to both reductive neurophilosopher and neuroscientist, he/she does not come from the natural world and thus real actuality but from the logical world and its mere possibilities. Unlike the situation for the reductive neurophilosopher, the logical world is not a fictive possibility, but rather a logical possibility. That logical possibility, however, has to be distinguished from those possibilities that are actually realized in our natural world. Hence, he/she would associate any kind of possibility with the logical world exclusively, whereas the natural world can only be characterized by actualities.

The philosopher presupposes linkage of the distinction between natural and logical worlds with the distinction between actuality and possibility. Natural world = actuality and logical world = possibility. Any cross-over linkage of, for instance, possibility with the natural world amounts simply to confusion. The concept of neural predisposition is then regarded as confused, since it links the natural world with possibility rather than actuality.

How does this affect the hard problem? The hard problem is considered to be merely a logical problem. It is about the logical possibility of consciousness rather than the natural actuality; that is, how consciousness is realized and manifested. Consequently, the hard problem in principle cannot be solved by neuroscience, since that would mean simply to confuse logical and natural worlds and thus possibility and actuality.

Neurophilosophical discussion 6

Explanatory gap problem and the brain's intrinsic activity

What about the explanatory gap problem? The explanatory gap problem consists in the gap between phenomenal and neuronal knowledge. Our knowledge about neuronal states does not include any knowledge about the phenomenal features of consciousness and vice versa. There is thus a gap between both types of knowledge. Why is there such a gap? We have no direct access to our brain's neuronal activity as such. Nobody ever perceived his/her own or others' neuronal states as such. We thus have only indirect access to the neuronal states (see also Chapters 1 and 4).

Imagine the following. If we had direct access to our own and others' neuronal states, we would be able to access and experience directly how our brain's neuronal states are transformed into phenomenal states and thus consciousness. We could then access our intrinsic activity and its spatiotemporal structure directly. Moreover, we would then be able to access directly the interaction of the intrinsic activity with extrinsic stimuli and how that makes the stimuli's association with consciousness possible. However, we do not possess this type of direct access, nor are we able to access our brain's intrinsic activity directly. We are also unable to access and observe directly the mechanisms underlying the interaction of the brain's intrinsic activity with extrinsic stimuli.

Since we have no direct access to either the intrinsic activity itself, or its interaction with extrinsic stimuli, there is a gap between our neuronal knowledge and our phenomenal knowledge. We have neural knowledge about the brain's intrinsic activity as observed in third-person perspective. We have phenomenal knowledge about our experiences in first-person perspective. In contrast, due to our lack of unmediated online, access to our own brain as brain and its neural states while being conscious, we have no direct knowledge at all about how the brain's intrinsic activity as neural activity is transformed into phenomenal states and consciousness.

Neurophilosophical discussion 7

Explanatory gap problem and the NPC

Why is there such a lack of direct access to our own brain and its neural states while being conscious? One may speculate that this may be because of the same unknown neuronal processes that predispose possible consciousness – the neuronal predispositions of consciousness. More specifically, the intrinsic activity itself, and its spatiotemporal structure, may be central in preventing us from directly accessing intrinsic activity and its interaction with the extrinsic stimuli.

If so, our brain's predisposition for possible consciousness comes at an epistemological price, namely, that we cannot directly access the brain's intrinsic activity. Instead we will have to develop the technological tools to overcome these limitations and to make at least indirect access to consciousness possible.

Epistemically, this implies that the brain's neural predisposition for consciousness is accompanied by a principal limitation in our knowledge, the impossibility of direct access to the brain's intrinsic activity in purely neuronal terms. If the intrinsic activity does, on the one hand, predispose consciousness and, on the other, prevent direct access to it as intrinsic activity, both possible consciousness and this epistemic limitation are closely linked. This would mean that consciousness is impossible without the limitation.

We would have no consciousness without the limitation; that is, the explanatory gap. Hence, the price we pay for having consciousness is to suffer from an explanatory gap. We could resolve the explanatory gap only by giving up consciousness. That, however, would mean that we would not be able to experience the joy of no longer suffering from the explanatory gap because then we would have simply no consciousness and thus no longer any experience. Nothing comes without a price; either way, nature in general and, more specifically, the possible linkage between consciousness and the explanatory gap, make us pay a price. This, however, is a rather speculative neuroepistemological assumption that needs to be investigated both empirically and epistemically in the future.

Neurophilosophical discussion 8

NPC and absent and inverted qualia

What about the absent qualia argument? This concerns the absence of qualia in the presence of the same brain with the same neuronal and behavioral states. Can qualia remain absent in the presence of the brain's intrinsic activity and its spatiotemporal structure? The stimulus must interact with the intrinsic activity in a certain way in order for it to be associated with consciousness. In this case qualia cannot remain absent, implying that the absent qualia argument is not empirically plausible.

If the brain's intrinsic activity does indeed presuppose consciousness and thus qualia, in the case of the 'right' extrinsic stimulus and its 'right' interaction with the brain's intrinsic activity, qualia cannot remain absent. In other words, the intrinsic activity makes the occurrence of qualia necessary if the 'right' stimulus occurs. Accordingly, the absent qualia argument would be empirically implausible. While it may still apply to the logical world, the absent qualia argument can no longer be feasible within our actual world.

If, in contrast, one considers only the intrinsic activity (without its interaction with extrinsic stimuli), qualia may still remain absent. Extrinsic stimuli may, for instance, induce almost 'normal' stimulus-induced activity, which might not interact with the intrinsic activity. In this case, qualia may remain absent. This would lend empirical plausibility to the absent qualia argument.

Is the vegetative state as defined by the absence of consciousness an instance of absent qualia? No, because in addition to the absence of qualia, there are also major changes in the brain and its neuronal activity. This violates the condition of the absent qualia argument which argues that the brain and its neuronal states remain exactly the same.

What about the inverted qualia? We recall from Chapter 13 that inverted qualia concern the dissociation between content and experience. The content may be the color red, but someone might experience 'blueness' even though there is no blue content.

Is this case empirically plausible? One would have to assume an abnormal interaction of the extrinsic stimulus with the brain's intrinsic activity. The extrinsic stimulus and its content red is then associated and interacts with those regions in the brain's intrinsic activity that process blue rather than red. The extrinsic stimulus and its content red may thus encounter an abnormally wired intrinsic activity in the brain, which then associates an abnormal experience, here 'blueness', with the extrinsic stimulus and its respective red content.

Neurophilosophical discussion 9

NPC and what it is like, and knowledge arguments

The 'what it is like' argument refers to the point of view from which experiences are made. This point of view makes subjective experience possible.

How does the 'what it is like' argument relate to the intrinsic activity and its spatiotemporal structure as a possible candidate for the NPC? The intrinsic activity and its spatiotemporal structure as described so far do not, in fact, bear any relationship to point of view and qualia. More technically, the intrinsic activity and its neuronal features are currently not necessary for the occurrence of phenomenal features and thus consciousness.

If the NPC do predispose possible consciousness, they must also show how they predispose and make point of view possible. The constitution of point of view will be the focus in the final part of this book, the part on the self and the brain.

What about the knowledge argument? The knowledge argument describes the principal difference between first- and third-person knowledge. Mary is a famous neuroscientist. She is supposed to have complete third-person knowledge about the brain and its neuronal mechanisms, while due to her color-blindness from birth, she completely lacks first-person knowledge of the phenomenal features of consciousness. Can she account for consciousness? 'No,' says the philosopher, 'and therefore consciousness cannot be reduced to the brain.'

What exactly is first-person knowledge? The first-person perspective presupposes some kind of self as the subject of experience, the one who is conscious and who makes or experiences consciousness. Accordingly, first-person knowledge concerns experience and thus consciousness. Most important, it concerns the experiences the own person makes, the own self. Hence, first-person knowledge is knowledge about the self, the own self and the self in general, as such. Accordingly, the question for the specific nature of first-person knowledge presupposes the problem of self, what a self is and how it is generated.

If now the intrinsic activity claims to be a possible candidate mechanism for the NPC, it must be able to address the question of the self. How is the brain's intrinsic activity and its spatiotemporal structure related to the self, the subject of experience, and its first-person knowledge? This will be the guiding question in the next part, where we will discuss the neurophilosophy of the self.

Take-home message

Chapter 13 focused on various philosophical arguments against reducing consciousness and its phenomenal features to the brain's neuronal features. This was complemented in Chapter 14 by discussing the recent neuroscientific suggestions for the neural correlates of consciousness (NCC), the sufficient neural conditions of actual consciousness. However, these various suggestions suffer from unspecificity. How can we establish a tighter and more specific link between the brain's neuronal features and the phenomenal features of consciousness? This is the focus in the present chapter. Here, we focused on the brain and its intrinsic activity. However, as we demonstrated, intrinsic activity on its own does not provide any link to consciousness. Furthermore, we had to further specify intrinsic activity, including, for instance, in spatiotemporal terms. Spatiotemporal structures may provide a neural predisposition of consciousness (NPC), the necessary neural conditions of possible consciousness. However, both the exact spatiotemporal structure of the brain's intrinsic activity, and its relation to consciousness, are currently unclear and remain subject to future research. What is clear, however, is if that is solved, the NPC may provide novel empirically-based answers to the various problems with reducing consciousness to the brain.

Summary

This chapter discusses the neuroscience of consciousness. The state or level of consciousness, rather than the contents of consciousness, have been the focus. This leads to the distinction between level-based and content-based NCC. The level of consciousness is impaired in neurological disorders of consciousness that include the vegetative state, anesthesia, epilepsy and sleep. Based on the data relating to these disorders, suggestions for the level-based NCC have been made that all focus on the brain's intrinsic activity as distinguished from its extrinsic stimulus-induced activity. These include the 'slow wave hypothesis' that argues in support of the central role of the slow frequency fluctuations as a hallmark feature of the brain's intrinsic activity in yielding consciousness. Another level-based NCC is assumed in the brain's global metabolism and energy demand, which is reduced in disorders of consciousness. The consideration of the brain's intrinsic activity suggests that it may possess certain unknown features that predispose and thus create the brain's tendency to yield consciousness rather than nonconsciousness. Therefore, the brain's intrinsic activity may signify what can be described as 'neural predispositions of consciousness' (NPC) as necessary neural conditions of possible consciousness. As such, the NPC must be distinguished from the NCC. The NCC refer to the sufficient neural conditions of actual consciousness. How can the brain's intrinsic activity serve as the NPC? One possible way is that the brain's intrinsic activity provides the form or organization of consciousness as manifested in its various phenomenal features. This is uncharted and novel territory that needs to be explored. It is clear, however, that the NPC may be able eventually to provide an answer to the philosophical arguments like the hard problem.

Revision notes

- What are disorders of consciousness? How can their clinical symptoms be described?
- Explain the distinction between content-based NCC and level-based NCC?
- How can we distinguish between intrinsic and extrinsic activity in the brain?
- What are low frequency fluctuations? How are they possibly related to consciousness?
- Does the brain's intrinsic activity require global metabolism and energy demand? If so, describe it in more detail. How is global metabolism altered in disorders of consciousness?
- What is the distinction between actual and possible consciousness?
- Why do we need to distinguish between neural correlates of consciousness (NCC) and neural predispositions of consciousness (NPC)?
- What is meant by the form of consciousness? And how is it related to the brain and its intrinsic activity?
- Why and how do the neural predispositions of consciousness (NPC) provide an answer to the philosophical problems of consciousness like the 'hard problem'?

Suggested further reading

- He, B. J. and Raichle, M. E. (2009) 'The fMRI signal, slow cortical potential and consciousness', *Trends in Cognitive Sciences*, 13(7), 302–9.
- Laureys, S. (2005) 'The neural correlate of (un)awareness: Lessons from the vegetative state', *Trends in Cognitive Sciences*, 9(12), 556–9.
- Northoff, G. (2013a/2013b) *Unlocking the Brain. Vol. I: Coding, Vol. II: Consciousness* (Oxford/New York: Oxford University Press).
- Northoff, G., Qin, P. and Nakao, T. (2010) 'Rest–stimulus interaction in the brain: A review', *Trends in Neurosciences*, 33(6), 277–84.
- Shulman, R. G., Hyder, F. and Rothman, D. L. (2003) 'Cerebral metabolism and consciousness', *Comptes Rendus Biologies*, 326(3), 253–73.

16
Conceptual, Phenomenal and Methodological Issues in the Investigation of Consciousness

Overview

So far we have discussed the concept of the mind and how it relates to the brain. In philosophy, various suggestions have been put forward to explain the mind–brain relationship in general terms. These suggestions are complemented by the various arguments against reducing consciousness to the brain. This leaves open for discussion the topic of how we might characterize consciousness itself. Among the wide-ranging discussions in philosophy and neuroscience about consciousness over the past 20–40 years, there has been a consensus that consciousness is 'located' at the junction between the mind and the brain. Therefore any discussion about consciousness must include conceptual, phenomenal and methodological issues. The present chapter focuses on the conceptual and phenomenal characterization of consciousness, as well as on the kinds of methodological strategies required to investigate consciousness empirically.

Objectives

- Understand the features that characterize consciousness as consciousness
- Provide a description of temporal and spatial continuity in consciousness
- Give the definition of unity and self-perspective and intentional organization
- Understand 'what it is like' and qualia
- Differentiate conceptually between consciousness and unconsciousness, including their different forms
- Definition of the nonconscious
- Understand the difference between the building block approach and the unified field model

Key concepts

'What it is like', phenomenal features of consciousness, unity of consciousness, spatial and temporal continuity, self-perspectival and intentional organization, first- and third-person perspective, unconscious, phenomenal and access consciousness, nonconscious, building block approach, unified field model.

Background 1

Investigation of consciousness: NCC and NPC

Consciousness is a multifaceted phenomenon that includes empirical, phenomenal, conceptual and epistemological (and metaphysical) dimensions. We therefore discussed the concept of consciousness in different ways. In the first chapter we presented arguments against the physical and ultimately neuroscientific explanation of consciousness. These included various arguments that pointed out phenomenal, epistemic, logical and explanatory features of consciousness that seem to resist any explanation in the brain's neuronal states. By developing these arguments, philosophy has put forward the required criteria that a neuroscientific explanation of consciousness should meet.

After discussing the philosophical issues in Chapter 13, we turned to the current neuroscientific findings in Chapter 14. More specifically, we discussed the current neuroscientific hypotheses for the neural correlates of consciousness (NCC), the sufficient neural conditions for consciousness. This was followed by a discussion about how the various NCC hypotheses relate to the different arguments as discussed in Chapter 13. This led us to conclude that the current NCC do not really fulfill the required criteria and reveal major shortcomings.

In Chapter 14, we discussed some alternative options presented by findings in current neuroscience. Instead of focusing mainly on stimulus-induced activity, we shifted our focus to the brain's intrinsic activity – its resting state activity. Rather than deciphering the sufficient neural conditions of actual consciousness, the NCC, the intrinsic activity might provide the necessary conditions for possible consciousness, which we described as neural predispositions of consciousness (NPC). While we currently lack detailed knowledge about the structure and organization of the brain's intrinsic activity, the search for the NPC may nevertheless provide a good starting point. Why? Because it promises to fulfill the criteria established by the various philosophical arguments in a more coherent way than the NCC.

Background 2

Investigation of consciousness: linkage between first- and third-person perspectives

Where does this leave us? The discussions in Chapters 14 and 15 clearly emphasized that we need to specify the neuronal mechanisms if we want to link them to consciousness. For instance, we do not know the spatial and temporal structure of the brain's intrinsic activity and whether or not it corresponds to the structure and organization of the spatiotemporal features in consciousness. In addition to the empirical features, we may also need to refine our conceptual definitions, our phenomenal descriptions and our methodological tools in a more differentiated way. This is the focus of the present chapter.

Thus far we have assumed conceptually that the concept of consciousness is homogeneous. There may, however, be different forms of consciousness with different features that may need to be distinguished. We also need to sharpen how consciousness differs from related concepts like the unconscious and the nonconscious.

In addition to conceptual refinement, we also need to detail our descriptions of the phenomenal features of consciousness. There may be other phenomenal features besides the phenomenal-qualitative features.

Finally, we also encounter methodological problems in the empirical research on consciousness. Consciousness is, by definition, tied to first-person perspective. It describes subjective experience and cannot be experienced in third-person perspective. This distinguishes consciousness from observation of the brain's neuronal states. Neural states can only be observed in third-person perspective.

When investigating the neuronal mechanisms of consciousness in neuroscience, we are thus confronted with the question: how can we account for something that occurs in first-person perspective by observation in third-person perspective? Can we link the third-person-based observations in neuroscience to the first-person-based experience of consciousness? And if yes, how?

Biography

Ned Block

In order to talk about consciousness clearly, we need to specify what we mean, conceptually, by the word consciousness. What, exactly, do we refer to when we speak about consciousness? This type of specification becomes even more urgent when considering the different concepts of consciousness as they are used in both philosophy and neuroscience.

The philosopher Ned Block from New York made a distinction between two different forms of consciousness – phenomenal consciousness and access consciousness (Block, 1995, 2007). In short, phenomenal consciousness is about experience itself and its phenomenal features. Access consciousness is about how we can access and become aware of our experiences, our phenomenal consciousness. Both phenomenal and access consciousness will be explained further below.

Concept of consciousness 1

Access consciousness

Let us start with access consciousness. Imagine that you are eating Chinese hot and sour soup. You usually like hot and sour soup, but today you experience it as particularly sour; it does not taste as good as it usually does. Here, in this example, the soup tasting sour is the content of your consciousness. This means that your consciousness is about the sour soup in the same way that your consciousness can also be about reading this book. Both reading this book and tasting the sour soup are the contents of your consciousness.

You become aware that the soup tastes sour to you. You access the contents of your consciousness, the sour soup. That in turn, makes it possible for you to become aware, or better meta-aware, of them as contents of your consciousness. Since you access the contents of your own consciousness and become consciously aware of them as the contents, Block calls this form of consciousness 'access consciousness'. *The concept of 'access consciousness' describes the ability to access the contents of consciousness and thereby to become aware of them as contents of consciousness.*

Once you are able to access the contents of your own consciousness, you are also able to detect, recognize and report them. For instance, you may now recall that the sour taste of your soup reminds you of a more or less similar sour taste you had when ordering lemon juice yesterday. For such a comparison to be possible, you must access the contents of your consciousness, identify and detect and recognize them as such, and then you need to compare them with previous contents. Access consciousness can be considered the door opener to these various processes.

The processes of access, detection, recognition and comparison require the involvement of various cognitive functions like working memory, attention, executive functions, etc. Access consciousness has thus been closely related to cognitive functions. As such, access consciousness is targeted by many approaches that focus on the NCC, the neural correlates of consciousness. Why? Because many of the experimental paradigms used in these studies require the subject to detect and report their conscious content. This is then taken as an index of consciousness as distinguished from those instances where the subjects remain unable to report the respective content.

For instance, many studies of the neural correlates of consciousness present subjects with stimuli signifying colors or objects. They present them in either a long duration (> 100ms) so that subjects are easily able to perceive these stimuli and to judge the kinds of content they see. This is compared to trials with shorter duration where subjects afterwards remain unable to detect, report and judge the kinds of content they perceived, implying unconscious rather than conscious processing. The main comparison thus consists in the difference between the ability to report and detect versus the inability to report and detect. Accordingly, what is investigated is thus access consciousness, the ability to access the contents of one's own consciousness, as compared to lack of access consciousness and inability to access the contents.

Concept of consciousness 2

Phenomenal consciousness

In addition to access consciousness, Ned Block also speaks of phenomenal consciousness. What does the concept of phenomenal consciousness mean? By using the term 'phenomenal', Block refers to experience itself, the experience of contents, as distinguished from access to the contents. *Therefore the concept of phenomenal consciousness describes this experience of contents by itself as distinguished from the access to the contents as associated with access consciousness.*

How can we now characterize the experience itself? The concept of experience implies its characterization by phenomenal features (see Part I). How can we now describe the phenomenal features of experience, and thus of consciousness? We already got to know one major phenomenal feature of consciousness in Chapter 13. In this chapter we discussed the 'what it is like' argument. The 'what it is like' argument is about a specific phenomenal feature of consciousness, the feeling of 'what it is like' to experience a particular content like the sour soup. The taste of the sour soup is experienced as 'sourness'. If the content of your consciousness is the color red of the book in front of you, your experience can be described by 'redness' (see Chapter 13).

Let's give another example. I am aware of sitting in front of the computer while typing these lines. I thus access the phenomenal contents of my consciousness. This is made possible by access consciousness. This needs to be distinguished from my experience of the very same computer and my writing, which is described by the 'what it is like' and ultimately qualia (as qualitative-phenomenal features) of experience and thus of phenomenal states. This is what Block calls 'phenomenal consciousness'. It refers to the phenomenal features like the 'what it is like' of our experience.

How are phenomenal and access consciousness related to each other? Block argues that both the experience of, and the access to, contents can dissociate from each other. We may be able to experience contents in consciousness while, at the same time, we may remain unable to access them as such. For instance, you may experience the sourness of your soup while simultaneously being unable to access it as the content of your consciousness.

How is this dissociation possible? There may be different neuronal mechanisms underlying the access-to and experience-of dimensions of consciousness. Access and phenomenal consciousness may be served by different neuronal mechanisms and systems. That makes their dissociation possible if, for instance, the neuronal mechanisms underlying access consciousness are deficient. Some authors (such as Crick and Koch, 2003) argue that in the case of visual contents, phenomenal consciousness may be related to early neural processing (50–200ms) in the primary visual cortex like V1 and V2, while later processing (200–500ms) in subsequent regions like V2–V5 and the prefrontal cortex may make it possible to access and detect the contents, thus access consciousness.

Above, we saw that, experimentally, most studies presuppose active access to and detection of the contents of consciousness, thus implying access consciousness. What would an experimental design for distinguishing phenomenal consciousness from access consciousness look like? To answer this question one may want to target those periods of pure experience or consciousness prior to the description and verbalization of that very same experience. How can we measure and quantify those periods of pure experience? We currently do not know how to measure and quantify the phenomenal features of a respective experience reliably.

The conceptual distinction between phenomenal and access consciousness is quite popular in both current neuroscience and philosophy. In addition, several other distinctions, especially regarding access consciousness, have been suggested by various authors. Because of space constraints, we will not have the time to discuss them in detail here.

What does the conceptual distinction between phenomenal and access consciousness imply for the neuroscientific investigation of consciousness? As indicated here, the conceptual distinction between phenomenal and access consciousness may correspond to a neuronal differentiation in the brain. If so, the distinction between phenomenal and access consciousness must be empirically plausible. However, more detailed neuroscientific investigations with experimental paradigms specifically targeting this distinction are necessary to further support this assumption.

Concept of consciousness 3

Consciousness and the unconscious

In addition to the distinctions within the concept of consciousness itself, we also need to distinguish it from the concept of the unconscious. What is the unconscious?

Notions of the unconscious can be traced back as far as to Plato and Aristotle, and have been elaborated ever since in both philosophical and psychological thought. You may also think of Sigmund Freud, the founder of psychoanalysis, as the master of the unconscious, who aimed at revealing the secret and hidden messages coded in our unconscious.

Whatever frameworks have been suggested in the past, unconscious states have been characterized most frequently by the supposedly hidden characteristics of a person's self (fate, temperament, soul, character) that need to be inferred, and cannot be accessed directly. These hidden characteristics were distinguished from those that were believed to be transparent, experienced directly, open to introspection, and thus accessible to consciousness.

The Californian philosopher John R. Searle (Searle, 2004, pp. 165–72), whom we have already encountered several times, distinguishes between different types of unconsciousness. He first speaks of the 'preconscious' – a state that is on the verge of becoming conscious thought, though not as yet conscious itself. It resembles what Sigmund Freud described as the 'preconscious system'.

Imagine you are standing in the kitchen and starting to cook. You focus on the different kinds of vegetables that lie in front of you, yet you are not yet conscious of the cooking pot standing behind the vegetables. However, the cooking pot is on the verge of entering your consciousness; it is in your preconscious. Now you suddenly become aware of the pot, which has thus moved from the preconscious to the conscious.

Another concept of the unconscious is the 'dynamic unconscious' – 'unconscious mental states function causally, even when unconscious' (Searle, 2004, p. 167). Unlike in the case of the preconscious, the dynamic unconscious is not on the verge of becoming conscious. Instead it remains unconscious. This corresponds, in some degree, to what Freud referred to as the 'dynamic or repressed unconscious', a term that describes how contents are blocked or suppressed in an attempt at preventing them from becoming conscious. It is important to note, however, that even dynamically unconscious states have the potential of becoming conscious.

How can we illustrate the dynamic unconscious? Imagine that you had a major trauma as a child when your whole family, including you, had a serious car accident where your brother died. That is well stored in your memories and in your dynamic unconscious. Now you observe another car accident while walking on the street.

What happens to you and what do you experience? You now become anxious and afraid, very much like when you experienced your own car accident as a child. Though you do not experience your own car accident as content in your consciousness, it nevertheless affects your experience of the current contents in your consciousness, the car accident you observe. And if things get worse, meaning that your anxiety increases, you may even re-experience your own car accident as content of your consciousness.

Searle's philosophical (and ultimately Freud's psychodynamic) distinction between the 'dynamic unconscious' and the 'preconscious' is mirrored in the more empirically- and neuroscientifically-based distinction between the 'subliminal' and the 'preconscious' made by the French neuroscientist Stanislav Dehaene (Dehaene *et al.*, 2006; Kouider and Dehaene, 2009; Dehaene and Changeux, 2011).

The concept of the 'subliminal' is supposed to describe neural processing where the stimulus remains unconscious. In this case, the stimulus cannot enter consciousness because it is simply too weak to induce the 'right' kind of neural processing, like the suggested 'ignition' of neural activity in a large-scale fronto-parietal network. This is different in the case of the 'preconscious', where the stimulus itself is strong enough. Instead it is the fronto-parietal network that is not ready because it is occupied with other stimuli (see Chapter 14 for extensive discussion of this 'global neuronal workspace' theory (GNW) by Dehaene and Changeux (2005, 2011)).

Concept of consciousness 4

Unconsciousness and nonconsciousness

Let's return to the philosopher John Searle and his concepts of the unconscious. The third concept of the unconscious is what Searle describes as the 'deep unconscious'. *The 'deep unconscious' refers to those states and contents that cannot, in principle, enter consciousness.* There is no way that the contents of the deep unconscious can ever become conscious and thus be experienced. This distinguishes the deep unconscious from both the system preconscious and the dynamic unconscious as discussed above. Both those forms of the unconscious can be characterized by contents that have the potential to become conscious.

In the case of the deep unconscious it remains impossible for these contents to become conscious. According to Searle, the reason for this is because the deep unconscious is not 'the sort of thing that can form the content of a conscious intentional state' (Searle, 2004, p. 168). Before going into the reasons why the content cannot become conscious, let us consider some examples of the deep unconscious.

Searle emphasizes this using the example of the computational rules that we follow unconsciously in acquiring language: while we can be preconscious or dynamically unconscious about the language and its letters, we remain deeply unconscious about the rules and principles of its universal grammar that guide our learning of the language. Hence, the rules that guide the acquisition of language or our construction of perception in the retina and the visual cortex are simply not the sorts of things we can become conscious of at all.

The same applies to some of our movements and actions. We do not experience the biomechanical parameters that guide our actions, like the joint angles, the velocity, etc. that are necessary to perform and experience actions. They remain unconscious and can never become conscious as such. They are thus examples of the deep unconscious.

The same holds true for the various biochemical processes related to drugs that we take. When, for instance, the biochemical substance GABA is modulated by a specific drug that targets it directly, we will never be able to experience the change in concentration of GABA, or the action of the drug itself, in our consciousness. The change in GABA concentration and the action of the drug simply cannot become contents that ever enter consciousness. They are simply not subject to experience or consciousness, and therefore remain deeply unconscious or, as Searle himself says, nonconscious.

Concept of consciousness 5

Concept of nonconscious

To what, though, does the concept of the 'nonconscious' refer? The concept of the nonconscious refers to neurobiological phenomena that remain nonconscious and cannot become the contents of consciousness at all: there are all sorts of things going on in the brain, many of which function crucially in controlling our mental lives but that are not cases of mental phenomena at all. So, for example, the secretion of serotonin at the synaptic cleft is simply not a mental phenomenon. Serotonin is important for several kinds of mental phenomena, and indeed some important drugs, such as Prozac, are used specifically to influence serotonin, but there is no mental reality to the behavior of serotonin as such. Let us call these sorts of cases the 'nonconscious'.

Another example of the nonconscious is the following: 'So, for example, when I am totally unconscious, the medulla, a part in my lower brain, will still control my breathing. This is why I do not die when I am unconscious or in a sound sleep. But there is no mental reality to the events in the medulla that keep me breathing: even when unconscious I am not unconsciously following the rule "Keep breathing"; rather, the medulla is just functioning in a nonmental fashion, in the same way that the stomach functions in a nonmental fashion when I am digesting food' (Searle, 2004, p. 168).

Considering this, we remain nonconscious of the bulk of the processes going on in our body and the brain. We can apparently become conscious only of the outputs or outcomes of these processes. Meanwhile, the underlying processes themselves, like the biochemical processes of digestion in the stomach, the biochemical modulation of drugs in the brain, and the neural activity in the brain itself as such, can never, in principle, enter consciousness.

Neurophilosophical discussion 1

Code and consciousness

Why is it that these contents cannot ever become conscious? The content may be coded and processed in simply the wrong format. It could be that, because of this wrong format, these types of contents are unable to enter consciousness. This reasoning is illustrated by Antti Revonsuo in an example about DNA (2006, p. 63): 'there is biological information coded in the DNA of our brain cells, but that type of information is in a totally nonconscious format. We will never be able to read it out just by reaching into our own minds and trying to bring it into consciousness. It is in a format unreadable at the phenomenal level.'

Consider an analogy to the computer. Sometimes people send us emails with attachments that we cannot open because we lack the specific program required for reading the file. These attachments are processed and coded in a format that our computer cannot read. In the same way, our consciousness cannot 'read or open' contents that are coded in the wrong format. This prevents us from accessing them and obstructs them from entering consciousness and unconsciousness.

Contents that are coded with the incorrect format cannot, in principle, enter consciousness. There is thus no principle difference between the 'deep unconscious' and the nonconscious. Why? Because both lead to the same outcome: these contents are unable to enter both unconsciousness and consciousness. Hence, the conceptual distinction does not correspond to a difference in the result. For this reason, both concepts are considered part of the nonconscious.

Neurophilosophical discussion 2

Code and the hard problem

Why is all this relevant in the current context of neurophilosophy and the brain? In addition to the neural mechanisms that distinguish between consciousness and unconsciousness, we may also need to search for those neuronal mechanisms that allow for the differentiation between nonconscious contents on the one hand, and unconscious and conscious contents on the other. We saw in the previous chapter that the current neuroscience of consciousness mainly targets those neuronal mechanisms that allow us to differentiate between consciousness and unconsciousness (see Chapter 13). This is reflected in the concept of the neural correlates of consciousness (NCC).

In contrast, the neuronal mechanisms that allow for the differentiation between nonconsciousness and consciousness remain open. While we discussed possible candidate mechanisms in the previous chapter, we currently lack the knowledge about the exact neuronal mechanisms.

Based on the above characterizations of the conscious/unconscious and the nonconscious, one may want to search for the kinds of format or coding the brain uses to process those contents that become unconscious and conscious. The next step would be to compare these formats. Solving this problem remains subject to future empirical investigation of the brain's neural code.

Why is this important to the various arguments philosophy raised against a neuroscientific explanation of consciousness? Recall the 'hard problem' from Chapter 13. The hard problem asks why there is consciousness rather than nonconsciousness. If we search for those neuronal mechanisms that allow for the differentiation between nonconsciousness and consciousness, we may be able to find a solution to the hard problem. The neuronal mechanisms in question might then provide an answer to why there is consciousness/unconsciousness rather than nonconsciousness. Hence, the rather complicated conceptual distinctions between consciousness, unconsciousness, and nonconsciousness, may open the door to addressing the hard problem in a truly neurophilosophical way.

Phenomenal features of consciousness 1

Qualia and what it is like

So far we have distinguished different concepts of consciousness – phenomenal and access consciousness – as well as the concept of consciousness from both unconsciousness and nonconsciousness. While understanding these distinctions is critical, it has not provided us with a definition of what consciousness actually is. Consciousness is such a basic phenomenon that any definition seems superfluous.

However, if we want to understand how consciousness is yielded, we need at least to determine somehow what it is we are searching for. Let me give a first tentative example before I provide a determination of consciousness.

Everybody knows what it is to be conscious. One experiences it continuously during the day and even during the night when dreaming. You are conscious while looking at your computer screen and reading the latest news of the world. More specifically, you perceive the news in a conscious mode. You feel certain emotions like anger about the latest news of some terrorist attacks. Moreover, you may even become aware of your own thoughts and cognitions that question the motivation of the terrorists. Ultimately, even your own self, the one who is reading these lines, enters consciousness, yielding what is called 'self-consciousness'.

The philosopher Thomas Nagel (Nagel, 1974) characterized consciousness by 'what it is like'. This is considered the standard definition. The concept of 'what it is like' describes the experience, and thus consciousness, that goes along with a particular quality, a phenomenal-qualitative feel also known as qualia. You experience the redness of the book's red color in terms of this phenomenal-qualitative feel. The concept of redness means that you have a quale of the color red. Such experience of redness, that is, the quale red, remains absent during mere observation of the book's red colored cover. Accordingly, the presence of the phenomenal-qualitative feel, the qualia, distinguishes experience from observation. Qualia may therefore be considered hallmark features of consciousness.

How is the phenomenal-qualitative feel possible? For that, Nagel argues, you need to take a particular point of view. What is a point of view? Most generally, a point of view anchors us as humans in a particular position when compared to the rest of the physical and biological world.

However, the particular point of view you are taking while reading these lines is also different from the points of view of other humans. In other words, the individual has a specific point of view. And it is in that specific point of view that you experience the world in terms of the qualitative-phenomenal feel, i.e. qualia.

How can we further characterize the point of view? The specific point of view you are taking while experiencing these lines is subjective. Subjective means that it is specific to you and your person – no other person can share your particular point of view. As a result, your associated experience remains private (as distinguished from public). In short, experience is essentially subjective.

Let us describe another example to distinguish experience from observation. You watch a movie with your friend. Though you watch the same pictures on the screen, each of you experiences different qualities in, for instance, emotions. Most important, you cannot experience the anger of your friend, you can only observe it. Conversely, your friend remains unable to experience your anxieties.

Experience, and thus consciousness, is subjective rather than objective. The subjective nature is the core of consciousness: we cannot even imagine consciousness without the subjective, since then we would no longer talk of consciousness at all. In other words, subjectivity is a defining, and thus intrinsic, feature of consciousness. This is why consciousness is such a hard nut to crack, especially for neuroscientists who, by default of their scientific investigation, focus on objectivity rather than subjectivity.

Phenomenal features of consciousness 2

Temporal and spatial continuity and unity

When reading these lines you experience a continuous flow of time (and space) – there is a smooth and continuous transition from the past to the present to the future. As you read, you are reminded of books you've read in the past on similar subjects. You are able to integrate these memories into your present experience of reading this book. You also anticipate the next pages that you will read in the near future and, going even further in time, you may already envision another book that you will write yourself.

The experience of your reading is thus embedded in a dynamic and continuous flow of time, extending from the past to the present to the future, and condensed in the present moment. This is what the American philosopher and psychologist William James (1890) described as the 'specious present' or 'dynamic flow'. The phenomenological philosopher Edmund Husserl referred to it as 'phenomenal time' (see Part I for details of both James and Husserl). The dynamic flow describes the organization of time as continuum (rather than as discontinuum) in consciousness.

The same holds true for space. Space is also experienced as homogeneous and continuous rather than heterogeneous and discontinuous. For instance, you do not experience the distinct parts of space around you as occupied by different kinds of furniture that are separate and isolated from each other. Instead, despite their differences in physical location, you nevertheless experience one homogeneous and continuous space in your consciousness. One may consequently want to speak of a spatiotemporal continuity as a central feature of our experience of time and space. This is also known as 'inner time and space consciousness'.

What about the experience of the body? As in your experience of time and space, there is also a homogeneous and continuous structure and organization in the experience of your own body. You experience your own body as a homogeneous and continuous whole rather than experiencing its different parts as isolated and separated. In addition, you experience your body subjectively as an integrated part of your environment, which in turn you also experience as structured and organized, rather than distinct and segregated. There is thus a certain structure and organization in our subjective experience and consciousness. This is described as 'phenomenal structure' (van Gulick, 2004).

Another important phenomenal feature of consciousness is *unity*. You do not experience the book lying in front of you as principally diverse and segregated from the table, the floor, the room, etc. Instead of such diversity, you rather experience the book in continuity and thus in relation to the table and the rest of the room including yourself.

One may now want to argue that there is some kind of spatial and temporal distinction. Imagine a book is lying on the table. Your focus is almost entirely on the book, which is thus the figure in your perception. However, you also perceive the table, though only in the background of your perception. Nevertheless, you experience them as unity, as a homogeneous and unified field in space and time, of which they are distinct aspects or parts. Hence, consciousness may be characterized by spatiotemporal unity as distinct from spatiotemporal diversity.

Phenomenal features of consciousness 3

Intentionality

There is more to consciousness than qualia, spatiotemporal continuity, and unity. When reading these lines, your experience is directed toward the book. Your consciousness is about something, the book and its introduction. Perception is always perception of something else, and that in turn shapes your experience. Hence, consciousness seems almost always to be directed toward something, an object or event.

Most important, this holds true even when the object or event remains absent in the real physical world. This is the case during dreams or auditory hallucinations in schizophrenia. This is called *intentional organization* or intentionality, which can be described as the directedness or 'aboutness' of conscious states. We have already encountered the concept of intentionality in Chapter 1 in the context of phenomenology. Phenomenological philosophy regards intentionality as a hallmark feature of mental states. Such directedness toward contents should be distinguished from the objects, events, or people themselves, which are meaningless without consciousness.

Why are these different phenomenal features of importance? One may assume that distinct neuronal mechanisms underlie the different phenomenal features of consciousness. For instance, the qualia may be subserved by different neuronal mechanisms from the ones that may underlie spatiotemporal continuity and unity. And finally, intentionality, the directedness towards contents, might be related to different neuronal mechanisms.

How does this relate to the current neuroscience of consciousness? As discussed in the previous two chapters, current neuroscience more or less focuses on consciousness as a whole. It does not distinguish the different phenomenal features and their possibly different underlying neuronal mechanisms. Why would it be an advantage to focus on the different phenomenal features? Because in doing so we might be able to specify our neuronal mechanisms. Let's go into more detail.

As discussed in Chapters 14 and 15, our current neuroscientific hypotheses lack both phenomenal and neuronal specificity; they are too unspecific and vague, meaning that they remain unable to capture the intrinsic features of consciousness. Hence the distinction between different phenomenal features may open the door for a more specific neuroscientific account of consciousness, in terms of both neuronal systems and phenomenology.

Phenomenal features of consciousness 4

Self-perspectival organization and first-person perspective

Consciousness is also about yourself. You might experience the reading of this book in a completely different way than your friend. Consciousness is thus always tied to the perspective of your specific self, which provides the particular perspective from which you experience the reading of this book.

There is always a subject of experience, the one who experiences the event, object, person, or place. Without a subject, experience and consciousness would be impossible. All your experience is centered on your self. This is what philosophy calls 'self-perspectival organisation'. It is considered a phenomenal hallmark feature of consciousness. However, the very same subject of experience can also become a content of the experience itself. This yields what philosophers call self-consciousness. In other words, the self, or subject, can become the content of its own experiences.

In Chapter 13, we saw that some phenomenological philosophers, like Merleau-Ponty, regard the body as essential in providing this self-perspectival organization. We experience our experiences on the basis of our own bodies. In this model, it is our body that provides us with the point of view and an anchor within the world. The body and its sensorimotor functions are thus considered central for consciousness by those thinkers who argue in favor of the embodiment of consciousness and the mind (see Part I and Chapter 7).

Why is this relevant for neuroscience? As a scientist, you might wonder about consciousness and discard it as being unsuitable for scientific investigation. Why? Because science only considers objective features that can be investigated independently of self-perspectival organization as being viable. This, however, is not the case with consciousness. Consciousness is intrinsically tied to self-perspectival organization. Without self-perspectival organization there would be no consciousness. Accordingly, instead of being objective, consciousness seems to be inherently subjective.

This also affects our knowledge of consciousness, which consequently seems to be subjective rather than objective, and thus not suitable for science. The apparently inherent subjective character of consciousness thus poses a problem, not only in terms of our possible knowledge, but also as a seemingly unsuitable project for experimental investigation and observation.

This brings us to a final point, namely that scientific investigation is based on third-person observation, independent of specific first-person perspective. This seems to conflict with the investigation of consciousness, precisely because it cannot be observed in third-person perspective. I remain unable to experience subjectively your subjective experience, even if I am doing exactly the same thing as you at the same time in the same context. This raises several methodological problems, which will be discussed below.

Methodological issues in consciousness research 1

First-person versus third-person access

How can we investigate consciousness? It seems clear that consciousness, as tied to the first-person perspective, needs a special method that cannot be completely covered by neuroscience. Let us sketch the situation in more detail. The phenomenal features of consciousness occur in first-person perspective, therefore they may need to be investigated in first-person rather than third-person perspective. This is the claim phenomenology makes as it aims to reveal the features of subjective experience and thus of consciousness.

If we now want to investigate consciousness and its phenomenal features scientifically, we are confronted with a dilemma. Scientific investigation presupposes observation in third-person perspective. At the same time, consciousness is defined by phenomenal features that are accessible only in first-person perspective. In short, neuronal explanation is quantitative, objective and third-person based. This distinguishes it from phenomenal descriptions that are qualitative, subjective and first-person based. In order to investigate consciousness neuroscientifically, we need to link both neuronal and phenomenal explanations, as well as their seemingly opposing features.

What can we do? Either we claim that consciousness, as a first-person based phenomenon, cannot be investigated scientifically, or we argue that the first-person phenomenal features of consciousness are nothing but third-person based neuronal functions of the brain. The latter strategy is followed by many Anglo-American philosophers and neurophilosophers.

Alternatively, one may pursue an intermediate path between the two, and take the first-person based phenomenal features of consciousness seriously in their own right. While doing so, one also aims to link neuronal features to the phenomenal features. Neurophenomenology advocates this as a methodological strategy. Neurophenomenology aims to develop methodological strategies for the empirical investigation of first-person phenomenal features, as described in phenomenology, in relation to third-person based neuronal mechanisms (see Chapters 1 and 4 for details).

First-person based phenomenal features of consciousness and third-person based neuronal features of the brain are then no longer considered to be mutually exclusive. Instead, they are regarded as complementary and as informing each other. This might ultimately lead to what has been described as 'first-person neuroscience'.

Methodological issues in consciousness research 2

First-person neuroscience

What is 'first-person neuroscience'? 'First-person neuroscience' can be defined as a methodological strategy that aims to link subjective first-person experience systematically to third-person observation of neuronal states (Varela and Shear, 1997; Northoff and Heinzel, 2006; Northoff et al., 2007). The development of such methods distinguishes first-person neuroscience from neuroscience as it is commonly practiced (in the third-person). Third-person neuroscience relies exclusively on third-person observation of neuronal states, more or less independently of subjective experience in first-person perspective.

Since phenomenal states are assumed to be central, this approach has also been termed as 'critical phenomenology' by the British psychologist Max Velmans (Velmans, 2000). Here, the term 'critical' describes the aim to include and link both first-person based phenomenal features and third-person based neuronal features. This critical approach to the phenomenal element of consciousness may correspond roughly to the neuroscientific elements.

How can we realize first-person neuroscience as the linkage between first- and third-person perspective in our methodological strategy? The linkage between subjective experience and neuronal states requires two steps:

(1) Subjective experience needs to be evaluated systematically, including objectification and quantification of subjective data. This 'science of experience' is

a necessary precondition for any linkage between subjective experience and neuronal states. This is different from psychology where psychological states are not considered in subjective terms.
(2) The systematically objectified and quantified subjective data then enable links to be made between analogous data about neuronal states. To do this, special methodological strategies need to be developed; this is the core of what is called 'first-person neuroscience'. To make such phenomenal–neuronal linkage possible and systematic, specific rules for the validity of this type of linkage needs to be developed.

Why do we want to link first- and third-person perspective systematically, rather than just reverting to third-person based observation? The systematic examination and evaluation of subjective experience allows us to preserve its richness and complexity on the one hand, while on the other allowing us to quantify its main characteristics objectively. The objectification and quantification of subjective first-person data allows for scientific investigation and, consequently, for the establishment of a 'science of experience'. This science of experience can be described as an investigation of subjective experience and its first-person based phenomenal features in third-person perspective (Varela, 1996).

This makes it clear that the methodological distinction between first- and third-person neuroscience may not be only of mere academic interest, but also be highly relevant to how we can find the neural mechanisms underlying consciousness. If so, the future development of first-person neuroscience may be the methodological convergence between the brain and consciousness, and more generally between neuroscience and philosophy.

Taken together, the investigation of the neuroscientific mechanisms underlying consciousness requires a special methodological strategy. This is because the description of phenomenal features is first-person based, while the explanation of the brain's neuronal activity presupposes the third-person perspective. One way to link both is to pursue what has been called first-person neuroscience, which relates first-person based phenomenal features of consciousness directly to the third-person based explanations of the brain's neuronal activity.

Hence, in addition to future neuroscientific research, we may also need to develop novel methods for investigating consciousness, like first-person neuroscience. The research on consciousness thus takes a combined conceptual, phenomenal, neuronal and methodological effort, and may therefore be considered a truly neurophilosophical problem.

Methodological issues in consciousness research 3

'Building block approach'

How can we consider the first-person based phenomenal features of consciousness in our methodological strategy of investigating consciousness? Recall that unity is one central phenomenal feature that describes the linkage and continuity of space and time.

How can we investigate consciousness empirically, given the phenomenal features of spatial and temporal continuity and unity? One way to investigate consciousness empirically is to start with specific functions; for example, visual function. Conscious and unconscious visual states can be compared with each other by investigating the spatial and temporal course of neural activity in unconscious and conscious visual stimuli. This can be accompanied by the investigation of altered conscious visual states using experiments that use blind sight and binocular switching.

The specific example of visual consciousness is then taken as a starting point to infer the features of consciousness in general as they occur across the different functions (like auditory, tactile, cognitive, affective, motor, etc.). The resulting neuroscientific theory is thus established and built on the basis of different functions, the so-called 'building blocks', as Searle (2004) calls them. By establishing theories about one particular block, like the visual block, this approach aims to put all the different blocks together and thereby to develop an overall neuroscientific theory of consciousness. Because it is based on different blocks of consciousness, Searle describes this approach as the 'building block approach'.

Conceptually, the building block approach presupposes that consciousness is, as Searle (2004, p. 106) says, atomistic in that it consists in 'more-or-less independent conscious units'. What are these conscious units? Each unit may correspond to a particular function like auditory, visual or affective function and its respective neuronal mechanisms.

Methodological issues in consciousness research 4

'Unified field model'

According to Searle, this 'building block approach' is, however, not in accordance with phenomenology. It is not that we have an isolated consciousness of red, independent of all other colors, or an isolated taste of beer, independent of its color, its smell, etc.: 'the unconscious subject would suddenly have a conscious experience of red and nothing else' (Searle, 2004, p. 108). Instead, we experience the color red in the context of the surrounding colors and noises, or the experience of tasting the beer in relation to its smell, color, etc. – the experience of red and the taste of the beer surface as aspects or parts of an always already unified field of consciousness.

The object's red color thus does not induce consciousness per se, but rather modifies an already existing 'qualitative subjective unity' – the conscious field: 'to put the point very crudely, a conscious experience of red can only occur in a brain that is already conscious. We should think of perception not as creating consciousness but as modifying a pre-existing conscious field. Again, consider dreams. Like many people, I dream in colour. When I see the colour red in a dream, I do not have a perceptual input that creates a building-block of red. Rather the mechanisms in the brain that create the whole unified field of dream consciousness create my experience of red as part of the field' (Searle, 2004, p. 108).

Consciousness may then no longer be conceptualized as the mere collection of different building block functions, but rather as a unified field with 'qualitative subjective unity'. Such qualitative subjective unity may then provide the very basis and foundation for the different functions and their association with consciousness as different surface features.

If one wants to investigate this underlying qualitative subjective unity by itself in a direct way, one needs a research strategy other than the building block approach.

Why? Because the building block approach focuses only on the surface features – the different functions – while neglecting their supposedly underlying qualitative subjective unity. Following Searle, this requires a methodological strategy he describes as the 'unified field model'.

What is the 'unified field model'? The 'unified field model' does not consider consciousness as freshly created each time it occurs. Instead, our single conscious experiences are considered as modifications or surface features of an already pre-existing consciousness: 'But I like to think of it this way: you are not creating a new consciousness; you are modifying the pre-existing conscious field. On the unified-field model we should think of perceptual inputs not as creating building-blocks of consciousness but as producing bumps and valleys in the conscious field that has to exist a priori to our having the perceptions' (Searle, 2004, p. 108; see also Noe, 2004).

How is Searle's assumption of a unified field approach related to the phenomenal features of spatiotemporal continuity and unity described above? As illustrated above, spatiotemporal continuity and unity do not occur as mere collections of the time and space related to different functions or different blocks.

Instead, time and space and their unity provide the underlying ground upon which the different functions, like the visual, auditory and tactile, rely. If this is the case, the approach to consciousness in terms of blocks will likely miss out on the underlying ground. In order to capture the 'ground' itself – the spatiotemporal continuity and unity – one may therefore need a different approach: one that is based on the different functions or the blocks themselves.

This is where Searle's unified field approach becomes relevant. This may allow for the characterization of the spatiotemporal continuity and unity that occurs prior to subsequent distinction into different functions or blocks of consciousness. One may argue consequently that the investigation of the phenomenal features of spatiotemporal continuity and unity requires a unified field approach, rather than the building block approach.

Methodological issues in consciousness research 5

'Unified field' and the brain's intrinsic activity

Why is the unified field approach relevant in the current neurophilosophical context? Recall Chapter 15, in which we characterized the brain by intrinsic activity. Intrinsic activity seems to show a certain spatiotemporal structure and organization; but the details of how such a structure and organization is generated remain unclear at this point in time. The brain's intrinsic activity must be considered as a whole, rather than in isolated elements or blocks related to different functions like visual, auditory, etc. Hence, the brain's intrinsic activity may, by itself, be regarded as the kind of field Searle targets in his concept of the 'unified field model'.

But we have to be careful. In order for the brain's intrinsic activity to provide the kind of unified field Searle has in mind, we need to show the following. First, we need to demonstrate that the field of the brain's intrinsic activity is really unified with regard to space and time. This means that the brain's intrinsic activity must independently show some kind of spatial and temporal continuity and homogeneity.

The brain's intrinsic activity may then be regarded as a spatiotemporally unified field. This has yet to be demonstrated.

Second, we need to demonstrate that the spatiotemporally unified field of the brain's intrinsic activity is essentially subjective – that it provides a point of view. Only then can the spatiotemporally unified field of the brain's intrinsic activity come close to what Searle means by the 'unified field model' that presupposes the essentially subjective nature of the unified field. Even more than spatiotemporal unity, the subjective nature of the field of the brain's intrinsic activity will need to be explored in the future.

This methodological approach to the brain and its neuronal activity in terms of a 'unified field' has to be distinguished from the building block approach. In the case of the building block approach, different functions and their respective stimuli are the main focus of investigation. The guiding question here is: what are the neuronal processes related to extrinsic stimuli that bring the stimuli into consciousness? This must be distinguished from the question guiding the unified field approach: how is it possible for the brain's intrinsic activity to assign consciousness to the different extrinsic stimuli?

The focus of the building block approach is thus on extrinsic activity in the brain. The focus in the unified field approach is on the brain's intrinsic activity and the kind of spatiotemporal structures it provides for the extrinsic stimuli to be processed. Hence, the methodological distinction between a building block approach and a unified field model has crucial relevance for how to investigate consciousness in neuroscience.

Take-home message

The previous chapters investigated the neural correlates and neural predispositions of consciousness. While this provided insight into the possible neuronal mechanisms underlying consciousness, it left open for discussion certain conceptual, phenomenal and methodological issues. These were the focus of this chapter. The distinction between different forms of consciousness as well as their distinction from both unconsciousness and nonconsciousness is a central conceptual issue. Phenomenal features are about the features that signify experience, and thus consciousness; these include qualia, spatiotemporal continuity, unity, intentionality and self-perspectival organization. Finally, different methodological approaches were discussed, including first-person neuroscience and the unified field model. On the whole, this makes clear that the investigation of consciousness is a truly transdisciplinary and neurophilosophical enterprise. It links and connects conceptual, phenomenal, methodological and empirical issues. Without considering the variety of these different issues, any explanation of consciousness may remain incomplete. In short, to solve the puzzle of consciousness, we need a truly neurophilosophical effort.

Summary

Consciousness is one of the main features, if not the central hallmark, of the mind. But what is consciousness? Consciousness is a multifaceted phenomenon that includes empirical, phenomenal, conceptual and epistemological (and metaphysical) dimensions. While the metaphysical and epistemological dimensions of consciousness were discussed in the first and second parts of this book, I left open its conceptual, phenomenal and methodological characterization. These were the focus in this chapter. Conceptually, different forms of consciousness can be distinguished: phenomenal consciousness describes the 'what it is like' and thus the subjective experience itself. Meanwhile 'access consciousness' concerns the awareness of the contents in consciousness. Finally, the concept of consciousness in general must be distinguished from the concept of the unconscious. In the unconscious, contents are not conscious, though they potentially can become conscious. This distinguishes the concept of the unconscious from the nonconscious, where the contents are neither conscious, nor have the possibility of ever becoming conscious. What about the phenomenal characterization of consciousness – its first-person based experiential features? Philosophers like Thomas Nagel assume the phenomenal-qualitative character of conscious states, the 'what it is like', to be central for consciousness. Besides the 'what it is like', consciousness may be characterized by other phenomenal features. These include unity as well as spatial and temporal continuity. In addition, consciousness can be characterized by self-perspectival and intentional organization and, most importantly, by subjective access and first-person perspective. These issues raise questions about appropriate methodological strategies for linking the first-person based phenomenal features of consciousness to the third-person based neuronal features of the brain. Hence, special methodological strategies like first-person neuroscience and the unified field model may be required to investigate the neural basis of consciousness.

Revision notes

- Why is the distinction between phenomenal and access consciousness both conceptually and empirically relevant?
- Describe the different forms of the unconscious.
- How is the unconscious distinguished from consciousness?
- What is the nonconscious and how is it distinguished from the unconscious?
- How can you characterize the 'what it is like' of consciousness? What is meant by phenomenal-qualitative features, e.g. qualia and point of view?
- What is unity and how is it manifested in consciousness?
- Define and describe self-perspectival and intentional organization.
- What is meant by spatial and temporal continuity? What are the terms philosophy uses to describe experience in consciousness?
- How can you distinguish first- and third-person perspective from each other?
- How can you distinguish first- from third-person neuroscience?
- What is the building block approach and how does it differ from the unified field model?

Suggested further reading

- Block, N. (1995) 'On a confusion about a function of consciousness', *Behavioral and Brain Sciences*, 18, 227–87.
- Nagel, T. (1974) 'What is it like to be a bat?', *The Philosophical Review*, 83(4), 435–50.
- Searle, J. (2004) *Mind: A Brief Introduction* (Oxford/New York: Oxford University Press), ch. 9.
- van Gulick, R. (2004) *Are There Neural Correlates of Consciousness?* (Exeter: Imprint Academic).
- Velmans, M. (2000) *Understanding Consciousness* (London: Routledge/Psychology Press).

Part V
Neurophilosophy of Self: From Consciousness to Self

The hallmark feature of our mind is that it is subjective. Subjective means that all people experience the same event in different ways, even though the event may be the same and experienced in the same context and at the same time. Subjectivity is the hallmark feature of the mind and thus of consciousness. Who is conscious? Who experiences something? It is the self that experiences something and that is conscious. Part V is about the self, the 'who' of experience and thus the subject of consciousness.

There has been an abundance of different concepts of self in the history of philosophy and most recently also in neuroscience. Discussing all the different concepts of the self would be beyond the scope of this book. We therefore focus only on those concepts that are empirically and thus neurophilosophically relevant, while leaving aside those that are only logically and thus philosophically important (see Chapter 17). Most important, we will confront reductive and non-reductive neurophilosophical concepts of the self. This will be followed by a discussion of the relationship between self and consciousness in both philosophy and neuroscience. Empirical material, like the loss of consciousness in vegetative state patients, will here be confronted with the conceptual reflections of the philosophers about the relationship between self and consciousness (see Chapter 18).

We have seen that the clinical disciplines of neurology and psychiatry contributed hugely to our understanding of the brain, especially in the early days of neuroscience (see Chapter 3). This continues until the present day, where in particular psychiatric disorders may provide insight on how the brain may work. Psychiatric disorders like schizophrenia and depression show changes in the mental features which are sometimes rather bizarre and strange. At the same time, recent neuroscience has demonstrated neuronal changes in these patients. These disorders can thus tell us about the relationship between neural and mental features, though in an indirect way. This is exploited in Chapter 19 where I discuss the neuronal mechanisms of self and self-consciousness in depression and schizophrenia (see Chapter 19).

Finally, Chapter 20 is about another problem closely related to the self, namely the encounter with other selves, which is conceptualized as intersubjectivity. Here again, empirical findings from mirror neurons are directly confronted with conceptual reflections to show how a neurophilosophical concept of intersubjectivity, as distinguished from both philosophical and neuroscientific ones, can be developed (see Chapter 20).

17
Brain and Self

Overview

We discussed consciousness and its neuronal mechanisms in Part IV, where we considered that consciousness always presupposes some kind of subject – a subject who experiences the respective object, event, or person in its consciousness. The subject of experience is considered central to experience, since without a subject, experience itself, and consciousness, would most likely remain impossible. What is the subject of experience? This has been discussed under the umbrella concept of the self. The self has taken on various meanings in past and recent philosophical debates. Most recently, the introduction of functional brain imaging allows us to investigate the neuronal mechanisms underlying our experience of a self. This raises questions about the relationship between the different philosophical concepts of the self and the neuroscientific findings related to it. This is the topic of this chapter.

Objectives

- Understand different philosophical concepts of the self
- Follow how these concepts resurface in the present philosophical discussion about the self
- Understand the different kinds of methods used experimentally to study the self
- Understand the main findings about the spatial and temporal patterns of neural activity related to self-specificity
- Discuss the missing neuronal, psychological, experimental, phenomenal and conceptual unspecificity of the current neuroscientific findings on the self

Key concepts

Self, mental substance, self-representation, empirical self, phenomenal self, minimal self, cortical midline structures, self-reference effect, gamma synchronization, neuronal, phenomenal, psychological, experimental and conceptual specificity.

Background 1

Conceptual determination of the self in philosophy

You're winning a game of tennis while your friend is watching. You feel pride. Who experiences that pride? You. You read these lines. You are the subject of the experience of interest or of boredom. Without you as the subject of this experience, you could not experience anything at all, not even boredom. This subject of experience has been described as the 'self'. Your 'self' makes it possible for you to experience things. In other words, it is a necessary condition for experience and thus also for consciousness. It is clear, therefore, that there is much at stake when it comes to the self.

The concept of self has been subject to intense philosophical discussion over the centuries. Different philosophers have suggested different concepts of self. Because of time and space constraints, we will only focus on those that are relevant in the attempt to map the interface between philosophical and neuroscientific accounts of the self.

There are four main concepts of self discussed in current philosophy. First is 'the mental self', which is based on our thoughts and a specific mental substance. Second is the 'empirical self' – this concept of the self represents and reflects the biological processes in one's body and brain. Third is the 'phenomenal self', from which originates our experience in consciousness. Our consciousness is accompanied by an awareness of our self, referred to as pre-reflective self-awareness or phenomenal self. Finally, and most recently, philosophers speak of a 'minimal self'. This concept of the self is based on our body and its physiological processes.

I will discuss each of these different concepts and how they relate to the brain in this chapter, but before doing this, I must shed some light on several related concepts. We experience our self in daily life during, for example, the act of perceiving certain objects, people or events in our environment. For example, while making a list of all the things you have to do today, you experience not only the act of thinking and writing, but an awareness and experience of your own self. Hence, your self as the very subject of experience seems to be part of that experience. In other words, your self is a content of your consciousness. This is described as self-consciousness. The concept of self-consciousness will be the focus of the next chapter.

However, there is more to the self than the self itself and our experience of it in self-consciousness. You wake up every morning. Every day. Every week, every year. Your body changes. You become older. You become wrinkled and your hair turns white. Despite all these bodily changes, you nevertheless have the feeling that you have the same self. You still experience your self as being the same self as 20 years ago.

You are the same person. There is thus a temporal dimension to your self that seems to be coherent and persistent over time. You and your self are continuous across time. The temporal dimension of your self has consequently been discussed under the umbrella of what is called 'personal identity' in philosophy. While our discussion will touch upon the temporal dimension of the self and thus upon personal identity, we will not discuss it explicitly.

In a world of over seven billion people, there are many, many selves: you, your friends, your family, etc. Most interestingly, you can relate to them – you can communicate with other selves and sometimes even feel their emotions as in, for instance, the grief someone might feel when they lose a loved one. Or you might experience pain when your friend's arm is broken. How is this possible? In philosophy, this is called 'intersubjectivity'. Intersubjectivity will be discussed in Chapter 20.

Finally, your self is not isolated from the rest of the world. You can share others' experiences and feel connected to the world. The world and its specific objects, people and events has meaning for you – you can relate to it more or less, and can appropriate it for your own self. How is such basic integration of your own self within the world possible? And how is that related to your brain and its neuronal mechanisms? This will be the focus in Chapter 20.

Background 2

Empirical investigation of the self in neuroscience

How can we investigate the self? In order to address the self experimentally, we need some quantifiable and objective measures that can be observed from the third-person perspective. How can we obtain such measures? Psychologists focusing on memory observed that items related to ourselves were better remembered than those unrelated (see Northoff et al., 2006). For example, as a resident of Ottawa, I recall the recent thunderstorm that wiped out several houses locally much better than a person who, say, living in Germany, only heard about it in the news.

There is thus superiority in the recollections of those items and stimuli that are related to one's self. This is described as the self-reference effect (SRE). The SRE has been well validated in several psychological studies. Most interestingly, it has been shown to operate in different domains. Not only in respect of memory, but also in relation to emotions, sensorimotor functions, faces, words, etc. In all these different domains (see below for details), stimuli related to one's own self, known as self-specific stimuli, are recalled much better than those that are unrelated to one's own self, known as non-self-specific stimuli.

How is the SRE possible? Numerous investigations (see, e.g., Klein and Gangi, 2010; Klein, 2012 for summaries) show that the SRE is mediated by different psychological functions. These range from personal memories, including autobiographical memories, through memories of facts (semantic memories) to those cognitive capacities that allow for self-reflection and self-representation. Hence, the SRE is by itself not a unitary function, but rather a complex, multifaceted psychological composite of functions and processes.

How can we link the SRE to the brain? Before the introduction of functional imaging techniques such as fMRI at the beginning of the 1990s, most studies conducted focused on the effect of dysfunction or lesions in specific brain regions caused by brain tumors or stroke. These revealed that lesions in the medial temporal regions that are central in memory recall, such as the hippocampus, change and ultimately abolish the SRE effect.

With the introduction of brain imaging techniques such as fMRI, we could then transfer the experimental paradigms of comparing self- and non-self-specific stimuli to the scanner and investigate the underlying brain regions. The basic premise here is that, if self-specific stimuli are recalled better than non-self-specific stimuli, they must be processed by the brain in a different way. This might be, for instance, by higher degrees of neural activity and/or different regions.

This led to the investigation of numerous experimental designs of SRE-like paradigms in the fMRI scanner. For example, subjects were presented with trait adjectives that were either related to themselves (such as, for me, my hometown of Ottawa) as opposed to unrelated (Sydney, an unrelated city for me). In other tests, the participant was presented with images of his/her own face and these were compared with faces of other people. Autobiographical events from the subject's past were also compared with those from other people. One's own movements and actions could be compared with those of other people, implying what is called ownership (e.g. my movements) and agency ('I myself caused that action').

The stimuli belonged to different domains such as memory, faces, emotions, verbal, spatial, motor or social. Most of the stimuli were presented either visually or auditorily, and the presentation of these stimuli was usually accompanied by an online judgment about whether the stimuli are related and personally meaningful or not to the research subject.

On the whole we can see that current neuroscience can investigate the self in various experimental ways using mainly functional brain imaging. However, any empirical research relies on certain presuppositions. This also holds true for current neuroscientific research on the self, which aims to reveal the neuronal mechanisms underlying our experience or sense of self. However, before examining the neuroscientific findings, we need briefly to shed some light on the concept of the self and how it has been defined in philosophical discussions.

Philosophical concepts 1

Mental self

What is the self? What must it look like in order to presuppose experience and be the subject of our experience? The self has often been viewed as a specific 'thing'. Stones are things, the table on which your laptop stands is a thing. And in the same way the table makes it possible for the laptop to stand on it, the self may be a thing that makes experience and consciousness possible. In other words, metaphorically speaking, experience and consciousness stand on the shoulders of the self.

However, another question is whether the self is a thing or, as philosophers such as Descartes suggest, a substance? A substance is a specific entity or material that serves as a basis for something like a self. For instance, the body can be considered a physical substance, while the self can be associated with a mental substance.

Is our self real and thus does it exist? Or is it just an illusion? Let us compare the situation to perception. When we perceive something in our environment, we sometimes perceive not a real thing, but an illusion that in reality does not exist. The question of what exists and is real is what philosophers call a metaphysical question. Earlier philosophers, such as Descartes, assumed that the self is real and exists.

However, Descartes also assumed that the self is different from the body. Hence, self and body exist, but differ in their existence and reality. Thus, from this perspective, the self cannot be a physical substance and is instead a mental substance. It is a feature not of the body, but of the mind.

However, the characterization of the self as a mental entity has been questioned. For example, the Scottish philosopher David Hume argued that there is no self as a mental entity. There is only a complex set or 'bundle' of perceptions of interrelated events that reflect the world in its entirety. There is no additional self in the world; instead there is nothing but the events we perceive. Everything else, such as the assumption of a self as a mental entity, is an illusion. The self as a mental entity and thus as a mental substance does not exist and is therefore not real.

To reject the idea of self as a mental substance and to dismiss it as mere illusion is currently popular. One major proponent of this view today is the German philosopher Thomas Metzinger (Metzinger, 2003). In a nutshell, he argues that through our experience we develop models of the self, so-called 'self-models'. These self-models are nothing but information processes in our brain. However, since we do not have direct access to these neuronal processes (e.g. all those processes and activities of the cells, neurons, in the brain), we tend to assume the presence of an entity that must underlie our own self-model. This entity is then characterized as the self.

According to Metzinger, the assumption of the self as a mental entity results from an erroneous inference from our experience. We cannot experience the neuronal processes in our brain as such. Nobody has ever experienced his/her own brain and its neuronal processes. Therefore, the outcome of our brain's neuronal processes, the self, cannot be traced back to its original basis, the brain, in our experience.

Where, then, does the self come from? We assume that it must be traced back to a special instance different from the brain. This leads us to assume that the mind and the self are mental entities rather than physical, neuronal entities originating in the brain itself. Metzinger argues that the self as a mental entity simply does not exist. Therefore, Metzinger (2003) concludes, selves do not really exist. Hence, the title of his book – *Being No One*.

Philosophical concepts 2

From the metaphysical to the empirical self

What is the self if not a mental entity? Current authors, such as Metzinger (2003) and Churchland (2002), argue that the self as a mental substance or entity does not exist. How do we come up with the idea of a self or the self-model, as Metzinger calls it? The model of our own self is based on summarizing, integrating and coordinating all the information from our own body and brain.

What does such integration look like? Take all that information together, coordinate and integrate it, and you have a self-model of your own brain and body and their respective processes. In more technical terms, our own brain and body are represented in the neuronal activity of the brain. Such representation of one's own brain and body amounts to a model of your self. The self-model is therefore nothing but an inner model of the integrated and summarized version of your own brain and body's information processing. The self is thus a mere model of one's own body's and brain's processes.

The original mental self, the self as mental substance or entity, is, in this line of thinking, replaced by a self-model. This implies a shift from a metaphysical discussion of the existence and reality of the self, to the processes that underlie the representation of body and brain as a self-model. Since this representation is based on the coordination and integration of the various ongoing processes in the brain and body, it is associated with specific higher-order cognitive functions such as working memory, attention, executive function and memory, among others.

What does this imply for the characterization of the self (presupposing a broader concept of self beyond the self as mental substance)? The self is no longer characterized as a mental substance but as a cognitive function. Methodologically, this implies that the self should be investigated empirically rather than metaphysically.

We therefore need to search for the cognitive processes underlying the special self representation. The self is consequently no longer an issue of philosophy, but rather one of cognitive psychology and ultimately of cognitive neuroscience. According to this model, the self is no longer a metaphysical matter, but a possible subject of empirical investigation.

Philosophical concepts 3

Phenomenal self

One of the problems one encounters is that a substance or meta-representation cannot be experienced as such. Nobody ever experienced a mental substance or a meta-representation in consciousness because we simply cannot be conscious of them. Therefore, instead of speculating about something that lies beyond the scope of our experience, why not start with experience itself, and thus with consciousness. Rather than looking at what lies 'outside' our consciousness, like a substance or meta-representation, the self may be found within consciousness itself.

However, this localization is denied in phenomenological philosophy precisely because it focuses on consciousness itself and what lies 'inside' our experience. More specifically, phenomenological philosophy is interested in investigating the structure and organization of our experience, and thus of consciousness. It focuses on how our experience is structured and organized, and reveals phenomenal features as we experience them from the first-person perspective.

How does the phenomenal approach determine the self? Currently, it is argued that the self is an integral part of experience itself (Northoff, 2012a, 2012b, 2012c). The self is always present and manifested in the phenomenal features of our experience such as intentionality (i.e. the directedness of our consciousness towards specific contents), qualia (i.e. the qualitative character of our experience; what it is like), etc. Without these features the self would remain impossible.

Consequently, phenomenological philosophers such as Zahavi (2005) consider the self to be an inherent part of consciousness itself. Here, the self is supposed to be always accompanied by some kind of consciousness of the external world, even if we are not aware of the self being part of that experience. Phenomenological philosophers therefore speak of what they call pre-reflective self-awareness (or pre-reflective self-consciousness).

The concept of pre-reflective self-consciousness contains two main terms, 'pre-reflective' and 'self-consciousness'. Pre-reflective means that the experience of the self does not stem from any reflection or cognitive operation. Instead, it is already always there as an unavoidable part of our experience as such. The self is thus pre-reflective. It is simultaneously an inherent part of our experience and thus of our consciousness. The self is consequently no longer outside our consciousness, but an integral part of it, hence the second term, self-consciousness. Such an approach suggests an intimate and intrinsic link between self and consciousness.

Characterizing the self in terms of self-consciousness implies a significant shift. The self is no longer metaphysical, as Descartes proposed. Nor is it empirical, as advocated by Hume and others such as Metzinger and Churchland. Instead, the self is part of experience and of consciousness itself, and can therefore be characterized as the 'phenomenal self'. Such a phenomenal self is open to systematic investigation of the phenomenal features of our experience, which would complement the metaphysical, empirical and logical approaches to the self.

Philosophical concepts 4

Minimal self

How can we describe the pre-reflective self-consciousness in more detail? It is always already there in every experience so that we cannot avoid it or separate it from the experience. The self is always present in our consciousness and thus in our subjective experience. Even if we do not focus on the self as such, we cannot avoid or remove its presence. Hence, the term pre-reflective self-consciousness describes an implicit or tacit experience of our self in our consciousness.

Since the self as pre-reflectively experienced is the basis of all phenomenal features of our experience, it must be considered essential for any subsequent cognitive activity. Such a basic and fundamental self occurs in our experience before any reflection. For instance, when reading the lines of this book, you experience the contents and, in addition, you also experience your self as reading these lines.

Hence, your immediate experience and consciousness comes with both the content and your own self, since the experience of such self occurs prior to any reflection and recruitment of higher-order cognitive functions. This is why this concept of self is a sort of minimal version of the self. Current phenomenological philosophers such as Gallagher (2000) or Zahavi (2005) therefore speak of a 'minimal self' when referring to the self as implicitly, tacitly and immediately experienced in consciousness.

How can we describe the concept of the 'minimal self'? The minimal self refers to a basic form of self that is part of any experience. As such, it is not extended across time as it is in the experience of the self as a continuity across time in personal identity. Instead the minimal self describes a basic sense of self at any given moment in time, but does not yet provide a link between different moments in time and thus continuity across time.

How can such continuity across time be constituted? Cognitive functions such as memories, and autobiographical memories in particular, may be central. In this model, the self may become more complex. One might speak of a cognitive, extended or autobiographical self, as does, for example, the Portuguese-American neuroscientist, Antonio Damasio (see e.g. Damasio, 1999, 2010).

Another important feature of the minimal self is that, although we experience it, we may not be aware of it as such. This means that we might not be able to reflect upon it in order to gain knowledge of it. We are, to put it in technical terms, only pre-reflectively aware of the minimal self. In contrast to such pre-reflective awareness, there is no reflective awareness of the minimal self. How can we become reflectively aware of the minimal self? For that to be possible, the different moments or points in time need to be integrated and, as philosophers say, represented. For such representation to occur, cognitive functions are needed which make it possible to link together the different time points.

Finally, the minimal self may also occur prior to verbalization and thus linguistic expression. Rather than being tied to specific linguistic concepts, as is the case with more cognitive concepts of the self, the minimal self must be considered pre-linguistic. It is an experience, a sense of self that can barely be put into terms of concepts. We can experience it as self but are not really able to describe such experiences in terms of concepts and thus articulate them in a linguistic way.

Thus, the minimal self is pre-linguistic and pre-conceptual and will therefore not be affected speculatively by second-language acquisition. It is the kind of experience, an implicit sense of self, which subjects will most likely take with them as more or less stable when moving to a new country where they have to acquire a new language. However, at the same time, the minimal self provides the essential basis upon which more cognitive forms of self are developed. These are, then, central and instrumental in providing the ability to learn a second language.

Philosophical concepts 5

Social self

How does the self interact with other selves? So far, we have described the self in an isolated and purely intra-individual way. However, in daily life, the self is not isolated from others but always related to other selves. This is called inter-individualism rather than intra-individualism. This raises questions about what is described as the 'problem of other minds' or, more generally, questions concerning intersubjectivity. Here we will give a brief description of the problem of intersubjectivity. It will, however, be discussed in greater detail in Chapter 20.

How can we make the assumption of attributing mental states, and thus self and mind, to other people? Philosophy has long relied on what is called the 'inference by analogy'. What is the 'inference by analogy'? 'Inference by analogy' goes like this. We observe person A to show the behavior of type X. And we know that in our own case the same behavior X goes along with the mental state type M. Since our own behavior and that of person A are similar, we assume the other person A to show the same mental state type M we experience when exhibiting behavior X.

What inference do we draw here? There is similar or analogous behavior between ourselves and the other person. In addition, my own behavior is associated with a particular mental state. Since the other person shows the same behavior, I infer that he/she also has the same mental state. Hence, by indirect inference and analogy via our own case, we claim to obtain knowledge of the other person's mental state. How can we make such an inference? We may make it on the basis of our own mental states and their associated behavior. And what we do may also hold true for the other person, who in the same way attributes mental states to us by inferring them from the comparison between our behavior and their own mental states.

Why do we make such inferences? Because it seems to be the easiest and best way for us to explain other people's behavior. The assumption of mental states thus seems to be the best explanation for your behavior. The 'inference by analogy' may thus be considered an inference to the best possible explanation.

The inference by analogy describes intersubjectivity in a cognitive and ultimately linguistic way when attributing mental states and a self to other people. There might be, however, a deeper level of intersubjectivity. We also feel the other person's mental states when sharing the emotional pain, for example, that one's spouse experienced when his/her father died. Such sharing of feeling is described as empathy and sheds light on a deeper pre-cognitive and pre-verbal dimension of intersubjectivity. This has been emphasized especially in phenomenological philosophy (see, for instance, Zahavi, 2005).

However, both, empathy and the attribution of mental states to another person are puzzling: despite the fact that we do not experience the other's mental states and consciousness, we nevertheless either share them (as in empathy) or infer them (as in inference by analogy). We have no direct access to another person's experience of a self and its mental states in first-person perspective but nevertheless share their mental states and assume that they have a self. How is that possible?

This is where we need to introduce yet another perspective. There is first-person perspective – tied to the self itself and its experience or consciousness of objects, events or people in the environment. Then there is the third-person perspective – this perspective allows us to observe the objects, events or people in the environment from the outside, rather than from the inside. The picture is not complete.

What is the second-person perspective? The second-person perspective has initially been associated in philosophy with the introspection of one's own mental states. Rather than actually experiencing one's own mental states in first-person perspective, the second-person perspective makes it possible to reflect and introspect about one's own mental states. An example of this is when you ask yourself whether the voice you heard was really the voice of your good friend (see also Schilbach et al., 2013).

The second-person perspective thus allows us to put the contents of our consciousness as experienced in first-person perspective into a wider context, the context of oneself as related to the environment. In other words, the second-person perspective makes it possible to situate and integrate the purely intra-individual self with its first-person perspective into a social context. This transforms the intra-individual self into an inter-individual self. Another way of thinking of the second-person perspective is to call this concept of the self, the 'social self'.

How can we define the concept of the social self? The concept of the social self describes the linkage and integration of the self into the social context of other selves. This shifts the focus from experience or consciousness in the first-person perspective to the various kinds of interactions between different selves as associated with the second-person perspective. As we have already indicated, there may be different kinds of social interactions, including affective pre-cognitive and more cognitive ones that involve meta-representation, as described above.

Neuroscientific findings 1

Spatial patterns of neural activity during self-specific stimuli

How can we relate the various philosophical concepts of the self to the neuroscientific findings of self-reference? Above, we discussed that psychology, and later neuroscience, quantified the self in terms of the self-reference effect (SRE). The SRE describes the different impact of self-referential and non-self-referential stimuli on psychological (e.g. reaction time, recall, etc.; see above) and neural (e.g. degree of activity, regions, etc.; see below) measures. Below we want to highlight briefly some of the main findings of recent imaging studies on the self-reference effect.

What results did the various imaging studies yield in fMRI? Two different kinds of regions showed up. First, one could see that the regions specific for the respective domains like emotions or faces were recruited. For instance, there is a region in the back of the brain that specifically processes faces (as distinguished from, say, houses); this is called the fusiform face area. This region is obviously active during the presentation of faces, no matter whether it is one's own face or another person's face. Importantly, clear differences between self- and non-self-specific stimuli could not be observed in these domain-specific regions in most studies (see Northoff et al., 2006).

What about other regions that are not specific to particular domains (known as domain-independent regions) involved in the neural processing of the self? Meta-analyses of the various studies demonstrated the involvement of a particular set of regions in the middle of the brain. These regions include the perigenual anterior cingulate cortex (PACC), the ventro- and dorsomedial prefrontal cortex (VMPFC, DMPFC), the supragenual anterior cingulate cortex (SACC), the posterior cingulate cortex (PCC) and the precuneus. Since they are all located in the midline of the brain, they have been coined 'cortical midline structures' (CMS).

The self-specific stimuli – those that were personally relevant for the subjects – induced higher neural activity in these regions than non-self-specific stimuli, or those that remained irrelevant und unrelated to the person. This was observed in the various domains for faces, trait adjectives, movements/actions, memories and social communication. Therefore, the CMS seem to show a special significance for the self and self-reference.

However, there is also some differentiation within the CMS. The self-specific stimuli may be presented in different ways to the subject in the scanner. If subjects have to make judgments requiring cognitive involvement, the dorsal and posterior regions such as the SACC, DMPFC and PCC are recruited to a stronger degree. If, in contrast, stimuli are merely perceived without any judgment, and thus without any cognitive component, the ventral and anterior regions such as the VMPFC and PACC are highly involved.

This led to the assumption that the different regions mediate different aspects of self-reference. The ventral and anterior regions, such as the PACC and VMPFC, may be more involved in the representation of the degree of self-reference in the stimulus. However, dorsal regions, such as the SACC and the DMPFC, may be related to monitoring and reflection of the stimulus and its self-reference when we become aware of the stimulus as self-specific.

Finally, the posterior regions, such as the PCC, may be implicated in integrating the stimulus and its degree of self-reference into the autobiographical memory of the respective person. These regions seem to be implicated in the recall and retrieval of, in particular, personally relevant and autobiographical information from the past of that person. Thus, it can be concluded that specific regions in the midline of the brain, the cortical midline structures, seem to be involved in the neural processing of self-reference or attributing personal relevance or self-relevance to stimuli.

Neuroscientific findings 2

Temporal patterns of neural activity during self-specific stimuli

In addition to the spatial patterns of self-reference, its temporal patterns have also been investigated using EEG. Again, self-specific and non-self-specific stimuli have been compared with each other while the subjects undergo EEG measurement. This has revealed early changes during self-specific stimuli at around 100–150ms after stimulus onset.

More specifically, self-specific stimuli induced different electrical activity changes at 130–200ms after their onset when compared to non-self-specific stimuli. This was accompanied by later changes at around 300–500ms. Hence, the temporal pattern between self- and non-self-specific stimuli shows both early and late differences.

In addition, different frequencies of neural activity were investigated. The neural activity oscillated rhythmically in different frequency ranges in the fluctuations of neuronal activity.

One frequency often induced by stimuli is gamma frequencies in the range of 30 to 40 Hz. Interestingly, some EEG (and MEG) studies observed higher power in the gamma range in anterior and posterior midline regions during self-specific stimuli than non-self-specific stimuli. The question, though, is whether such an increase in gamma power is specific to self-specific stimuli since it can also be observed in other functions independent of self-reference (see below).

Neuroscientific findings 3

Social patterns of neural activity during self-reference

How can we investigate the social nature of the self described earlier? Various studies have been conducted to investigate different kinds of interaction between different selves. Pfeiffer *et al.* (2013) and Schilbach *et al.* (2013) distinguish two different methodological approaches. One investigates social cognition, the cognition of mental states in other people, from the third-person perspective. Here, social cognition is investigated in an 'offline' mode. More recently this 'offline' methodological strategy has been complemented by an 'online' mode. In the 'online mode', social interaction is investigated from the 'inside', by taking on the perspective of the interacting selves (rather than the observer's point of view).

Besides conducting several studies, the same group has recently investigated the neural overlap between emotional processing, resting state activity, and social-cognitive processing (Schilbach *et al.*, 2012). They conducted a meta-analysis including imaging studies from all three kinds of investigations – resting state, emotional and social-cognitive. In a first step they analyzed the regions implicated in each of the three tasks. This yielded significant recruitment of neural activity, especially in the midline regions like the ventro- and dorsomedial prefrontal cortex and the posterior cingulate cortex (bordering the precuneus). In addition, neural activity in the temporo-parietal junction and the middle temporal gyrus was observed.

In a second step, they overlaid the three tasks, emotional, social-cognitive and resting state, in order to detect commonly underlying areas. This indeed revealed the midline regions, the dorsomedial prefrontal cortex and the posterior cingulate cortex, to be shared among emotional and social-cognitive tasks and resting state activity. Based on this neural overlap, the authors concluded that there may be an intrinsically social dimension in our neural activity which might be essential for consciousness of both our own self and other selves. If this is true, it will have radical consequences, not only for the concept of the self, but also for consciousness in general.

Neurophilosophical discussion 1

Different forms of specificity

So far we have covered philosophical approaches to the concept of the self. We also discussed neuroscientific findings about self-reference. Now, the question is how both philosophical concepts and neuroscientific findings are related to each other. *This requires what one may describe as a neurophilosophical discussion. A neurophilosophical discussion relates empirical findings in neuroscience directly to concepts in philosophy.*

How are the neuroscientific findings about self-reference related to the philosophical concepts of self? Are the philosophical concepts of self empirically plausible and thus compatible with the neuroscientific findings of self-reference? In order to address these questions one should start by investigating the degree of specificity for the self of the neuroscientific findings. One can thus speak of the neuronal specificity of the cortical midline structure for the self. The concept of neuronal specificity describes the quest for the exclusive association of a particular neuronal measure, like the activity of a certain region or network, with exclusively one specific function.

And they may also be discussed in the context of psychological functions associated with the self and thus psychological specificity. Furthermore, one may question the ability of the experimental designs and measures to really tap into the self. This is called experimental specificity.

One may also raise the question of whether the results really reflect the experiential and thus phenomenal features related to the self. Experience may, for instance, be confounded by features that are not directly related to one's own self. One may thus want to speak of phenomenal specificity. Finally, one may want to discuss how the results relate to the different concepts of the self, and whether they correspond exclusively to one specific concept. If they do, this would imply conceptual specificity.

Neurophilosophical discussion 2

Neuronal specificity of midline regions

Let's start with neuronal specificity. The concept of neuronal specificity describes whether the spatial and temporal patterns of neural activity observed in studies about self-specificity are really specific to the self. We roughly distinguished two kinds of different regions, the domain-specific regions and the domain-independent regions.

Domain-specific regions are those that are related to the processing of a content in specific modality (e.g. sensory) or domain (e.g. verbal, sensory, motor, etc.). Depending on the stimuli and/or the task, domain-specific regions were activated in the imaging data described above.

Are these domain-specific regions specific for the self? No, because the imaging data show that the very same regions are also recruited when applying stimuli that are not related to the self at all. For example, you are shown an image of a house in Brazil. For you, a resident of Canada (with no connection to Brazil), this image has no degree of self-specificity or self-relatedness to you. It nevertheless activates your fusiform face area.

The self-specialist may now want to argue that at least the degree of neural activity in the fusiform face area or other domain-specific areas may be different between self- and non-self-specific stimuli. However, empirical data are not clear. While some studies report some difference, though small, in sensory regions, the majority of studies did not apparently observe differences between self-specific and non-self-specific stimuli in these domain-specific regions. Hence, it seems as if the domain-specific regions like the sensory and motor cortex remain unspecific for the self. This implies neuronal unspecificity.

What about the domain-independent regions like the cortical midline structures? There has been much discussion whether these regions are specific for self-specific stimuli as distinguished from non-self-specific stimuli. Is the self 'located' in the midline regions? While initial enthusiasm was in support of the theory that the midline regions are specific for the self, recent investigations have implicated the same set of regions in a variety of different functions.

Let me be more specific. Tasks requiring the need to understand other people and their mental states – mind-reading as described in the theory of mind in psychology – strongly recruit the midline regions. Emotional stimuli and emotional tasks also lead to strong activation in the midline regions. In addition, various kinds of social tasks that require social exchange and reciprocity also recruit these regions. Finally, day-dreaming or mind-wandering and other forms of introspection also recruit these regions.

The involvement of the midline regions in various functions other than self-reference sheds some doubt on the neuronal specificity of the midline structure for the self. Hence, even the domain-independent regions like the midline regions do not seem to show any specificity for the self.

The same diagnosis of neuronal unspecificity is also true of the reported gamma synchronizations. Gamma synchronization is not specific to the self but has been observed in a variety of different functions including sensorimotor, working memory, attention and episodic memory retrieval. Hence, there is neuronal unspecificity in both a temporal and spatial sense with regard to the self.

Neurophilosophical discussion 3

Psychological and experimental specificity

Most of the fMRI above compared self-specific versus non-self-specific stimuli, such as a grand piano for a professional pianist compared to a saw for a carpenter. In addition to the mere perception, subjects were required to make a judgment after each stimulus, as to whether it was self- or non-self-specific. This raises questions about what exactly the study is measuring – the perception or the judgment of the stimulus? Is it capturing the effect of the stimulus itself, or the task related to that stimulus?

Most likely the results reflect a mixture between stimulus- and task-related effects. This, therefore, casts some doubt on whether the midline regions show psychological specificity for the self. The judgment about self-specificity requires various cognitive functions such as attention, working memory and autobiographical memory retrieval. Some authors, such as the French neuroscientist Dorothée Legrand, argue that the midline regions may be more related to what she describes as 'general evaluation function', than being specific to the self and self-specific stimuli.

What about when research investigates the self in relation to more basic functions such as movements and actions? Even when subjects perform some motor tasks, we face the same confusion of different functions: the self's components, such as ownership (my own movement), as well as agency (whether I am the agent of the movement), may be confounded by the neural mechanisms underlying the execution of the movement/action by the person.

Such psychological unspecificity highlights the need in neuroscience to specify the experimental design and measures. We need measures that are specific to the self as distinguished from the various associated sensorimotor, affective and cognitive functions. We also need experimental designs to segregate stimulus-related effects and task-related effects. For example, we might do this by spacing perception and judgment temporally apart from each other.

Neurophilosophical discussion 4

Self-specificity and other functions

We also need to discuss the relationship between the self and other functions. Recent imaging studies demonstrated strong neural overlap between the self and reward, the self and emotions, and the self and decision-making. For example, when receiving a reward such as money in relation to specific stimuli, regions of the reward system like the ventral striatum (VS) and the ventromedial prefrontal cortex (VMPFC) become active (Northoff, 2012a, 2012b, 2012c). These same regions are also active when the same stimulus is conceived of as self-specific, rather than non-self-specific, by the respective subject.

The same effects can be observed in emotions where emotional and self-specific stimuli have been shown to overlap, especially in the anterior midline regions. Finally, the same effect can be observed in decision-making: if external cues are provided when making a decision (such as a higher or lower price of the same kinds of apples), lateral cortical regions become active. If, in contrast, no such external cues are provided, we need to come up with some internal criterion to guide our decision about which apples to purchase (Nakao et al., 2012). Such an internal criterion can only stem from our self. Studies comparing both kinds of decision-making show the predominant involvement of the midline regions in internally guided decision-making, when compared to externally guided decision-making (Nakao et al., 2012).

Together, this neural overlap between the self and other functions such as reward, emotions and decision-making raises questions about the relationship between them. Different models could be imagined. Self- and self-specificity could be an independent function, just like attention, working memory, emotion, sensorimotor, etc. However, in that case, one would expect specific regions in the brain and specific psychological functions to subserve specifically, and exclusively, self-specificity. However, at this point in time, this cannot be supported empirically.

Finally, one could also suggest that self and self-specificity are basic functions that underlie and provide the basis for all other functions – sensorimotor, affective, cognitive, and social. In this sense, self and self-specificity would occur prior to the recruitment of the other functions. Self-specificity would then always be present, making its involvement and manifestation in the various functions unavoidable.

Rather than searching for self-specificity in relation to specific functions, such as language, one would then need to look for more basic functions that must occur prior to sensorimotor, affective or cognitive functions.

One could, for instance, imagine that the strong involvement of the self in language acquisition requires the recruitment of midline regions. This involvement of midline structures may be implicitly presupposed in many of the tasks or paradigms described above when presenting self-relevant and non-self-relevant words, known as trait adjectives. While the linguistic tasks themselves seem to involve the lateral cortical regions more, their degree of activity may nevertheless be dependent upon the midline regions and their high resting state activity. Hence, future studies should investigate the relationship between midline regions and the lateral networks implicated in language, which may correspond psychologically to the relationship between self and language.

Neurophilosophical discussion 5

Phenomenal specificity of self-reference

The assumption of self-specificity brings us back to the concept of the self as a 'minimal self' (see above). To recap, the minimal self describes a basic sense of self that occurs immediately and is always already part of our experience of the world. The question now is how the concept of the minimal self is related to the neuroscientific results discussed above. To answer this question, we have to shed light briefly on the experience of the minimal self as manifested in pre-reflective self-consciousness.

Various phenomenal features such as qualia and first-person perspective characterize our consciousness. If the minimal self is part of any experience (rather than being outside of it), the self should be manifested in these phenomenal features too. What experiential and thus phenomenal features does the self add? One may assume that the self, first and foremost, makes possible the generation of qualia. Without self, there is no point of view and hence, in our experience, no qualitative features.

Phenomenological philosophers assume that the special contribution of the self consists in what they describe as 'belongingness' or 'mineness' (Gallagher, 2000; Zahavi, 2005): the contents of our experience are experienced as belonging to a particular self; they are experienced as 'mine'. For instance, I experience my friend's laptop on which I write for a while as my laptop, though I do not own it. This goes along with the experience of a feeling of belonging to my self. However, such experience is not possible for the person sitting beside me who, though looking at the same laptop, does not experience any relation to the self. Instead, she/he may experience mineness or belongingness of the CD lying besides the laptop because she/he is a composer and it is a CD of her/his work.

This relation to the own self is particularly important when one needs to acquire a second language. The foreign language will appear as totally strange, as having no relation to one's own self and thus no self-relevance will be detected in any of the words. Why? Because none of the new words are as yet associated with any experiences in specific contexts and situations.

The words thus do not yet elicit any sense of relation to the self. However, once one immerses oneself more and more into a new culture or learning context and gains new experiences, the novel words will become associated with self-relevance, thus inducing a sense of self. In short, the novel language will increasingly become associated with one's own self and become part of it. It is to be supposed that this self-relevance of language may facilitate the acquisition and learning of the new language.

Neurophilosophical discussion 6

Phenomenal specificity and phenomenal limits

In order to account for phenomenal specificity, neuroscience needs to demonstrate which neuronal mechanisms underlie the experience of mineness and belongingness. We also need to distinguish those that underlie other phenomenal features of experience, including intentionality, unity, first-person perspective, qualia and spatiotemporal continuity.

One would therefore require distinct experimental measures and designs for each of these phenomenal features. Only then would we be able to achieve phenomenal specificity and to distinguish clearly the phenomenal or minimal self from phenomenal consciousness. In short, we need to distinguish experimentally between self-specific and non-self-specific phenomenal measures.

However, the phenomenological philosopher may want to raise the following question: is such phenomenal specificity with the experimental distinction between self-specific and non-self-specific phenomenal measures really possible at all? The minimal self is considered part of the experience and thus of consciousness more generally. Any consciousness of the world goes along with an experience of the self in a pre-reflective way. And the opposite holds true too. Any experience of the self is part of an experience of the world. Both the experience of self and experience of the world are thus intrinsically linked.

What does this intrinsic link between the experience of self and the experience of the world imply for the phenomenal specificity of the self? It means that we will remain unable to clearly segregate experimental measures for the minimal self from those of our experience in general. Why? Because these phenomenal features are always already 'infected' by the self – they are encoded and ingrained into the self. Hence, the requirement of maximal experimental and phenomenal specificity may have reached its phenomenal limits. If so, we may be forced to acknowledge that there may be limitations to what we can and cannot investigate experimentally when it comes to the minimal self.

Neurophilosophical discussion 7

Minimal self and body

What about self and body? We experience our own body as our own. The body plays a central role in the concept of the minimal self in a dual way. First, because of its basic and minimal character, the minimal self is supposed to be subserved by functions of brain and body. The sensorimotor functions of the body that link it with the environment are thus supposed to yield and constitute the minimal self.

Phenomenologically-minded neuroscientists therefore consider the minimal self to be embodied and embedded (see Chapters 1 and 7 for details on embodiment and embeddedness).

This leads us to the characteristic feature of the body, namely, that it can be experienced in consciousness. The body is not only an objective body that can be observed from a third-person perspective. This is the body that both the neuroscientist and the doctor investigate. It can also be experienced from a first-person perspective. This is the body we consciously experience, also known as the 'lived body'.

The lived body is my body as distinguished from others' bodies. Hence, we experience the lived body in relation to our self – in terms of mineness and belongingness. Thus, the experience of the body, the lived body, may be regarded as the first and most fundamental manifestation of the phenomenal or minimal self. Our self in its most basic and minimal form is thus essentially a bodily self.

This relationship to the self is also reflected in what we described earlier as ownership and agency. Ownership describes the fact that I experience my body as mine, rather than some other body. Neuroscientifically, the ownership of the body has been associated with neuronal activity in specific regions of the brain such as the sensory cortex and the parietal cortex. The parietal cortex mediates the spatial position of the body in the world.

Agency is the experience that it is I, rather than some other person, that causes action and movement. I, my self, am the agent of the lines I am currently writing here on my laptop. Neurally, regions such as the premotor cortex and the motor cortex have been associated with agency; these are regions that are implicated in generating movement and action in general.

How is the experience of such a bodily self mediated? In determining this, sensorimotor function is considered to be central, especially the coordination and integration between sensory and motor circuits in the brain. For instance, when generating an action and movement, a copy of this sensorimotor coordination, a so-called efference copy that signals a forward model, is sent to the sensory cortex.

Why? By receiving an efference copy of the intended and to-be-performed action, the sensory cortex can prepare itself for, and thus 'anticipate', potentially incoming sensory stimuli. It can thus predict more easily the next sensory state on the basis of what the motor cortex is currently doing. Through this process sensory and motor functions become intrinsically linked together and provide the integration of the body within the environment, known as embeddedness.

Neurophilosophical discussion 8

Body and proto-self

Sensorimotor functions are not only mediated by cortical regions such as the motor and the sensory cortex. In addition, they are already processed in subcortical regions like the periaqueductal gray, the superior and inferior colliculi, and the basal ganglia like the pallidum, the caudate and the subthalamic nucleus. In addition to the sensorimotor functions, these regions are central in regulating and controlling the vegetative and thus the inner visceral or homeostatic functions of the body. This in turn is central in eliciting emotions.

How are these regions related to the minimal self? Investigations that did not include a strong cognitive or task-related component like a judgment (see above) demonstrated neural activity in these subcortical regions during self-specific stimuli. Because of their involvement in various functions, these regions are definitely not specific for self-referent stimuli. What this shows is that they nevertheless participate in constituting a self, a minimal or phenomenal self.

Contemporary neuroscientists like Jaak Panksepp and Antonio Damasio do, therefore, speak of a bodily self or 'proto-self' that occurs prior to the minimal or phenomenal self. They call this the 'core or mental self'. These subcortical regions seem to coordinate and integrate the inputs from the body at each moment in time. This allows the body to experience itself as one's own body in the most basic way. Panksepp goes so far as to characterize the term self as 'Simple Ego-type Life Form', to indicate the basic and most fundamental nature and relevance of the self for the body as a biological organism.

Neurophilosophical discussion 9

Difference between the concepts of self and self-reference

The concept of conceptual specificity focuses on the question: do the neuronal findings really reflect the self, or some other function? First and foremost, we must see what exactly is investigated in the imaging studies. Remember that all experimental paradigms are based on the self-reference effect. This effect assumes the distinct processing of stimuli (e.g. items, objects, people or events) that are related to the self compared to those that are not related to the self.

For instance, a picture of Ottawa has a specific self-reference to me since I live there, while to you, as a resident of Australia, it has no self-reference whatsoever. Due to such difference in self-reference, the picture of Ottawa will be processed differently by your brain and mine.

The self-reference effect presupposes the distinction between the self and a specific content to which the self may refer or not. What is investigated experimentally in the paradigms described above is thus not so much the self itself, but rather the degree of reference of a particular content to the self, known as self-reference. Some neuroscientific authors do therefore also speak of self-related or self-referential processing, which describes the processes that are assumed to constitute the relation of a particular content to the self.

Why is this important? It means that the experimental paradigms do not target and measure the self as such, but rather the degree of the relation of a specific content to the self. For instance, the degree of neural activity in the midline structures reflects the degree of self-reference, rather than the self itself. The same holds true for the gamma oscillations, which at best correspond to the degree of self-reference, rather than to the self itself.

This means that the various empirical findings remain unspecific with regard to the concept of the self itself. They tell us about self-reference as the relationship of particular content to the self, but not about the self. This means that the empirical findings are conceptually unspecific with regard to the concept of self. How can we resolve this conceptual unspecificity? In order to close the gap between the concepts of self and self-reference, we may need to shift our focus from the self to self-consciousness. This will be discussed in the next chapter.

Neurophilosophical discussion 10

Self as brain-based neurosocial structure and organization

What does this imply for the self? Our self may be considered as linked intrinsically to the body. This is called the embodied self. Furthermore, since it is based on self-reference, our self may also be linked intrinsically to the environment. This is called the embedded and social self. Our self cannot consequently be regarded as an entity located somewhere in the brain and isolated from both body and environment. Instead, our self seems to be intrinsically social, as suggested by the advocates of the concept of a social self (see above).

What does this intrinsically bodily and social nature imply for the conceptual characterization of the self? Our self may be described as structure and organization, rather than as an entity – be it mental or physical. Such structure and organization needs to develop through childhood and adolescence with persistent changes even throughout adulthood. Despite all the changes there may also be persistence and continuity across time, which then accounts for what can be described as identity. Identity may describe the persistence and continuity of self over time which, in an exploratory study, has recently been associated with the midline structures and their high intrinsic activity.

We can also see that this concept of self as structure and organization is embodied and embedded. Hence, the virtual structure of the self spans the brain, body and environment. At the same time, this virtual structure is dependent upon the respective environmental context. Freud's characterization of the ego as structure and organization surfaces here in a more specific way; namely, the ego consists of a relationship between the brain, the body, and the environment. This means that it can be determined in an intrinsically relational way. Future investigation might link the different features Freud attributed to the ego to the self.

What do we mean by the concepts of structure and organization? The structure must be virtual in that it spans the physical boundaries of the brain, the body, and the environment. Does this mean that we have to revert to a mental structure and organization that is distinct from the physical structure and organization of the brain? No! The results from neuroscience clearly link the self with neuronal processes related to both intra-individual experiences and inter-individual interaction. There is thus a neuronal basis for the distinct aspects of the self within the context of the brain, the body, and the environment. We therefore reject the mental characterization of the structure and organization that is supposed to define the self.

How can we define the concepts of structure and organization in a more positive way? One way is to characterize structure and organization as social. This distinguishes it from mental or physical features. The social characterization would then be the underlying basis that links and integrates the purely physical and the purely mental. The self would then be based on the brain, but would also extend beyond it to the body and the environment. This means that, conceptually, we need to characterize the concept of the self as brain-based, rather than brain-reductive (as the proponents of the empirical self tend to do). The brain-based nature of the self also excludes both mind- and consciousness-based approaches to the self.

If the social characterization of the structure and organization as related to the self is indeed basic and fundamental, one would assume that our brain's neural activity is intrinsically neurosocial: the brain cannot avoid including the social environmental context in the encoding of stimuli into its own neural activity. The neural activity is thus, by default, neurosocial rather than merely neuronal. This is supported by the neural overlap between resting state activity and the neural activity changes during emotional and social-cognitive tasks, as described above.

Whether the brain encodes its neural activity in an intrinsically neurosocial way remains unclear at this point. What is clear is that the exact characterization of the brain's neural activity will be essential if we are to develop a truly neurophilosophical, brain-based (rather than brain-reductive), and neurosocial (rather than merely neuronal) concept of the self.

Take-home message

We all have consciousness. And we all have a self who experiences consciousness. This 'who' addresses the question of the subject who experiences consciousness. Who is this who, the self? In this chapter, we addressed this question specifically. The concept of self has long been discussed in philosophy. Most notably among the various philosophical definitions for the self is the concept of self as a mental substance. Another concept describes the empirical self as self-representation. Yet another concept is the phenomenal self that is always already part of any experience. Finally, there is the minimal self that is closely related to the body. In addition, the self has also been investigated in psychology and neuroscience by comparing self-specific versus non-self-specific stimuli. Both types of stimuli lead to different psychological and neural effects. How are these psychological and neural measures of the self-specific stimuli related to the different concepts of self as discussed in philosophy? To answer this question, one needs to investigate the empirical findings with regard to their specificity for the self and their neuronal, psychological, experimental and phenomenal specificity. This sheds some doubt on a direct one-to-one correspondence between the philosophical concepts of self and the neuronal and psychological measures of self-specific stimuli. Finally, questions about the conceptual difference between self and self-reference must be raised, which makes any direct linkage between conceptual suggestions and empirical findings problematic.

Summary

While the previous part focused on consciousness, the present chapter aims to target yet another central feature of the mind, the self as the subject of all our experience, and hence of consciousness. More specifically, the focus is on different concepts of the self and how they are related to recent findings about neural mechanisms related to the self-reference of stimuli. I first introduce different basic concepts of the self as they are currently discussed in philosophy. The first concept of self is the self as a mental substance, which was introduced originally by Descartes. This is rejected by current and more empirically-oriented concepts of the self, where the idea of a mental substance is replaced by assuming specific self-representational capacities. These self-representational capacities represent the body's and brain's physical, neuronal states in a summarized, coordinated and integrated way. As such, the self-representational concept of the self must be distinguished from the phenomenological concept of self that is supposed to be an integral part of experience and thus of consciousness. This phenomenal self resurfaces in the current debate as the 'minimal self' – a basic sense of self in our experience that is supposed to be closely related to both brain and body. Current neuroscience investigates the spatial and temporal neural mechanisms underlying those stimuli that are closely related to the self when compared to the stimuli that show no relation or reference to the self. This is described as the self-reference effect. When comparing self-specific versus non-self-specific stimuli, neural activity in the middle regions of the brain, the so-called cortical midline structures, is increased. Moreover, increased neuronal synchronization in the gamma frequency domain can be observed. The question is how specific these findings are for the concept of self as discussed in philosophy. Neuronal specificity describes the specific and exclusive association of the midline regions with the self. This is not the case, since the same regions are also associated with a variety of other functions. This goes along with the quest for the psychological and experimental specificity of psychological functions and experimental paradigms and measures used to test for the self. One may also raise the issue of phenomenal specificity: the concept of phenomenal specificity refers to whether the phenomenal features of the self, including 'mineness' and 'belongingness', are distinguished from other phenomenal features like intentionality or qualia. Finally, one may discuss the question of conceptual specificity that targets the distinction between the concepts of self-reference and self.

Revision notes

- What is the self and why is it relevant?
- Characterize the concept of the mental self.
- Why is the concept of the self as a mental substance rejected in the current debate, and by what is it replaced?
- What is self-representation?
- What is the phenomenal self?
- How does the phenomenal self resurface in the current debate about the self?
- What are the experimental methods of testing the self empirically in psychology and neuroscience? What is the self-reference effect?
- What are the main results in the spatial and temporal pattern of neural activity?
- Are the midline regions neuronally specific for the self?
- What is psychological and experimental specificity, and are they fulfilled in the case of the self?
- Explain phenomenal specificity and its relationship to possible phenomenal limits.
- Does the self-reference effect really target and account for the self itself? If not, why not?

Suggested further reading

- Damasio, A. R. (1999) 'How the brain creates the mind', *Scientific American*, 281(6), 112–17.
- Legrand, D. and Ruby, P. (2009) 'What is self-specific? Theoretical investigation and critical review of neuroimaging results', *Psychological Review*, 116(1), 252–82.
- Metzinger, T. (2003) *Being No One* (Cambridge, MA: MIT Press). *This is a central book about one of the current concepts of self that denies its existence as such. Instead the self is considered to be an illusion. The book provides a nice overview of the current discussion.*
- Northoff, G., Heinzel, A., de Greck, M., Bermpohl, F., Dobrowolny, H. and Panksepp, J. (2006) 'Self-referential processing in our brain – a meta-analysis of imaging studies on the self', *Neuroimage*, 31(1), 440–57. *This paper gives an overview of empirical findings on the self in imaging studies and how they relate the different concepts of self discussed in both psychology and philosophy.*
- Zahavi, D. (2005) *Subjectivity and Selfhood: Investigating the First-person Perspective* (London: MIT Press). *This book presents a phenomenological philosophical viewpoint on the self, the self as pre-reflective self-consciousness. Excellent overview with many links to the philosophy of mind.*

18
Brain and Self-consciousness

Overview

In the previous chapter, we discussed different concepts of self. This included the concept of the self as a mental substance, as a higher-order cognitive function, as a minimal self, and as an integral part of our experience, known as the phenomenal self. Neuroscientific findings have demonstrated the association of particular spatial and temporal patterns of neural activity with the self-reference of extrinsic stimuli. How are philosophical concepts of self and neuroscientific discoveries about self-reference related to each other? Because they concern different concepts – self versus self-reference, and different domains – metaphysical versus empirical, there seems to be a gap between our philosophical and neuroscientific explanations of the self. This explanatory gap may be closed and bridged by self-consciousness as a common functional pathway of both the self and stimuli. This is the topic in the present chapter.

Objectives

- Understand the explanatory gap in the context of the self
- Distinguish between direct and propositional awareness
- Explain the immunity to error through misidentification
- Understand the empirical findings about self and rest
- Discuss how the brain's intrinsic activity could be related to the formal or transcendental concept of self
- Understand the concept of the neural predisposition of self

Key concepts

Self, self-consciousness, explanatory gap (s), direct and propositional awareness, immunity to error through misidentification, intrinsic activity, neural overlap, vegetative state, prediction of consciousness, pre-reflective self-consciousness, neural predisposition of self.

Background 1

Philosophical and neuroscientific approaches to the self

In the previous chapter, we introduced various concepts of the self. One example includes a concept first put forward by Descartes: the concept of a mental self as a specific mental substance or property. Most contemporary philosophers no longer assume a specific mental substance; instead, most understand the self as an empirical self, based on a specific form of representation known as self-representation. Self-representation assumes the brain's neural processes that represent features of the body and the environment are re-represented by themselves in additional neural activity patterns. In this model, the self is no longer characterized by a specific mental substance or property but rather by self-representation.

An alternative concept of the self was developed in phenomenological philosophy, the branch of philosophy that focuses on the structure and organization of consciousness (see Part I for details). The phenomenological approach assumes the self to be an integral part of our experience. Any experience is not only about the respective content – a person, object or event from the environment – but must necessarily also include the self that experiences that consciousness.

Consequently, the self is always *already* part of our experience, even if the experience is not directly about the self, or self-reflective about the self as self. This is what the phenomenological philosophers describe as 'pre-reflective self-consciousness'. In other words, the self and consciousness are intrinsically linked: they necessarily co-occur. Hence, when experiencing the laptop in front of me, not only do I experience the laptop but I also, simultaneously, experience my own self in a pre-reflective way. In short, my experience includes the laptop as the content of my experience, and my self as the subject of that very same experience.

The phenomenal self as a concept resurfaces in the concept of the minimal self, an approach to the self suggested by contemporary phenomenological philosophers. The minimal self concerns the minimal necessary conditions that allow for experience of the self. The body and its sensorimotor functions are considered central for the minimal self, which therefore focuses mainly on the subjective experience of one's own body as the minimal basis of any self.

In addition to these philosophical concepts of self, we also discussed various empirical findings in neuroscience about the neuronal mechanisms of the self-reference effect (SRE). The self-reference effect describes the specific psychological (e.g. reaction time, recall, etc.) and neural (e.g. activity level, functional connectivity, etc.) effects of self-specific stimuli. Self-specific stimuli are those stimuli that are specifically related to the self as distinguished from non-self-specific stimuli. By testing for SRE, neuroscience aims to tap (indirectly) into the self and the specific contents it links and associates with itself.

The recent research demonstrated strong involvement of a particular set of regions in SRE, namely the cortical midline regions. Whether the neural activity in these regions is neurally, experimentally, and psychologically specific for the self and SRE remains open, however.

Background 2

'Explanatory gap (s)'

How are these neuroscientific findings about self-reference related to the philosophical concepts of the self? Beyond neuronal, psychological, experimental and phenomenal unspecificity (see Chapter 17), one of the main problems is the conceptual difference between self and self-reference. The concept of self-reference as used in neuroscience describes the relationship of a particular stimuli, be it an event, person or object, to the self. The guiding question is: how are the stimuli referred or related to the self? Thus here the self is approached indirectly via stimuli and their degree of self-referentiality.

The concept of self is different. It concerns exactly that to which the stimuli refer; the degree of reference of the stimuli to the self is measured in terms of their degree of self-referentiality (or self-specificity). Rather than concerning how the stimuli are related to the self, the guiding question here is: what is that very self to which the stimuli are referred or related to? Unlike in the empirical approach, the self is approached in a direct way.

Philosophical and neuroscientific approaches differ conceptually. Each attempts to understand consciousness through either the concept of self or the concept of self-reference, respectively. Equally, they differ in their relationship to the self as being either direct or indirect. Finally, they also presuppose different kinds of questions that relate back to each other's underlying domain. Neuroscience raises 'how' questions: how does something like the self-reference of stimuli function; and how is it processed in such a way that it presupposes the empirical domain – the domain of observation? Philosophy asks 'what' questions – what is the existence and reality of the self; and what does it imply about the metaphysical domain – the domain of existence and reality.

So far we have highlighted the three main differences between philosophical and neuroscientific approaches. These are (i) a difference in concepts; namely self versus self-reference; (ii) a difference in approach – either direct or indirect; and (iii) a difference in domains – metaphysical versus empirical. These differences point to an explanatory gap between our philosophical and neuroscientific explanations of the self. *The concept of explanatory gap (s) describes principal differences between philosophical and neuroscientific explanations of the self, namely the differences in the concepts, the approach and the domains presupposed in the respective explanations.*

You may recall that we have already encountered an explanatory gap, in Part IV. There, I outlined the fact that philosophers often speak of an 'explanatory gap problem'. This gap concerns the differences in our explanations of neuronal and phenomenal states: these states are explained in a parallel rather than an integrative way (see Chapter 13).

As such, the explanatory gap problem concerns the relationship between neuroscientific and phenomenological explanations of consciousness. In order to distinguish between them, we can speak of an explanatory gap (c) and an explanatory gap (s). In contrast to the explanatory gap (c), the explanatory gap (s) is not about consciousness, but about two different explanations, philosophical and neuroscientific, of the self.

Background 3

Explanatory gap (s) and self-consciousness

How can we bridge the explanatory gap (s) – the gap between neuroscientific and philosophical explanations of the self? This question may be reformulated in the following way: where and how can we find a bridge between the self – its nature and thus its reality and existence – and the stimuli, including their degree of reference to the self? In order to address that question one may want to investigate how the self-reference of stimuli is manifested and how that manifestation is then related to the self itself.

Neuroscience investigates the processes and mechanisms underlying the self-reference of stimuli by comparing the psychological and neural effects of self- and non-self-specific stimuli. This is usually done by letting subjects judge or evaluate the stimuli as being either self- or non-self-specific (see Chapter 17 for details). This type of judgment, however, presupposes some kind of experience of the stimulus and its respective degree of self-specificity. In other words, before being able to judge and evaluate the stimuli's degree of self-specificity, the subject must experience and thus be conscious of them.

However, the subjects must not only be conscious of the stimuli themselves, but also of their self against which they measure the stimuli's degree of self-specificity. If the subjects are not conscious of themselves, they cannot relate the stimuli to their own self. This, in turn, would make any attribution of higher or lower degrees of self-specificity to the stimuli impossible. Metaphorically speaking, self-consciousness itself might provide the glue or bridge between the self itself and the self-reference of the stimuli.

What is self-consciousness? Self-consciousness is manifested in various facets. When we experience, for instance, the movements and actions we initiate, one can speak of 'agency'. Or when we experience and become aware that the legs belonging to our body are our legs. This is called 'ownership'. Self-consciousness is also a given when we recognize the face in the mirror as our own face. This is called 'self-recognition' or 'self-identification'.

These various forms of self-consciousness describe manifestations of the self in consciousness. The self may be manifest in our consciousness of our own face, called self-recognition, in our experience of our own body, called ownership, and in our experience of movements and actions, called agency. While this description paints a picture of the manifestation of self-consciousness, it leaves open for discussion the relationship between self and consciousness.

How are self and consciousness related to each other? Is the self intrinsically tied to consciousness, so that there is no self outside the realm of consciousness? This is what many philosophers thought, and still think, as we will see below. Or is there a self prior to and thus independent of consciousness? This is what certain neuroscientific findings seem to suggest, as we will discuss later in this chapter. Before moving on to these topics, however, we need to review the different forms of self-consciousness.

Philosophical concepts 1

Indirect and direct access to the self in consciousness

How can we access our self as self? 'In and through our consciousness', you may want to respond. Very simple; the easiest way is to become aware of the self and its features. For instance, 'I know that it is me, my self, that is writing these lines. I know that it is my self rather than some other self or another person that is writing this book.' Another example would be the following: 'I am aware that it is me, my self, that talks slowly and nicely to my shouting neighbor. I am calm and relaxed, this is me, my self, rather than some other person's self.' Yet another example: 'I believe that the image of a person in front of me is not Barack Obama, but rather George W. Bush. Who believes this? It is me, my own self.'

Here, I access my self by reflecting and introspecting about what I am doing, believing, etc. I access my self in an indirect way via my perception, beliefs and knowledge, which are tied to and expressed in language and propositions. Therefore, indirect access to the self in consciousness is called 'propositional awareness' in current philosophy of mind. I access my self in terms of concepts.

In addition to propositional awareness as indirect access to the self, one may also have a more direct access. This is described as direct awareness in the context of self-consciousness. Direct awareness refers to the awareness of one's own self independent of its beliefs, knowledge, perceptions, etc. about some objects, events, or persons in the world. Direct awareness is, for instance, described by the French philosopher Descartes in his assumption: 'I think, therefore I am'. One's own thoughts are assumed to provide direct access to the self itself; in short, one's thoughts are one's self.

In order for direct awareness of the self to be possible, one must assume a self that is independent of its beliefs, perceptions, knowledge, etc. While Descartes assumed the self to be a mental substance, his successor David Hume denied the existence and reality of such a kind of self. According to Hume, there is no self separate and distinct from its own perceptions, beliefs and knowledge.

The assumption of a separate and distinct self is nothing but an illusion for Hume, because we are not able to find any kind of mental substance or property besides what Hume describes as a 'bundle of perceptions'. The self remains elusive in this model. This is what philosophers describe as the 'elusiveness thesis'. If, however, the self is elusive, direct awareness of the self remains impossible too. Why? Because one cannot become aware of something that is absent.

Philosophical concepts 2

Reflective and pre-reflective self-consciousness

What about the experience of our self in experience? Recall from Chapter 1 that phenomenological philosophy focuses on experience and thus consciousness itself, aiming to reveal its structure and organization. Following their focus on experience, phenomenological philosophers assume the self to be present in consciousness. The self is assumed to be part of all experience, whether we focus on the self or not; the self is always already there.

One may now distinguish between two different forms of experience with regard to the self. The self may be, by itself, the content of experience. I focus on myself as the content of experience. Here, my perception and cognition are about myself – as in when I see myself in a mirror or perceive my own body. This is called 'reflective self-consciousness'.

The more interesting case in phenomenological philosophy is the assumption that the self is always already part of any experience and thus of consciousness in general. For instance, while I experience the birds singing in the trees and notice how the wind shifts the leaves, I also experience my own self. Why? Because without the self, any experience of singing birds would remain impossible. My experience of the singing birds thus always already includes my own self as the subject of that very experience – as the one who experiences the singing birds. If I did not have a self, any such experience would remain altogether impossible. This is what phenomenological philosophers describe as 'pre-reflective self-consciousness'.

Let us describe 'pre-reflective self-consciousness' in further detail. Before I reflect upon my experience of the trees and the birds and how it implicates my self, my self is already present in the experience itself. In pre-reflective self-consciousness, the self is present, though not in an explicit way, as it is in reflective self-consciousness. Instead, here the self is accessed in an implicit and non-thematic way. This type of implicit and non-thematic presence of the self in our experience is described as 'self-givenness'. This term refers to the presence and thus the givenness of our self in any experience in either an implicit or explicit way.

How is my self manifested in my experience? All my experiences about the tree, the birds and the wind are centered around my self and directed towards it. This is experienced as what is described as 'mineness' and 'belongingness'. The concepts of mineness and belongingness describe the experience of a relationship between the subject or self and the contents of its experience.

For instance, the experience of the text I currently see in front of me is characterized by a high degree of mineness and belongingness. It is 'my' text and 'belongs' to me because I wrote it, and it reflects my ideas, the ideas of my self. Put purely empirically, there is thus a high degree of self-specificity of the text which is manifest phenomenologically in the experience of mineness and belongingness.

What kind of self do we pre-reflectively encounter in our experience? This self is not a mental substance in the Cartesian sense, nor is it an illusion, as Hume and many of the current advocates of an empirical self claim. Instead, the phenomenologist philosophers argue that the kind of self we encounter in our pre-reflective self-consciousness is part of the structure or organization of experience. In this model, the self no longer consists of a particular substance, property or feature but, instead, in some kind of structure and organization. What is the structure and organization, and how can we define it in more detail? This question warrants further investigation. It will be discussed below, as well as in Chapters 15 and 16.

Philosophical concepts 3

No 'immunity to error through misidentification' with regard to the contents in our consciousness

How can we further characterize self-consciousness? One essential criterion is whether we can err about our own self or not. If I experience my own self as part of my experiences, I may also be mistaken in that it is really my self in particular who has these experiences. Rather than my self, it may be another person's self who has these experiences. This might happen, for example, when I empathize so closely with my partner's pain that I feel the pain myself. In this case, I might be prone to confuse my own self as the subject of experience with my partner's self, who is the actual subject of experience.

Alternatively, I may mistake the subject of my experiences and no longer associate them with my own self, but rather with another self. This is what happens in schizophrenia – a psychiatric disorder typically characterized by delusions, thought disorders and hallucinations. Schizophrenic patients sometimes experience themselves as another person, like Jesus or the Prime Minister of Canada, to whom they then attribute their experiences. Jesus or the Prime Minister, rather than their own self, becomes the subject of their experiences and thus their self.

Can we really err about our own self in self-consciousness? Or are we immune to making such errors? This is debated strongly in current philosophy of mind under the heading of 'immunity to error through misidentification', which will be summarized briefly below.

Think about yourself while you read these lines. You think about the concept of self-consciousness and have many thoughts about it as a concept. Now you are asked by your fellow student: 'What are you thinking about?' You will answer: 'I'm thinking about self-consciousness!' 'Are you sure?' your fellow student asks. 'Wait a minute,' you may reply, 'sorry, you're right: I was not really thinking about self-consciousness, but rather about my next vacation.'

What is happening here? You were introspecting about your own thoughts. At first you identified your thoughts as thoughts about self-consciousness. But this turned out to be false, known as a misidentification. Instead, your thoughts were really about your next vacation. This is called 'error through misidentification'. An example of this is when you are sitting on a train and think it is moving when in fact it is the train on the track next to you that moves. You look out of the window and see the movement and become confused about which train it is that has begun to leave the station.

What exactly does the error consist in here? Here, the error through misidentification concerns the contents of our thoughts and perceptions. We can misidentify the contents we presume to perceive and, in doing so, be deceived. In other words, we are not immune to erring about the contents of our experience.

Philosophical concepts 4

'Immunity to error through misidentification' with regard to the own self or subject of our consciousness

We have so far discussed that we can err and thus misidentify the contents of our consciousness. In contrast, we left out the subject of consciousness, the self or I that experiences the contents of its consciousness and is conscious of them. In other words, we are now shifting our focus from the contents of consciousness to the subject of consciousness.

Can we misidentify and thus err about our own self that experiences the contents in consciousness? You said: '*I* think about self-consciousness'. But you misidentified self-consciousness as a false positive content of your thoughts because you were really thinking about something else, namely, your next vacation. Can you also misidentify the subject of your own experiences, your self? Can you misidentify the person who thinks these thoughts? 'No, that is impossible,' the philosophers say. 'We cannot be mistaken about the person or subject who thinks these thoughts or perceives certain contents like the moving or non-moving train.' If so, misidentification of our self as the subject of experience remains impossible. Error remains consequently impossible. This is called 'immunity to error through misidentification' in current philosophy of mind.

To put it differently, we remain immune to the misidentification of our own self. When I say 'I think about self-consciousness', I might misidentify the content – self-consciousness – but I cannot misidentify the 'I'. The 'I' is my self and cannot refer to any other self because my introspection and reflection cannot provide access to any other self.

In sum, we are not immune to error with regard to the contents of our consciousness which we can easily misidentify. In contrast, such error through misidentification seems to remain impossible when it comes to the self or subject of consciousness, the I.

Philosophical concepts 5

Self as subject and object

How can we further specify 'immunity to error through misidentification'? To do this we need to go briefly into the linguistic use of the first-person pronoun with regard to the self. One may use the pronoun 'I' in different ways, as a subject or as the object of a sentence.

We can use 'I' to describe the self as subject. 'I think about self-consciousness' or 'I perceive rain outside'. Here, 'I' is the subject that describes your first-person perspective. As described above, there can be no error with regard to the 'I' who thinks and perceives. More technically, there is immunity to error through misidentification relative to the first-person pronoun and its use as the subject of experience.

Alternatively, we can also use the first-person pronoun as the object of a sentence. Consider 'I feel that *I* am doomed to failure' or 'I sense that *I* will win the lottery'. The first 'I' in each sentence uses the 'I' in the role as subject. The second 'I' in each sentence uses the 'I' in the role as object rather than subject. Hence, I refer to myself, my 'I', as an object, rather than as the subject.

What about error through misidentification in both cases? There is immunity to error through misidentification in the first use of 'I' as subject. However, there is no longer immunity in the case of 'I' as object. As an object, I may indeed be mistaken; it might not be my 'I' that is doomed to failure, but rather my neighbour's 'I' because of his financial troubles.

The mistaking of one's self as object can occur, for instance, in psychiatric patients with depression, who relate all kinds of negative feelings to their own self. The severely depressed patient may say: '"I" feel that "I" am doomed to failure and death'. However his 'I' is not doomed to failure and death. Instead, the patient makes an error through misidentification in the use of 'I' as object.

In the case of severe depression I might misidentify my self as a false positive object of my experiences. Why? Because when one uses 'I' in the role of an object, it is treated like a content of experience. And since one can be mistaken about the contents of experience, one can also err about one's own 'I', once it is used in the role of an object. In short, there is no immunity to error through misidentification relative to the use of the first-person pronoun as object.

Let's summarize. We can apparently not err when it comes to the identification of the subject of our experiences. We seem to remain immune to such error through misidentification when it comes to our own self – the 'I' and its role as subject. It must be me, my self or 'I', that writes these lines, thinks these thoughts, hears the birds and perceives the green trees. This we know and there we cannot err. In short, there is immunity to error through misidentification relative to the first-person pronoun and its use in the role as subject. However, such immunity is no longer a given once one uses the 'I' in the role as object rather than as subject.

Disorders of consciousness 1

Vegetative patients: self-specificity in vegetative state

We have already encountered patients in a vegetative state (VS) in the previous part about consciousness. Patients with VS suffer from dysfunction in their brain. This leads to the apparent loss of consciousness with no observable behavioral reactions. What about the self in these patients? We have already discussed how self-reference can be investigated in general by comparing the neural effects of self-specific stimuli (like one's own name) and non-self-specific stimuli (like another person's name) (see Chapter 17).

The same thing can be tested in patients in a vegetative state (VS) using functional brain imaging. Recent imaging data shows that they react to cognitive stimuli when being asked to imagine playing tennis or asked to navigate mentally through their house. This goes along with neural activity changes in the same regions that healthy subjects use to perform the same tasks (see Chapter 16). Inwardly, in their brain, these patients show neural activity, whereas their outward behavior shows no signs of consciousness.

What about stimuli related specifically to the self, like one's name? One study by the Chinese scientist Pengmin Qin (Qin and Northoff, 2011) played each respective VS patient his/her own name and then another person's name. The brain's activity in the fMRI was compared. As in healthy subjects, VS patients showed neural activity in the auditory cortex and in various parts of the midline regions like the VMPFC, DMPFC, SACC and PCC when listening to their own name.

Did the VS patients differ from healthy subjects? The difference in signal degree between hearing one's own and another's name was significantly smaller in the VS patients when compared to healthy subjects. These results suggest that the VS patients are still able to elicit neural activity in response to hearing their own name. However, they are not as able to differentiate and distinguish between different kinds of stimuli.

Disorders of consciousness 2

Self-specificity and consciousness in vegetative state

How is this neural activity related to consciousness? Consciousness is measured with a typical scale, the revised consciousness recovery scale (CRS-R). The degree of signal differentiation between hearing one's own name versus other names predicts the degree of consciousness: the better one's own name and others' names could be differentiated in their underlying neural signals, the higher the degree of consciousness. Higher degrees of signal differentiation are accompanied by higher degrees of consciousness. For example, patients in a minimally conscious state (MCS) are 'more conscious' in comparison to patients in VS.

These results were confirmed and extended in another fMRI study in a separate group of VS patients by another scientist from China, Zirui Huang (Huang et al., 2013). Here the patients were listening to autobiographical and heterobiographical questions ('Are you from Ottawa?'; 'Are you from Sydney?') based on prior interviews with their relatives. Again, the patterns as described above were observed with neural activity in midline regions and less signal differentiation between auto- and heterobiographical questions. Most important, the degree of signal differentiation predicted the degree of consciousness, as in the previous study.

In addition, they also investigated the resting state activity when the eyes of the patients were closed. This revealed decreased resting state activity in various measures in exactly those regions where decreased signal differentiation was observed. More specifically, functional connectivity – the degree of synchronized activity between anterior (PACC, VMPFC) and posterior (PCC) midline regions – was reduced in VS. The same could be observed in the degree of variability – the degree of standard deviation of the resting state activity across time – that was significantly reduced in the patients.

Analogous findings were reported in EEG studies in VS. These studies investigated specific electrophysiological responses and waves, known as event-related potentials, in response to hearing their own versus other names. This elicited different early potentials between 100 and 300ms (like a negative wave around 100ms, the N100, the MisMatchNegativity (MMN) at around 200ms, and a positive wave around 300ms, the P300) when hearing their name compared to other names. VS patients showed analogous responses. However, these early responses were smaller in those patients with lower degrees of consciousness.

On the whole, these initial studies show the spatial and temporal pattern of neural activity underlying self-reference to be preserved to a certain degree in VS patients. Most important, the degree of signal differentiation between self-specific versus non-self-specific stimuli predicted the degree of consciousness in VS and MCS patients.

Neurophilosophical discussion 1

Is the self still present in vegetative patients?

What do these findings imply for the relationship between self and consciousness? To answer this, we have to sketch the experimental situation in more detail.

During brain scanning, self-specific and non-self-specific stimuli are presented. Applied to healthy subjects, this experimental scenario is confronted with different variables. This scenario presupposes the self, to which the respective stimuli are related and measured against. In addition to the self and the self- (and non-self-)specific stimuli, the experimental scenario also implicates consciousness, since healthy subjects are supposed to experience the stimuli and thus exhibit conscious behavior. This in turn enables them to judge the stimuli as self- or non-self-specific, which in turn is supposed to require self-consciousness.

What does the same experimental scenario look like in patients in VS? As in the previous experiment, self- and non-self-specific stimuli are applied by the experimenter. What about the self? One would assume that since there is differential activity between self- and non-self-specific stimuli, some kind of self must be present. Why? The stimuli cannot be measured and related to anything in the absence of a self. This means that the observed neuronal differences cannot be accounted for. There must be some kind of standard that serves as a measure to distinguish the different neuronal effects of the various stimuli. This standard may reflect what philosophers call the self.

Philosophers usually associate the concept of self with some kind of experience and thus with consciousness. Our self can exist and be real only if we can experience it and thus be conscious of it as self. This is what philosophers describe as self-consciousness (see above). Our self must be open to experience and consciousness, otherwise it cannot exist and be real. In short, no consciousness, no self.

This assumption is in conflict with the observations made in patients in VS. Consciousness remains absent by definition. So how can there be a self, if there is no consciousness in these patients? Following the philosophers' definition of the self being tied to consciousness, the absence of consciousness should go along with the absence of self. This belief is called into question indirectly by the observation of neuronal differentiation between self-specific and non-self-specific stimuli in VS patients.

Even though the neuronal measurement of self- and non-self-specific stimuli touches only indirectly upon the self, the neural processing of the stimuli nevertheless presupposes some kind of self as a standard or measure against which the different stimuli can be compared. Otherwise any kind of neuronal differentiation between self- and non-self-specific stimuli remains impossible.

Neurophilosophical discussion 2

Possible relationship between self and consciousness in vegetative state

What does this imply? Two possibilities. First that one may assume the VS patients do indeed show some kind of consciousness. The original definition of VS as loss of consciousness may then need to be revised. This is indeed what many neuroscientists suggest on the basis of these imaging findings. Conceptually, this implies that we can leave behind the original definition of the self as being tied to consciousness. What is revised in this case is not the definition of the concept of self and its relation to consciousness, but rather the definition of the vegetative state itself.

However, one may also go the reverse way. Rather than revising the definition of VS to include consciousness, one may revise the conceptual definition of the self. More specifically, one may dissociate the concept of self from the concept of consciousness. In contrast to the implicit presupposition in philosophy of linking the self to consciousness, one may assume the concept of self to be prior to and thus occur independently of consciousness. The concepts of self and consciousness are then no longer intrinsically linked to each other as presupposed in philosophy.

If the concepts of self and consciousness are no longer intrinsically linked, both may dissociate from each other. This dissociation would then be assumed to be manifest empirically in VS as an occurrence of self (i.e. as indirectly manifest in the neuronal effects of self-specific stimuli) without concurrent manifestation of consciousness, while at the same time the definition of VS as loss of consciousness can be maintained. Hence, rather than revising the definition of VS, we revise the definition of the concept of self.

How can we decide which of the ways described above is better and more plausible? Different criteria for plausibility, empirical (e.g. neuronal and psychological), phenomenal and conceptual-logical may be applied, as suggested in the methodology of neurophilosophy (see Part I). This will tell us what exactly is still preserved and what is damaged in VS. However, for the time being, we only know that something is still preserved in VS, since otherwise the findings described above could not have been observed.

Neurophilosophical discussion 3

Propositional awareness and vegetative state

We have discussed different forms of self-consciousness. These include direct versus indirect propositional awareness, pre-reflective versus reflective self-consciousness, and use of the self in different roles as subject and object (see above). What do the findings in VS tell us about these different forms of self-consciousness? Let us start with direct versus indirect propositional awareness. We may access our self in an indirect way via our beliefs and knowledge that we express in concepts and propositions. Or we may access the self in a more direct way, via our own thoughts and their respective propositions.

What does this imply for VS? One may now be inclined to argue that VS patients do not possess the kind of cognitive capacities that are necessary to yield beliefs and knowledge, and thus ultimately concepts and propositions. This in turn makes it impossible for them to have indirect propositional awareness of their own self.

What about direct awareness? Unlike in the case of indirect propositional awareness, no beliefs and knowledge need to be generated. Instead, the access to the self is more direct, as manifest in specific features like one's thoughts. What are these specific features in the case of VS? Is the direct access to one's own self and our thoughts still preserved in VS patients? If so, these specific features, like one's own thoughts, could serve as a standard or measure against which the various stimuli can be set and compared with the consecutive neuronal differentiation between self- and non-self-specific stimuli.

The advocates of propositional awareness assume that access to the self occurs in conceptual and propositional terms. These conceptual or propositional capacities seem to be disturbed in VS. This, though, raises the question: is there a more basic form of awareness, such as pre-conceptual and pre-propositional awareness? This will be the focus in the next section.

Neurophilosophical discussion 4

Pre-reflective self-consciousness and vegetative state

We recall from above and Chapter 17 the concept of pre-reflective self-consciousness that postulates the self to be always already part of any experience. The self comes quasi with the experience and is henceforth an integral part of it. What this means is that the self is intrinsically linked to consciousness independently of whether consciousness is about the self or some other completely unrelated content, like watching a foreign film by an unknown director, for example. This must be distinguished from reflective self-consciousness, where the self becomes the content by itself in consciousness.

What about pre-reflective and reflective self-consciousness in VS patients? One may now want to argue that VS patients lack the cognitive capacities to reflect upon their own self (though even that may be debated, given the findings of neural activity during cognitive stimulation; see Part IV). This would make reflective self-consciousness impossible.

In contrast, pre-reflective self-consciousness may still be possible in VS. Why? The neuronal differentiation between self-specific and non-self-specific stimuli may be related to some kind of pre-reflective self-consciousness in these patients. Such pre-reflective self-consciousness could still be present even in the absence of consciousness. One would consequently assume that the pre-reflective and pre-cognitive aspects of neural processing might have yielded the neuronal differentiation between self- and non-self-specific stimuli. This might be empirically plausible.

However, this interpretation may conflict with the conceptual definition in phenomenological philosophy. The conceptual definition assumes the pre-reflective component to be intrinsically tied to experience and consciousness. We experience our self in a pre-reflective way in consciousness, implying that the pre-reflective self is an integral and inherent part of consciousness. In short, the pre-reflective self is and consists in consciousness – pre-reflective self-consciousness.

That, however, contradicts the findings in VS. VS patients are by definition non-conscious. The philosophical definition of pre-reflective self-consciousness implies that the absence of consciousness should accompany the absence of a pre-reflective self and pre-reflective self-consciousness. However, this would make it unclear how and why the VS patients are able to differentiate neuronally between self- and non-self-specific stimuli. In other words, the findings in VS argue against the intrinsic linkage between the pre-reflective self and consciousness.

What about the immunity against the error of misidentification in VS? VS patients seem to remain immune to this error. They were able to show different levels of neural activity in response to self-specific stimuli like hearing one's own name or autobiographical scenes as distinguished from non-self-specific ones. One may thus assume that, at least neuronally, VS patients seem to remain immune to error through misidentification.

Whether such neuronal immunity applies to the use of the first-person pronoun in both its roles as subject and object remains open for discussion. To understand this fully, one would require more complex paradigms than the mere presentation of self- and non-self-specific stimuli. More complex tasks, where subjects are required to use the 'I' as both subject and object, would need to be tested in the future.

Neurophilosophical discussion 5

Pre-reflective self-consciousness and the brain's intrinsic activity

Empirical observation indicates some pre-reflective process related to the self that occurs seemingly independently of consciousness. How can we further tap into such earlier pre-reflective processes? We recall the seemingly central role of the brain's intrinsic activity. The brain's intrinsic activity describes the neural activity in the brain that cannot be traced to any extrinsic stimuli from outside the brain. Providing a certain yet unknown structure and organization, the brain's intrinsic activity may predispose consciousness and thus be a necessary condition of possible consciousness, also known as a neural predisposition of consciousness (NPC) (see Chapter 16).

What about the role of the intrinsic activity in the self and more specifically in processing self-specific stimuli? In the previous chapter we discussed the findings on self-reference where self- and non-self-specific stimuli were applied. This demonstrated the involvement of cortical midline regions. Interestingly, the same regions, the cortical midline structures, also show high neural activity in the resting state – the absence of extrinsic activity. This suggests high intrinsic activity in these regions. Recent studies have now observed neural overlap between the high intrinsic activity and the activity during self-specific stimuli. I here want to highlight some of the most relevant studies and describe them in further detail.

To put it differently, we may want to search for some kind of feature in the brain's intrinsic activity that somehow serves as a measure and standard against which extrinsic stimuli and their degree of self-specificity can be determined. This yet unknown feature in the brain's intrinsic activity may then predispose the neuronal (and psychological and phenomenal) differentiation into self- and non-self-specific stimuli. Before further elaborating the concept of neural predisposition in the context of the self, we need to discuss the empirical evidence for a relationship between self-specificity of extrinsic stimuli and the brain's intrinsic activity.

Neuroscientific findings 1

Neural overlap between self and rest

Antoine d'Argembeau is an excellent researcher from Belgium. He focuses on memory and its alterations, as in Alzheimer's disease for example. Additionally, especially because autobiographical memory implicates the self, he is also interested in investigating the neuronal mechanisms that underlie self-specificity using functional imaging.

D'Argembeau *et al.* (2005) conducted an H2O PET investigation. Subjects were tested under four conditions: thinking/reflection about their own personality traits; thinking/reflection about another person's personality trait; thinking/reflection on social issues; and a pure rest condition where subjects could relax. This allowed them to compare self- and non-self conditions, as well as to investigate the relation between self-conditions and the resting state.

What were their results? The VMPFC showed significant increases in regional cerebral blood flow (rCBF) during the self-conditions when compared to the non-self condition and the social condition. In addition, they compared all three task-related conditions – thinking/reflection about their own personality traits; thinking/reflection about another person's personality trait; thinking/reflection on social issues, against rest. This yielded increased rCBF in the DMPFC and the temporal regions, while no differences were observed in the VMPFC. Conversely, the rest condition showed rCBF increases in the large medial fronto-parietal and posterior medial network with again no differences in the VMPFC.

The separate account of self and rest allowed the authors to compare directly both conditions with each other. This yielded strong overlap in the VMPFC between both conditions, rest and self-specificity, that showed similar rCBF increases. In contrast, other and social conditions induced rCBF decreases in the same region.

Post-scanning subjective measures demonstrated that self-referential thoughts were most abundant in the self-condition, while they were diminished in the other three conditions. The authors therefore correlated the post-scanning measures of self-referential thinking with the rCBF changes. This yielded a positive relationship in the VMPFC. The higher the rCBF in the VMPFC, the higher the degree of self-referentiality in the thoughts subjects reported themselves (across all four conditions as described above).

This strong association or regional overlap between self and rest, especially in the same regions, was further confirmed by recent meta-analyses by Pengmin Qin (Qin and Northoff, 2011). Pengmin Qin conducted a meta-analysis of human imaging studies on the self compared to non-self. Most important, he also included studies on the resting state to compare their neural activity patterns to the one during self- and non-self-specific stimuli. This allowed him to compare directly the resting state in the default-mode network with the regions recruited during self- and non-self-specific stimuli.

What were his results? They confirm the ones described above – they demonstrate regional overlap between self and rest. More specifically, the regional activities during self-specific stimuli and those during the resting state overlapped, especially in the PACC extending to the VMPFC. Meanwhile, no such regional overlap with the resting state was observed in the conditions of familiarity and non-familiarity in either the PACC or any other region.

The overlap between rest and self is further supported by a recent MEG (magnetoencephalography) study by Lou et al. (2010). Lou investigated judgment of self-related words and focused on three main regions: precuneus, thalamus/pulvinar and anterior midline regions (including VMPFC, DMPFC and PACC). Using Granger causality analysis (which allows one to test for the direction of functional connectivity), he observed that the three regions were bidirectionally connected to each other (i.e. showing high degrees of statistical covariance in their signal changes).

Most interestingly, the increase in functional connectivity occurred 900ms before stimulus onset, while it was further enhanced by the stimulus itself for the first 900ms. Such functional connectivity was strongest in the gamma range between 30 and 45 Hz before and after stimulus onset, and strongest in the self-condition after stimulus onset.

Taken together, these studies show strong overlap between high resting state activity and stimulus-induced activity, as elicited by high self-specific stimuli in anterior and posterior midline regions, including both cortical and subcortical (albeit with limited evidence) regions. Hence the resting state activity in the anterior regions of the inner ring seems to be closely related to self-specificity, though in ways that are unclear.

Neuroscientific findings 2

Prediction of self-specificity by the resting state

How can we gather further empirical support for the assumed neural overlap between resting state activity and self-specificity in the PACC? To answer this, we conducted an intracranial study in collaboration with neurosurgeons in Toronto (Lipsman et al., 2013). We first investigated nine patients with depression, who underwent deep brain stimulation in the subgenual part of the anterior cingulate cortex. We measured the cell firing rates and their frequency fluctuations during self- and non-self-specific stimuli (i.e. own name versus other names presented visually) as well as during the resting state (i.e. long baseline and intertrial intervals). To test for regional specificity, we undertook the same measurements in the subthalamic nucleus (STN) in patients with Parkinson's disease.

What kinds of firing rates would one expect? Since a stimulus is applied when presenting the patient with his/her own name, one would expect a stimulus-related increase in the firing rates. This was indeed the case in non-self-specific stimuli. There we observed a significant increase in the firing rates. In contrast, this was not the case for self-specific stimuli, i.e. one's own name: we did not observe any significant change in the cells' firing rates during self-specific stimuli when compared to the preceding baseline. Even more interesting was the fact that this was specific for the subgenual cingulate cortex. In contrast, we could not observe such patterns in the subthalamic nucleus, where both self- and non-self-specific stimuli induced significant increases in the cells' firing rates.

Since previous studies demonstrated the association of increased gamma power with self-specificity (see Chapter 17), we then focused our subsequent analyses on gamma power. This demonstrated increased gamma power in the subgenual cingulate during self-specific stimuli and rest, compared to non-self-specific stimuli. Most interestingly, we observed increased gamma power in the resting state preceding the onset of self-specific stimuli. This was not the case in non-self-specific stimuli.

This raises the question of whether the increased gamma power in the preceding resting state is related to the expectation of a self-specific stimulus. Or, conversely, whether it is related to the ongoing spontaneous fluctuations in the resting state's gamma power independent of the presentation of subsequent stimuli. If the latter holds true, one would expect high resting state gamma power to be predictive of the subsequent stimuli's degrees of self-specificity. High gamma power in the preceding resting state should then predispose the subjects to assign a high degree of self-specificity to the stimulus.

This hypothesis was tested in another three intracranial patients from Berlin, Germany, who underwent another paradigm (Lipsman et al., 2013). Instead of the stimulus itself being self- or non-self-specific (i.e. own name versus another name), we let the subjects themselves decide about the degree of self-specificity. For that, we presented pictures of different faces from the famous Ekman series. After an initial perception period, subjects had to judge the degree of self-specificity of the pictures, whether it was high or low self-specific (while controlling for emotions and race in subsequent behavioral analysis).

The local field potentials showed significantly higher gamma power during those stimuli that were rated by the subjects as high self-specific when compared to the ones rated as low self-specific. The higher the gamma power induced by the stimulus, the higher the degree of self-specificity assigned to the stimuli by the subjects themselves. This confirms the association of self-specificity with high gamma power in the PACC during stimulus-induced activity.

We plotted the degree of gamma power in the preceding resting state interval – the intertrial interval that precedes the onset of the stimulus. Interestingly, we could see that up to 800ms pre-stimulus onset, gamma power was already significantly higher in those trials where the subsequent stimulus was assigned a high degree of self-specificity. Meanwhile, low pre-stimulus gamma power predicted a low degree of self-specificity of the subsequent stimulus. Hence, the stimulus' degree of self-specificity (rather than its emotion or race) was predicted by the degree of gamma power in the preceding resting state periods.

Taken together, these findings go beyond the previous studies that showed regional overlap between resting state activity and self-specificity in anterior regions like the PACC. They show that the degree of self-specificity of stimuli is predicted by the preceding resting state activity level, namely its gamma power. The preceding resting state's neuronal measures, like the gamma power (and possibly others too), must consequently contain some information related to self-specificity. Otherwise the observed prediction of the stimuli's degree of self-specificity by the preceding resting state activity would be impossible.

Critical reflection 1

Location of the self in the brain's intrinsic activity

These early findings raise many questions, especially with regard to the philosophical definition of the self and its relationship to consciousness. Therefore let's imagine a dialogue between a philosopher (PH) and a neuroscientist (NS) discussing the self and its relation to the brain's intrinsic activity.

PH: I am really confused. Your experiments show strong neural overlap between intrinsic activity and self-specificity. And, even stronger, you demonstrate the resting state activity to predict the degree of self-specificity that is assigned to the subsequent stimulus.

NS: Yes, that is true. Why is that confusing?

PH: How can the self already be present in the resting state when it is not yet experienced? In short, how can something that is absent already be present?

NS: Could you please explain?

PH: In the resting state intervals prior to the onset of the stimulus, you do not experience anything. There is no experience of self yet. Experience, and thus consciousness, only begin once the stimulus is presented, starting from stimulus onset. Then you may indeed experience a self as part of that experience in a pre-reflective way, as the phenomenological philosophers assume.

NS: I can understand that. That is not at all confusing.

PH: But in your data, you interpret it as if the self is already present in the resting state, prior to the stimulus onset. This confuses the presence of the self in the experience of the stimulus with the absence of the self in the preceding resting state. In short, you confuse absence and presence of self.

NS: There can be no self during the resting state?

PH: Yes, you are correct, there can indeed be no self during the resting state. The self is intrinsically tied to experience and consciousness, which begins only during stimulus onset.

NS: That, though, does not exclude the possibility that there is a self already present in the resting state. It may not be the kind of phenomenal or minimal self with its pre-reflective self-consciousness you seem to presuppose when talking about the absence of the self in the resting state.

PH: There is no other self than the self in experience and its pre-reflective self-consciousness. Accordingly, there is no self at all in the resting state. You confuse the self with the brain. These are two different categories: the self belongs to the category of people, while the brain is a mere object. Since people are not mere objects, you cannot identify both as you do when locating the self in the brain.

Critical reflection 2

Neural predispositions of self (NPS) and the brain's intrinsic activity

NS: I still do not fully understand your categorical rejection. Category error, you say! But if the findings demonstrate that the brain's intrinsic activity is somehow related to self-specificity, who cares about categories. They are merely conceptual and logical anyway; what counts are the empirical findings.

PH: How do you imagine the relationship between self and intrinsic activity? It is not an easy question. I agree you show these findings of self-specificity that must indeed presuppose some kind of self as an intrinsic measure or standard for the extrinsic stimuli, while at the same time, one does not experience such self – one is not conscious of it and this implies that there is no such self.

NS: I still think that the conclusion you draw is too strong and therefore not justified. The only thing the data imply is that there is no self as self-consciousness in the brain's intrinsic activity. However, even that may be doubted.

PH: Why?

NS: Consider dreams. Isn't your brain in a true resting state, with its intrinsic activity at its fullest display, while sleeping and dreaming?

PH: Sure. That does not tell us anything about the self, however. Dreaming only tells us about the brain and its intrinsic activity.

NS: But don't you experience your self in your dreams? The contents associated with your self resurface in your dreams. Sigmund Freud, the first psychoanalyst, described this process in full detail. There is thus some kind of self even in dreams.

PH: You and your self must have interesting dreams!

NS: Even that irony tells much about your real self. Anyway let's be serious again. The brain's intrinsic activity does not harbor a full-fledged self, in the organization of self-consciousness. But at the same time the brain's intrinsic activity seems to contain some information about the self as encoded and ingrained in its resting state. Are we in agreement on that?

PH: Yes, though the second assumption of information about the self being encoded in the intrinsic activity remains rather mysterious to me. That needs to be further investigated and detailed empirically and conceptually.

NS: I must admit that, empirically, I am currently not able to provide you with more details. That is a subject for future research.

PH: And conceptually?

NS: Conceptually, the intrinsic activity seems to make possible or predispose in some way the neuronal (and psychological and phenomenal) differentiation between self- and non-self-specific stimuli. What about the neural predispositions of self (NPS)?

PH: As in the case of consciousness, the NPS would then describe the necessary but non-sufficient conditions of the possible neuronal (psychological and phenomenal) differentiation between self- and non-self-specific stimuli.

NS: Sounds correct to me!

Critical reflection 3

Neural predispositions of self (NPS) and the explanatory gap

PH: The concept of neural predispositions of self reminds me of another philosophical concept of self.

NS: Which one?

PH: The German philosopher Immanuel Kant spoke of a purely formal and abstract self, called a transcendental self. The transcendental self is devoid of any content and cannot be experienced as such. It must consist in a specific structure or organization that constitutes the perspective of the self, called self-perspectival organization. But we must in some way presuppose it, since without it any kind of self, including self-consciousness and thus the experience of a self, would remain impossible. This is a transcendental self in the sense of Kant. Such a transcendental self is outside of and prior to any experience, i.e. consciousness. However, despite being outside experience, the transcendental self is indispensable for experience. More recently, the American philosopher John Searle suggested a similar notion of the self in order to explain consciousness.

NS: While I do not have all the details, what you say sounds plausible to me. The self lies, metaphorically speaking, dormant in the brain's intrinsic activity. You may then want to speak of a truly neuro-transcendental self, rather than a merely transcendental self. A neuro-transcendental self is awakened by the extrinsic stimuli for which it serves as a measure or standard for determining the latter's degree of self-specificity.

PH: Interesting thought, to describe Kant's transcendental self as a dormant self. What would Kant have said to that?

NS: I do think that a dormant self in the brain's intrinsic activity – what you and Kant call transcendental self – might provide a key to bridging the explanatory gap (s).

PH: The explanatory gap (s) describes the gap between the concepts of self and self-reference, and thus between philosophy and neuroscience. Is that correct?
NS: Yes.
PH: How and why?
NS: By providing a predisposition of the possible self (rather than a correlate, i.e. a sufficient condition of the actual self), the brain's intrinsic activity requires the extrinsic stimuli to 'bring its predisposition into life', so to speak. The interaction between the brain's intrinsic activity and the extrinsic stimuli may then allow us to measure the degree of self-specificity of the latter by the former. Meanwhile, the intrinsic–extrinsic interaction may then allow the brain to yield consciousness in association with the determination of the stimuli's degree of self-specificity.
PH: I agree that self-consciousness would then indeed serve as a bridge that is fed from both sides – the (transcendental or dormant) intrinsic self and the extrinsic stimulus. However, as I understand them, the empirical findings are not really there yet. You say that we currently lack empirical data to explain how extrinsic stimuli and intrinsic self interact with each other, and how their interaction is related to consciousness. Is that correct?
NS: Correct. And you are right; more empirical findings, targeted specifically towards intrinsic–extrinsic interaction in both healthy and VS subjects need to be collected to further support my assumption.

Take-home message

The previous chapter discussed and compared philosophical concepts of self with recent neuroscientific findings about self-reference. The main focus here was on pointing out the discrepancies between philosophical concepts and neuroscientific findings. One central conceptual problem was the difference between self and self-reference. While self-reference concerns stimuli and their relation to the self, the concept of self is about the existence and reality of the self. There is thus an explanatory gap in our explanation of self and self-reference. This is called the explanatory gap (s) in this chapter. How can we close and bridge that explanatory gap (s)? The current chapter discusses different forms of self-consciousness that might mediate between the self and the self-referent stimuli. This leads to recent neuroscientific findings that show a strong relationship between the brain's intrinsic activity and self-specificity. Those findings lead us to conclude that there must be some kind of information about self-specificity ingrained and encoded in the brain's intrinsic activity. Conceptually, one may therefore speak of a neural predisposition of self. This neural predisposition may refer to a rather formal and abstract notion of the self that is devoid of any content. This abstract and formal self comes close to what Kant described as the transcendental self.

Summary

The previous chapter discussed the philosophical conceptions about the self and how they are related (or not) to the empirical findings about self-reference. There is a clear gap between the metaphysical (i.e. philosophical) explanations of the self and the empirical explanations about self-reference. This led us to formulate what I describe as an 'explanatory gap' in the context of the self, i.e. explanatory gap (s). How can we bridge such an explanatory gap (s)?

To answer this question, we investigated how we can access the self in our experience and thus in consciousness. This raises the question of self-consciousness and its epistemological features. Recent philosophy of mind distinguishes between direct and propositional awareness. Direct awareness provides direct access to the self, while propositional awareness is more indirect – we access the self via propositions about features or properties of the self.

Another epistemological feature of consciousness is that we apparently cannot misidentify our own self as the subject of experience. This is described as immunity to error through misidentification. While we can misidentify the objects – the contents of our experience (including our own self when taken as object or content of experience – we cannot misidentify our own self as the subject of our experience. How are these epistemological features of self-consciousness related to the brain and its neuronal mechanisms? Recent findings show neural overlap between the brain's intrinsic activity and the neural activity during self-specific stimuli over non-self-specific stimuli. Moreover, it seems that the degree of neural activity during self-specific stimuli predicts the degree of consciousness, as suggested by the findings in patients in a vegetative state.

On the whole, these findings suggest some kind of as yet unclear relationship between three different variables: the self, self-consciousness, and the brain's intrinsic activity. How are these three related to each other? Different possible models of their relationship are discussed, including the dissociation of the concept of self from that of consciousness. What seems to be clear is that the brain's intrinsic activity appears to encode and ingrain some kind of information about self-specificity, though exactly how or what is still unclear. This leads us to assume what can be called a 'neural predisposition of self' that describes the necessary (as distinguished from sufficient) neural conditions of the possible (rather than actual) constitution of a self. This neural predisposition of self (NPS) may in turn contribute to closing the explanatory gap between self and self-reference, and thus between philosophical and neuroscientific accounts of the self.

Revision notes

- What is the explanatory gap (s)? Why is there this explanatory gap? How does it compare to the explanatory gap (c) in the context of consciousness?
- How can you distinguish between direct awareness and propositional awareness? Are they associated with different forms or types of self-consciousness?
- What is immunity to error through misidentification? Why does it apply only to the use of the first-person pronoun as subject and not to its use as object?
- Explain the main empirical findings about self and rest. What do the findings show? What can we infer from the neural overlap between intrinsic activity and self-specific neural activity?

- How can the formal or transcendental self be related to the current empirical findings about self and rest?
- What are the findings on self-specificity in patients in a vegetative state? How are they related to their level of consciousness?
- What are the neural predispositions of self?
- Why could Kant's concept of a transcendental self eventually be relevant in the current neurophilosophical discussion?

Suggested further reading

- Bermudez, J. L. (1998) *The Paradox of Self-consciousness* (Cambridge, MA: MIT Press).
- Gallagher, S. and Zahavi, D. (2010) 'Phenomenological approaches to self-consciousness', *Stanford Encyclopedia of Philosophy* (Winter 2010 edition), Edward N. Zalta (ed.). Available at: http://plato.stanford.edu/archives/win2010/entries/self-consciousness-phenomenological/. The citation above refers to the version in the following archive edition: Winter 2010 (substantive content change).
- Huang, Z., Dai, R., Wu, X., Liu, D., Hu Jm Gao, L., Tag, W., Mao, Y., Jin, Y., Wu, X., Zhang, Y., Lu, L., Laureys, S., Weng, X. and Northoff, G. (2013) 'The self and its resting state in consciousness. An investigation of the vegetative state', *Human Brain Mapping*, in press.
- Qin, P. and Northoff, G. (2011) 'How is our self related to midline regions and the default-mode network?', *Neuroimage*, 1(57), 1221–33.
- Shoemaker, S. (1968) 'Self-reference and self-awareness', *Journal of Philosophy*, 65, 555–67.

19
Abnormalities of Self and Brain in Psychiatric Disorders

Overview

So far we have discussed the concepts of self and self-consciousness, and how they are related to the brain. The regions in the midline of the brain and their intrinsic activity seem to be central in mediating a basic sense of self – the experience or consciousness of our self in pre-reflective self-consciousness. However, the neuroscientific findings, as well as the neurophilosophical assumptions, must be regarded as preliminary. How can we further support them? To provide more support, we should revert to a methodological strategy that has been prevalent throughout the history of neuroscience: we should observe abnormal alterations of mental features like consciousness and self in neurological and psychiatric patients. These observations of the neural abnormalities in psychiatric patients tell us indirectly about the neuronal mechanisms in the healthy brain. In the context of consciousness, we discussed patients in a vegetative state as characterized by their loss of consciousness. Now, in the context of the self, we will discuss psychiatric disorders like schizophrenia and depression, where the patients experience major abnormalities in their self while simultaneously showing changes in their brain function. This is the topic in the present chapter.

Objectives

- Understand the role of disorders in order to explain the neuronal mechanisms underlying mental features like self and consciousness
- Describe and learn about depression and schizophrenia, and their abnormalities in the self
- Learn about the neural abnormalities in depression
- Understand the neurophilosophical significance of the neural and phenomenal changes in depression and schizophrenia for the determination of the concept of self in general

Key concepts

Psychiatric disorders, depression, schizophrenia, self, self-consciousness, intrinsic activity, embodied and relational self, immunity against error through misidentification.

Background 1

Self and self-consciousness

So far, we have discussed different concepts of self (mental, empirical, phenomenal, minimal) and how they are related to recent neuroscientific findings about the self-reference of stimuli. The neuroscientific findings suggested that the cortical midline structures are recruited during self-specific stimuli. However, the direct relationship between the neuroscientific findings and philosophical concepts was hampered by the neuronal, experimental, psychological and phenomenal unspecificity of the empirical data.

The main difference between philosophical concepts and neuroscientific findings consists of a conceptual difference – the difference between the concepts of self and self-reference, while the concept of self denotes existence and reality, and thus the metaphysical domain, while the concept of self-reference concerns stimuli that are related to the self; this presupposes the empirical domain – the domain of observation. Hence, philosophical concepts and neuroscientific findings concern different referents, self and stimuli, as well as different domains, metaphysical and empirical (see Chapter 1 and, especially, Chapter 2 for details).

How can we link self and self-reference, and thus philosophical concepts and neuroscientific findings? It is possible that self-consciousness might provide one possible bridge. Self-consciousness describes the experience or consciousness of one's own self. The self thus becomes the content of consciousness in the same way as another person, an object or an event can become the content of consciousness (see Chapter 18).

How is the concept of self-consciousness related to the concepts of self and self-reference? The concept of self, as understood in the philosophical debate, is usually associated with consciousness. If we can experience the self and thus be conscious of it, only then can the self be real and existent. Depending on the type of access to the self, different forms of self-consciousness – direct and indirect as well as pre-reflective versus reflective – can be distinguished from each other. Moreover, we seem to remain immune to any error when it comes to identifying our own self in the role of subject of experience. This is described as immunity against error through misidentification (see Chapter 18).

Background 2

Self and brain

How are these different forms of self-consciousness related to the neuroscientific findings about self-reference? We first demonstrated that self and consciousness can possibly dissociate from each other, as in patients in a vegetative state (VS). As already noted, VS is defined by the absence of consciousness. Despite the apparent absence of consciousness, these patients' brains nevertheless show neuronal differentiation between self-specific and non-self-specific stimuli. Because the distinction between self-specificity and non-self-specificity seems to presuppose some kind of self as a measure or standard, the existence and reality of a self might be able to dissociate from its association with the consciousness that remains absent in these patients. Alternatively, one may assume that these VS patients do not really lose their consciousness and possess some kind of self-consciousness.

How could the self occur prior to and independently of consciousness? To answer this question, one may want to discuss the findings in healthy subjects in more detail. Such findings demonstrate a strong neural overlap between the regions implicated in the brain's high resting state activity and the regions recruited in particular during self-specific stimuli. Even more important is that the spatial and temporal pattern of the brain's resting state activity, called its intrinsic activity, seems to predict whether one will experience an extrinsic stimuli as self-specific or non-self-specific (see Chapter 18).

This led us to the conclusion that the brain's intrinsic activity must contain – ingrain and encode – some information about the self. This information might in turn function as a standard for the assignment of self-specificity to extrinsic stimuli. This is a rather daring hypothesis, which at this point must be regarded as tentative. It is daring for several reasons. First, more empirical findings with a higher degree of specificity (neuronal, phenomenal, experimental, psychological) are needed. Second, it goes against the assumption of the self as being associated with consciousness.

Third, it may have major consequences for our definition of subjectivity, for the concept of self, and the core nucleus of consciousness in general. Fourth, by suggesting a specific relationship between the brain and the self (and thus between the brain and subjectivity), neurophilosophy will be able to make a major contribution to both philosophy and neuroscience. Should this hypothesis turn out to be valid, it will shed a completely new light on one of the core problems in philosophy: the problem of subjectivity. It will likewise contribute to neuroscience by revealing the neural mechanisms of consciousness. In short, it will solve the problem of the neural correlates of consciousness (NCC). In other words, there is much at stake – high gains and high losses – when discussing the relationship between the self and the brain.

Background 3

Psychiatric disorders and the self

How can we obtain more empirical support for this hypothesis? To answer this question, we may want to go back briefly to the beginnings of neuroscience.

As we demonstrated in Part I, investigation into the brain, including its relationship to mental features, started at the end of the nineteenth century. At this time, the major way of finding out how the brain mediates mental features like language, consciousness, etc. was the observation of clinical patients in neurology. For example, patients with a lesion in the Broca region of the prefrontal cortex showed severe disturbances when speaking and using language. From both the symptoms and the location of the lesion in the brain, it was inferred that this region is specifically involved in speaking and the use of language. This is described as a clinical-anatomical or lesion-based method. This later led to the development of neuropsychology as a separate field within neuroscience (see Chapter 3 for details).

The clinical-anatomical observations in neuropsychology were later complemented by the introduction of functional brain imaging techniques like fMRI. Rather than inferring indirectly from the location of the lesion to the neural mechanisms underlying the mental function in question, these techniques allowed direct online visualization of the neural activity related to specific mental functions (like the use of language, emotional feeling, etc.) (see Chapter 3 for details).

However, as we saw in the case of self-reference, the use of functional brain imaging is still hampered by major problems of unspecificity (see Chapter 17). One may thus want to combine them with the old methodological strategy of inferring neuronal mechanisms underlying mental features from their alteration in pathological cases. This strategy was already in place when we discussed the case of the vegetative state in both the contexts of consciousness (see Chapter 16) and self-consciousness (Chapter 18).

How can we use the same strategy to lend further empirical evidence to the assumption that some information about the self and its self-specificity is ingrained and encoded in the brain's resting state activity? 'Very simple,' you may say, 'just provide me with pathological cases where we can observe abnormal changes in both the self and the brain's intrinsic activity.' Hence, rather than looking for disorders of consciousness, we now look for disorders of self and see how they are related to the brain and its intrinsic activity. Let's turn to discussing psychiatric disorders, namely depression and schizophrenia.

Schizophrenia 1

Basic disturbance of the self

What is schizophrenia? Schizophrenia is a complex mental disorder, a psychiatric disorder, where patients show bizarre symptoms. They can show positive symptoms like hearing voices that nobody else hears. These are called auditory hallucinations. Often, they feel paranoia, that they are being persecuted and followed by others, and perceive reality in a delusional way. These are called delusions. Their thoughts are often disturbed and chaotic – this is described as thought disorder.

The most bizarre symptom is that they may experience themselves as another person or self. For instance, patients may experience themselves as Jesus, Nefertiti, or some other person, and they behave accordingly. These patients may then dress like Jesus and grow a beard. Or they put on cosmetics in very much the same style as Nefertiti and behave and walk like a queen. In addition to these so-called positive symptoms, they also show negative symptoms, with blunted affect, emptiness of thought, anhedonia (i.e. the absence of pleasure), and lack of initiative and activity. For the sake of simplicity, we will focus mainly on the positive symptoms and what they tell us about the self in general. How are these strange symptoms possible?

Where do these bizarre symptoms come from, and what is their underlying cause? At the beginning of the twentieth century, when brain imaging had not yet been developed, psychiatrists in Germany and Switzerland like Emil Kraepelin and Eugen Bleuler assumed abnormality of the self to be basic in schizophrenia. Unlike in our times, these psychiatrists had to rely on nothing but clinical observation. And, based on this, they assumed an abnormal change of the self to be fundamental in schizophrenia.

More specifically, Kraepelin (1913, p. 668) characterized schizophrenia as 'the peculiar destruction of the inner coherence of the personality' with a 'disunity of consciousness' ('orchestra without a conductor'). Bleuler (1911, p. 58; 1916) also pointed out that schizophrenia is a 'disorder of the personality by splitting, dissociation' where the 'I is never completely intact'.

A German contemporary of Bleuler and Kraepelin, Berze (1914) even referred to schizophrenia as 'basic alteration of self-consciousness'. Another famous German psychiatrist, Karl Jaspers (1963, p. 581), also noticed 'incoherence, dissociation, fragmenting of consciousness, intrapsychic ataxia, weakness of apperception, insufficiency of psychic activity and disturbance of association, etc.' to be basic as unifying 'central factors' in schizophrenia.

Schizophrenia 2

Subjective experience of the abnormal self

How is such basic disturbance of the self manifested in the patients' subjective experience of their own self? The early descriptions of a disrupted self are complemented by current phenomenological accounts that focus predominantly on the experience of one's own self in relation to the world. Josef Parnas (Parnas et al., 2001; Parnas, 2003) describes nicely how 'presence' is altered in schizophrenia. The experience of the world and its objects are no longer accompanied by a pre-reflective self-awareness.

Let me specify. One's own self, the self that experiences the experience of the world, is no longer included in that very experience: 'The prominent feature of altered presence in the pre-onset stages of schizophrenia is disturbed ipseity, a disturbance in which the sense of self no longer saturates the experience. For instance, the sense of mineness of experience may become subtly affected: one of our patients reported that this feeling of his experience as his own experience only "appeared a split-second delayed"' (Parnas 2003, p. 225).

The patients remain unable to refer to themselves in their experience of the world. It is as if the experience of the world is no longer their own experience of their own self. Because one's own self is absent in their experience of the world, schizophrenic patients become detached, alienated and estranged from their own experience. This detachment from their own self makes it impossible for them to experience their experiences as subjective.

The experiencing self is consequently no longer affected by its own experiences, which Sass (2003) describes as a 'disorder of self-affectivity': the self is no longer experienced as one's own and, most important, is no longer experienced as the vital center and source of experience, actions, perceptions, thoughts, etc. This reflects what Sass (2003) calls the 'diminished self-affection', meaning that the self is no longer affected by its own experiences.

If, however, the self is no longer affected by its own experience, the self stands apart from the objects and the events in the world that are experienced. A gulf, a phenomenological distance as Parnas says (2003, p. 225), opens up between the world and the self. The objects and events of the world no longer intuitively make sense and are thus no longer meaningful to the experiencing subject. One's own self becomes almost objective and mechanical in experience and perception of the world.

Schizophrenia 3

Abnormal intrinsic activity in the cortical midline structures

Recall from the previous chapters that self-specific stimuli were associated specifically with neural activity changes in the cortical midline structures, the regions in the midline of the brain's cortex (see Chapter 17). Interestingly, these regions also show extremely high levels of neural activity in the resting state, during the absence of specific tasks or stimuli. Most important, these regions' high levels of resting state activity seem to overlap with and predict the neural activity changes during self-specific stimuli (see Chapter 18).

Taken together, these findings suggest a close relationship between the brain's intrinsic activity in the midline regions and self/self-specificity. The intrinsic activity may then be supposed to contain, encoded and ingrained, some information about the self and self-specificity. If so, one would now expect schizophrenia and its abnormalities of self to be characterized by abnormalities in both resting state and stimulus-induced activity during self-specific stimuli. This is exactly what the present results suggest.

Various recent studies have investigated the midline regions as part of the default-mode network (DMN) in schizophrenia (see Kühn and Gallinat, 2012, for a recent review). Recent imaging studies in schizophrenia reported abnormal resting state activity and functional connectivity, especially in the anterior cortical midline structures (aCMS). One study (Whitfield-Gabrieli *et al.*, 2011) demonstrated that the aCMS (and posterior CMS, like the PCC/precuneus) show decreased activity changes during short-term memory tasks. The decreases in activity concerned predominantly negative activity changes that are described as task-induced deactivation (TID) in fMRI. This was observed in both schizophrenic patients and their relatives when compared to healthy subjects. This is indicative of decreased task-related suppression and possibly increased resting state activity.

Furthermore, the very same schizophrenic subjects also showed increased functional connectivity – temporal synchronization across different regions – of the aCMS with other posterior regions of the CMS, the PCC. Both functional hyper-connectivity and decreased signal changes (e.g. TID) correlated with each other. A higher degree of TID, that is, deactivation, was accompanied by a higher degree of functional connectivity. Finally, both decreased TID and increased functional connectivity in aCMS correlated with the positive symptoms like auditory hallucinations and delusions in these patients (as measured with the so-called PANS scale).

Decreased TID in aCMS were also observed in an earlier study that likewise investigated working memory (Pomarol-Clotet *et al.*, 2008). Similar to the study described above, the researchers had subjects perform a working memory task and observed abnormally decreased TID in aCMS in schizophrenic patients when compared to healthy subjects. Similar to the study by Whitfield-Gabrieli *et al.* (2011) described above, they also observed abnormal task-related activation in the right dorsolateral prefrontal cortex in schizophrenic patients. Other studies also reported abnormal TID in aCMS, as well as abnormal functional connectivity in schizophrenic patients from aCMS and posterior CMS to the insula.

In addition to TID and functional connectivity, other abnormal measures of resting state activity are the temporal features – more specifically fluctuations or oscillations in certain temporal frequencies. For example, Hoptman *et al.* (2010) demonstrated that low frequency fluctuations in the resting state were increased in the aCMS (and the parahippocampal gyrus) in schizophrenic patients. In contrast, other regions like the insula showed decreased low frequency fluctuations. Abnormally increased low frequency oscillations (< 0.06 Hz) in the aCMS (and posterior CMS regions and the auditory network) and their correlation with positive symptom severity were also observed in another study on schizophrenic patients (Rotarska-Jagiela et al., 2010).

Schizophrenia 4

Abnormal neural activity during self-specific stimuli

Besides its resting state activity, the brain can also be characterized by stimulus-induced activity known as extrinsic stimuli. As we saw in Chapter 18, one can use self-specific stimuli (like one's own name) and investigate how they impact neural activity in the brain. Comparison can then be made between their effects and those of non-self-specific stimuli (like another person's name). This comparison has been performed in healthy subjects but recently also in some studies with schizophrenic patients as participants.

A recent imaging study by Holt et al. (2011) showed that abnormal anterior-to-posterior midline connectivity is related to self-specificity. These researchers investigated schizophrenic patients during a word task where subjects had to judge trait adjectives according to their degree of self-specificity (and two other tasks: other-reflection (i.e. relation of that word to another person) and perception-reflection (i.e. a word printed in upper- or lower-case letters).

What about their results? Schizophrenic patients showed significantly elevated activity in posterior midline regions like the mid- and posterior cingulate cortex during self-reflection. Meanwhile, signal changes in the anterior midline regions like the medial prefrontal cortex were significantly reduced when compared to healthy subjects. Finally, functional connectivity was abnormally elevated from the posterior to the anterior midline regions in schizophrenic patients. Analogous results of altered midline activity with an altered relationship, that is, an imbalance between anterior and posterior midline regions, have also been observed in other studies on self-specificity in schizophrenia (see Taylor et al., 2007; Menon, 2011).

Taken together, these results demonstrate that patients with schizophrenia have abnormal resting state activity, especially in the anterior and posterior midline network (see Kühn and Gallinat, 2012, for a recent meta-analysis). The same network also shows alterations in the balance between anterior and posterior midline regions when probing for self-specific stimuli.

Now we can raise the question of whether these abnormalities in stimulus-induced activity are related to the resting state abnormalities described above. Unfortunately, at this time studies testing the linkage between resting state abnormalities and self-specific stimuli still need to be conducted on patients with schizophrenia. These results are needed in order to support the assumption that the information about the self in the brain's intrinsic activity may serve as a measure for the assignment of self-specificity to extrinsic stimuli.

To put it differently, the resting state abnormalities may signify what earlier psychiatrists described as the 'basic disturbance of the self' present in schizophrenic patients. Meanwhile the stimulus-induced activity changes that are associated with self-specific stimuli may correspond to the abnormal subjective experience of the self as described by Parnas and other current psychiatrists.

Neurophilosophical discussion 1

Schizophrenia and the concept of the mental self

We showed that schizophrenia can be characterized as a 'basic disturbance of the self' accompanied by corresponding neural abnormalities. To what kind of concept of self does this refer? This is a matter of neurophilosophical discussion that also illuminates a particular model of the general relationship between the brain and the self.

The first philosophical concept of the self we discussed in Chapter 17, was the concept of a mental self. The concept of the mental self assumes that the self presupposes a specific existence and reality that is non-physical and mental. The philosopher Descartes assumed a specific mental substance, and current philosophers following his line of thought suggest specific mental properties or features to account for a self that cannot be reduced to the physical properties of the brain.

How does this concept of the mental self relate to the observations in schizophrenia? If there is indeed a specific mental, non-physical feature accounting for the self, neuroscientific investigation should not reveal neural abnormalities to be specifically related to the abnormal self in schizophrenia. This raises the following question: are the observed abnormalities in resting state and stimulus-induced activity specific for the self, as distinguished from other functions?

The neural investigation of the self in schizophrenia suffers from the same problems of unspecificity (phenomenal, neural, psychological, experimental) as discussed in Chapter 17. This is further amplified with regard to, in particular, the resting state, whose abnormalities in schizophrenia have not yet been related (directly or indirectly) to the abnormalities of the self in these patients. Hence there is no direct and plausible evidence for the abnormal self in schizophrenia to be associated exclusively with the brain and its neural abnormalities.

This fact leaves open the possibility that some non-physical feature (inside or outside the brain) might help to account for the abnormal self in schizophrenia. The definition of the self by a specific mental, non-physical, feature would then be based on negative evidence (the absence of any empirical findings of such a feature in the brain). In this model, the self would then be characterized as non-physical. This suggests that any empirical evidence in favor of a mental characterization of the self can only be negative (i.e. absence of positive neuronal findings) and indirect (via the non-physical nature).

Neurophilosophical discussion 2

Schizophrenia and the concept of the empirical self

What about the concept of an empirical self as signified by a special form of representation called self-representation (see Chapter 17)? Self-representation is supposed to be based on higher-order cognitive functions that allow the represented features to be re-represented again as represented features. One would consequently assume these higher-order cognitive functions and the subsequent self-representation to be disturbed in schizophrenia. This would explain the 'basic disturbance of self'. There have been many theories developed about some higher-order cognitive deficit in schizophrenia.

How do the neuronal findings described above relate to this assumption? The abnormal stimulus-induced activity during self-specific stimuli might be compatible with the disturbance in higher-order cognitive functions. In order to process stimuli as self-specific, they need to be represented as self-specific stimuli processed by the brain of the respective self; this meta-representation or self-representation of the stimuli would then allow the yielding of a self, the empirical self. Additionally, the respective higher-order cognitive functions are disturbed in schizophrenia, leading consecutively to abnormal self-representation. This might be manifested in the abnormal stimulus-induced activity during self-specific stimuli.

The self's characterization as a higher-order cognitive function and its disturbance in schizophrenia would conflict with the assumption of the abnormal self as a 'basic disturbance' that underlies all subsequent functions, including sensorimotor and cognitive functions. The self cannot be a higher-order cognitive function while at the same time basic to and underlying the same higher-order cognitive functions.

Would the resting state abnormalities be a better candidate for the 'basic disturbance of self' in schizophrenia? The resting state can indeed be regarded as very basic for any subsequent functions, including both sensorimotor and cognitive functions. If so, one would expect the resting state abnormalities in schizophrenia to be related to the 'basic disturbance of self'. Empirically, one would thus assume the resting state abnormalities to predict the 'basic disturbance of self'. This remains to be demonstrated. If it were, one would then also need to specify what kind of information about the self is encoded and ingrained in the brain's intrinsic activity, and in what ways.

Neurophilosophical discussion 3

Schizophrenia and pre-reflective self-consciousness

Phenomenological philosophy assumes what is described as 'pre-reflective self-consciousness'. The concept of 'pre-reflective self-consciousness' describes our self as the very subject of experience, as always already part of any kind of experience, irrespective of whether that experience is about the self or some other content. The experience of any kind of content, like an object, person or event, is possible only when there is a subject, a self, who experiences it (see Chapter 18).

How is the inclusion of the self as part of experience manifested in the experience itself? The phenomenological philosophers argue that, because of the inclusion of the self in our experience, we experience all contents as being related to the self. This is manifested in our experience by the terms 'mineness' and 'belongingness'. These terms signify the experiential relationship of the contents of experience to the experiencing subject, the self (see Chapters 17 and 18).

What about mineness and belongingness in schizophrenia? As described above, these are indeed disturbed in schizophrenia. Schizophrenic patients no longer experience other people, objects or events as related to themselves. They often do not experience mineness or belongingness; in short, the contents of experience are no longer experienced in a subjective way. Instead, the contents of experience are experienced as foreign and in a purely objective way.

How is this related to the neuronal findings described above? We currently do not know. The neuronal findings on stimulus-induced activity during self-specific stimuli do not directly measure the subjective experience and thus the pre-reflective self-consciousness by itself. It may be included in the measurement, but no specific phenomenal measure was included in the experimental designs. In other words, the neuronal findings remain unspecific with regard to the phenomenal features of mineness and belongingness. Future studies are thus needed to specifically investigate the subject of the phenomenal features of mineness and belongingness, associated with self-specific stimuli.

To put it differently, it is not sufficient only to investigate the neural and psychological effects of the presentation of self-specific stimuli compared to non-self-specific stimuli. Instead, we need to target the subjective-experiential and thus phenomenal features associated with the self-specific stimuli. Rather than comparing self-specific versus non-self-specific stimuli, one may then want to compare the self-specific stimuli that are experienced in terms of mineness and belongingness with those that are not experienced in this way – as non-mineness and non-belongingness.

Neurophilosophical discussion 4

Schizophrenia and the concept of the minimal self

In addition to the mental self and the empirical self, we also discussed the concepts of the phenomenal self and the minimal self. The concept of the phenomenal self is characterized as the subjective experience of a self as part of that very experience. In other words, the subject of an experience, the self, is experienced as part of that same experience.

The concept of the minimal self extends this premise and argues that, for any such experience of the self, we do not need higher-order cognitive functions. Instead, the body and its sensorimotor functions are considered to be central and it is presumed that these occur prior to any linguistic functions. The concept of the minimal self can thus be characterized as sensorimotor- and bodily-based, while at the same time pre-linguistic and pre-cognitive.

How does this relate to the findings described above? One may assume a disturbance of the minimal self in schizophrenia. If so, these patients should show sensorimotor abnormalities and abnormal changes in the respectively underlying neural mechanisms, as in, for instance, sensory and motor cortices. This is indeed the case, and has been demonstrated.

Especially in the sensory cortices like the auditory cortex, schizophrenic patients seem to show severe abnormalities. The observed neural abnormalities during self-specific stimuli as reported above would then be linked to sensorimotor, rather than higher-order, cognitive functions. In order to support that assumption, one would need to conduct a study where the self-specific and non-self-specific stimulus processing interacted with sensorimotor functions. One would then assume this interaction and its underlying neural correlates to be abnormal, and to account for the abnormal 'minimal self' in schizophrenia.

How, though, does the minimal self relate to the resting state abnormalities? The resting state abnormalities occur prior to the recruitment of any specific functions, like sensorimotor and cognitive functions. Why? Because usually the resting state – the brain's intrinsic activity – does not recruit sensory and motor functions by itself. The minimal self thus cannot be associated with the resting state, since that would leave out exactly what is supposed to characterize the minimal self – the sensorimotor functions.

What kind of concept of self would the resting state – if it could be demonstrated as being associated with the self – imply? A concept of self based on the brain's intrinsic activity can neither be characterized by sensorimotor nor by cognitive functions. Why? Because these functions are simply not yet present and recruited in the resting state. The resting-state-based concept of self may thus need to be even more minimal than the concept of the minimal self.

In other words, the resting-state-based concept of self would need to occur prior to any subsequent sensorimotor and cognitive functions. This very basic concept of self could then eventually account for what earlier psychiatrists described as the 'basic disturbance of self' in schizophrenia. This concept of self would then be based on the brain's intrinsic activity, rather than some specific functions related to extrinsic stimuli. However, such a basic concept of self would need to be elaborated in much more detail, both conceptually and empirically.

Neurophilosophical discussion 5

Schizophrenia and 'immunity against error through misidentification'

One hallmark of self-consciousness is that we apparently cannot err about it. We can be mistaken about the contents of our experience, as is the case when you think your own train is moving when in reality it is another train that moves on the adjacent track. Here, you are mistaken about the contents of your experience; you misidentify them, meaning you are not immune to error through misidentification.

It is different when it comes to your own self, however. You cannot be mistaken that it is *you*, your *own* self, who experiences reading this book. It must be you, your own self and no other person's self that experiences reading this book. You thus cannot misidentify your own self as the subject of experience. This means that you cannot assume and experience another person's self. This is what philosophers call 'immunity against error through misidentification' (see Chapter 18).

One further differentiation is necessary, though. We can use the concept of self, the first-person pronoun, 'I', in either the role of subject or as object. If you use the 'I' in the role of subject, as, for instance, in 'I think', 'I perceive', etc., you designate the self who experiences the actions. Using the 'I' in the role of subject in this sense is supposed to provide you with immunity against error through misidentification, as described above.

This is different once you use the 'I' in the role of an object: 'I think I am doomed to failure.' While the first 'I' in 'I think' signifies the 'I' in the role of subject, the second 'I' in 'I am doomed to failure' uses the 'I' in the role of object. There is no longer immunity against error through misidentification because one can be mistaken about who is doomed to failure – whether it is one's own self, the 'I', or some other person's self.

How does all this apply to schizophrenia? Schizophrenic patients suffer from delusions in that they assume the contents of their experience (like objects, people or events) show strange meanings. For instance, they often feel they are being followed and persecuted by other people, though in reality this is not the case. They thus attribute false positive meaning to objects, events and people, which leads them to be mistaken about the contents of their experience. There is thus indeed no immunity against error through misidentification in schizophrenia.

How about the use of the 'I' in the role of subject? As noted earlier, schizophrenic patients can experience themselves as another person's self. They take on, for instance, the self of Jesus, Nefertiti or some other person. It is from the perspective of that person that they experience their own activities. In other words, the subject of their experiences, the 'I' in the 'I perceive' and the 'I think', changes and is replaced by another person's 'I'. Rather than their own selves, it is now the 'I' of the other person who experiences the activities. 'I' no longer stands for one's own self, but for some other self. There is thus no longer immunity to error against misidentification in the use of the first-person pronoun as subject.

Neurophilosophical discussion 6

Self as subject and object in schizophrenia

One may now be inclined to argue that the 'I' in the role of the subject remains untouched in schizophrenia. Instead, it is only the 'I' in the role of the object that is altered. This may indeed be the case in patients suffering only from auditory hallucinations and delusions without any ego-disturbances (see below). Here, their 'I' hears voices and is persecuted by others: 'I experience voices and feel that I am being followed by others'.

Accordingly, the second 'I' in 'I hear voices and I am followed by others' may indeed indicate misidentification of their own 'I' in the role of object, while the first 'I', in the role of subject, seems to be more or less preserved at this stage. There is thus immunity against error through misidentification in the use of the 'I' as subject.

This immunity may hold true as long as the schizophrenic patients experience themselves as preserved and intact; this is manifested in the experience of the auditory hallucinations and delusions as experienced by and thus related to their own self. The auditory hallucinations and delusions are thus still experienced in terms of mineness and belongingness. This is exactly what makes them so painful and difficult to cope with for the patient.

That, though, may change once the schizophrenic patient starts to suffer from ego-disturbances, namely that she or he experiences her/himself as another person's self. There is consequently no longer experience of any mineness and belongingness. This, in turn, entails that the immunity of error through misidentification no longer applies. Schizophrenic patients may then err and thus misidentify their own self; that is, their 'I' who experiences the contents and thus the subject of their own consciousness.

Neurophilosophical discussion 7

Immunity and the brain's intrinsic activity

How is all this related to the brain? Schizophrenia can be characterized by abnormalities in the brain's intrinsic activity. While it is still unclear, these may be related to the 'basic disturbance of self'. One could postulate that the lack of immunity against error through misidentification in the use of 'I' as subject may be related to the abnormalities in the intrinsic activity. While this neurophilosophical assumption is at best tentative at this point, it nevertheless remains to be explained in detail.

Based in particular on the findings in healthy subjects reported in Chapter 18, one may assume that the brain's intrinsic activity contains, encoded and ingrained, some set of information about the self (and self-specificity). Though it is currently unclear, that information may signify something about the self as a subject of experience reflecting the use of 'I' in the role of subject.

If that information is by itself severely altered, due to changes in the brain's intrinsic activity, misidentification with regard to the subject of experience may become possible. What exactly needs to change in the brain's intrinsic activity in order to make such misidentification possible remains unclear at this point. We thus need to explore the brain's intrinsic activity in much more neuronal detail, and see how that is related to information about self and self-specificity.

How is abnormal intrinsic activity manifested neuronally? The abnormalities in the intrinsic activity entail its abnormal interaction with the extrinsic stimuli. This leads to what can be described as abnormal rest–stimulus interaction. Even if self-specific stimuli like one's own name are presented, they can no longer interact in a normal way with the intrinsic activity.

This in turn may make it impossible to assign to them a high degree of self-specificity with regard to the self as an experiencing subject. Instead, other originally non-self-specific stimuli like the name Jesus are assigned abnormally high degrees of self-specificity. Rather than one's own name, it is then the name Jesus that is experienced in terms of mineness and belongingness.

Depression 1

Increased self-focus

In addition to schizophrenia, depression is another major disorder in psychiatry. Major depressive disorder (MDD) is a psychiatric disorder that is characterized by extremely negative emotions, suicidal thoughts, hopelessness, diffuse bodily symptoms, lack of pleasure, ruminations and enhanced stress sensitivity. The self is also altered in these patients, showing an 'increased self-focus', which will be illustrated with a quote from another paper:

> She sat by the window, looking inward rather than looking out. Her thoughts were consumed with her sadness. She viewed her life as a broken one, and yet she could not place her finger on the exact moment it fell apart. 'How did I get to feel this way?' she repeatedly asked herself. By asking, she hoped to transcend her depressed state; through understanding, she hoped to repair it. Instead, her questions led her deeper and deeper inside herself – further away from the path that would lead to her recovery. (Cited by Treynor et al., 2003, p. 247)

This description of a depressed patient shows three crucial characteristics of the self which will be conceptualized as increased self-focus, association of the self with negative emotions, and increased cognitive processing of the self. Let us start with increased self-focus. Like our patient, almost all depressed patients look inward rather than outward, and focus very much on themselves, while no longer being able to shift their focus to others but always see arrows pointing towards the self. Social–psychological theory describes self-focused attention as a focus on internal perceptual events (i.e. information from those sensory perceptions that react to changes in bodily activity).

The self-focus may also involve an enhanced awareness of one's present or past physical behavior (i.e. heightened cognizance of what one is doing or what one is like). In addition to such increased self-focus, the depressed patient's focus is often also on their own body. Depressed patients show heightened awareness of their own body which results phenomenologically in the subjective perception of diffuse bodily symptoms. The increased self-focus may therefore be accompanied by what I call 'increased body-focus'.

The increased self-focus and body-focus imply that the depressed person's attention is no longer focused on their relations to the environment and environmental events, as in healthy individuals, but rather on themselves as the prime focus, with the environment shifting into the background. Thus the increased self-focus is associated with what one may call 'decreased environment-focus'.

Depression 2

Decreased environment-focus

The concept of the decreased environment-focus describes how the depressed patient's subjective experience and perception are no longer directed toward the environment and its respective people and events. Instead, his/her subjective perception and experience are directed towards his/her own body and own cognition, resulting in 'increased self-focus'. This means that, in depression, the balance between environment-focus and self-focus is shifted unilaterally toward the self at the expense of the environment.

This is also supported by recent empirical data. Empirical research clearly indicates that in depression there is heightened self-focused attention. A variety of studies that have assessed self-focused attention with diverse measures and methodologies all converge on the finding of an increased, and perhaps prolonged, level of self-focused attention in depression (Ingram, 1990). What remains unclear, though, is whether this increased self-focus is purely explicit and thus conscious, or whether it is already present at an implicit, and thus unconscious, level.

Another characteristic is the attribution of negative emotions to the self. The self is associated with abnormal sadness, guilt, mistakes, inabilities, death, illness, etc. which may ultimately result in paranoid delusions.

Finally, there is also increased cognitive processing of the self. The patient described above typically suffers from increased cognitive processing. She/he thinks about her/himself and her/his mood and tries desperately to discover the reasons for her/his depression, but as a result only sinks deeper and deeper into the depressed mood. This cognitive processing of the self is described as rumination, and is often considered to be a method of coping with negative mood that involves increased self-focused attention and self-reflection.

Depression 3

Neural abnormalities

Meta-analyses of all imaging studies in human MDD that had focused on resting state activity yielded hyperactive results in several midline regions like the PACC, the ventromedial prefrontal cortex (VMPFC), thalamic regions like the dorsomedial thalamus (DMT) and the pulvinar, pallidum/putamen, and midbrain regions like the ventral tegmental area (VTA), substantia nigra (SN), amygdala, the tectum and the PAG. In contrast, resting state activity was hypoactive and thus reduced in the dorsolateral prefrontal cortex (DLPFC), the posterior cingulate cortex (PCC) and adjacent precuneus/cuneus.

Involvement of these regions in MDD is further corroborated by the investigation of resting state activity in animal models of MDD. Reviewing evidence for resting state hyperactivity in various animal models yielded diverse participating brain regions – the anterior cingulate cortex, the central and basolateral nuclei of the amygdala, the bed nucleus of the stria terminalis, the dorsal raphe, the habenula, the hippocampus, the hypothalamus, the nucleus accumbens, the PAG, the DMT, the nucleus of the solitary tract, and the piriform and prelimbic cortex (Alcaro et al., 2010). In contrast, evidence of hypoactive resting state activity in animal models remains sparse, with no clear results.

Taken together, these findings indicate abnormally high resting state activity in extended subcortical and cortical medial regions of the brain. This has led authors to assume dysfunction in the limbic system in depression, and more specifically in the 'limbic-cortico-striato-pallido-thalamic circuit'. This in turn leads to abnormalities in the reciprocal interactions between medial prefrontal and limbic regions.

Most important, it seems as if there is an imbalance in the distribution of resting state activity: this seems to be abnormally high in the medial regions, both cortically and subcortically. Meanwhile, resting state activity is abnormally low in more lateral regions like the dorsolateral prefrontal cortex.

How are these resting state abnormalities related to the increased self-focus? Some recent studies applied self-specific stimuli using either emotional pictures or words, known as trait adjectives, in depressed patients, using fMRI. They observed that the depressed patients showed abnormal stimulus-induced activity in response to the self-specific stimuli (when compared to the non-self-specific).

Behaviorally, depressed patients attributed abnormally high degrees of self-specificity to, in particular, the negative emotional stimuli when compared to healthy subjects. This abnormally high degree of self-specificity, as well as the depression symptoms' severity, correlated with the degree of abnormal activity in the midline regions during self-specific stimuli. This means that the abnormal midline stimulus-induced activity in these patients seems to be closely related to their assignment of abnormal degrees of self-specificity to extrinsic stimuli, as well as to their depressive symptoms.

We have to be careful, though. These data do not show the resting state abnormalities themselves to be related to the increased self-focus. Instead, it is rather stimulus-induced activity in response to extrinsic self-specific stimuli that is abnormal in these patients. The direct linkage between resting state activity and increased self-focus remains to be demonstrated. Furthermore, it remains unclear how, for instance, the lateral cortical resting state hypoactivity contributes to the decreased environment-focus in these patients.

Neurophilosophical discussion 8

Embodied and relational concept of self

What can we learn about the concept of self from depression? Depression can be characterized by an increased self-focus accompanied by a decreased environment-focus. The patients experience their own self as being abnormally strong, while at the same they feel disconnected from the environment. Both increased self-focus and decreased environment-focus seem to be coupled and linked to each other in a negative, reciprocal way: if one increases, the other decreases and vice versa.

What does this reciprocal dependence between self-focus and environment-focus in our experience tell us about the concept of self? Even the subject of experience, the self, cannot be considered independent and isolated from its environment. Instead of being detached from the environment, the self, the subject of experience, seems to be deeply embedded in that very same environment.

Based on these observations, one may want to characterize the self by embeddedness into the environment, known as an embedded self. The concept of embeddedness describes the integration and linkage of the self to its respective context, the environment. Is such integration a necessary, and thus intrinsic, feature of the self as the subject of experience? We currently do not know. If it is the case, the self as subject of experience would be defined by its relation to the environment and thus by embeddedness. The absence of embeddedness would then entail the absence of the self. This must be regarded as a tentative hypothesis that requires investigation in both conceptual and empirical terms.

Conceptually, this implies that the concept of self must be defined as embedded in the environment. However, the exact characterization of the relational self remains unclear at this point. What constitutes and makes possible the relationship between self and environment, and thus the self as the subject of experience? This remains to be investigated in the future, both empirically and conceptually.

Neurophilosophical discussion 9

Conceptual and empirical characterization of the brain's intrinsic activity

How does all this relate to the brain? Depression can be characterized by an imbalance in resting state activity levels between midline and lateral cortical regions. One may now want to hypothesize that the medial–lateral resting state imbalance may be related to the imbalance between self and environment in experience. However, more findings are necessary to support the assumption that the neuronal resting state balance corresponds to the phenomenal balance between self and environment in experience. In short, we need to investigate the relationship between neuronal and phenomenal balances.

On the whole, depression can be characterized by dysbalance between self and environment in experience, a phenomenal dysbalance. This suggests that the concept of self cannot be considered in an isolated way, but rather in relation to the environment. One may thus opt for an embedded and relational concept of self as the subject of experience.

At the same time, depression shows a neuronal dysbalance in the resting state activity between midline and lateral regions. Is the resting state's balance between midline and lateral regions central for the embedded and relational self, the subject of experience? If this is the case, the resting state's dysbalance between midline and lateral regions should be accompanied by changes in the self and its relation to the environment. This is indeed the case in depression, as described above in the sections on increased-self focus and decreased environment-focus. Most important, this tells us that the intrinsic activity of the brain must contain some information about its own relationship to the environment.

This has major implications in both the empirical and conceptual regard. Empirically, it implies that we would need to investigate how the brain's intrinsic activity couples and links itself to the environment independent of and prior to the recruitment of specific functions, e.g. sensorimotor, cognitive, etc. These functions all presuppose some kind of subject, the subject that experiences the respective functions. Therefore, the linkage of the brain's intrinsic activity to the environment must occur prior to and independent of these functions in order to make possible the constitution of a subject of experience as embedded and relational.

This also has an important conceptual implication for how we should characterize the brain's intrinsic activity. This may then no longer be considered to be purely intrinsic as opposed to the extrinsic environment. Instead, the seemingly intrinsic activity may be characterized by information about the apparently extrinsic environment. In other words, the conceptual boundary between intrinsic and extrinsic becomes rather blurred and thus empirically implausible.

Take-home message

The first two chapters in this part discussed how the self and self-consciousness are related to the brain and its intrinsic and extrinsic activity. One major suggestion from this neurophilosophical encounter consisted in the assumption that the brain's intrinsic activity must contain, encoded and ingrained, some information about the self. This is at best a tentative hypothesis and pure speculation at worst. How can we garner further empirical and conceptual support for this neurophilosophical hypothesis? To answer this question, one may want look for insight in cases showing abnormal mental features in neurologic or psychiatric disorders. For that purpose we discussed the psychiatric examples of schizophrenia and depression that show abnormalities in both self and intrinsic activity. Conducting detailed investigation and neurophilosophical discussion provides further support for the assumption of a more basic concept of self that may be closely related in an as yet unclear way to the brain's intrinsic activity. This details and shifts not only the definition of our philosophical concepts of self and self-consciousness, but also provides new ideas for future experimental investigation of the relationship between the brain and the self. Hence there is much we can learn from psychiatric disorders like schizophrenia and depression. Understanding these disorders can help us to achieve a better picture of the relationship between the brain and the self, and can aid us in developing a truly neurophilosophical model.

Summary

The previous chapters in this part focused on how the philosophical concepts of self and self-consciousness are related to recent neuroscientific findings on the self-reference of extrinsic stimuli. Different concepts of self and self-consciousness were distinguished and tested for their empirical plausibility. This raised the question of a more basic concept of self that exists prior to and independently of any particular functions, e.g. sensorimotor, affective, cognitive, etc. This corresponded to the observation of neuronal overlap and prediction of self-specific neural activity by the resting state, the brain's intrinsic activity. In this case, one would assume some information about the self and self-specificity to be contained in the brain's intrinsic activity. How can we furnish further conceptual and empirical support for this neurophilosophical assumption? To answer this we looked at pathological cases in clinical psychiatry that show abnormalities in both self and brain.

Schizophrenia is a complex psychiatric disorder that can be characterized by auditory hallucinations, delusions and ego-disturbances. Early psychiatrists assumed a 'basic disturbance of self' to underlie the various symptoms in schizophrenia. Recent brain imaging studies demonstrate abnormalities in the intrinsic activity, its spatial and temporal pattern, as well as abnormal stimulus-induced activity in response to self-specific stimuli. This has major implications for the concepts of self and self-consciousness. The abnormal self in schizophrenia may not be explained by traditional philosophical concepts like the mental and the empirical self. The 'disturbance of self' seems to be more basic in schizophrenia – prior to and independent of any particular function, e.g. sensorimotor, cognitive, etc. This might correspond with the abnormalities in the resting state in schizophrenia, which need to be further specified in the future. One would, for instance, expect that the resting state abnormalities predict the self-abnormalities in schizophrenia. If so, one may indeed assume a more basic concept of self as ingrained and encoded in the brain's intrinsic activity. However, even that encoded and ingrained self may still be prone to error through misidentification, since patients can experience their self as another person's self. There is thus no immunity against error through misidentification even in the use of the 'I' in the role as subject.

Depression can be characterized by an abnormally increased self-focus and a decreased environment-focus in the experience of self and environment. This is accompanied by a resting state imbalance between medial and lateral cortical regions. Conceptually, this implies that even the self as the subject of experience is always already linked and integrated to its respective environmental context. In other words, the self as the subject of experience is embedded and relational. Most important, some information about this kind of embedded and relational self and its relation to the environment seems to be already contained in the brain's intrinsic activity. This topic needs to be researched empirically in further detail. The fact that the intrinsic activity possibly contains some information about the extrinsic environment implies that the distinction between intrinsic and extrinsic activity may become blurred.

Revision notes

- Why do we go back to psychiatry to learn something about the relationship between brain and self?
- What is this methodological strategy called that was applied in the past? Please give examples.
- Define and describe schizophrenia and its symptoms.
- What is meant by a 'basic disturbance of self' in schizophrenia?
- Explain the neuroscientific findings in schizophrenia in both resting state and stimulus-induced activity.
- What are the implications of schizophrenia for the concept of the phenomenal and minimal self?
- Why does schizophrenia lend support to the assumption of a most basic concept of self prior to and independent of specific functions, e.g. sensorimotor, cognitive, etc.
- What about the misidentification of the self in schizophrenia? Does this violate what philosophers describe as immunity to error through misidentification?
- Why is it necessary to distinguish between the role as subject and object with regard to schizophrenia?
- What is depression?
- Describe increased self-focus and decreased environment-focus.
- How is resting state activity changed in depression?
- How is abnormal resting state activity related to increased self-focus?
- What is the concept of an embedded and relational self?
- Why may the conceptual distinction between intrinsic and extrinsic become empirically implausible when the intrinsic activity contains some information about the self and its relationship to the environment?

Suggested further reading

- Alcaro, A., Panksepp, J., Witzak, J. and Northoff, G. (2010) 'Is subcortical-cortical midline activity in depression mediated by GABA-A receptors and glutamate? A cross-species translational approach', *Neuroscience & Biobehavioral Reviews*, 34(4), 592–605.
- Northoff, G. (2007) 'Psychopathology and pathophysiology of the self in depression – neuropsychiatric hypothesis', *Journal of Affective Disorders*, 104, 1–14.
- Northoff, G. (2011) *Neuropsychoanalysis in Practice* (Oxford/New York: Oxford University Press), chs 10–12 on depression and schizophrenia.
- Treynor, W., Gonzalez, R. and Nolen-Hoeksema, S. (2003) 'Rumination reconsidered: A psychometric analysis', *Cognitive Therapy and Research*, 27, 247–59.
- Wiebking, C., Bauer, A., de Greck, M., Duncan, N. W., Tempelmann, C. and Northoff, G. (2010) 'Abnormal body perception and neural activity in the insula in depression: An fMRI study of the "material me"', *World Journal of Biological Psychiatry*, 11(3), 538–49.

20
Brain and Intersubjectivity

Overview

So far we have focused on two essential features of the mind, consciousness and self. Consciousness describes our ability to experience our self and the world. The concept of self concerns the subject of experience, namely that any experience is experienced by a subject. There is no experience, and thus consciousness, without a subject. Where does this subject come from? This was the guiding question of Chapters 17 and 18. In addition to consciousness, the subject – the self – shows further ability – the ability to relate to other people, or other selves. This raises the question: how can the self relate to others? This is called the 'self–other relationship'. The ability of the self to relate to other selves is described by the concept of intersubjectivity in philosophy. Intersubjectivity and its relationship to the brain is the focus in this chapter.

Objectives

- Understand the self–other relationship and the problem of intersubjectivity
- Explain inference by analogy
- Distinguish between different forms of intersubjectivity, including, for example, pre-reflective and reflective intersubjectivity
- Explain the difference between theory-theory and simulation-theory
- Account for the main neuroimaging findings in theory of mind and mentalizing
- Understand mirror neurons and the respective empirical data
- Explain the concept of embodied simulation
- Account for the relationship between embodied simulation and intersubjectivity

Key concepts

Intersubjectivity, self–other relationship, embodied simulation, theory-theory, simulation-theory, inference by analogy, mirror neurons, empathy, shared manifold.

Background

Problem of other minds

Throughout this book, we have investigated the mind and its various features. In Part II we focused on the metaphysics of the mind, its existence and reality, and how this relates to the body. This was discussed as the mind–brain problem. This focus was extended further in Part III, in which we discussed the different possible explanations of the mind.

We then shifted our focus from the mind itself to its features in Parts IV and V. Part IV discussed consciousness as a hallmark feature of the mind. In addition to several arguments against reducing consciousness to the brain (see Chapter 16), we also characterized consciousness by various phenomenal features that signify our experience as experience. These phenomenal features include spatiotemporal continuity, unity, intentionality, self-perspectival organization, and first-person perspective (see Chapter 17 for details). We discussed the recent empirical suggestions for the necessary and sufficient neuronal mechanisms, known as neural correlates and predispositions, that underlie consciousness and its phenomenal features.

Consciousness describes experience by a subject. Experience, and thus consciousness, seem to be tied intrinsically to a subject: without a subject, any experience or consciousness seems to be impossible. What is this subject, the subject of experience? This led us to the question of subjectivity. Subjectivity is the focus in Part V.

Chapter 17 discussed the concept of self and the metaphysics of self. Different philosophical suggestions for a concept of self were presented, including the mental self, the empirical self as a self-representational self, the phenomenal self, and the minimal self. These were compared and contrasted with the recent neuroscientific findings on self-reference. This was followed by a discussion of how empirical findings and philosophical concepts can be related to each other. This revealed the neuroscientific data to be unspecific in several regards (neuronal, phenomenal, experimental, psychological) (see Chapter 17).

In Chapter 18, we discussed the concept of self-consciousness, and how the self can access consciousness. Different forms of access to one's own self in consciousness were discussed, including direct awareness and indirect propositional awareness, as well as immunity to error through misidentification. These topics were then put into the context of recent imaging findings. This revealed a central role for the brain's intrinsic activity as a neural predisposition, or the necessary conditions of possible self-consciousness. The assumption of a possible relationship between self-consciousness and the brain's intrinsic activity was further supported by discussing abnormalities in both self and brain in psychiatric disorders like depression and schizophrenia (see Chapter 18).

In addition to accessing our own self in consciousness, we can also access other people's selves and their respective consciousnesses. While you sit beside me and read these lines, I sense that you are bored by what you read and that you want to put this book behind you as soon as possible. You now feel detected and exposed in your feelings and intentions. I was apparently able to read your mind. How is that possible? The question of reading other people's thoughts is described as the problem of other minds or intersubjectivity in philosophy (and at least as heavily in psychology, as we will see below). This is the focus in the present chapter.

Philosophical suggestions 1

Problem of intersubjectivity

In addition to access and becoming aware of itself as manifested in self-consciousness, the self is also able to become aware of other selves. For instance, while watching a movie I am aware that you are sitting beside me and, based on the kind of facial grimace you exhibit when violence breaks out, I can feel and share the emotional pain you experience. How is this possible? My self is able to reach out to your self as the other and to establish some kind of self–other relation. This self–other relationship is often subsumed under the concept of intersubjectivity, also known as the 'problem of other minds' in philosophy.

What exactly is the problem of other minds about? Let me give an example. Your friend is sitting beside you while you are watching a movie on TV. You laugh more or less at the same time, though your friend is more serious and not so much taken by the comedic appearance of one of the characters. You assume that, despite his seriousness, he nevertheless understands the plot and the comical situation in which the actors find themselves. You thus presuppose that he has the capacity of logical reasoning and understanding. You also assume that he has certain experiences and is thus conscious. In other words, you assume that he has certain mental states and thus, in short, a mind.

How do you come to the conclusion that he has mental states and a mind? All you can observe is how he sits in front of the movie watching it. You can, for instance, see his body and his behavior, his smiling and laughing, his shoulders shrugging and his serious facial expression. In contrast, you cannot experience his experience and thus share his consciousness: you do not experience what he experiences in his first-person perspective, such as a feeling of boredom, a feeling of happiness, or some thoughts about his recent loss of his job as a movie maker in Hollywood.

How is it, then, nevertheless possible for you to assume that he has mental states and thus a mind? You could also, by the same token, assume that he has no mind whatsoever, that he is a zombie who is acting purely mechanically without any experience or consciousness. However, you do not do that. Instead, you assume that he has a mind, mental states, and is conscious. And, most important, in the same way that you assume he has mental states, you also attribute mental states to any other person. This is the problem of other minds or intersubjectivity, as discussed in philosophy.

Biography 1

John Stuart Mill

How do we come to and make the assumption of attributing mental states and minds to other people? Philosophy has long relied on what is called 'inference by analogy', developed by the British philosopher (and economist) John Stuart Mill (1806–73). He was born in London to a father who taught him classical philosophical literature from a very young age. Mill started to learn Greek at the age of three and was reading Plato and other Greek classics by his eighth birthday.

Mill made contributions to social, economic and political philosophy and was also a politically active Member of Parliament for some years. He was also one of the first outspoken advocates of women's rights. In this text, we shall focus on his development of inference by analogy.

Philosophical suggestions 2

Inference by analogy

What is 'inference by analogy'? It goes like this: you observe another person A to show behavior of type X. And you know that in your own case the same behavior, type X, is accompanied by the mental state type M. Since your own behavior and the behavior of person A are similar, you assume the other person A experiences the same mental state type M that you experience.

You thus infer from the analogy of behavior between you and the other person that this other self experiences the same mental state as yourself. Hence, by indirect inference, you claim to obtain knowledge of the other person's mental state. Why do we make these types of inference? We make them precisely because it seems to be the easiest and best way for us to explain others' behavior. Only by assuming and inferring that you show mental states, can I explain your behavior. The assumption of mental states thus seems to be the best explanation for your behavior. 'Inference by analogy' may thus be considered an inference to the best possible explanation.

But how can you apply the concepts used to describe one's own mental states to those that you suggest occur in other people? You must have some kind of criteria upon which you assign and associate the concept that describes a particular mental state with the behavior you observe in another person.

For instance, you observe that the person in front of you retracts her leg and shows grimaces on her face. You assume that she is in pain and thus associate the mental concept 'pain' with the behavior you observe. The criterion for applying mental concepts to other people is thus their behavior. What kind of criteria one should apply is a question that is strongly debated in contemporary philosophy of mind.

Philosophical suggestions 3

Self and other in phenomenological philosophy

We have already encountered phenomenological philosophy several times in this book. Phenomenological philosophy focuses on consciousness itself, and more specifically on the structure and organization of our experience in first-person perspective (see Chapter 1). The structure and organization of experience is manifest in certain phenomenal features which we discussed in the context of consciousness (see Chapter 14). Finally, the structure and organization of experience is also assumed to include reference to a self, a phenomenal or minimal self that is always already part of any experience in an implicit way, known as a pre-reflective self-consciousness (see Chapters 17 and 18).

How did phenomenological philosophers like Edmund Husserl, Martin Heidegger, John-Paul Sartre and Merleau-Ponty as well as contemporary philosophers including Dan Zahavi (see his excellent book on subjectivity and selfhood – Zahavi, 2005) and Shaun Gallagher address the problem of other minds? They considered the argument of 'inference by analogy' to be based on false positive presuppositions. 'Inference by analogy' presupposes (i) that we have direct and immediate (and complete) access to our own mental states; and (ii) that there is no such direct and immediate access to the mental states of other people. Both assumptions may be doubted.

(i) The access to our own mental states may not be as direct and immediate as assumed. Often we access our own mental states only in conjunction with both our own body and others in our social environment. We thus, at least in part, access our own self via the other and therefore in an indirect and mediated way. This is the case when you start to laugh at a scene in a movie only after seeing and hearing your friend laugh. Why is this example relevant for understanding the ability to access our own mental states? Here, it is the other person and the social context – both of you watching a movie – that opens the door to accessing your own mental state. Hence, the argument of inference by analogy seems to overestimate our ability to access our own mental states in a direct and immediate way.

(ii) What about the other assumption: that we lack direct access to the other's mental states? There are many situations in our daily lives where we seem to access the other person's mental states directly.

Let me give you the following example. For the first time, you feel that you are in love. You find the other person to be beautiful, intelligent and extremely attractive. The other person seems to send you similar signals. You feel for the first time that your feelings are reciprocated. Both of you look at each other with deep love and mutual affection in your eyes. You understand each other without saying a single word and can instead communicate just by your emotional feelings, which strongly overlap and are shared by the two of you. Such sharing of emotional feelings is described by the concept of 'empathy'. And by means of empathy, you are strongly connected and linked to the other via the emotional feelings you share with the respective other person – empathy must thus be considered an instance of what philosophers call 'intersubjectivity'. What does this case tell us about access to others' mental states? In this case you have access to the mental states of the other without making any inferences. Hence, the argument of inference by analogy underestimates our ability to access other people's mental states in a direct and immediate way.

Philosophical suggestions 4

Empathy as affective form of intersubjectivity

How is it possible that by merely looking at each other and without saying anything you sense the other's mental states? The body is most likely central. You experience your body as your body that is specific for you and your self; this reflects the concept of the 'lived body', as discussed in Parts I and II. At the same time, the very same body, which you experience as your body, is located in the environment and provides you with an interface between your self and the other's self. Via your body, you have access to the environment, including the other. This is what is described as 'embodied subjectivity'.

How does the concept of 'embodied subjectivity' relate to the concept of intersubjectivity? Thanks to embodied subjectivity, your own experience of your self and your body – the lived body – is not as isolated from the other and the social environment (as is assumed in the argument of inference by analogy). The experience of my own self is closely tied to the other; put metaphorically, the other is present in my own self. More technically, there is 'internal otherness', the other is within the self, or alterity (i.e. other) within ipseity (i.e. self). The subjectivity of the self is here in direct interface with the subjectivity of the other. This highlights an intricate linkage between intrasubjectivity and intersubjectivity, and thus between your own self and the other self within your own body.

How is this intricate linkage between intrasubjectivity and intersubjectivity manifested in our experience and thus in consciousness? Here, phenomenological philosophers emphasize affective and motivational as well as sensorimotor functions.

What is empathy? As a preliminary definition, empathy concerns the sharing of emotional feelings. The concept of empathy concerns affective-motivational and sensorimotor functions that are the means used to reach out to others. As such, affective-motivational and sensorimotor functions may be regarded as manifestations of a primary and more basic sense of intersubjectivity. This complements the more cognitive account of inference by analogy that is based predominantly on cognitive and linguistic functions to conceptualize others' mental states (see above). The inference by analogy can thus be characterized as a cognitive form of intersubjectivity.

While the cognitive inferential knowledge of the other is not denied, phenomenology argues that there may already be some kind of pre-cognitive sharing between self and other as prior to and at a more basic level of intersubjectivity: rather than the cognitive functions of the mind as in inference by analogy, the affective-motivational and sensorimotor functions of the body are considered to be central here. One may thus want to speak of an affective or pre-cognitive (and pre-conceptual and pre-linguistic) form of intersubjectivity as distinct from a cognitive form of intersubjectivity.

There may be essentially two different forms of intersubjectivity. The cognitive form of intersubjectivity provides mediated and indirect access to the other's mental states via the recruitment of cognitive functions. This is reflected in what philosophy describes as 'inference of analogy.' Meanwhile, the affective form of intersubjectivity allows for more direct and immediate access to others' mental states via shared emotional feelings as described in the concept of empathy.

Philosophical suggestions 5

Determination of the concept of empathy

How can we define the concept of empathy in more detail? This concept was coined in the German language as 'Einfuehlung'. Originally, the term 'Einfuehlung' was used in the context of esthetic philosophy where it described the ability to 'feel oneself into' works of art as well as nature and its esthetic objects and events.

However, the German philosopher Theodor Lipps (1851–1914) made use of the term in a broader context: he detached the concept of empathy from its exclusively esthetic context and shifted it into a more psychological context, the ability to feel and understand other people's emotional and cognitive states. As such, the concept of empathy could be linked directly to the concept of intersubjectivity as discussed in philosophy. Empathy today describes our understanding and sharing of other people's sensory and emotional (and cognitive) states.

How can we distinguish empathy from related concepts like imitation, recognition, sympathy and emotional contagion? Imitation concerns only the mere simulation of another person's movements and gestures. For example, imagine an infant in the arms of its mother. The baby sees the mother smile and imitates this by smiling back. If, in contrast, the mother is yawning, the baby make similar movements with its mouth. However, and this is important, you may not feel the same emotions and feelings while imitating, which distinguishes imitation from empathy.

Another concept is recognition. You may recognize the other's movements and their intentions, which may not go along with shared emotional feelings. For example, you may recognize that the other person is desperately trying to find a diamond necklace that has fallen into a river. You watch her movements and gestures as she feels around the riverbank on her hands and knees. This, though, does not imply that you feel and share her emotional feeling of despair.

What about sympathy? The concept of 'sympathy' is defined as social affinity with another person. This is, for instance, the case when you as an outsider sympathize with the group of soccer fans who cry out over the penalty the referee gave against their team. You might not share their emotional feelings, but you could sympathize with them because the penalty was unfair and wrong. Hence, you do not share their emotional feelings, but you understand wholly why the team is upset. You are thus sympathetic, but not emphatic.

What about emotional contagion? Despite the fact that you are not a soccer fan, the distress of the soccer fans nevertheless affects you strongly. You suddenly become aware that a few tears are running down your cheeks. You have no idea where that is coming from, because you don't really care about soccer, the penalty or the team the soccer fans love so much. Their crying is contagious and affects you. This is called emotional contagion.

The difference between empathy and 'emotional contagion' lies in the awareness of the origin of the shared emotion: in the case of emotional contagion one is not aware that the actual emotional state is induced by another person. In the case of empathy, one is well aware of the other person, as well as of the shared emotional feelings.

Finally, one needs to distinguish empathy from what is described as 'theory of mind' (ToM). Theory of mind concentrates on the cognitions of a person used to understand the mental state of the other, while empathy focuses on the emotional feelings of the other. For example, when you and your spouse see a couple arguing to the point that it looks as though they will break up, you may look at your beloved and ask him/her what he/she thinks. (And, of course, you hope that he/she thinks that this will never happen to the two of you, and that your romantic love will endure for ever!) You thus want to get a glimpse of the other's cognition and mental state. This is described by theory of mind (ToM).

Psychological theories 1

Theory-theory (TT)

How can we explain the cognition of other people's mental states? This is an empirical question that touches upon psychology, where it is discussed under the heading of 'theory of mind'. The concept of theory of mind describes the ability to attribute mental states to other people and to explain and predict their behavior in terms of their intentions, beliefs and desires. Even though the other person's mental states are not directly observable, nor can be experienced by us, we nevertheless attribute them to the other person in order to explain and predict their behavior.

How is an explanation and prediction of others' mental states possible, and what are the underlying mechanisms? There are two main competing theories in current psychology: theory-theory (TT) and simulation-theory (ST). Let's start with theory-theory.

Theory-theory argues that the ability to explain and predict another person's behavior and mental states requires a theory about the mind and mental states in general. We need to have a particular theory of the mind in order to be able to assign mental states to another person. Only if we have some kind of theory of the mind in general will we be able to attribute beliefs, desires and intentions to other people. Examples of a possible theory of the mind include common sense or folk psychology (see Part III for more details). Hence, the cognition of others' mental states is linked to the availability of some kind of theoretical knowledge about the mind.

Theory-theory is often closely linked to functionalism. Recall Part II. Functionalism describes the assumption that the mind consists in causal relations between input (the stimulus), internal state (the mental state), and output (the behavior). This describes a particular theory of the mind in terms of causal relations. You deploy this theory in your encounters with others. For instance, you know that smiling reflects happiness: this is one of your pieces of knowledge about the mind. This knowledge allows you infer a specific mental state from the presence of a particular input and observable behavior; for example, you give your loved one flowers and they smile at you. You recognize the smile as a behavior that indicates happiness and pleasure.

Where does this theory about the mind and theoretical knowledge come from? There are two main sides to this current discussion. One side argues that the theoretical knowledge about the mind is learned and acquired, while the other assumes it to be innate.

Let's start with learning and acquisition. Consider the acquisition of scientific theories. We explain the data, predict new data and theories from other experiments, falsify our predictions, and develop new predictions by correcting the old ones. 'Learning by doing', one may want to call this.

Learning by doing is now assumed to operate in the case of the acquisition of theoretical knowledge about the mind and mental states. For example, we learn and acquire knowledge about behavior and about how the mind operates by being exposed to others and their behavior. This leads to what is described as 'theory formation', which provides the knowledge base for any theory of mind and its subsequent application when attributing mental states to others.

Those who oppose the idea that we learn and acquire knowledge of the mind and behavior argue that knowledge of the mind is innate and somehow already contained in our mind when we are born. How is innate knowledge about the mind possible? The supporters of the innateness thesis argue that there is some kind of theory of mind module that is genetically and evolutionary pre-configured in our brains. This is also called 'nativism'.

Psychological theories 2

Simulation-theory (ST)

An alternative to theory-theory is simulation-theory. Simulation theory argues that we do not theorize about mental states when attributing them to another person. Instead, we simulate them on the basis of our own mental states. This simulation allows us to attribute mental states to the other person.

For example, two years ago, when you broke your leg roller-skating, you experienced serious pain. Now, when you observe another person falling down in more or less the same way you did two years ago, what happens? This scenario led you to simulate your own mental states associated with your experience two years ago. Since her situation is similar, you project the pain you felt on to her, and attribute a mental state of pain to her. Your simulation of her mental states thus makes it possible for you to attribute a mental state to her.

What is the exact difference between theory-theory and simulation-theory? In the case of theory-theory, the attribution of mental states is based on knowledge as manifested in a theory of the mind in general. The cognition of others is thus knowledge-driven. This is different in the case of simulation-theory. Unlike in theory-theory, the attribution of mental states is not based on knowledge and theorizing. Rather, it is based on a particular process, simulating others' mental states on the basis of your own mental states. The attribution of mental states to others is thus no longer knowledge-driven, but process-driven.

There is also another important difference. The involvement of knowledge and theory implies concepts and propositional attitudes. The theorizing thus takes place on a personal level. This is different in the simulation theory. Unlike in theory-theory, the simulation may not need to involved concepts and propositional attitudes. The simulation may thus be characterized as pre-conceptual, pre-linguistic and pre-cognitive, rather than conceptual, linguistic and cognitive as in theory-theory. This may, for instance, be the case in empathy. The sharing of emotional feelings may be based on the simulation of one's own affective states and emotional feelings that are then projected on to the other person.

However, the difference between theory-theory and simulation-theory might not be as drastic as it seems. One can opt for less radical versions of simulation-theory that argue for the involvement of some kind of general knowledge about mental states during simulation. In this case, theory-theory and simulation-theory are not wholly mutually exclusive, but might rather be fruitfully combined.

Neuroscientific findings 1

Theory of mind and brain imaging

Imaging of the brain allows us to investigate the various processes related to our ability to represent and attribute affective and cognitive mental states to the other. This has been done in various studies with a range of tasks that require the participants to detects another subject's goals, mental states, intentions and emotions. We are thus mentalizing, which allows us to generate theories about the other's beliefs, desires and emotions to understand and predict their behavior.

How can we further describe the process of mentalizing? Research in psychology has distinguished three main mentalizing processes: (i) representing cognitive and affective states, i.e. representation; (ii) attributing these mental states either to the self or the other, i.e. attribution; and (iii) applying these mental states to the observed behavior in the other person, i.e. application. All three subprocesses – representation, attribution and application – have been tested using functional imaging of the brain. The results show that they seem to be associated with different neural networks.

The representation of affective mental states, the affective theory of mind, has been associated with limbic subcortical and cortical regions. These include subcortical regions like the amygdala and the striatum, as well as cortical regions like the orbitofrontal and ventromedial prefrontal cortex (OFC, VMPFC), the perigenual anterior cingulate cortex (PACC), and the ventral temporal pole (TP).

Meanwhile, the representation of cognitive mental states seems to be mediated by neural activity in dorsomedial and lateral prefrontal cortex (DMPFC, DLPFC), dorsal anterior cingulate cortex (ACC), and the inferior prefrontal cortex. Posterior regions like the superior temporal sulcus (STS), the temporo-parietal junction (TPJ), and the posterior cingulate cortex (PCC) are also thought to play a central role here.

What about the second subprocess: the attribution of mental states to self or other? This presupposes the ability to represent the mental states of our own self and the other. In this case, anterior medial regions like the PACC and the VMPFC are assumed to be central. In addition, posterior regions like the PCC, STS, TPJ and parietal cortex are also assumed to be especially important for distinguishing mental states related to own self and those associated with other people. We can see that these regions closely overlap with those implicated in the representation of affective and cognitive mental states.

Neuroscientific findings 2

Empathy and brain imaging

Research on theory of mind focuses mainly on cognitive functions and how they are related to the attribution of mental states in other people. What about affective functions? These may be central in empathy. Recent experimental research investigated empathy by using tasks to elicit empathy while subjects undergo brain scanning. One may distinguish between two different experimental approaches to empathy, an affective and a cognitive approach.

The affective approach presupposes empathy as an involuntarily elicited sharing or matching between one's own feelings and those of the perceived other. Here, empathy is induced experimentally by passively observing others' affective states without explicit instruction. For instance, the passive viewing of others' pain activated the neural circuit of self-experiencing pain, which included the anterior cingulate cortex (ACC) and the anterior insula. Most important, the activation of these brain regions is correlated with the post-scan subjective ratings of pain intensity and personal traits of empathy.

How do these findings relate to the postulated process of simulation? Empathy is thought to be based on the simulation of the other's feelings in one's own self. One may now associate the neural activity in these regions, the ACC and insula, with the process of simulation underlying the generation of empathy. However, this kind of inference might be problematic, as we will discuss below.

In addition to the affective approach, one may also pursue a more cognitive strategy to empathy. The cognitive approach focuses more on the voluntary effort we make during empathic reaction. Here, the researchers test for empathy by explicitly asking the subject to imagine or evaluate others' emotional states. For example, our empathic reaction to others' pain, which induces the activation of ACC and anterior insula, is modulated by top-down attention and cognitive appraisal.

Despite researchers' recent attempts to distinguish affective and cognitive subprocesses in empathy, and investigate the interaction between them, it is strongly suggested that both affective and cognitive processes of empathy are ultimately based on a perception–action coupling (Preston and de Waal, 2002). The perception of the other's emotional state is assumed automatically to elicit the related somatosensory and subsequent emotional-affective processing in one's own self.

The somatosensory processes raise questions about the involvement of somatosensory brain regions in the different approaches to empathy. The activation of the somatosensory cortex in the empathy of pain and tactile experience is supported by a series of studies using transcranial magnetic stimulation (TMS), event related potentials (ERP), magnetoencephalography (MEG), and fMRI.

Taken together, these findings show the involvement of different brain regions in empathy, including the ACC, the insula, and sensory and motor regions. However, this leaves open the exact mechanisms and processes by means of which such neural activity, and ultimately the shared emotional feelings, are generated. This brings us back to the concept of simulation that supposedly underlies empathy.

The concept of simulation describes the representations of another person's feelings in one's own representational apparatus. How can this process of simulation take place and be realized in the brain? One would assume the existence of specific regions or cells that are especially in charge of such simulation during the observation of others' movements and emotional feelings. This leads us to what has been described as mirror neurons and mirroring mechanisms in recent neuroscience.

Biography 2

Giovanni Rizzolatti and Vittorio Gallese

Mirroring mechanisms were first assumed when a specific subset of neurons in the ventral premotor cortex (F5) of the macaque monkey were found to be active as one monkey observed another monkey's movement. This was reported by the Italian researchers Rizzolatti and Gallese in Parma, Italy. Giovanni Rizzolatti (1937–) was originally born in Kiev in the former Soviet Union, in what is now the Ukraine. His family left the Ukraine and went to Italy, where he studied and detected the mirror neurons in monkeys.

He was joined by his student, Vittorio Gallese (1959–), who studied in Parma and, like Rizzolatti, is still working there. While Rizzolatti focuses on monkeys, Gallese also reaches out to humans and is especially interested in the more philosophical notions of the self, especially the bodily self and how the self is intrinsically connected with other selves.

Neuroscientific findings 3

Mirror neurons and goal-orientation

What are mirror neurons? These so-called 'mirror neurons' show activity not only when the monkeys performed a goal-oriented movement and action, but also when observing other individuals – monkeys or humans – executing a similar action. Neurons that are active during both execution and observation of action were also observed in other regions of the brain, including the posterior parietal cortex, more specifically the inferior parietal lobule, and the superior temporal sulcus. These regions are not only active during the planning and execution of one's own movements and actions, but also during the observation of the other's movements and actions.

This has been demonstrated not only in monkeys but also in humans. Analogous regions have also been observed in humans in functional imaging. During the observation of other people's actions, neural activity was shown in premotor and parietal areas, the human homologue of the areas in which mirror neurons were originally described in monkeys (see Gallese, 2007; and Rizzolatti and Fabbri-Destro, 2008, for reviews).

As such, it is more the action and its goal-orientation than movement itself that is crucial for inducing neural activity changes in these neurons. Mirror neurons in F5 are also active when the monkey hears the sounds usually associated with the to-be-mirrored action, even if they are not able to see the movement associated with the respective action and its goal-orientation. The neurons respond only to the action-associated sounds, but not to other sounds, including interesting and novel sounds. These neurons have therefore also been called 'audiovisual mirror neurons'. These findings indicate that it is not so much the observed movement itself that is coded in these neurons' neural activity, but rather the goal or intention of the action that is associated with the respective movement.

These findings have been further substantiated in a variety of studies on grasping and reaching, where subjects were asked to transport their hands to a particular location in space without either reaching or grasping the objects themselves. Even though the actual movements of grasping or reaching were not executed, mirror neurons were nevertheless demonstrated as active when the subjects observed the movement of the hand.

It was the Italian researcher Marco Iacoboni who originally tested the involvement of the mirror system in coding intentions. Iacoboni, who now researches and teaches in Los Angeles, conducted an interesting study, which will be described briefly. Subjects were asked to watch three different stimuli: hand grasping without a context (e.g. objects), context only (e.g. a scene containing only objects but no hand), and hand grasping acts embedded in contexts (e.g. objects).

When compared to the first two conditions, the third condition – hand grasping embedded in contexts – yielded a significant increases in signal changes in the prefrontal cortical mirror system. More specifically, these concerned signal changes in the inferior frontal gyrus close to Broca and the ventral premotor cortex (Iacoboni et al., 2005). The increased neural activity during specifically the third condition suggests that the mirror neurons seem to code the intention of a movement rather than the movement itself and its mere physical characteristics. Such coding of the other's intentions may, for instance, help us to understand another person's intentions even when she/he has not yet performed or executed the respectively associated movement.

Neuroscientific findings 4

Mirror neurons and different regions and functions

In addition to prefrontal mirror mechanisms, studies in both humans and monkeys have demonstrated that other regions, like the posterior parietal cortex and the right inferior parietal lobule, are also crucially involved in mirroring the intentions of movements.

For example, neurons in the parietal cortex showed activity during the observation of an act (e.g. grasping an object) in the stages preceding the actual movement itself. These neurons thus seem to be sensitive to the overall action intention – the goal-orientation that consisted in reaching for the object and bringing it to the mouth. Hence, parietal neurons seem to be active even before the actual movement is observed, indicating that they must code the goal or intention of the to-be-executed movement. This indicates that intention detection and action prediction seem also to recruit neural activity in the parietal cortex.

Mirror neurons have been shown predominantly in the motor and action domain. This might enable us to understand the other person's intentions from their actions and movements. However, there is much more to human communication than merely understanding the intentions of the other person. There are, for instance, the other person's sensations and emotions, which we also understand and can empathize with. If the mirror mechanisms were to hold true in the domains of sensations and emotions, one would expect that the observation of another's sensations or emotions would induce neural activity in the same regions as those involved in the experience of the same sensations and emotions.

For instance, observing tactile experience like touching in other people has indeed been shown to induce neural activity in the same region, the somatosensory cortex, as when having the experience oneself. Similar observations have been made in the case of pain. The observation of pain induces neural activity in the somatosensory cortex, the anterior cingulate cortex, and the insula, all of which are also activated when one is in pain.

Another study investigated emotions, most specifically the emotion of disgust. Both the experience of disgust in olfaction, and witnessing the same emotion expressed by the facial mimicry of another person smelling something, activated the same regions – the left anterior insula and the anterior cingulate cortex, at the same overlapping locations (see Wicker et al., 2003). Hence there seems to be some regional sharing during the emotional sharing of disgust.

The neural overlap in the regions recruited during emotional experience and emotion observation/recognition is further supported by observations that deficits in emotion recognition are coupled or paired with deficits in emotional experience.

Patients with lesions in the insula showed deficits in both the experience and facial recognition of disgust. Analogous findings are reported for other emotions, like happiness, fear, and anger, though these emotions involve different regions (amygdala, ventral striatum, ventral tegmental area).

Taken together, there seems to be solid empirical evidence that observation and execution seem to recruit neural activity in similar regions. This apparently holds true not only for motor function, but also for other functions like emotion, pain, etc.

Neurophilosophical discussion 1

Anatomo-processual inference

How do these neuroscientific data about theory of mind, empathy and mirror neurons relate to the psychological theories of theory-theory and simulation-theory, and ultimately to the philosophical problem of other minds, or intersubjectivity? The central question here is whether, and how, the neuroscientific data provides evidence for the processes postulated in theory-theory (TT) or simulation-theory (ST). Let me focus on ST and the concept of simulation.

The main assumption of simulation and mirroring relies on the empirical findings of neural activity changes during observation in those regions that also mediate the execution of the same function (action, sensation or emotional feeling). From such anatomical overlap in regions like F5, inferior parietal lobule, superior temporal sulcus, insula, anterior cingulate, etc. (see above), the investigators infer mirroring and simulation as possible underlying processes to account for the involvement of the same region during observation.

Let us detail this inference. The guiding question is the following: what is the underlying process that makes such double involvement of the same regions during execution and observations possible? Since the very same regions are involved in execution, the investigators infer that these regions may mirror or simulate the execution of action, emotional feeling or sensation when one observes them in others. Hence the authors infer from the anatomical overlap between execution and observation to the supposedly specific underlying processes of mirroring and simulation.

One may consequently want to speak of an 'anatomo-processual inference'. The 'anatomo-processual inference' describes an inference from the anatomo-functional level to the processual level: from the fact that these regions are involved in both execution and observation, one infers the involvement of the processes of mirroring and simulation. More specifically, involvement of these regions in observation is assumed to be possible by mirroring and simulating the other's actions, feelings or sensations in an 'as if' mode – as if the observing person would execute the same function.

This anatomo-processual inference is crucial in the current neuroscientific and philosophical discussion about mirror neurons and empathy. Why? Because it provides the framework within which investigators interpret and conceptualize their findings. And that in turn is taken as the starting point for relating the empirical findings to psychological and philosophical theories about intersubjectivity and their concept of simulation. Accordingly, the anatomo-processual inference bridges the gap from the empirical findings to the more conceptual and theoretical assumptions.

Neurophilosophical discussion 2

Is the anatomo-processual inference empirically plausible?

Is the anatomo-processual inference empirically plausible or not? If not, what kind of empirical findings are needed to make the anatomo-processual inference empirically plausible? We may want to shed a brief light on the kinds of data that would be needed to infer from the anatomical regions to the process of simulation. First, the data should tell us about a specific process that may underlie observation; this criterion may be called the 'revelation of process'. And second, the findings should demonstrate the involvement of that process in both observation and execution.

What do the current data tell us? The current empirical findings demonstrate that observation of others' actions, sensations or emotions induces neural activity changes in the same regions that are activated during the execution of the same action, sensation or emotion by the observing person. More specifically, the data tell us something about the sensitivity of the regions in question, namely that they are sensitive to various observational variables like the intention of movements, emotional feelings, etc.

In contrast, the data themselves do not reveal any kind of process that may possibly underlie the involvement of these regions during observation. The identification of a specific process is not contained in the data themselves. Instead, the specific process is inferred by the investigator on the basis of his/her assumption that the specific process must underlie the observed data because otherwise these could not be generated. He/she thus assumes that the same data could not have been yielded in the absence of that process. The first criterion is thus based on the assumption and interpretation of the data, rather than on the data themselves.

In addition to lacking the process, the data also do not tell us anything about the relationship of the assumed process to both observation and execution. This puts the second criterion, the involvement of that process in both execution and observation, into doubt.

On the whole, both the specific process and its involvement in both execution and observation are based on the assumption and interpretation of the data, rather than being contained in the data themselves. Despite the seemingly compelling evidence, the above-mentioned criteria for a reliable and valid inference might nevertheless not be met. Therefore we currently cannot exclude processes other than simulation, nor different processes underlying observation and execution.

One could, for instance, imagine processes other than simulation and/or completely independent processes as potentially underlying execution and observation. The same regions or even one single neuron (at the cellular level) may mediate different processes that might remain independent of each other; though recruiting the same regions or same neuron observation and execution should then be regarded as different functions that can no longer be traced back to the same underlying process.

Neurophilosophical discussion 3

Lack of necessity in the inference from anatomy to processes

Where is the problem in our empirical data that prevents us from inferring directly from anatomy to processes? The problem is that nothing in the current data tells us anything about the relationship between particular anatomical structures and specific processes. All we know is that both observation and execution induce neural activity changes in the same regions; this, however, does not imply any information about the processes underlying observation and their relationship to those that mediate execution.

Hence, to make inferences from the anatomical structures to specific processes is to go beyond the current data. This implies that, at present, we apparently remain unable to make empirically plausible inferences about the process that may possibly underlie execution and observation from our data to simulation.

What would we need to do to make the anatomo-processual inference more empirically plausible? We need experimental designs that interfere directly with the assumed process itself, rather than targeting only observation and execution as the outcomes or results of that very same process. This means that we need strategies to experimentally manipulate the process of simulation. Then we could observe how that affects and impacts the neural activity in those common regions that are supposed to underlie both observation and execution.

This reverses the experimental design. Rather than observation and execution themselves, the process itself is then the independent variable – the variable that is manipulated experimentally. Observation, execution, and especially the neural activity in the respective regions, are then the dependent variable – the variable that is supposed to be impacted by the manipulation of the supposedly underlying process.

Neurophilosophical discussion 4

Different concepts of simulation

How are the empirical findings and the concept of simulation related to the philosophical issue of intersubjectivity? One may assume intuitively that empathy and simulation imply intersubjectivity. In that case, one would need to show the relationship between the empirical findings related to simulation and the philosophical definitions of intersubjectivity. Vittorio Gallese also took on this project, and in addition to his empirical-experimental work, also has a strong theoretical-philosophical and specifically phenomenological interest.

Gallese (2003, 2006, 2007) distinguishes between 'standard simulation theory' and 'embodied simulation'. 'Standard simulation theory' is characterized by explicit cognitive processing where the observer adopts the other's perspective, generates the hypothesized or pretended mental state, and infers from that to the other's mental state. Following Gallese, these cognitive processes of hypothesizing the other's state, generating one's own state and inferring by matching, can account for what is commonly called simulation, or 'standard stimulation'.

This explicit cognitive processing is called into question by Gallese, though, and his concept of embodied simulation. 'Embodied simulation' describes a mandatory, nonconscious and pre-reflexive mechanism to reveal other people's mental states that is not the result of conscious, deliberate and cognitive efforts. The mere perception of others' emotions, actions or sensations automatically activates the very same neuronal mechanisms in the observer that also underlie the execution of the same behavior by the observer him/herself.

Here, the emphasis is on 'automatic'. This means that the observer cannot avoid the instantiation and recruitment of the activation in the respective brain regions, and ultimately the sharing of emotional feelings with the other. Let us detail that in the following.

For instance, the observation of action in another person automatically activates those regions that are responsible for executing and performing the same action by the observer him/herself. The case is similar with sensations and emotional feelings. This is described as 'automatic sharing'. By internally simulating the other person's behavior in an automatic and non-inferential way, one brings one's own body into the same (physiological and phenomenal) state as the one that is observed. Because of the central role of the body in this simulation, Gallese speaks of 'embodied simulation'.

In addition to the body, embodied simulation can be characterized by the involvement of subjective experience. By simulating the other's state automatically in one's own body, the corresponding subjective experience is induced.

Let me give an example. When observing another person's facial expression of, for instance, disgust, the regions underlying the generation of the emotional feeling of disgust, the insula and the anterior cingulate cortex, are activated in your brain. This in turn induces the feeling of disgust in you as the observing person. This makes it possible for you to understand and access the other person's emotions, to which you might react by instantiating another emotion or simulating and re-enacting the same emotion. Not only do you know what the other person is feeling, but you understand and may even experience the very same emotional feeling.

Neurophilosophical discussion 5

Embodied simulation and intersubjectivity

According to his account of embodied simulation, Gallese (2003, p. 171) distinguishes between two types of personal identity, i-identity and s-identity. The concept of i-identity describes the individual solipsistic identity where the self is considered independently of other selves in an isolated way. This contrasts with the s-identity, the intersubjective identity, which allows one to identify one's self as part of a larger community of other selves presupposing an intersubjective embedding of one's self. Rather than focusing on the i-identity, as is so often the case in current cognitive neuroscience, Gallese is much more interested in the s-identity.

How does the concept of embodied simulation relate to that of the s-identity as the intersubjective embedding of the self? The embodied simulation provides a shared or similar subjective experience between different persons. Such shared or similar experience provides us with a certain tacit or implicit knowledge about the other person. Gallese calls this knowledge 'implicit certainties', which he regards as constitutive of the s-identity and its intersubjective embedding.

These implicit certainties concern action, feelings and sensations. In the case of empathy, for instance, these implicit certainties consist in the fact that we do not only grasp and feel another person's emotional feelings, but also, at the same time, experience the other person as self – as a human being and co-species inhabiting a common world (see Gallese, 2003, pp. 175–6; Thompson, 2001, p. 19).

Self and other are then no longer regarded as opposites but as extensions of the same underlying 'correlative and reversible system self/other'; the observing self and the observed self are then 'part of a dynamic system governed by reversibility rules' (Gallese, 2003, p. 176). The 'correlative and reversible system self/other' allows us to constitute what Gallese calls the 'shared manifold of intersubjectivity'.

What exactly does Gallese mean by the concept of 'shared manifold of intersubjectivity'? The 'shared manifold' between self and other provides the basis for acquiring some knowledge about the other selves. The shared manifold thus allows for the creation of models of the self–other relationship, and to detect some kind of coherence, regularity, and predictability between different selves. This makes it possible for us to constitute our s-identity as the fundamental basis of social cognition, which may be essential in creating social communities and co-species bonds.

In this sense, the shared manifold is instantiated at the neuronal level in the brain and the body with, for instance, mirror neurons enabling 'supramodal intentional, emotional and sensitive shared spaces'. The shared manifold in this threefold sense allows us to constitute a 'we-centric space' with self–other identity or s-identity, which first and foremost makes social cognition and subsequent self–other distinction, like i-identity, possible.

Neurophilosophical discussion 6

Relationship between intra- and intersubjectivity

What does this entail for the philosophical concept of intersubjectivity and its relationship to intrasubjectivity? Remember that philosophy has long presupposed 'inference by analogy' to access and gain knowledge other people's mental states. 'Inference by analogy' describes how we can access other people's mental states only indirectly via our own mental states. This means that intersubjectivity necessarily presupposes intrasubjectivity. Intrasubjectivity is thus primary, while intersubjectivity builds upon it and must be regarded as secondary (see the left part of Figure 20.1).

If, in contrast, Gallese's assumption of a 'shared manifold' holds true and agrees with the neuronal workings of our brain, the concept of 'inference by analogy' may no longer be considered empirically plausible. This is so because Gallese's concept of the 'shared manifold' bases intersubjectivity on what he describes as 'embodied simulation'. It also occurs prior to the constitution of any distinction between self and other.

This determination of the concept of intersubjectivity differs from its traditional philosophical definition, where it is based on intrasubjectivity and the difference between the self and the other. Gallese thus reverses the relationship between intrasubjectivity and intersubjectivity: philosophy considered intersubjectivity an outcome of a primary and basic intrasubjectivity, whereas he assumes intersubjectivity to be more basic and fundamental than intrasubjectivity (see the right part of Figure 20.1).

This also reverses the sequence of philosophical problems. Usually the problem of intrasubjectivity and its question of the self is considered to be the primary and basic problem. Once the question of the self is solved, we can discuss how the self is related to the other. Hence the question of intersubjectivity comes second and builds upon the more basic and primary issue of intrasubjectivity. The solution to the question of intersubjectivity thus depends on the solution to the question of intrasubjectivity.

530 Neurophilosophy of Self

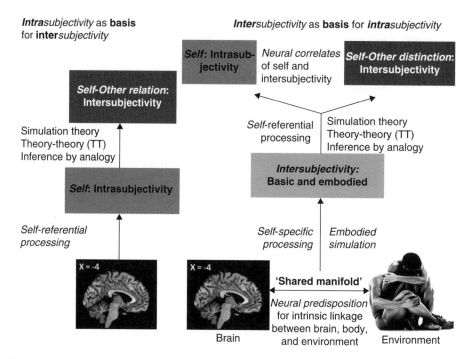

Figure 20.1 Models of the relationship between intra- and intersubjectivity
The figure schematically illustrates two different models of the relationship between self and other.

Left part: In this case the self and thus intrasubjectivity including self-referential processing (on the empirical side) provide the basis for any intersubjectivity as relation between self and other. The self is supposed to be based on the brain (lower and middle part). Specific mechanisms like simulation, theory-theory and inference by analogy (right middle upper) then provide the step from the self to the other by establishing self–other relations (upper part).

Right part: In this case a basic intersubjectivity is assumed, which is supposed to stem from the involvement of the body and its interface to others, as indicated in the picture on the lower right. This intrinsic linkage between brain and body is assumed to be due to a neural predisposition by the brain that provides a specific yet unknown mechanism by means of which brain and body can be linked intrinsically (lowest part). Such intrinsic linkage between brain and body allows for embodied simulation, the simulation of the other's body including its sensorimotor, affective and cognitive states (see lower middle part). That leads to a primary, basic and embodied intersubjectivity, as indicated in the middle part. From there on, self and other are distinguished (upper part) by cognitive functions as debated in simulation theory, theory-theory (TT) and inference by analogy (upper right).

The introduction of the 'shared manifold' reverses the sequence of philosophical problems. The problem of intersubjectivity is the most basic problem. It needs to be solved first before addressing the problem of intrasubjectivity.

What does this reversal of sequence imply for 'inference by analogy'? Inference by analogy presupposes a distinction between the self and the other. First, I have to access and know my own mental states. Then I can infer from my mental states to those of another person. Here, intersubjectivity presupposes intrasubjectivity.

Let me explain this relationship in more detail. The inference by analogy by means of which intrasubjectivity is able to account for intersubjectivity presupposes a more basic form of intersubjectivity. Gallese calls this the s-identity. It is the s-identity that provides the basis for accessing our own self and its mental states as distinct from other selves. This means that the s-identity can be considered a necessary condition or predisposition for the possibility of inferring others' mental states as described in inference by analogy.

Accordingly, inference by analogy may then be considered necessary to establish a cognitive relationship between self and other. Analogously, one may then also establish a sensorimotor or affective relationship with the other. All three – cognitive, sensorimotor and affective relationships – are then based on and predisposed by the 'embodied simulation' and the 'shared manifold'.

Critical reflection 1

Neural predisposition of intersubjectivity

How is intersubjectivity constituted in this basic and foundational sense? We saw that it is manifest in all basic functions, including sensory, motor and emotional functions, as well as cognitive functions. The constitution of intersubjectivity may thus be predisposed by the structure and organization of the brain. In short, it might be a 'more general and basic endowment of our brain' (Gallese, 2003, p. 41). If so, there must be some kind of neural predisposition ingrained in the brain itself that provides the necessary conditions for the possible constitution of intersubjectivity.

What are the neuronal mechanisms that provide the neural predispositions of intersubjectivity? The concepts of embodied simulation and shared manifold rest on two main assumptions – the involvement of sensorimotor functions and subjective experience or consciousness. Thus sensorimotor functions and subjective experience, along with their underlying neuronal mechanisms, might be the sufficient conditions of the shared manifold (and, thus, the basic form of intersubjectivity).

What makes it possible for sensorimotor functions to realize and implement the shared manifold? And why does the shared manifold seem unavoidably (and thus necessarily) to be accompanied by subjective experience and consciousness? The association of subjective experience with the shared manifold suggests that the neural predispositions of intersubjectivity might overlap with the neural predispositions of consciousness. This overlap between the neural predispositions of both intersubjectivity and consciousness may then make the association of the shared manifold with subjective experience possible.

The question of the neural predispositions of intersubjectivity thus leads us back to the neural predispositions of consciousness. Recall that in Part IV we assumed the brain's intrinsic activity to predispose the possible occurrence of consciousness in its still unclear spatiotemporal structure and organization (see Chapter 16).

How is this spatiotemporal structure and organization of the brain's intrinsic activity related to intersubjectivity? One would assume that the brain's intrinsic activity and its particular spatiotemporal structure must be related to the environment (though it is still unclear how). There must be some kind of invisible relationship between the environment and the brain's intrinsic activity such that the latter's spatiotemporal structure and organization spans across the boundaries of the brain to the environment in a virtual way. The exact neuronal mechanisms of such a relationship are unclear at this time.

Critical reflection 2

Attuning between brain and environment

How can we further illustrate the need for a more basic linkage between the brain and the environment? This is the point where the sensorimotor functions may come in. The possible constitution of a shared manifold by sensorimotor functions might presuppose a prior and more basic relationship between the brain and the environment. This basic relationship between the brain and the environment might enable the sensorimotor functions to become attuned to the world.

What exactly is meant by attuning the sensorimotor functions to the world? There is much more to movement than the simple movement of your leg. The simple movement of your leg is part of an overall action and has certain direction and goal-orientation. For example, the movement of your leg might be coordinated with the movement of your other leg. Both separate movements are part of an overall movement to take a step. And this step might be part of an overall action directed toward picking up your dinner from a local restaurant. Thus the movement of my right leg cannot be considered as being isolated from any of the other movements that are part of this goal-orientation. To do so would be to dissociate and detach the movement from its context. What 'more' is there to the movement? It is physical, of course. But at the same time the movement is also attuned to the rest of the body and to the world. To speak metaphorically, the movement is part of a more basic coordinated system spanning in a virtual way the organism, its brain and body, and the rest of the world. As such, a single movement must somehow already be attuned to the world. Vittorio Gallese, the Italian researcher responsible for developing the research on mirror neurons, would probably say that a single movement is always already part of the 'shared manifold'. In short, a single movement is part of the basic, prior and foundational intersubjectivity spanning the brain, the body and the world.

In the same way that our actions are not simply a collection of isolated movements, our perceptions are not just a collection of single isolated sensations. Instead, our perceptions can be characterized by 'internal connections' between the different sensations and their respective stimuli. And, most important, those connections are tailored and attuned to the world somehow and let us perceive meaningful contents, be they objects, people or events.

Taken together, both sensory and motor functions cannot be considered in complete isolation. Instead, they seem to be always already attuned to the world. This presupposes some kind of basic, foundational and prior relationship between the brain and the environment. This basic relationship and its underlying neuronal mechanisms may consequently provide the neural predisposition of the shared manifold and thus of intersubjectivity and ultimately even consciousness. It is left to future research to demonstrate this.

Take-home message

The previous chapter focused on the self, its relationship to the brain, to consciousness and its abnormal changes in psychiatric disorders like schizophrenia and depression. This neglected one central hallmark feature of the self, however, namely that it is able to relate to other selves. We are able to share others' emotional feelings and to decipher others' thoughts and mental states. How is this possible? In philosophy, this is addressed through the concept of intersubjectivity, while psychology has suggested different mechanisms like simulation-theory (ST) and theory-theory (TT). In neuroscience, specific neurons in the brain that are recruited during observation of others' actions and emotions have been observed, called mirror neurons. Based on the mirror neurons, the possibility of intersubjectivity thus seems to be deeply ingrained and encoded in the functioning of the brain. This raises not only empirical questions for the exact neuronal mechanisms that underlie intersubjectivity, but also shows philosophical issues like the relationship between intra- and intersubjectivity in a novel light.

Summary

The previous chapters focused on the self as the subject of experience. This demonstrated specific networks and regions that seem to be essential in processing self-reference. In addition to the subjective experience of one's own self, we can also relate to another person and his/her mental states. How is this self–other relationship, and thus intersubjectivity, possible? This is the topic of the present chapter.

Philosophy's concept of 'inference by analogy' describes our ability to infer other's mental states from our own. This has been associated with cognitive mechanisms. Phenomenological philosophy, in contrast, shifts the focus toward pre-cognitive mechanisms in establishing the self–other relationship. Here intersubjectivity is based on sensorimotor and affective functions. This leads to the assumption of a more basic, primary and foundational intersubjectivity. What kinds of mechanisms underlie this type of intersubjectivity?

Psychology has discussed two main approaches for the possible mechanisms underlying the self–other relationship. Theory-theory (TT) assumes the constitution of a theory about the mind in general. This theory then serves as the basis for inferring and relating to others' mental states. Simulation-theory (ST) suggests the simulation of others' mental states on the basis of our own mental states. Here, the construction of a theory about the mind is not important. Both TT and ST are well reflected in current neuroscience. TT-based imaging investigates different subprocesses like representing, attributing, and applying, and their respective neuronal mechanisms. ST is more related to the detection of mirror neurons that have been observed in both humans and monkeys. These mirror neurons become active when the respective person or animal observes the same action or feeling in another person or animal. What exactly do these mirror neurons do? Many researchers today assume that they simulate the other's actions or feelings in the brain.

The empirical plausibility of this anatomo-processual inference is questioned and discussed critically. Finally, the neurophilosophical implications of the empirical findings for the conceptual distinction between intra- and intersubjectivity are discussed and put into a larger perspective. The chapter concludes with a discussion about the neural predispositions of a more basic, primary and foundational form of intersubjectivity. This may be related to the brain's intrinsic activity and its still unclear link to the environment.

Revision notes

- Explain inference by analogy.
- What are the different meanings of the concept of intersubjectivity in philosophy?
- How must the concept of empathy be distinguished from other related concepts like sympathy, emotional contagion, and theory of mind?
- Explain the difference between theory-theory and simulation-theory, including their respective features.
- What kinds of neurons and which regions in the brain are active when observing another person's actions and movements?
- Are movements' physical features encoded in the neural activity of neurons and regions? If not, what is encoded in these neurons' neural activity during observation?
- What are the different meanings of the concept of simulation?
- How are the different concepts of simulation related to the concept of intersubjectivity?
- What is the anatomo-processual inference? Why is it relevant in the present context?
- What is the difference between intra- and intersubjectivity?
- Explain the requirements for the neural predispositions of intersubjectivity.

Suggested further reading

- Abu-Akel, A. and Shamay-Tsoory, S. (2011) 'Neuroanatomical and neurochemical basis of theory of mind', *Neuropsychologia*, 49, 2971–84.
- Carruthers, P. (2006) *The Architecture of the Mind: Massive Modularity and the Flexibility of Thought* (Oxford/New York: Oxford University Press).
- Gallese, V. (2003) 'The roots of empathy: The "shared manifold" hypothesis and the neural basis of intersubjectivity', *Psychopathology*, July/August, 36 (4), 171–80.
- Gallese, V. (2007) 'Embodied simulation: From mirror neurons to interpersonal relations', *Novartis Foundation Symposium*, 278, 3–12 (discussion: 12–19, 89–96).
- Goldman, A. (2006) *Simulating Minds: The Philosophy, Psychology, and Neuroscience of Mindreading* (Oxford/New York: Oxford University Press).

Epilogue: Is the Brain a Door Opener?

Philosophy is about the world. Metaphysics raises the question of what exists and is real in the world (that is, in all logically possible worlds), while epistemology focuses on how we can know about those worlds. This is already difficult in itself, as thousands of years of philosophizing since the ancient Greeks have shown. Even more puzzling is the fact that we, as humans, are part of that very same world whose existence and reality we aim to capture in philosophy. How can we describe our own existence and reality as humans within the world we live in? We have a body that includes a brain. Both body and brain can be described in purely physical terms, similar to the rest of the world and its various objects. This is the easy part.

What about the hard part? The hard part is to account for our various mental features like self, consciousness, emotions, free will, etc. Why and how can these mental features come into existence and reality in a world that seems to be purely physical? The answer that Western philosophers, especially in early modernity in the fifteenth and sixteenth centuries, gave is simple and clear: we must assume a mind with mental features, a separately existing entity that is as real as the physical features in the world.

However, the assumption of a mind conflicts with science. Science relies on observation of the world. We can observe physical features like force, gravitation, atoms, quanta and many others. None of these physical features as they are observed in science include the kind of mental features that characterize humans, though. There is no self or consciousness in force, gravitation or other physical measures. Nor can we observe these mental features in the molecules of chemistry or the genes at issue in biology. Nothing remotely resembling mental features similar to the idea of mind can be observed in the world of science. Scientists consequently say to philosophers: 'All that exists and is real in the world are physical, chemical and biological features. Mental features, let alone a mind, simply do not exist.'

'No mind? No mental features? Objection, my dear fellow of science,' the philosopher will almost certainly disagree! The scientist confuses empirical and metaphysical matters: science is about the empirical world; that is, the things we can observe. In contrast, philosophy is about the metaphysical world; that is, what exists and is real in any possible world. Accordingly, even if something cannot be observed empirically by scientific means, it may nevertheless be real in metaphysical terms. The rejection of the mind and its mental features in science consequently rests on a confusion: namely, on conflating empirical and metaphysical issues. Hence, philosophy is not science, and science is not philosophy. To conflate both is to compare apples and oranges.

Now it's the scientist's turn again: 'Then, dear philosopher, you have to explain how the existence and reality of the mind and its mental features is related to the body and brain, including their physical features.' By throwing the ball back into the metaphysical territory, the scientist raises the question of what is described as the *mind–brain problem* in philosophy. The mind–brain problem is a metaphysical problem that asks for the existence and reality of the mind, and how this mind is related to the rest of the physical world, including brain and body.

How does the mind–brain problem fare in our times? The mind–brain problem is ultimately about ourselves as humans and our being; that is, existence and reality, in the world. If the mind really exists in the world, we as humans must be special compared to the rest of the merely physical world (unless one assumes the world itself to be mental or psychical rather than physical, as is assumed in panpsychism). If, in contrast, there is no mind, humans are not special at all and ultimately nothing but a chunk of mere – though probably rather complex – physical matter. Hence, there is much at stake for us as humans in the mind–brain problem, a continuously lingering problem that still awaits its solution at the beginning of the twenty-first century.

This is the background against which neuroscience, the science of the brain, has developed in the past 20–30 years. By exploring the brain and its various neural mechanisms, neuroscience ventures more and more into the realm of mental features like self, consciousness, emotions, and free will. Specific neural mechanisms are detected and postulated to underlie particular mental features – this is the search for the neural correlates of mental features as exemplified by the concept of the neural correlates of consciousness (NCC).

Does the progress in neuroscience allow us finally to replace the search for the mind's metaphysical existence and reality in philosophy with the empirical investigation of the neural correlates of mental features in neuroscience? Like its predecessors in physics, chemistry, and biology, neuroscientists enthusiastically reject the assumption of the mind's existence and reality: as in physical, chemical and biological measures of the world, no mental features can be observed in the neural features of the brain. There is thus no mind in the world as distinguished from brain and body as part of the merely physical world.

The neuroscientist's claim can be made even stronger. The existence and reality of a mind in the world, and consequently the mind–brain problem, are nothing but illusions. The search for the mind's metaphysical existence and reality would better be dominated by and validated against the empirical observation and investigation of the brain and its neural features. Mind is here reduced to and eliminated in favor of the brain, while, at the same time, philosophy is dominated by neuroscience. This is called neurophilosophy these days; that is, a brain-reductive and reductive (in disciplinary terms) neurophilosophy.

Such reductive neurophilosophy considers neuroscience rather than philosophy as the new 'Leitwissenschaft'. The one-sided domination of philosophy by neuroscience as 'Leitwissenschaft' entails a shift in the kinds of questions raised. Philosophy asks for the 'why' of existence, reality, and norms (including their respective features, e.g., the 'what'…): Why does something exist? Why do we know something? Why are there ethical norms? etc. Neuroscience, in contrast, focuses more on the 'how' as related to mechanisms and functions: How does the brain's neural activity generate sensory and motor functions? How do the brain's neural mechanisms give rise to consciousness and other mental features? By accepting neuroscience as its Leitwissenschaft and role model, reductive neurophilosophy shifts from 'why' to 'how questions': instead of focusing exclusively on the 'why' as isolated from the 'how' as philosophy does, reductive neurophilosophy only considers the 'how' as detached from the 'why'. How does the philosopher stand in relation to such reductive neurophilosophy? Traditional philosophy is about the mind in the same way that neuroscience is about the brain. Mind and brain are not only different but also coupled to different methods of investigation, primarily *rational-argumentative* in philosophy and predominantly *observational-experimental* in neuroscience. To eliminate mind in favor of brain is hence to apply the wrong method (observation and experiments) to something (the mind) that requires a different method (rational argument).

Accordingly, reductive neurophilosophy confuses not only metaphysical and empirical matters but also different and mutually exclusive methods of investigation. Neuroscience, being confined to empirical matters and observation, will not in principle be able to solve the metaphysical puzzle of the mind–brain problem; this is the methodological territory of philosophy. Put differently, the philosopher may say that the reductive neurophilosopher confuses 'how questions' and 'why questions'.

How can we solve this deadlock? Let us introduce a metaphor of different rooms with separate doors. Metaphorically, the proponents of philosophy enter the door on the left and claim to have arrived in the metaphysical room: there is a big table in the middle called mind, while its surrounding chairs and the rest of the furniture correspond to mental features that are designed in orientation on the rational-argumentative method. In contrast, the proponents of neuroscience and their reductive neurophilosophical allies argue that we need to take the door on the right, which leads to an empirical room: the table in the middle of the room is here occupied by a big brain surrounded by various kinds of neural furniture that are designed to work well with the observational-experimental method.

As in a real house, the occupants of both right and left rooms know of each other's existence. That makes matters worse, though. The reductive neurophilosophical ally of neuroscience claiming the whole house for himself and believing that he/she has entered the correct room, the room of neuroscience, claims the left room of the philosopher for himself: 'Tear down the walls, demolish the table of the mind and put a brain instead, and replace the rather antique and old-fashioned mental furniture by the modern, fancy neural designer furniture.'

That, however, is to reckon without one's host. In addition to the doors on left and right, there is also a door in the middle. What kind of room does the door in the middle open into? The neuroscientist may argue that it opens in to his room, the room on the right. What if, however, the middle door opens into a separate room? A room that has neither a mind nor a brain in the middle, a room where we can find furniture that is different and designed in a rather peculiar way, containing elements of both rational-argumentative and observational-experimental methods.

What does such a room look like? There is a brain but it is rather small and no longer in the middle of the room. Instead, there is a huge carpet that is also a world map embracing the rest of the room, including the brain and the antique and modern furniture. There are pictures of the table and the mind hanging on the wall. All these are scattered about the room with no apparent strategy, but still brought together by the common ground the carpet and its world map, i.e. the (natural) world. Finally, there are walls with doors to both left and right rooms, the rooms of philosophy and neuroscience.

This middle room is where the world, that is, the carpet, serves as a basis and center. The brain is here only a part of the world rather than being at the very center as in the right room, the room of neuroscience. The neurophilosophy of the middle room is consequently no longer brain-reductive but *brain-based* in the same way that it is based on the world, that is, *world-based*. Moreover, the placement of both antique and neural furniture suggests a combination of rational-argumentative and observational-experimental styles, i.e. methods. There is no longer any need to replace philosophy with neuroscience. Brain-based neurophilosophy is consequently non-reductive rather than reductive like its neuroscientific sibling.

Accordingly, the door to the middle room offers the entrance to a room, non-reductive neurophilosophy, that is different from both the left room of philosophy and the right room of neuroscience and its ally reductive neurophilosophy. How is the middle room different? Unlike the right and left rooms, the middle room, due to its position in the middle, cannot avoid combining 'why' and 'how' questions. This distinguishes the middle from both right and left rooms, where both kinds of questions are detached and isolated from each other.

The focus in this book was on this middle room; that is, on linking and combining 'why' and 'how' questions in a systematic way, which amounts to what I here described as non-reductive neurophilosophy. The aim in this book was to show how such a non-reductive neurophilosophical approach (that is, the middle room) can contribute to problems like the mind–brain problem (Part II), explanation (Part III), consciousness (Part IV), and self (Part V) that are usually dealt with in either the left or right room, philosophy or neuroscience.

We have to be careful, however. By considering the brain in the context of philosophy, I only opened the door to the middle room while not really entering the room itself, leaving its furniture, its designs and style open and undetermined. For instance, I did not discuss how the brain contributes to the world to make possible mental features while at the same time being part of that very same world and its physical features. Metaphorically speaking, I have left the exact placement of the brain in relation to the carpet, the world, open. One may thus want to speak of a world–brain problem here.

Philosophy is about the world. This was the original hallmark feature of the left room, the room of philosophy, before it placed the mind as a table in its center. The right room of neuroscience, in contrast, is about the brain rather than either mind or world. Due to its position in the middle, the middle room is able to link both world and brain and thus to tackle the world–brain problem: what and how can the brain contribute to the world such that what philosophers describe as mental features can occur in the world?

The focus shifts here from the mind–brain problem in the left room to the world–brain problem in the middle room. This is made possible by shifting from the brain and the problem of its neural mechanisms, as dealt with in the right room of neuroscience, to the world itself including its relationship to the brain in the middle room. Now it is up to you, as students and teachers, to walk through the door opened by the brain and to explore the novel room, the middle room of non-reductive neurophilosophy, and the various problems and questions it harbors. This, I hope, will enable you to develop and establish truly neurophilosophical furniture as distinct from both the philosophical furniture in the room to your left and the neuroscientific furniture in the room to your right.

References

Alcaro, A., Panksepp, J., Witzak, J. and Northoff, G. (2010) 'Is subcortical-cortical midline activity in depression mediated by GABA-A receptors and glutamate? A cross-species translational approach', *Neuroscience & Biobehavioral Reviews*, 34(4), 592–605.

Baars, B. J. (2005) 'Global workspace theory of consciousness: Toward a cognitive neuroscience of human experience', *Progress in Brain Research*, 150, 45–53. doi:10.1016/S0079-6123(05)50004-9.

Baars, B. J. (2007) 'The global workspace theory of consciousness', in M. Velmans and S. Schneider (eds), *The Blackwell Companion to Consciousness* (Oxford: Blackwell).

Baars, B. J., Ramsoy, T. Z. and Laureys, S. (2003) 'Brain, conscious experience and the observing self', *Trends in Neurosciences*, 26(12), 671–5.

Barnes, J. (ed.) (1984) *The Complete Works of Aristotle*, Volumes I and II (Princeton: Princeton University Press).

Bechtel, W., Mandik, P. and Mundale, J. (2001a). 'Philosophy meets the neurosciences', in W. Bechtel, P. Mandik, J. Mundale et al. (eds), *Philosophy and the Neurosciences: A Reader* (Malden, MA: Blackwell). ISBN 9780631210450.

Bechtel, W., Mandik, P., Mundale, J. and Stufflebeam, R. (eds) (2001b) *Philosophy and the Neurosciences: A Reader* (New York: Basil Blackwell).

Bennett, M. and Hacker, P. (2003) *Philosophical Foundations of Neuroscience* (Oxford/New York: Blackwell).

Bennett, M., Dennett, D., Hacker, P. and Searle, J. (2007) *Neuroscience and Philosophy. Brain, Mind, and Language* (New York: Columbia University Press).

Berkeley, G. (1948–1957) *The Works of George Berkeley, Bishop of Cloyne*, A. A. Luce and T. E. Jessop (eds.) (London: Thomas Nelson), 9 vols.

Bermudez, J. L. (2005) *Philosophy of Psychology* (London: Routledge).

Berze, J. (1914) *Die primäre Insuffizienz der psychischen Aktivität ihr Wesen, ihre Erscheinungen und ihre Bedeutung als Grundstörung der Dementia praecox und der Hypophrenien überhaupt* (Leipzig: Deuticke).

Bickle, J., Mandik, P. and Landreth, A. (2006) *The Philosophy of Neuroscience.* Journal [serial on the Internet]. Available at: http://plato.stanford.edu/entries/neuroscience/.

Bickle, J., Mandik, P. and Landreth, A. (2010) 'The philosophy of neuroscience', *Stanford Encyclopedia of Philosophy* (Summer 2012 edition), Edward N. Zalta (ed.). Available at: http://plato.stanford.edu/archives/sum2012/entries/neuroscience/. The citation above refers to the version in the following archive edition: Summer 2012 (minor correction).

Bleuler, E. (1911) *Dementia praecox, or: The Group of Schizophrenias* (New York: International Universities Press).

Bleuler, E. (1916). *Lehrbuch der Psychiatrie* (11th edn) (Berlin/Heidelberg/New York: Springer).

Block, N. (1980) 'Are absent qualia impossible?', *The Philosophical Review*, 89, 257–82.

Block, N. (1986) 'Advertisement for a semantics for psychology', in P. French, T. Euhling and H. Wettstein (eds), *Studies in the Philosophy of Mind. Midwest Studies in Philosophy* (Minneapolis, MN: University of Minneapolis Press).

Block, N. (1990) 'Inverted Earth', in J. Tomberlin (ed.) *Action Theory and the Philosophical Perspectives*, 4, 53–79.

Block, N. (1995) 'How can we find the neural correlate of consciousness?', *Trends in Neuroscience*, 19(11), 456–9.

Block, N. (2007) 'Consciousness, accessibility, and the mesh between psychology and neuroscience', *Behavioral and Brain Sciences*, December, 30(5–6), 481–99; discussion 499–548. doi: 10.1017/S0140525X07002786.

Block, N. and Fodor, J. A. (1972) 'What psychological states are not?', *Philosophical Review*, 81(2), 159–81.

Botteril, G. and Carruthers, P. (1999) *The Philosophy of Psychology* (Cambridge: Cambridge University Press).
Breidbach, O. (1997) *Die Materialisierung des Ichs – Eine Geschichte der Hirnforschung im 19. und 20. Jahrhundert* (Frankfurt am Main: Suhrkamp).
Brook, A. (1994) *Kant and the Mind* (Cambridge/New York: Cambridge University Press).
Buszáki, G. (2006) *Rhythms of the Brain* (Oxford/New York: Oxford University Press).
Carruthers, P. (2005) *Consciousness: Essays from a Higher-Order Perspective* (Oxford: Oxford University Press). ISBN 0-19-927736-2.
Chalmers, D. (1996) *The Conscious Mind* (Oxford/New York: Oxford University Press).
Chalmers, D. (2010) 'The singularity: A philosophical analysis', *Journal of Consciousness Studies*, 34, 34–46.
Chomsky, N. (1959) 'A review of B. F. Skinner's verbal behavior', *Language*, 35(1) 26–58.
Churchland, P. S. (1986) *Neurophilosophy. Towards a Unified Mind–Brain Science* (Cambridge, MA: MIT Press).
Churchland, P. M. (2002) 'Self-representation in nervous systems', *Science*, 12(296), 308–10.
Clark, A. (1997) *Being There* (Cambridge, MA: MIT Press).
Clark, A. (2000) *Mindware: An Introduction to the Philosophy of Cognitive Science* (New York: Oxford University Press). ISBN 978-0-19-513857-3.
Cooper, J. M. (ed.) (1997) *Plato: Complete Works* (Indianapolis, IN: Hackett).
Cowan, W. M., Harter, D. H. and Kandel, E. R. (2000) 'The emergence of modern neuroscience: Some implications for neurology and psychiatry', *Annual Review of Neuroscience*, 23, 343–91. Review.
Crick, F. (1994) *The Astonishing Hypothesis: The Scientific Search for the Soul* (New York: Simon & Schuster).
Crick, F. and Koch, C. (2003) 'A framework for consciousness', *Nature Neuroscience*, February, 6(2), 119–26. Review.
Damasio, A. R. (1999) 'How the brain creates the mind', *Scientific American*, 281(6), 112–17.
Damasio, A. R. (2010) *The Self Comes to Mind* (New York: Viching).
D'Argembeau, A., Collette, F., Van der Linden, M., Laureys, S., Del Fiore, G., Degueldre, C., Salmon, E. (2005) 'Self-referential reflective activity and its relationship with rest: A PET study', *NeuroImage*, 25(2), 616–24. doi:10.1016/j.neuroimage.2004.11.048.
Davidson, D. (1980) *Essays on Actions and Events* (Oxford: Clarendon Press).
de Graaf, T. A., Hsieh, P. J. and Sack, A. T. (2011) 'The "correlates" in neural correlates of consciousness', *Neuroscience and Biobehavioral Reviews*, 36(1), 191–7. doi:10.1016/j.neubiorev.2011.05.012.
Dehaene, S. and Changeux, J. P. (2005) 'Ongoing spontaneous activity controls access to consciousness: A neuronal model for inattentional blindness', *PLOS Biol*, 3(5), e141.
Dehaene, S. and Changeux, J. P. (2011) 'Experimental and theoretical approaches to conscious processing', *Neuron*, 70(2), 200–27. doi:10.1016/j.neuron.2011.03.018.
Dehaene, S., Changeux, J. P., Naccache, L., Sackur, J. and Sergent, C. (2006) 'Conscious, preconscious, and subliminal processing: A testable taxonomy', *Trends in Cognitive Sciences*, 10(5), 204–11. doi:10.1016/j.tics.2006.03.007.
Dennett, D. C. (1987) *The Intentional Stance* (Cambridge, MA: MIT Press).
Dennett, D. C. (1988) 'Quining qualia', in A. Marcel and E. Bisiach (eds), *Consciousness in Modern Science* (Oxford: Oxford University Press).
Dennett, D. C. (1993) *Consciousness Explained* (Boston, MA: Back Bay Books).
Edelman, G. M. (1993) 'Neural Darwinism: Selection and reentrant signaling in higher brain function', *Neuron*, 10(2), 115–25.
Edelman, G. M. (2003) 'Naturalizing consciousness: A theoretical framework', *Proceedings of the National Academy of Sciences of the United States of America*, 29 April, 100(9), 5520–4.
Edelman, G. M. (2005) *Wider than the Sky. The Phenomenal Gift of Consciousness* (New Haven, CT: Yale University Press).
Edelman, G. M. and Tononi, G. (2000) *A Universe of Consciousness: How Matter becomes Imagination* (New York: Basic Books).
Fodor, J. A. (1983) *The Modularity of Mind. An Essay on Faculty Psychology* (Cambridge, MA/London: MIT Press).

Freud, S. (1895) *Project for a Scientific Psychology*, Standard edition, 1 (London: Hogarth Press).
Gallagher, I. I. (2000) 'Philosophical conceptions of the self: Implications for cognitive science', *Trends in Cognitive Sciences*, 4(1), 14–21.
Gallagher, S. (2005) *How the Body Shapes the Mind* (Oxford/New York: Oxford University Press).
Gallagher, S. (2009) 'Neurophenomenology', in T. Bayne, A. Cleeremans and P. Wilken (eds), *Oxford Companion to Consciousness* (Oxford: Oxford University Press), pp. 470–2.
Gallagher, S. and Zahavi, D. (2008) *The Phenomenological Mind* (London: Routledge), ch. 2.
Gallese, V. (2003) 'The roots of empathy: The "shared manifold" hypothesis and the neural basis of intersubjectivity', *Psychopathology*, July–August, 36(4), 41–80.
Gallese, V. (2006) 'Intentional attunement: A neurophysiological perspective on social cognition and its disruption in autism', *Brain Res*, 24 March, 1079(1), 15–24. Review.
Gallese, V. (2007) 'Embodied simulation: From mirror neurons to interpersonal relations', *Novartis Foundation Symposium*, 278, 3–12 (Discussion: 12–19, 89–96).
Gazzaniga, M. (2008) *Cognitive Neuroscience* (Cambridge, MA: MIT Press).
He, B. J. and Raichle, M. E. (2009) 'The fMRI signal, slow cortical potential and consciousness', *Trends in Cognitive Sciences*, 13(7), 302–9.
He, B. J., Snyder, A. Z., Zempel, J. M., Smyth, M. D. and Raichle, M. E. (2008) 'Electrophysiological correlates of the brain's intrinsic large-scale functional architecture', *Proceedings of the National Academy of Sciences USA*, 105(41), 16039–44. doi:10.1073/pnas.0807010105.
Hebb, D. (1949) *The Organization of Behaviour* (New York: John Wiley).
Holt, D. J., Cassidy, B. S., Andrews-Hanna, J. R., Lee, S. M., Coombs, G., Goff, D. C., Gabrieli, J. D. et al. (2011) 'An anterior-to-posterior shift in midline cortical activity in schizophrenia during self-reflection', *Biological Psychiatry*, 69(5), 415–23. doi:10.1016/j.biopsych.2010.10.003.
Hoptman, M. J., Zuo, X. N., Butler, P. D., Javitt, D. C., D'Angelo, D., Mauro, C. J. and Milham, M. P. (2010) 'Amplitude of low-frequency oscillations in schizophrenia: A resting state fMRI study', *Schizophrenia Research*, 117(1), 13–20. doi:10.1016/j.schres.2009.09.030.
Hornsby, J. (1997) *Simple Mindedness: A Defence of Native Naturalism in the Philosophy of Mind* (Cambridge, MA: Harvard University Press).
Huang, Z., Dai, R., Wu, X., Liu, D., HuJm Gao, L., Tag, W., Mao, Y., Jin, Y., Wu, X., Zhang, Y., Lu, L., Laureys, S., Weng, X. and Northoff, G. (2013) 'The self and its resting state in consciousness. An investigation of the vegetative state' *Human Brain Mapping*, in press.
Husserl, E. (1980) *Ideas Pertaining to a Pure Phenomenology and to a Phenomenological Philosophy – Third Book: Phenomenology and the Foundations of the Sciences*, trans. T. E. Klein and W. E. Pohl (Dordrecht: Kluwer).
Husserl, E. (1982 [1913]) *Ideas Pertaining to a Pure Phenomenology and to a Phenomenological Philosophy – First Book: General Introduction to a Pure Phenomenology*, trans. F. Kersten (The Hague: Nijhoff).
Husserl, E. (1989) *Ideas Pertaining to a Pure Phenomenology and to a Phenomenological Philosophy – Second Book: Studies in the Phenomenology of Constitution*, trans. R. Rojcewicz and A. Schuwer (Dordrecht: Kluwer).
Iacoboni, M., Molnar-Szakacs, I., Gallese, V., Buccino, G., Mazziotta, J. C. and Rizzolatti, G. (2005) 'Grasping the intentions of others with one's own mirror neuron system', *PLoS Biol.*, March, 3(3), e79. Epub 2005 Feb 22.
Ingram, R. E. (1990) 'Self-focused attention in clinical disorders: Review and a conceptual model', *Psychology Bulletin*, 107(2), 156–76.
Jackson, F. (1986) 'What Mary didn't know', *The Journal of Philosophy*, 83, 291–5.
Jackson, F. (1990) 'Epiphenomenal qualia', in W. G. Lycan (Ed.), *Mind and Cognition. A Reader* (Oxford: Basil Blackwell), pp. 519–47.
Jackson, F. (1993) 'Appendix A (For Philosophers)', *Philosophy and Phenomenological Research*, LIII(4), 899–903.
James, W. (1890) *The Principles of Psychology*, 2 vols, Dover Publications, 1950, Vol. 1: ISBN 0-486-20381-6, Vol. 2: ISBN 0-486-20382-4.
Jaspers, K. (1963) *General Psychopathology* (Chicago, IL: University of Chicago Press).

Kant, I. (1998) *The Critique of Pure Reason*, ed. by P. Guyer (Cambridge: Cambridge University Press).
Kaplan-Solms, K. and Solms, M. (2000) *Clinical Studies in Neuro-Psychoanalysis: Introduction to a Depth Neuropsychology* (London: Karnac Books).
Kim, J. (1979) 'Causality, identity, and supervenience in the mind–body problem', in *Midwest Studies in Philosophy*, Bd.IV: Studies in Metaphysics, 31–49.
Kim, J. (1982) 'Psychophysical supervenience', *Philosophical Review*, 41, 51–70.
Kim, J. (1984) 'Concepts of supervenience', *Philosophy and Phenomenological Research*, 45, 153–76.
Kim, J. (1985) 'Supervenience, determination, and reduction', *The Journal of Philosophy*, 82, 616–18.
Kim, J. (1993) 'Supervenience as a philosophical concept', in *Supervenience and Mind* (Cambridge, MA: MIT Press), pp. 131–60.
Kim, J. (2005) *Physicalism, or Something Near Enough* (Princeton, NJ: Princeton University Press).
Kitcher, P. (1992) 'The Naturalists Return', *Philosophical Review*, 101(1), 53–114.
Klein, S. B. (2012) 'Self, memory, and the self–reference effect: An examination of conceptual and methodological issues', *Personality and Social Psychology Review*, 16(3), 283–300. doi:10.1177/1088868311434214.
Klein, S. B. and Gangi, C. E. (2010) 'The multiplicity of self: Neuropsychological evidence and its implications for the self as a construct in psychological research', *Annals of the New York Academy of Sciences*, 1191, 1–15.
Knobe, J., Buckwalter, W., Nichols, S., Robbins, P., Sarkissian, H. and Summers, T. (2012) 'Experimental philosophy', *Annual Review of Psychology*, January, 10(63), 81–99. Epub 2011 Jul 29.
Koch, C. (2004) *The Quest for Consciousness: A Neurobiological Approach* (Greenwood Village, CO: Roberts & Company).
Kouider, S. and Dehaene, S. (2009) 'Subliminal number priming within and across the visual and auditory modalities', *Experimental Psychology*, 56(6), 418–33. doi:10.1027/1618-3169.56.6.418.
Kraepelin, E. (1913) *Ein Lehrbuch fur Studierende und Arzte* (Leipzig, Germany: Barth).
Kriegel, U. and Williord, K. (eds) (2006) *Self-Representational Approaches to Consciousness* (Cambridge, MA: MIT Press).
Kripke, S. (1972) *Naming and Necessity* (Cambridge, MA: Harvard University Press). ISBN 0-674-59845-8 and reprints 1972.
Kühn, S. and Gallinat, J. (2012) 'Resting-state brain activity in schizophrenia and major depression: A quantitative meta-analysis', *Schizophrenia Bulletin*. doi:10.1093/schbul/sbr151.
Lashley, K. S. (1949) 'Persistent problems in the evolution of mind', *Quarterly Review of Biology*, 24(1), 28–48.
Lashley, K. (1950) 'In search of the engram', *Society of Experimental Biology*, Symposium 4, 454–82.
Legrand, D. and Ruby, P. (2009) 'What is self-specific? Theoretical investigation and critical review of neuroimaging results', *Psychological Review*, 116(1), 252–82. doi:10.1037/a0014172.
Leibniz, W. (1998) *Philosophical Texts*, ed. and trans. by R. S. Woolhouse and Richard Francks (Oxford: Oxford University Press).
Lewis, C. M., Baldassarre, A., Committeri, G., Romani, G. L. and Corbetta, M. (2009) 'Learning sculpts the spontaneous activity of the resting human brain', *Proceedings of the National Academy of Sciences USA*, 106(41). doi:10.1073/pnas.0902455106.
Lipsman, N., Nakao, T., Kuhn, A. A., Bajbouj, M., Huebl, J., Merkl, A., Kanayama, N., Krauss, J. K., Anderson, A., Giacobbe, P., Hamani, C., Hutchison, W. D., Dostrovsky, J. O., Womelsdorf, T., Lozano, A. M. and Northoff, G. (2013) 'Neuronal gamma band synchronization within human subcallosal cingulate cortex is related to self-relevance', *Cortex*, in revision.
Llinás, R. (2002) *I of the Vortex: From Neurons to Self* (Cambridge, MA: MIT Press).
Lloyd, D. (2011) 'Mind as music', *Frontiers in Psychology*, 2(63).
Logothetis, N. K. (2008) 'What we can do and what we cannot do with fMRI', *Nature*, 12 June, 453(7197), 869–78. doi: 10.1038/nature06976. Review.

Logothetis, N. K., Murayama, Y., Augath, M., Steffen, T., Werner, J. and Oeltermann, A. (2009) 'How not to study spontaneous activity', *NeuroImage*, 45(4), 1080–9. doi:10.1016/j.neuro image.2009.01.010.
Lou, H. C., Gross, J., Biermann-Ruben, K., Kjaer, T. W. and Schnitzler, A. (2010) 'Coherence in consciousness: paralimbic gamma synchrony of self-reference links conscious experiences', *Human Brain Mapping*, 31(2), 185–92. doi:10.1002/hbm.20855.
Lurija, A. R. (ed.) (1973) *The Working Brain* (New York: Basic Books).
Maandag, N. J., Coman, D., Sanganahalli, B. G., Herman, P., Smith, A. J., Blumenfeld, H., Shulman, R. G. and Hyder, F. (2007) 'Energetics of neuronal signaling and fMRI activity', *Proceedings of the National Academy of Sciences USA*, December, 18, 104(51), 20546–51. Epub 2007 Dec 13.
MacDougall, D. (1907) 'Hypothesis concerning soul substance, together with experimental evidence of the existence of such substance', *Journal of the American Society for Psychical Research*, May.
Martin, C. (1988) *The Philosophy of Thomas Aquinas: Introductory Readings* (London: Routledge & Kegan Paul).
McDowell, J. (1994) *Mind and World* (Cambridge, MA: Harvard University Press).
McGinn, C. (1991) *The Problem of Consciousness* (London: Blackwell).
McGinn, C. (1999) *The Mysterious Flame. Conscious Minds in a Material World* (New York: Basic Books).
McGinn, C. (2004) *Consciousness and Its Objects* (Oxford/New York: Oxford University Press).
Menon, V. (2011) 'Large-scale brain networks and psychopathology: A unifying triple network model', *Trends in Cognitive Sciences*, 15(10), 483–506. doi:10.1016/j.tics.2011.08.003.
Merleau-Ponty, M. (1962 [1945]) *Phenomenology of Perception*, trans. Colin Smith (London/ Routledge).
Metzinger, T. (2003) *Being No One* (Cambridge, MA: MIT Press).
Moreno, J. D. (2003) 'Neuroethics: An agenda for neuroscience and society', *Nature Reviews*, February, 4(2), 149–53.
Morton, A. (1980) *Frames of Mind* (Oxford: Oxford University Press).
Nagel, E. (1961) *The Structure of Science: Problems in the Logic of Scientific Explanation* (New York/ Burlingame: Harcourt, Brace & World).
Nagel, T. (1974) 'What is it like to be a bat?', *The Philosophical Review*, 83(4), 435–50.
Nagel, T. (2000) 'The psychophysiological nexus', in P. Boghossian and C. Peacocke (eds), *New Essays on the A Priori* (Oxford: Clarendon Press).
Nakao, T., Ohira, H. and Northoff, G. (2012) 'Distinction between externally vs. internally guided decision-making: Operational differences, meta-analytical comparisons and their theoretical implications', *Frontiers in Neuroscience*, 6(31). doi:10.3389/fnins.2012.00031.
Noe, A. (2004) *Action in Perception* (Cambridge, MA: MIT Press).
Northoff, G. (2000) *Das Gehirn. Eine neurophilosophische Bestandsaufnahme* (Paderborn: Mentis).
Northoff, G. (2001) *Personale Identität und operative Eingriffe in das Gehirn* (Paderborn: Mentis).
Northoff, G. (2004) *Philosophy of the Brain* (Amsterdam: John Benjamins).
Northoff, G. (2007) 'Psychopathology and pathophysiology of the self in depression – neuropsychiatric hypothesis', *Journal of Affective Disorders*, 104(1–3), 1–14.
Northoff, G. (2011) *Practice of Neuropsychoanalysis* (Oxford/New York: Oxford University Press), see especially chs 1–3.
Northoff, G. (2012a) *Das undisziplinierte Gehirn. Was nun Herr Kant?* (Munich: Isiriana/Random House).
Northoff, G. (2012b) 'Immanuel Kant's mind and the brain's resting state', *Trends in Cognitive Sciences*, 16(7), 356–9.
Northoff, G. (2012c) 'Autoepistemic limitation and the brain's neural code: Comment on "neuroontology, neurobiological naturalism, and consciousness: A challenge to scientific reduction and a solution" by Todd E. Feinberg', *Physics of Life Reviews*, 9(1), 38–9.
Northoff, G. (2013a) *Unlocking the Brain. Vol. I: Coding* (Oxford/New York: Oxford University Press).
Northoff, G. (2013b) *Unlocking the Brain. Vol. II: Consciousness* (Oxford/New York: Oxford University Press).

Northoff, G. and Bermpohl, F. (2004) 'Cortical midline structures and the self', *Trends in Cognitive Sciences*, 8(3), 102–7.

Northoff, G. and Heinzel, A. (2006) 'First-person neuroscience: A new methodological approach for linking mental and neuronal states', *Philosophy, Ethics, and Humanities in Medicine*, 1(1), E3. doi:10.1186/1747-5341-1-3.

Northoff, G. and Qin, P. (2011) 'How can the brain's resting state activity generate hallucinations? A "resting state hypothesis" of auditory verbal hallucinations', *Schizophrenia Research*, 127(1–3), 202–14.

Northoff, G., Heinzel, A., de Greck, M., Bermpohl, F., Dobrowolny, H. and Panksepp, J. (2006) 'Self-referential processing in our brain – a meta-analysis of imaging studies on the self', *Neuroimage*, 31(1), 440–57.

Northoff, G., Bermpohl, F., Schoeneich, F. and Boeker, H. (2007) 'How can we bridge between psychoanalysis and neuroscience? First-person neuroscience', *Psychotherapy and Psychosomatics*, 76, 141–53.

Northoff, G., Hayes, D. and Duncan, N. W. (2010a) 'The brain and its resting state activity – experimental and methodological implications', *Progress in Neurobiology*, 92(4), 593–600.

Northoff, G., Qin, P. and Nakao, T. (2010b) 'Rest–stimulus interaction in the brain: A review', *Trends in Neurosciences*, 33(6), 277–84.

Northoff, G., Wiebking, C., Feinberg, T. and Panksepp, J. (2011) 'The "resting-state hypothesis" of major depressive disorder – a translational subcortical-cortical framework for a system disorder', *Neuroscience and Biobehavioral Reviews*, 35(9), 1929–45.

Nussbaum, M. C. and Rorty, A. O. (1992) *Essays on Aristotle's De Anima* (Oxford: Clarendon Press).

Owen, A. M. (2013) 'Detecting consciousness: A unique role for neuroimaging', *Annual Review of Psychology*, 64, 109–33. doi: 10.1146/annurev-psych-113011-143729. Epub 2012 Oct 2. Review.

Panksepp, J. (1998) *Affective Neuroscience* (Oxford: Oxford University Press).

Parnas, J. (2003) 'Self and schizophrenia: A phenomenological perspective', in T. Kircher and A. David (eds), *The Self in Neuroscience and Psychiatry* (Cambridge: Cambridge University Press), 105–28.

Parnas, J., Vianin, P., Saebye, D., Jansson, L., Volmer-Larsen, A. and Bovet, P. (2001) 'Visual binding abilities in the initial and advanced stages of schizophrenia', *Acta Psychiatrica Scandinavica*, 103(3), 171–80. Available at http://www.ncbi.nlm.nih.gov/pubmed/11240573.

Petitot, J. and Varela, F. (eds) (1999) *Naturalizing Phenomenology* (Stanford, CA: Stanford University Press).

Pfeiffer, U. J., Timmermans, B., Vogeley, K., Frith, C. D. and Schilbach, L. (2013) 'Towards a neuroscience of social interaction', *Frontiers in Human Neuroscience*, 7(22). doi: 10.3389/fnhum.2013.00022. eCollection 2013.

Pomarol-Clotet, E., Salvador, R., Sarro, S., Gomar, J., Vila, F., Martinez, A., McKenna, P. J. (2008) 'Failure to deactivate in the prefrontal cortex in schizophrenia: Dysfunction of the default mode network?', *Psychological Medicine*, 38(8), 1185–93. doi:10.1017/S0033291708003565.

Popper, K. and Eccles, J. (1989 [1977]) *Das Ich und sein Gehirn* (Munich: Piper).

Preston, S. D. and de Waal, F. B. (2002) 'Empathy: Its ultimate and proximate bases', *Behavioral and Brain Sciences*, February, 25(1), 1–20; discussion 20–71.

Putnam, H. (1975) *The Meaning of "Meaning". Mind, Language and Reality: Philosophical Papers*, Vol. 1 (Cambridge: Cambridge University Press).

Putnam, H. (1982) *Reason, Truth and History* (Cambridge: Cambridge University Press).

Qin, P. and Northoff, G. (2011) 'How is our self related to midline regions and the default-mode network?', *Neuroimage*, 1(57), 1221–33.

Qin, P., Di, H., Liu, Y., Gong, Q., Duncan, N. W., Laureys, S. and Northoff, G. (2010) 'Anterior cingulate activity and the self in disorders of consciousness', *Human Brain Mapping*, 31, 1993–2002.

Quine, W. v.O. (1951) 'Two dogmas of empiricism', *The Philosophical Review*, 60, 20–43.

Quine W. v.O. (1969) 'Epistemology naturalized', in W. Quine (ed.), *Ontological Relativity and Other Essays* (New York: Columbia University Press).

Raichle, M. E. (2009) 'A brief history of human brain mapping', *Trends in Neurosciences*, 32(2), 118–26. doi:10.1016/j.tins.2008.11.001.
Rey, G. (1997) *Contemporary Philosophy of Mind: A Contentiously Classical Approach* (Oxford: Blackwell).
Revonsuo, A. (2006) *Consciousness. The Science of Subjectivity* (Cambridge, MA: MIT Press).
Rizzolatti, G. and Fabbri-Destro, M. (2008) 'The motor system and its role in social cognition', *Current Opinions in Neurobiology*, April, 18(2), 179–84. doi: 10.1016/j.conb.2008.08.001. Epub 2008 Aug 20. Review.
Rosenthal, D. M. (2005) *Consciousness and Mind* (Oxford: Clarendon Press).
Roskies, A. (2002) 'Neuroethics for the new millennium', *Neuron*, 35, 21–3.
Rotarska-Jagiela, A., van de Ven, V., Oertel-Knochel, V., Uhlhaas, P. J., Vogeley, K. and Linden, D. E. (2010) 'Resting-state functional network correlates of psychotic symptoms in schizophrenia', *Schizophrenia Research*, 117(1), 21–30. doi:10.1016/j.schres.2010.01.001.
Rowlands, M. (2010) *The New Science of the Mind: From Extended Mind to Embodied Phenomenology* (Cambridge, MA: MIT Press).
Russell, B. (1910) *Philosophical Essays* (London: Longmans, Green).
Ryle, G. (1949) *The Concept of Mind* (Chicago: Chicago University Press).
Sadaghiani, S. and Kleinschmidt, A. (2013) 'Functional interactions between intrinsic brain activity and behavior', *Neuroimage*, 15 October, 80, 379–86. doi: 10.1016/j.neuroimage.2013.04.100. Epub 2013 May 3.
Sass, L. A. (2003) 'Self-disturbance in schizophrenia: Hyper reflexivity and diminished self-affection', in T. Kircher and A. David (eds), *The Self in Neuroscience and Psychiatry* (Cambridge: Cambridge University Press), 128–48.
Schilbach, L., Bzdok, D., Timmermans, B., Fox, P. T., Laird, A. R., Vogeley, K. and Eickhoff, S. B. (2012) 'Introspective minds: Using ALE meta-analyses to study commonalities in the neural correlates of emotional processing, social and unconstrained cognition', *PLoS One*, 7(2), e30920. doi: 10.1371/journal.pone.0030920. Epub 2012 Feb 3.
Schilbach, L., Timmermans, B., Reddy, V., Costall, A., Bente, G., Schlicht, T. and Vogeley. K. (2013) 'Toward a second-person neuroscience', *Behavioral and Brain Sciences*, August, 36(4), 393–414. doi: 10.1017/S0140525X12000660.
Schlicht, T. (2007) *Erkenntnistheoretischer Dualismus. Das Problem der Erklärungslücke in Geist-Gehirn-Theorien* (Paderborn: Mentis).
Schopenhauer, A. (1818/1819) *The World as Will and Idea. Vols I and II*, Dover edition 1966 (London: Dover).
Searle, J. (2004) *Mind: A Brief Introduction* (Oxford/New York: Oxford University Press).
Searle, J. R. (1999) 'The future of philosophy', *Philosophical transactions of the Royal Society of London*, 29 December, 354(1392), 2069–80.
Shoemaker, S. (1984) *Identity, Cause, and Mind: Philosophical Essays* (Oxford/New York: Oxford University Press).
Shulman, R. (2012) *Brain and Consciousness* (Oxford/New York: Oxford University Press).
Singer, W. (1999) 'Neuronal synchrony: A versatile code for the definition of relations?', *Neuron*, 24(1), 49–65, 111–25.
Singer, W. (2009) 'Distributed processing and temporal codes in neuronal networks', *Cognitive Neurodynamics*, 3(3), 189–96. doi:10.1007/s11571-009-9087-z.
Skinner, B. F. (1938) *The Behavior of Organisms* (New York: Appleton-Century-Crofts).
Solms, M. and Turnbull, O. (2003) *The Brain and the Inner World: An Introduction to the Neuroscience of Subjective Experience* (New York: Other Press).
Taylor, S. F., Welsh, R. C., Chen, A. C., Velander, A. J. and Liberzon, I. (2007) 'Medial frontal hyperactivity in reality distortion', *Biological Psychiatry*, 61(10), 1171–8.
Thompson, E. (2001) *Between Ourselves: Second Person Issues in the Study of Consciousness* (New York: Imprint Academic).
Thompson, E. (2007) *Mind in Life: Biology, Phenomenology, and the Sciences of Mind* (Cambridge, MA: Harvard University Press).
Tononi, G. (2004) 'An information integration theory of consciousness', *BMC Neuroscience*, 5, 42. doi:10.1186/1471-2202-5-42.

Tononi, G. (2008) 'Consciousness as integrated information: A provisional manifesto', *The Biological Bulletin*, 215(3), 216–42.

Tononi, G. (2012) 'Integrated information theory of consciousness: An updated account', *Archives Italiennes de Biologie*, December, 150(4), 293–329. Review.

Tononi, G. and Koch, C. (2008) 'The neural correlates of consciousness: An update', *Annals of the New York Academy of Sciences*, 1124, 239–61. doi:10.1196/annals.1440.004.

Treynor, W., Gonzalez, R. and Nolen-Hoeksema, S. (2003) 'Rumination reconsidered: A psychometric analysis', *Cognitive Therapy and Research*, 27, 247–59.

Tye, M. (2009) *Consciousness Revisited. Materialism without Phenomenal Concepts* (Cambridge, MA: MIT Press).

van Eijsden, P., Hyder, F., Rothman, D. L. and Shulman, R. G. (2009) 'Neurophysiology of functional imaging', *NeuroImage*, 45(4), 1047–54. doi:10.1016/j.neuroimage.2008.08.026.

van Gulick, R. (2004) *Are There Neural Correlates of Consciousness?* (Exeter: Imprint Academic).

Varela, F. (1996) 'Neurophenomenology: A methodological remedy for the hard problem', *Journal of Consciousness Studies*, 3, 330–49.

Varela, F. and Shear, J. (eds) (1997) *The View from Within: First-person Approaches to the Study of Consciousness* (Exeter: Imprint Academic).

Velmans, M. (2000) *Understanding Consciousness* (London: Routledge/Psychology Press).

Walter, H. (2001) *Neurophilosophy of Free Will: From Libertarian Illusions to a Concept of Natural Autonomy* (Cambridge, MA: MIT Press). ISBN 1-58811-417-1.

Watson, J. B. (1913) 'Psychology as the behaviorist views it', *Psychological Review*, 20, 158–77.

Whitfield-Gabrieli, S., Moran, J. M., Nieto-Castanon, A., Triantafyllou, C., Saxe, R. and Gabrieli, J. D. (2011) 'Associations and dissociations between default and self-reference networks in the human brain', *NeuroImage*, 55(1), 225–32. doi:10.1016/j.neuroimage.2010.11.048.

Wicker, B., Ruby, P., Royet, J. P. and Fonlupt, P. (2003) 'A relation between rest and the self in the brain?', *Brain Research Reviews*, October, 43(2), 224–30.

Wundt, W. (1893) *Principles of Physiological Psychology*, trans. E. B. Titchener (London: Allen). Translation of Wundt, 1874 [New York, 1904].

Zahavi, D. (2005) *Subjectivity and Selfhood: Investigating the First-person Perspective* (London: MIT Press).

Zeki, S. (2008) 'The disunity of consciousness', *Progress in Brain Research*, 168, 11–18. doi: 10.1016/S0079-6123(07)68002-9.

Index

Notes: **bold type** = extended discussion or term emphasized in text;
f= figure; n = footnote; t = table.

a posteriori knowledge   43, 57–60, 62, 64f, 65f, 164–5
a priori method   43, 57, 58f, 60, 64f, 65f, 112–13, 164
abnormal self   **497**
absent qualia   353, **358t**, 380, **398–9**
  'absent' concept   360
  neuroscientist and philosopher (dialogue)   **359**, 360, **398–9**
'absent qualia argument'   172, **357–9**, 361, 398
  NPC and   **422–3**
  refutation   **359–60**
  see also inverted qualia
'abstract' (concept)   **62**
access consciousness   389, 396f, 397, 426, **428–9**, 444
  definition   **428**
  relation to 'phenomenal consciousness'   430
  sufficient neural conditions   427
  see also cognitive functions
action potentials   9, 303
actual consciousness   402, 408–9, 420, 424
  necessary and sufficient conditions   410, 418
  see also 'possible consciousness'
'actuality' versus 'possibility'   **420–1**
affective form of intersubjectivity
  empathy   **516–17**
affective functions   334, 336, 342, 397, 407, 441, 462–3, 510, 517, 521–2, 533
  and consciousness   392, 393, 394–5, 396f
  see also emotions
affective neuroscience (Panksepp)   393
affective states   530n, 531
affective theory of mind   521
'agency'   451, 461, **465**, 474
'all bachelors are unmarried' (analytic sentence)   **48–9**, 112
Ameriks, K.   206
amygdala   80, 507, 521, 525
'analytic *a posteriori*' (Kripke)   165
analytic naturalism (*a priori* physicalism)   59, 60, 61, 98, 100–1
analytic sentences   57, 58f, 60

'analytic' versus 'synthetic' sentences   43, 52, 64f, 65f, 112–13
  a priori versus *a posteriori* knowledge (continuum)   **50–1**
  continuum   **49–50**
  dichotomy   **48–9**
anatomo-processual inference   **525–7**, 533
  definition   525
  empirical plausibility   **526**, 527
  'guiding question'   525
  lack of necessity   **526–7**
anatomy   77, 78, 89, 306
anesthesia   390, 403, 407–8, 411, 424
anterior cingulate cortex (ACC)   338–9, 412, 486, 507, 521–2, 524–5, 528
anterior cortical midline structures (aCMS)   341, 498–9
anthropology   88, 89, 311f
Aquinas, St Thomas   133, 138
  mind and body   **134–5**, 136
  mind and God   **133–4**
Aristotle   3, 24, 44, 73, 130, 153, 431
  concept of mind   **132–3**, 135
attention   72, 106, **390–1**
  'central role in consciousness'   391
auditory cortex   342, 415, 479, 502
auditory functions   335, 441
'auto-phenomenology'   273, 274
'automatic sharing' (Gallese)   528
'autonomous mind' (Bermudez)   258
awareness   292–3, 294, 425

Baars, B. J.   386
'basic disturbance of self'   *see* schizophrenia
behavior   456, 515, 519
  see also mental states
behaviorism   27–8, 72, 155, 172, **174–5**, 180–1, 184, 341
  empirical version (methodological behaviorism)   **157**, 158, 160
  objections (Chomsky)   **159**
  rejection of inner mental states   **157–8**
'being in world'   102, 235
beliefs   158, 159–60, 174, 251, 475

547

'belongingness'   463–5, 469, 476, 501–2, 504–5
Bennett, M. R.   **144**, 160, 228, 365
Berger, H.   **82**
Berger wave (alpha wave)   82
Berkeley, G.   151
Bermudez, J. L.   251–2, 296
  'autonomous mind'   258
Berze, J.   496
'bi-stable perception' (Kleinschmidt)   416
'binding' (of stimuli)   389, 406
'binding by synchronization'   384
biological naturalism (Searle)   183, **192–4**, 206, 210, 213
  and brain   **194–5**
  versus identity theory   193–4
  'phenomenal plausibility'   197
biology   46, 56, 59–60, 89
Bleuer, E.   75, **496**
Block, N.   295, **428**, 429–30
  converse spectrum inversion   361–2
body   **235–6**, 240–1, 355–7, 363, 393, 395, 404, 436, 452–3, 467–9, 472, 506, 535
  and minimal self   464–5
  in phenomenology   32
  and proto-self   465–6
  see also   lived body
body-world connection (Merleau-Ponty)   235–6
bootstrapping (Churchland)   116–18, 124
  'circularity'   115
  with concepts as outputs: definition   113–14
  with concepts as outputs: example   114–15
Botteril, G.   296
brain   ix–xi, 54, 535
  baseline metabolism   407, 408
  biological naturalism   **194–5**
  'category error in non-reductive neurophilosophy'   33–4
  clinical disorders   74–6
  concepts   231f
  'continuously active'   416
  'degradation'   176–7
  demystification   15, 16
  'design features' (Northoff)   324–6
  domains   11–13
  'door opener'   18, 535–8
  dualism and   (Sherrington and Penfield)   145–6
  'elevation'   177–8
  empirical characterization   17, 329–50
  empirical versus metaphysical domains   218
  'encoded information about self'   509

energy budget   341, 342–3
energy demand   **406–8**
  and environment   532
  extrinsic activity   347
  extrinsic features   **176–7**
  formal-syntactic properties   278, **289–90**
  hard problem   **7–8**, 9
  'internal criterion'   55
  intrinsic epistemic features   368
  intrinsic features   155, **177–8**, 378, **389–90**, 400
  intrinsic features and 'hard problem'   377
  intrinsic–extrinsic interaction model required   343–4
  investigation in philosophy   **11–13**
  'at junction between disciplines'   **34–5**
  knowledge (dark spots and explanatory gap problem)   9
  'lived brain' versus 'objective brain'   33
  localization of function   332–3
  metabolic consumption rate   341
  mode of operation   349–50
  neural activity (formal-syntactic structure)   278, 297
  neuronal features   380
  neurons as basic functional unit   73–4
  'nothing but car'   331–2
  'nothing but category error'   33
  past times   2–3
  philosophical arguments against linkage with consciousness   402
  philosophical concept(s)   **10–11**
  philosophy   **15–16**
  psychons and (Eccles)   **146**
  purpose   347
  relation to body during consciousness   235–6
  representation of world   98
  resemblance to music   290
  self and   **494–5**
  spatiotemporal activity   290–1
  structure and function   333
  'structure and organization'   332, 378, 379
  'sub-personal vehicle of mind'   286
  see also   philosophy of brain
brain: characterization   287
  mental   108–9
  physical   107–8
  'as whole'   331–2
brain: intrinsic activity   **342–8, 404–8**
  abnormal   505
  conceptual and empirical characterization   508–9
  and consciousness   **405–6**, 416–18
  definition   **404–5**

dialogue between neuroscientist (NS) and a brain (BR)   345–8
encoded concept of self   510
encoded information about self   495, 498, 501, 505
explanatory gap problem   421
guiding question   443
'immunity against error through misidentification'   503–5
intrinsic activity (concept)   346–7
intrinsic activity and consciousness   347–8
Kant's consciousness and brain's consciousness   344–5
Kant's mind and brain's intrinsic activity   344
location of self   487–8
'need to talk to brain itself'   345–6
NPS   488–9
pre-reflective self-consciousness   483–4
purpose   342–4
same as 'resting state'   405, 503
schemata   417
spatial characterization   412–13
spatiotemporal structure   416, **417–18**, 419f
temporal characterization   413–14
'unified field approach'   442–3
see also 'resting state activity'
brain: intrinsic versus extrinsic view   340–2, 343
history of neuroscience   340–1
interaction   342
present findings   341–2
brain: intrinsic features   324–6
examples   325
intrinsic observer-related intrusions   325–6, 330
brain: role
neural correlates   5–6
neural predispositions   5–6
brain: views   340–5
intrinsic activity   342–5
intrinsic versus extrinsic view   340–2
brain and body   233, 234–5
concepts (Merleau-Ponty)   234
embodiment and sensorimotor functions   234–5
brain at border between metaphysical and empirical domains   231f, 233
'deep interior of brain' (Nagel)   223–5
brain coding   326, 425
brain in epistemological domain
brain as subject and brain as object of cognition   226–7
Schopenhauer and brain   225–6

brain and function   332–9
cognitive functions   335–7
localization versus holism   336–9
modularity and modules   332–6
sensory functions   333–5
'brain as functioning'   230–1, 231f, 232–3, 241–2, 244
brain functions   332–3, 336–7, 349, 398, 524–5
neural overlap   338–9
brain imaging (1980s –)   73–4, 86, 95, 226–7, 289, 307, 403, 405, 460, 462, 510
empathy   521–2
theory of mind   520–1
see also   fMRI
brain and intersubjectivity   6, **18**, 29, 37, 55, 447, **512–34**
chapter focus   450, 512, 533
further research required   531, 532
problem of other minds   512–13
brain and intersubjectivity: critical reflection   531–2
attuning between brain and environment   532
neural predisposition of intersubjectivity   531
brain and intersubjectivity: neurophilosophical discussion   525–31
anatomo-processual inference   525–7
embodied simulation and intersubjectivity   528–9
intrasubjectivity: relationship with intersubjectivity   529–31
simulation concepts (Gallese)   527–8
brain and intersubjectivity: neuroscientific findings   520–5
empathy and brain imaging   521–2
mirror neurons and different regions and functions   524–5
mirror neurons and goal orientation   523–4
Rizzolatti and Gallese (biography)   522–3
theory of mind and brain imaging   520–1
brain and intersubjectivity: philosophical suggestions   513–18
determination of concept of empathy   517–18
empathy as affective form of intersubjectivity   516–17
inference by analogy   514, 515, 516
Mill (biography)   514
problem of intersubjectivity   513–14
self and other in phenomenological philosophy   515–16

brain and intersubjectivity: psychological theories 518–20
  'learning and acquisition' 519
  simulation theory (ST) 520
  theory-theory (TT) 518–20
brain lesions 74, 95, 113–14, 129, 233–4, 336–9, 370, 372, 392, 403, 409, 451, 495, 525
  see also 'lesion-based method'
brain metabolism 406–8
brain in metaphysical domain 218–20
  access to, and knowledge of, property P (McGinn) 219–20
  property P and brain (McGinn) 218–19
brain networks 332–3, 336–40, 349, 413
'brain as object of cognition' (Schopenhauer) 95, 212–13, **226–9**, 231f, 231–2
'brain as observed' 229–30, 231f, 231–2, 241, 244
brain and observer 316–22
  brain-based concepts 317, **318–22**, 330
  encoding of multiple data and facts into one concept 317–18
  GABA and glutamate 319–20
  indirectness in investigation of mind and brain 316–17, 330
  observer-based concepts 317, 318, **319–22**, 330
  resting state and stimulus-induced activity 321–2
  stimuli and their origins 320–1
  technological devices 316–17, 330
'brain without oscillations' 389–90
brain and philosophy: neurophilosophy 16–17, 91–124
  'central question' 98
  chapter focus 91, 93, 123
  ideas for future research 117–22
  philosophy, mind, naturalism 92–3
  see also brain-reductive neurophilosophy
brain regions 313, 332–3, 336–40, 349, 384, 400–1, 403, 405, 407, 412–15, 451, 457–8, 459–60, 523, **524–5**, 528
  and cognitive functions 82–3
  'task-positive' versus 'task-negative' 413
  see also cortical regions
brain and self 18, 29, 447, **449–70**, 471, **513**
  chapter topic 449, 490, 512
  neurophilosophical discussion 460–8
  neuroscientific findings 457–9
  philosophical concepts 452–7
brain and self: background 449–52
  conceptual determination of self in philosophy 449–50

empirical investigation of self in neuroscience 451–2
brain and self: philosophical concepts 452–7
  empirical self 453–4
  mental self 452–3
  metaphysical self 453–4
  minimal self 455–6
  phenomenal self 454
  social self 456–7
brain and self: neurophilosophical discussion 460–8
  body and proto-self 465–6
  data deficiencies/further research required 461, 462, 463, 467
  difference between concepts of self and self-reference 466–7
  different forms of specificity 460
  minimal self and body 464–5
  neuronal specificity of midline regions 460–1
  phenomenal specificity and phenomenal limits 464
  phenomenal specificity of self-reference 463–4
  psychological and experimental specificity 461–2
  self as brain-based neurosocial structure and organization 467–8
  self-specificity and other functions 462–3
brain and self: neuroscientific findings 457–9
  self-reference: social patterns of neural activity 459
  self-specific stimuli: spatial patterns of neural activity 457–8
  self-specific stimuli: temporal patterns of neural activity 458–9
brain and self-consciousness 18, 29, 447, **471–92**, 494
  chapter focus 450, 471, 490, 512, **513**
  data deficiencies/further research required 472, 485, 488–91
  disorders of consciousness 479–80
brain and self-consciousness: background 471–4
  'explanatory gap (s)' 472–3
  'explanatory gap (s)' and self-consciousness 473–4
  philosophical and neuroscientific approaches to self 471–2
brain and self-consciousness: critical reflections 487–90
  dialogue between philosopher and neuroscientist 487–90

location of self in brain's intrinsic activity **487–8**
NPS and brain's intrinsic activity **488–9**
NPS and 'explanatory gap (s)' **489–90**
brain and self-consciousness: philosophical concepts **474–9**
  contents in our consciousness **476–8**
  'immunity to error through misidentification' **476–7**
  indirect and direct access to self in consciousness **474–5**
  own self or subject of our consciousness **477–8**
  reflective and pre-reflective self-consciousness **475–6**
  self as subject and object **478–9**
brain and self-consciousness: neurophilosophical discussion **480–4**
  brain: intrinsic activity **483–4**
  pre-reflective self-consciousness **482–4**
  see also vegetative state
brain and self-consciousness: neuroscientific findings **484–7**
  neural overlap between self and rest **484–5**
  prediction of self-specificity by resting state **486–7**
brain stem 341, 342, 409, 419n
'brain as subject of cognition' (Schopenhauer) 95, 212, 213, **226–9**, 231f, 232–3
'brain in vat' (Putnam) 212, 213, **237–9**, 244, 345
  perceptions **238**, 239
brain and world **236–40**
  'evil demon' **236–8**
  semantic or content internalism versus externalism **238–40**
brain-based approach 95, 155, 180, **215–17**, 278, 297–8, 351
  bottom-up 278, 301
  mind-brain problem 178, **212–45**
brain-based concepts 302, 317, **318–22**, 327, 329, 330, 347
brain-based explanations **315–16**
brain-based neurophilosophy 91, **101**, 102, 123, **538**
brain-based problems
  'explanatory gap' **379**
'brain-mind differentiation problem' **243**
brain-paradox (Schopenhauer) 95, 213, **227–9**
brain-reductive approach 215, 309
brain-reductive concepts 302, 327, 330

brain-reductive explanations **314–16**
brain-reductive neurophilosophy 91, **99**, 99n, **100**, 101, 102, 123, 536, **538**
  see also explicit neurophilosophy
brain's consciousness **344–5**
Breidbach, O. 94
Broad, C. D. 139, 270
Broca, P. 74–5
Broca region 336, 495, 524
Brook, A. 206
Brown, T. G. 341
building block approach (Searle) 426, **440–1**, 442
  guiding question 443
'bundle theory of self' (Hume) **201–2**, 452, 475
Bunge, M. 139
Buszáki, G. **323–4**, 326, 389–90

C-fibers 59, 60, **161**, **164–6**, 271
camera **255**, **257**
'capacities' (McDowell and Hornsby) 260, 261
cars **331–2**, 416
Carnap, R. 156
Carruthers, P. 292, 296, 336, 534
'categories' (Kant) 221
category error/mistake 28, 100, 488
  brain in phenomenology **33–4**
  non-reductive neurophilosophy **33–4**
'causal closure' **139–40**
  of physical causes **129–30**
cave allegory (Plato) **130**
  skyscraper as modern cave (metaphorical comparison) **131**
cell firing rates 486
cell recording 74
cellular level (of explanation) 307, 312
Chalmers, D. 7, 145, 191, 369, 371, **372–4**, 399
  'dualist' 186
  'hard problem' **143**
  zombies and informational dualism **142–4**
Changeux, J. P. 386
chemistry 46, 56, 59
children 431
  acquisition of knowledge (Quine) 53
'Chinese room argument' (Searle) 179
Chomsky, N. **159**
Churchland, Patricia Smith 91, 94, 98–9, 119n, 168, 197, 208, 254, 295, 360, 365, 453
  'bootstrapping with concepts as outputs: definition **113–14**

Churchland, Patricia Smith – *continued*
  'bootstrapping with concepts as outputs: example **114–15**
  explicit neurophilosophy **96–7**
  neuroepistemology **106–7**
  versus Searle **117**
Churchland, Paul M. 96, 168, 254
Churchland (both) 273
  arguments against folk psychology **268–9**, 271
  characterization of brain 286
  linguistic versus vectorial representation **287–8**
'Churchland' (unspecified) 298, 454
Clark, A. 234
clinical disorders **74–6**
code (contents of consciousness)
  and consciousness **433–4**
  and hard problem **434**
cogito ergo sum 135, 205, **344–5**, 475
  'epistemic-metaphysical inference' **200–1**
  reasoning 'smashed by Kant' **200**
cognition 107
  'brain as object' versus 'brain as subject' 95
  and knowledge **94–5**
  mind-dependent versus mind-independent 230
cognitive functions 235, 334, **335–6**, 342, 377, 382, 384, 387–9, 396–7, 399, 401, 407, 429, 461–3, 503, 509–10, 517, 521–2, 530n, 531
  attention and representation 391
  brain regions and 82–3, **336–7**
  and consciousness **83–4**
  higher-order 86, 87, 453, 455, 500–2
  neural basis 83
  neurons and **81–2**
  relationship with consciousness **373–4**, 375, **392–3**, **394–5**, 396f
  vegetative states **391–2**
  working memory and attention **390–1**
  see also access consciousness
cognitive neuroscience 72–3, 85–8, 93, 297, 337, 341, 343, 373–4, 453
  see also critical neuroscience
cognitive psychology 69, **72–3**, 83, 85–6, 89, 175, 297, 303, 333, 337, 373, 453
  see also folk psychology
cognitive science 19
  'umbrella discipline' **72–3**
  see also 'philosophy and science'
colors 357–9, 360–2, **366–7**, 383, 391, 423, 429, 435, 441
coma 403, 412, 419n

commensurability 249, **256**, 303–4, 309
common sense 251, 252, 269, 303, 519
conceivability argument 353, **358t**, **369–70**, 380
  non-reductive neurophilosophy **372–3**
  reductive neurophilosophy **371**
  zombies **371–3**
concept of self 490
'concept-argument iterativity' **120–1**
'concept-fact iterativity' 92, **118–19**, 122–4, 323
  versus 'concept-argument iterativity' **120–1**
  as trans-disciplinary methodological strategy **119**
concepts **85–6**
  neuroscience **302–28**
  unilateral adaptation to facts (Searle) **115–16**
  'unilateral inference from facts' **112–13**
conceptual dualism (Bennett and Hacker) **144**
'conceptual feedback output' 118, 119f, 120
'conceptual feedforward input' **117–18**, 119f, 120
  see also feedforward connections
conceptual plausibility 43, **117**, 125–6, 153, 185, 188–9, 270
conceptual propaedeutic (Searle) **116**, 118
'conceptual re-entrant loop' **118–19**, 120, 121
conceptual specificity 449, 460, 469
conceptual-linguistic method (philosophy) 43, 48, 53, 57, 63, 65, **70**, 71, 92–3, 100, 102, 112, 120
  hallmark features (Quine) 62
conceptual-logical methodology 47, 63–7, **96–7**
conceptualization in neuroscience **313–14**
'conditions of possibility of' (McDowell and Hornsby) 260
'connectome' concept 81
conscious contents 387, 402
conscious states
  'aboutness' or 'directedness' 437
  phenomenal-qualitative character 444
'conscious units' (Searle) 441
consciousness 17–18, 29, 117, 235–6, **351–445**
  'always presupposes a subject' 449
  as basic function 397
  'biological feature' (Searle) **192–3**
  'central dimension' 383

Index 553

'central hallmark of mind'   443, 513
characterization in relation to other functions   396f
from clinical disorders of brain to   75–6
cognitive functions and   83–4
conceptual difficulties   481–2
conceptual logical account   353, 380
core phenomenal features   171
definition   443–4
diabetes and   411–12
disorders   402–4
empirical-experimental account   353, 380
energy and   407–8
explanation   358t
'extra ingredient'   398, 399, 400
'hallmark feature of mind'   354, 395
hallmark features   364–6, 447
hard problem of brain   7–8
as higher- or lower-order function   394–6, 396f
and higher-order representation   291–2
indirect and direct access to self   474–5
intrinsic activity and   347–8, 416
'irreducibility to brain'   380
knowledge of   358t
'located' at junction between mind and brain   426
'location in brain'   356–7
logical possibility   358t
low-frequency fluctuations and   405–6
from mind to consciousness   351–445
'necessary' conditions   408, 418
neural correlates   305, 409
'neural prerequisites'   409
neuronal processes (Searle)   194–5
neurophilosophy   351–445
neuroscientific explanation (arguments against)   427
phenomena and   30–1
'phenomenal hallmark features'   38, 70
philosophical arguments against linkage with brain   402
psychology and   71–2
qualitative and subjective features   294–7, 300–1
'quantitative and objective'   294–5
relationship with cognitive functions   392–3
'result of empirical mechanism' (Searle)   193–4
and self (VS patients)   481–2
'self-representational'   293
sensorimotor functions and   76–7
spatiotemporal structure and   417–18

split-brain patients   114
'standard definition'   435
subject   477–8
subjective aspects   374
subjective nature   411, 438
'sufficient' conditions   408, 418, 437
'transitive parts' (James)   72
unconscious and   39–40
consciousness: arguments about explanation   373–7
brain: easy and hard problems   374–5
Chalmers   373
easy problems   373–5
explanatory gap problem   375–7
hard problem   374–5, 377
consciousness: arguments about logical possibility   369–71
natural and logical worlds   370–1
philosopher-neuroscientist dialogue   370–1
zombies and conceivability argument   369–70
consciousness: arguments about phenomenal features   357–62
absent qualia argument   357–9
absent qualia argument: refutation   359–60
converse spectrum inversion   361–2
inverted spectrum argument   360–1
consciousness: concept   428–33
access consciousness   428–9
consciousness and unconscious   430–2
nonconsciousness (concept)   433
phenomenal consciousness   429–30
unconsciousness and nonconsciousness   432
consciousness: conceptual, phenomenal and methodological issues   18, 351, 426–45, 513
Block (biography)   428
chapter focus   426, 427, 443, 444
concept of consciousness   428–33
further research required   434, 442–3, 531
methodological issues   438–43
neurophilosophical discussion   433–4
phenomenal features of consciousness   434–8
consciousness: investigation
linkage between FPP and TPP   427–8
NCC and NPC   427
consciousness: neurophilosophical discussion   433–4
code and consciousness   433–4
code and hard problem   434

consciousness: phenomenal features 434–8
  intentionality 437
  qualia and 'what it is like' 434–5
  self-perspectival organization and FPP
    437–8
  temporal and spatial continuity and unity
    436
consciousness: reduction to brain (arguments
    against) 17, 353–81
  arguments about explanation of
    consciousness 373–7
  'central arguments' 380
  chapter focus 353, 380
  consciousness: reduction to
    physical-material basis (contrary
    philosophical arguments) 358t
  knowledge of consciousness: arguments
    364–7
  location problem (mental states) 354–7
  mental states 354–5
  metaphorical comparison (intrinsic
    features in heart and brain) 378
  neurophilosophical discussion 362–4,
    367–9, 371–3, 377–9
  physical states 354
'consciousness about contents' 383
consciousness recovery scale – revised
    (CRS-R) 479
consciousness research: methodological
    issues 438–43
  building block approach 440–1
  first-person neuroscience 439–40
  first-person versus third-person access
    438–9
  'unified field' and brain's intrinsic activity
    442–3
  unified field model 441–2
'consciousness-meter' 387
constructionist approach 335
content externalism 212, 213, 238–40
content internalism 238–40
content–context relationship 361–4
  empirical plausibility of qualia 363–4
contents (of consciousness) 278, 382, 400,
    402, 406, 409, 419f, 424, 429–30, 444,
    476–7, 478, 494, 501
  based on extrinsic stimulus input to brain
    404
  non-conceptual representation 295–6
  same as 'phenomenal content' 383
contextualization 313–14, 312–13
'contextualization model' 302, 310–12,
    315, 327, 330
  definition 311

continuity 249, 252, 275, 276
continuous changes (James) 72
'continuity hypothesis' 321
converse spectrum inversion (Block) 361–2
cooking 431
  correspondence to mind–brain relationship
    199
  see also food
cooperative naturalism 43, 63, **64–6**, 101,
    102, 123, 308
  see also naturalism
Copernicus, N. 44, 46
copy number variants (CNV) 80
'core or mental self' (Damasio and Panksepp)
    466
'correlative and reversible system self/other'
    (Gallese) 529
cortex 77, 336, 386, 396, 401, 415
cortical midline structures (CMS) 84, 342,
    413, 449, **458**, 460, 467, 469, 484, 494
  abnormal intrinsic activity 497–9
  anterior and posterior 412, 459
  neuronal specificity 460–1
  self-specificity and other functions 462–3
  see also midline regions
cortical regions 396, 409, 413, 485, 507–10
  see also brain regions
Craver, C. **312–13**
Crick, F. **384–5**, 430
critical neuroscience 69, **310**, 313–14
  see also neuroscience
'critical phenomenology' 439

d'Argembeau, A. 484
Damasio, A. R. 455, 466
Davidson, D. 167
Darwin, C. R. 46
data 320, 330
  versus 'facts' **317–19**
death 372, 431, 456, 478
'decreased environment-focus' **506**,
    508–10
  see also 'increased self-focus'
'deep interior of brain' (Nagel) 212, 213,
    223–5, 230, 232–3
'deep unconscious' (Searle) **432**, 433–4
Default-Mode Network (DMN, 2001–)
    338–9, 341–2, 405, 413, 416, 485, 498
definition problem 183, **186–8**, 193, 195,
    198
  see also mental–physical dichotomy
'degradation of brain' **176–7**
Dehaene, S. 386, 431–2
Dennett, D. C. 197, **252–3**, 273, 295, 360, 365

different levels of explanation 253
  'intentional stance' **253**, 258
  irreducibility of folk psychology (Dennett) 253–4
  'personal level' versus 'sub-personal level' 254
  'physical stance' 253
deoxyribonucleic acid (DNA) 79, 219, 384, 434
depression 79–80, 82, 324, 337, 447, 478, 486, 493, **505–9**, **510**, 511
  brain: intrinsic activity (conceptual and empirical characterization) 508–9
  decreased environment-focus 506
  embodied and relational concept of self 508
  increased self-focus 505–6
  neural abnormalities 507–8
  neurophilosophical discussion 508–9
derivability 249, **256**, 276, 303, 304, 309
Descartes, R. 24
  'assumption of mental substance' 58
  Cartesian doubt 135
  Cartesian dualism 202
  contested by Kant 199–201
  'epistemic-metaphysical inference' and mental approaches 200–1
  epistemological-metaphysical fallacy **196f**, 196n
  'evil demon' 237
  existence 135–6
  interactive dualism **136–7**, 139, 141
  living room and kitchen (metaphorical comparison) 26
  mental states 142
  metaphysical assumptions **199–200**
  metaphysical dualism between mind and body **24–5**, 26
  methodological dualism between introspection and observation 25–6
  mind and existence 135–6
  self as 'substance' 452
  substance dualism 137–8
'design features' 332, 348
design level 254
'design stance' 253
  same as 'design level' 253
designators,
  'rigid' versus 'non-rigid' (Kripke) 163–5
diabetes 410–12
  and consciousness 411–12
  glucose and stimulus-induced activity 412
  insulin and resting state activity 410–12
'diminished self-affection' (Sass) 497

direct awareness 471, 491
disciplines 25–6, 62
  versus domains **46–7**, 97
  subject matter and method of investigation 13–15
disgust **524–5**, 528
'disorder of self-affectivity' (Sass) 497
disorders of consciousness **479–80**
  vegetative state: self-specificity and consciousness **479–80**
DMT 507
'domain' concept 13, 56
domain linkage 91, 106, 111, 116, **120**, 148, 194, 267, 306f
domain monism 61, **99–100**, 102, 111, 120, 123, 148, 209, **213**, 214, 217, 274, 309
domain pluralism 61, **101**, **102**, 111, 120, 123, 209, **213**, 214, 274
domain problem/s 183, **196f**
  mind-brain problem **208–9**
domain-independent regions (of brain) 458, 460, **461**
domain-method constellations **47–8**
domain-specific regions (of brain) **460–1**
domain-specificity 329, 333, 335, 349
domains 33–4, 216–18, 242–3, 258, 307–8, 311f, **332**, 348, 451
  versus disciplines **46–7**, 97
  inferences between **188–90**
  neurophilosophy **105–11**
  subject matter and method of investigation 13–15
domains and methods 43, 58f
  dissociation 97
dopamine 77, 78–9, 82, 415
dormant intrinsic self 489, 490
dorsolateral prefrontal cortex (DLPFC) 498, 507, 521
dorsomedial prefrontal cortex (DMPFC) 412, 458–9, 479, 484–5, 521
'doubling of brain' *see* brain-paradox
'doubling of domains' 229
dreams/dreaming 30, 383, 390, 403, 416, 435, 437, 441, 488
Dretske, F. 296
dualism 3, 23, 125–7, **137–42**, 153, 155, 157–8, 168, 170, 174, 180, 184–6, 189, 215, 250
  appeal (Searle) 142
  epiphenomenalism and parallelism **140–2**
  mental causation problem **139–40**
  mind and body 24–5
  property dualism **138–9**
  substance dualism **137–8**

dualism: history   130–7
  allegory of cave (Plato)   130
  concept of mind (Aristotle)   132–3
  concept of mind (Plato)   131–2
  interactive dualism (Descartes)   136–7
  mind and body (Aquinas)   134–5
  mind and body (Descartes)   136
  mind and existence (Descartes)   135–6
  mind and God (Aquinas)   133–4
  skyscraper as modern cave (metaphorical comparison)   131
dualism and brain (Sherrington and Penfield)   145–6
dualism and science   147
'dynamic flow' (James)   417, 436
'dynamic localization' (Lurija)   338
'dynamic unconscious'
  Freud   431
  Searle   431, 432

easy problems (Chalmers)   373–5, 377, 380, 399
  and brain   374–5
  consciousness versus unconsciousness   374
  questions   374
  see also hard problem
Eccles, J. C.   146, 147
  explicit neurophilosophy   95–6
Edelman, G. M.   385, 386
'efference copy'   465
Egypt (ancient)   2–3, 73
Einfühlung (empathy)   517
Ekman series   486
electroencephalography (EEG)   82, 339, 413, 458–9, 480
'elevation of brain'   177–8
elimination   249, 250, 252, 275
elimination of mind   4–5
  and methodological strategy   297–8
'eliminative materialism' (Churchland)   98, 208, 250, 271, 272, 275
eliminative physicalism   168
eliminativism   4–5, 6, 16, 270, 294–5, 297–300
'elusiveness thesis' (Hume)   475
'embedded brain'   236, 345, 363–4
embedded mind   235
embedded self   467, 508–10
'embeddedness'   76, 95, 217, 237–40, 244, 465
embodied brain   236, 345
embodied self   493, 508
embodied simulation (Gallese)   512, 527–8, 529–31
  and intersubjectivity   528–9
'embodied subjectivity'   516–17

'embodiment'   76, 95, 103, 212, 217, 234–5, 237–40, 244, 465, 467
'emergence' or 'emergentism'   5, 139
emergent properties   139
emotional contagion (concept)   518
emotions/emotional feelings   350, 393, 435, 456, 459, 461–2, 465, 468, 486–7, 495, 506, 514, 518, 520, 524–6, 528, 533
  see also affective functions
empathy (*Einfühlung*)   18, 273, 456–7, 476, 512, 520, 525, 527–8, 534
  affective form of intersubjectivity   516–17
  brain imaging   521–2
  definition   517
  determination   517–18
  versus 'imitation'   518
empirical domain
  and metaphysical domain: differences   148–9
  and metaphysical domain: linkage   148
  mind-brain problem as 'border problem'   150
  see also observation
empirical elimination   271–2, 276
empirical approach/findings   2, 29, 43, 85–6, 192, 398, 447
'empirical implausibility'   184, 185, 186, 189, 270, 509
empirical investigation   426
  of self in neuroscience   451–2
empirical plausibility   43, 125, 153, 170, 189, 322–3, 352, 361, 368, 372, 422–3, 460, 482–3, 510, 529, 533
  of qualia as context–content relationship   363–4
empirical self, 449, 450, 453–4f, 468, 471–2, 476, 493, 500–1, 510
empirical-experimental investigation   273
empirical-metaphysical fallacy   196f, 196n, 205, 208
'empirical-metaphysical inference' (Kant)   108, 205–6, 207
  see also inference problem
empiricism   10–16, 18, 19, 33, 104–5
'enabling conditions' (McDowell and Hornsby)   260, 261
enabling conditions   419f, 419–20
  see also neural prerequisites
encapsulation   329, 335, 349
encoding   317–18, 468
energy and consciousness   407–8
energy demand   406–7
environment   240–1, 404, 436, 450, 452, 457, 464–5, 467, 472, 506, 508–9

environment–brain relationship 105
epilepsy 75, 82, 424
epiphenomenalism 127, **140–2**, 150, **167–8**
  see also physical causation
epistemic abilities (Kant) 206, 213, 229, 230, 376
epistemic domain (Kant) 183, 213
  'primary domain for mind–brain problem' 208
epistemic dualism (Kant/Schlicht) **206–7**
epistemic instruments 219, 220, 222
epistemic limitations (Kant) 183, **203–4**, 207–8, 221–3, 226, 230, 244, 422
  knowledge of mind (Kant) **198–9**, 203
  versus metaphysical assumptions (Kant versus Descartes) **199–200**
  and physical approaches **204–5**
epistemic modes (Schopenhauer) 227
'epistemic-metaphysical fallacy' **200–1**, 205, 221
'epistemic-metaphysical inference' (Kant) **200–1**, **205–6**, 207, 221, **367–9**
  see also inference problem
epistemological domain **225–7**
  naturalized 70
  versus phenomenal domain 36
'epistemological explanatory gap' **376–7**
epistemological limits 298, 324
epistemological naturalism 61, 92, **97–8**
'epistemological-metaphysical fallacy' **196f**, 196n
epistemology ix, 9, 11–15, 21, 23, 27, 40, **44**, 66–7, 238, 535
  explanatory gap problem **376–7**
  focus **24**
  FPP versus TPP 355
  Mind–brain problem **207–8**
  naturalized 43, 57
'epistemology naturalized' (Quine, 1969) **52–3**, 106
ethical domain 56, 98, 99f, 101f, 105–6, 111, 122–3
ethics 21, **24**, 47, 102, 311f
  see also neuroethics
event-related potentials (ERP) 480, 522
'evil demon' (Descartes) 237
examination anxiety (propositional attitude) 282
excitation–inhibition balance (EIB) 312, 320
execution (of action) **526–7**
experience 37–8, **475–9**
  always has a 'subject' **437–8**, 512, 513
  versus 'observation' 435
  subject of **449–50**
  subjectivity 435

'experiential similarity' **55–6**
'experiential states' 37
experimental specificity 382, **387–8**, 398, 449, 460, **461–2**, 468, 469, 500
explanation/s
  different levels (Dennett) 253
  versus metaphysics 258
  multi-level **312–13**
  neuroscience **302–28**
  'personal levels' versus 'sub-personal levels' 17, 247, **249–77**
  see also levels of explanation
explanatory context
  and 'contextualization model' **310–12**
  and critical neuroscience 310
explanatory continuity **261–6**
  commensurability and derivability 263
  continuity between personal and sub-personal explanations **261–2**
  explanations of anxiety at different levels 262
  philosophical functionalism **263–4**
  psychological functionalism **264–6**
explanatory dualism 258–9, **259**, 261
explanatory elimination **267–73**
  arguments against folk psychology (Churchland) **268–9**
  empirical versus metaphysical elimination **271–2**
  mental concepts **270–1**
  phenomenal elimination 272
  qualia **272–3**
  rejection of folk psychology (Churchland) 268
  theoretical posits and other properties in folk psychology **267–8**
explanatory gap 353, 378–9, 397, 473
explanatory gap problem (Levine) 9, 17, **358t**, **375–7**, 380–1
  as brain-based problem 379
  and brain's intrinsic activity **421**
  definition 375
  as epistemological or metaphysical claim **376–7**
  'gap between phenomenal and neuronal knowledge' **421**
  versus 'hard problem' 375, 376
  neuroscientist and philosopher (dialogue) **399–400**
  and NPC **421–2**
  questions raised 376
'explanatory gap (s)' 471, **472–3**, **490–1**
  definition 473
  NPS **489–90**
  and self-consciousness **473–4**

explanatory irreducibility 253–7
  different levels of explanation (Dennett) 253
  horizontal explanations 254–5
  incommensurability and non-derivability 256
  irreducibility of folk psychology (Dennett) 253–4
  metaphorical comparison (soccer with Messi) 255
  metaphorical comparison (soccer and camera) 255, 257
  normative-prescriptive and descriptive explanations (incommensurability) 257
  vertical explanations 254, 255–6
  see also irreducibility
explanatory models 258–9
explanatory reduction 144, 308–9
explanatory relevance 260, 261
explicit neurophilosophy 95–7
  Churchland 96–7
  Popper and Eccles 95–6
  see also neurophilosophy
'extended brain' 240–2, 244
extended mind 212–13, 235, 237, 240
exteroceptive stimuli 240, 320–1, 335, 342, 346, 348
extrinsic activity 405, 408
extrinsic observer-related intrusions 302, 323, 327
extrinsic stimuli (Hume) 201–2, 204–5, 224, 226, 321–2, 341–2, 344, 346–7, 349, 396f, 397, 404–5, 411, 417–18, 421–3, 443, 488–90, 499
  versus intrinsic features (Kant) 202–3
extrinsic stimulus-induced activity 347, 402, 414–15
'extrinsic view' of brain (Northoff) 329, 341, 349–50

Fabbri-Destro, M. 523
faces 458, 486
facts 85–6
  unilateral adaptation of concepts to (Searle) 115–16
  'unilateral inference' of concepts from 112–13
feedforward connections 385, 387–8
  see also 'conceptual feedforward input'
feedback circuits
  same as 're-entrant circuits' 385
'feedback output' 117–18
Feigl, H. 141, 160
Feyerabend, P. 270

films (motion pictures) 296, 483, **514**, 516
first-person access **438–9**
first-person knowledge 368
first-person neuroscience **439–40**, 443
  core 440
  definition **439**
'first-person ontology' (Searle) 116
first-person perspective (FPP) **38, 193**
  investigation of consciousness 427–8
  phenomenal feature of consciousness **437–8**
  see also point of view
'fixed domain location' **214**, 217
'flexible domain location' **214**, 217
floors 395–7
flowers 395–7
Fodor, J. A. 296, 335
  biography **284**
  LOT on sub-personal level **284–5**
  modules (characterization) **333**
  'structural isomorphism' **284–5**
  syntax and semantics: 'associated with sub-personal and personal levels' **285–6**
folk psychology, 249–50, 256–9, 266, 273–6, 303–4, 519
  arguments against **268–9**
  concepts (matching scientific psychology) 269–70
  definition **251**
  eliminative materialism 271–2
  explanatory elimination **267–9**
  interface problem (with scientific psychology) **251–2**
  irreducibility (Dennett) **253–4**, 257
  rejection **268**
  theoretical posits **267–8**
  see also mind, brain, science
food 521
  philosophical versus scientific approach **45**
  see also cooking
'force equals mass times acceleration' (synthetic sentence) **49**
formal language 278
formal-syntactic properties 288–9
  language, music, brain 289–90
free will **5, 29–30**, 136, 153, 159
Frege, G. 27
Freud, S. **39, 77–8**, 291–2, 337, 431, 467, 488
frontal cortex 74, 77
fronto-parietal network 432
functional connectivity 320, 326, 385–6, 485, 498, 499

functional magnetic resonance imaging (fMRI)   4, **82–3**, 87–8, 93, 336–7, 339, 346, 350, 354, 392, 406, 413, 451, 457, 461, 479–80, 495, 498, 507, 511, 522
'functional brain imaging'   69, 89, 156, 166, 303, 310, 316, 338, 452
'functional imaging techniques'   162
see also   brain imaging
functional magnetic resonance tomography (fMRT)   139
'functional relations'   173, 181
functional roles   278–80, 300
functionalism   5, 155, **172–6**, 177, **178–9**, 180–1, 184, **186**, 206, 213, 215–16, 250, 267, 276, 278–80, 299, 519
  'overcomes two central problems of behaviorism'   174
  versions   **175–6**
fusiform face area   458, 460–1

Galen   73–4
Galilei, G.   44, 46
Gallagher, I. I.   455
Gallagher, S.   234, 515
Gallese, V.   **522–3**, 531–2
  personal identity (types)   **528**
  simulation concepts   **527–8**
Gallinat, J.   498, 499
gamma fluctuations   405, 406, 414
gamma oscillations   **384–5**, 466
gamma power   **486–7**
gamma range   414, 459, 485
gamma synchronization   389, 449, 461
gamma-Aminobutyric acid (GABA)   79, 82, 312, **319–20**, 415, 432
  GABA-A receptors   511
Gangi, C. E.   451
genes   304, 307
  data deficiencies   80
  and neural activity   **79–80**
Germany   30, 75, 451, 496–7
global metabolism   403, 407, 419f
  and energy demand   402, 424
global neuronal workspace (GNW) (Dehaene and Changeux)   **386–7**, 432
global workspace   382, 395, 400, 408
  specificity   **388–9**
glucose   **406–7, 410–12**
glutamate   79, 82, 312, **319–20**, 415, 511
goal orientation   **523–4**, 532
God   **133–4**, 138, 141, 151, 156, 253
Goldstein, K.   75, 95, 234, 338, 341
Golgi, C.   74
Granger causality analysis   485
'grasping and reaching'   **523–4**

'gray pulpy mass' (Schopenhauer)   226–7
Greece (ancient)   **23–4**, 40, 156
grey matter   **74**
'groundless metaphysical speculation'   206, 209–10
gyri   459, 498, 524

H_2O (chemical formula)   144, 161, 164, 165, 194
H2O PET investigation   484
Hacker, P. M. S.   **144**, 160, 228, 365
hard problem (Chalmers)   17, **143**, 353, **358t**, 380–1, 397, 425
  and brain   **374–5**
  and brain's intrinsic features   **377**
  'cannot be solved by neuroscience'   421
  code and   **434**
  'empirical answer'   418
  'empirical problem'   420
  versus 'explanatory gap' problem   375, 376
  guiding question   374–5, 409
  'logical problem'   421
  neural predispositions and   **409–10**
  neural predispositions, prerequisites, and correlates of consciousness   **419f**
  neurophilosophical approach   **434**
  neuroscientist and philosopher (dialogue)   **399–400**
  (un)consciousness versus non-consciousness   374, 376, 409
  see also   easy problems
hard problem: NPC and   in
  neuroscience and reductive neurophilosophy   **418–20**
  non-reductive neurophilosophy   **418**
  philosophy   **420–1**
  see also   nonconsciousness
'hardware'   155, **173**, 174, **175**, 177, 180–1, 186, 206, 213, 215
Harvey, W.   16, 147
He, B. J.   405
heart   73, 104, 147, **176–7**, 178, 233, 240, 331–2, 342, 347–8, 379
  demystification   **16**
  engineering versus physiological concepts   9–10
  explanatory gap (metaphorical comparison)   **9–10**
  'hard problem'   **8–9**
heat   163, 168
Hebb, D.   **80–1**, 83
Hebb synapses   81
Hegel, G. W. F.   94, 225
Heidegger, M.   30, 515
Helmholtz, H. von   78

'hetero-phenomenology' 273
higher-order functions 394–6, **396f**, 397
higher-order perception (HOP) 293
higher-order representation 291–5
  and awareness 292–3
  and consciousness 291–2
  and phenomenal features 294–5
higher-order thought (HOT) 293
  'higher-order cognitive theories of consciousness' (HOT) **38**
hippocampus 84, 145, 342, 392, 412, 451, 507
Hippocrates 3, 73
history of philosophy **150–1**
holism 329, 349, 350
  see also localization versus holism
Holt, D. J. 499
Hoptman, M. J. 498
horizontal explanations 249, **254–6**, 276
Hornsby, J. 260
'how it works' knowledge 358t, 364–6, 375–6
Huang, Z. xi, 480
Hume, D. 344, 454
  'bundle theory of self' **201–2**, 452, 475
  contested by Kant **202–3**
  'elusiveness thesis' 475, 476
  empirical-metaphysical fallacy **196f**, 196n
Husserl, E. **30–1**, 38–9, 417, 436, 515

i-identity (Gallese) **528**, 529
'I am warm' (observation sentence) 53
Iacoboni, M. **523–4**
idealism (Berkeley) 151
identity 467
  contingent (empirical implausibility) **166**
  'contingent' versus 'necessary' **165–6**
  designators: 'rigid' versus 'non-rigid' (Kripke) **163–5**
  different forms in neurophilosophy **162–3**
  different forms in philosophy **161–2**
  rejection **164–5**
identity theory 155, 172, 174, 180–1, 184, 208
  versus biological naturalism **193–4**
  'dealt serious blow' by Kripke 166–7
identity theory in philosophy **160–1**
imagination 392, 393
imitation,
  versus 'empathy' 518
'immunity against error through misidentification' 471, **476–8**, 483, 491, 493–4, **503–4**
  brain's intrinsic activity **504–5**

implicit certainties (Gallese) 528
implicit neurophilosophy
  brain-paradox 95
  Schopenhauer **94–5**
incommensurability **256**
  'normative-prescriptive' versus 'descriptive' explanations **257**
incorporation model 302, **308–9**, 312, 314, 327, 330
  incorporation of other disciplines into neuroscience **309f**
incorporation naturalism **307–8**
  same as 'replacement naturalism' 64n, 98
'increased body-focus' (Northoff) 506
'increased self-focus' **505–6**, 508–10
  see also 'decreased environment-focus'
independent component analysis (statistical technique) 289
indirectness 330, 346
'indirectness of studies of mind and brain' 305, **316–17**
  definition **316**
individual solipsistic identity (Gallese) 528
'inference by analogy' (philosophy) **456**, 457, 512, 514, **515**, 516–17, 529–31, 533
'inference problem' 183, **188–90**, 195, **196f**, 198, 205
inferior parietal lobule 523, 524, 525
inferotemporal (IT) cortex 385, 387–8
'information' **191–2**
  see also Chalmers
information integration **385–6**, 406
informational dualism **142–4**, 150, 153
'informed consent' **110–11**
infra-slow fluctuations (ISFs) 413
inner mental states,
  rejection in behaviorism **157–8**
'inner time and space consciousness' 436
Input,
  conceptual propaedutic as (Searle) **116**
input–output 176, 287, 341
input–output relations **157–9**
intrinsic activity *see* 'brain: intrinsic activity'
insula 339, 498, 511, 521–2, 524–5, 528
insulin
  and resting state activity **410–12**
Integrated Information Theory (IIT) (Tononi) 385
intentional organization 426, 444
'intentional stance' (Dennett) **253**, 254
intentionality **31–2**, **254**
  'hallmark feature of mental states' 437
  phenomenal feature of consciousness **437**
  same as 'intentional organization' **437**

inter-individual interaction 467
inter-individualism versus intra-individualism 456, 457
interactive dualism (Descartes) 127, **136–7**, 139, 141, 146, 199
interface problem (Bermudez) 249, **252**, 276
　domain (empirical) 252
　folk versus scientific psychology **251–2**, 258
　versus 'mind–brain problem' 252
internal criteria 55
'internal otherness' 517
interoceptive input 335
interoceptive stimuli 240, 241, 320, 342, 346, 348
intersubjective identity (Gallese) 528
intersubjectivity 108, 273, 450, 456, **512–34**
　empathy as affective form **516–17**
　neural predisposition 531
　philosophical problem **513–14**
　pre-cognitive form 517
　pre-reflective versus reflective 512
　'shared manifold' (Gallese) 529, 530f, 530, 531
　see also 'self and other'
intertheoretic reduction **97–8**
intrasubjectivity 517, 533
　relationship with intersubjectivity **529–31**
intrinsic features (Kant) **203–4**
　versus extrinsic stimuli (Hume) **202–3**
intrinsic observer-related intrusions 302, 328, 322
'intrinsic view' of brain 329, 341, 350
introspection **37–8**, 477
　Descartes **25–6**
inverted qualia **358t**, 363, **398–9**
　NPC and **422–3**
'inverted qualia argument' 172
　neuroscientist and philosopher (dialogue) **398–9**
　see also qualia
inverted spectrum 353, **360–1**, 363, 380, 398
investigation (methodological strategy) 47
ipseity 517
　'disturbed ipseity' 497
irreducibility 249, **253–4**, 257–9, 261, 267, 275–6, 299, 353
　see also explanatory irreducibility
'isolated brain' 240, 241, 363

Jackson, F. **366–7**, 373
Jackson, H. 75, 337

James, W. **71–2**, 128, 191, 417
　'dynamic flow' or 'specious present' 436
Jaspers, K. 497

Kant, I. 6, 50, 94, 210, 213, 216, 225–7, 230, 329
　biography **198**
　'categories' 344
　'epistemic instruments' 199, 200
　'epistemic-metaphysical inference' and 'empirical-metaphysical inference' **205–6**
　'functionalist *avant la lettre*' 206
　versus Hume **202–3**
　inference problem **196f**
　'inner sense' versus 'outer sense' 207
　intrinsic features and epistemic limitations **203–4**
　versus McGinn (property P) **221–3**
　mind as subject of cognition **227–8**
　'mind-dependence' **204**
　neuroscientific investigation **344–5**
　stance on mind-brain problem (current interpretations) **206–7**
　transcendental self 489, 490
　transcendental 'unity' 345
Kant: epistemic limitations 207, 208, 209n
　in knowledge of mind **198–9**, 203
　versus 'metaphysical assumptions' (Descartes) **199–200**
　and physical approaches **204–5**
Kant's consciousness, and brain's consciousness **344–5**
Kant's mind, and brain's intrinsic activity **344**
Kaplan-Solms, K. 78
Kim, J. 167
Kitcher, P. 206
Klein, S. B. 451
Kleinschmidt, A. 416
Kleist, K. 75
knowledge 58f, **105–6**, 231n, 232, 233, 380, 475
　a priori versus *a posteriori* (continuum) **50–2**
　epistemic problem 316
　objective versus subjective 358t
　see also observation sentences
knowledge argument 353, **358t**, **366–7**, 373, 397
　'knowing how' versus 'knowing that' 367
　NPC and **423–4**
　refutation 367

knowledge of consciousness: arguments 364–7
 knowledge argument 366–7
 knowledge argument: refutation 367
 'what it is like' argument 364–5
 'what it is like' argument: refutation 365–6
knowledge-acquisition 53, 106, 208, 529
knowledge-justification 53, 106
Koch, C. 384–5, 387–9, 408, 430
Koehler, W. 338, 341
Kraepelin, E. 75, **496**
Kriegel, U. 293, 296
Kripke, S. 51, 57, 166
 biography 163
 non-rigid designators and rejection of identity 164–5
 rigid designators and identity 163–4
'Kripkenstein' approach 163
Kuhlenbeck, H. 75
Kühn, S. 498, 499

L-allele 80
language 27, 52, **53**, 179, 219, 278, 299–300, 495
 formal-syntactic properties 289–90
language-acquisition 53, 432, 456, 463–4
'language of thought' (LOT) (Fodor) 278, 286–7, 293, 299, 300, 333
 and brain 288–9, 291
 'structural isomorphism' (Fodor) **284–5**
 sub-personal level 284–5
 'taken to brain' 297
Lashley, K.S. 81, 337, 341, **417**
laughter 393, 514, 516
'Law of Equipotentiality' (Lashley) 337–8
'Law of Mass Action' (Lashley) 337–8
'learning by doing' **519**
Legrand, D. 461
Leibniz, G. W. 141
Leitwissenschaft 536
'lesion-based method' 81, 82, 495
 see also brain lesions
level of consciousness 402, 404, 409, 424
levels of explanation **304**, 329, 330, 331
 and neurophilosophy 266
 see also explanation
Levine, J. 375
Lewis, C. M. 415–16
life (Searle) 116
linguistic analysis 29, 92
 mental features 27–8
linguistic functions 502, 517
linguistic representation 287–8
linguistics 27, 40–1, 60, 73, 159, 175, 286, 455–6

linguistics of mind 27–8
Lipps, T. 517
'lived body' (Merleau-Ponty) 32, 213, **234**, 237, 394, 438, 465, 516–17
 see also mind–body problem
Llinás, R. 413
Lloyd, D. **288–91**
 brain and music (resemblance) 290
 formal-syntactic properties in language, music and brain 289–90
 LOT and brain 288–9, 291
 MOT and brain 290–1
localization (view of brain) 329, 349
localization versus holism **336–9**, 350
 cognitive functions in specific regions 336–7
 holism in history of neuroscience 337–8
 neural overlap of functions 338–9
 'reversion to holism' 339
'localization-based approach' (to brain) 336
location problem (mental states) **354–7**
 arguments against reduction of consciousness 357
 'consciousness in brain' 356–7
 definition 355
 'neurophilosophical problem' 355–6
 neuroscientist and philosopher (dialogue) 356–7
locus coerulus (LC) 77, 79
logic 52, 57, 222, 359, 361
logical behaviorism **158–60**, 176, 181
 objections 159–60
logical consistency 125–6, 187
logical plausibility 122, 125–6, 185, 187–9, 368, 372
logical possibility (of consciousness) 369–71
logical world 57, 353, **372**, 418, 420–2
Logothetis, N. K. 82, 405
Lou, H. C. 485
low-frequency fluctuations 320, 402, **404–5**, 419f
lower-order functions **394–6**, 396f, 397
Lurija, A. R. 338

Maandag, N. J. 415
MacDougall, D. **3**, 147
machine/computer functionalism 175, 176, 179
magnetoencephalography (MEG) **83**, 459, 485, 522
magnetoresonance spectroscopy (MRS) 82
major depressive disorder (MDD) 505, 507
'Mary, superscientist' (Jackson) 358t, **366–7**, 373, 423

'mass action' 337–8
'massive modularity' (Carruthers) 336
'material me' 511
materialism 125–6, 185, 215, 217, 250, 258
 definition 156
 same as 'physicalism' 4, 156
 reductive 271
 supervenience 'non-reductive form' 168
McDowell, J. 260
McGinn, C. 107–9, 212, 228–32, 244, 362, 400
 access to, and knowledge of, property P (McGinn) 219–20
 biography 218
 brain in metaphysical domain 218–20
 versus Kant (property P) 221–3
 versus Nagel 224–5
 property P and brain (McGinn) 218–19
meaning 17, 160, 178–9, 180, 244, 247, 278–301
memory 72, 84–5, 335–6, 451, 455, 458, 461
 'autobiographical' 73
 'declarative' versus 'non-declarative' 84
 types 84
 'working' 73
Menon, V. 499
mental approaches 200–1
mental causation 127, 129, 139–40, 160, 168, 174
mental concepts 9, 10, 213
 elimination (neurophilosophy) 270–1
 versus physical concepts 220
mental escapes 150–2
 see also panpsychism
mental features 29–30, 250–1
 'do not exist' (scientific viewpoint) 535
mental properties
 Descartes versus Hume 344
mental representation 282, 300
 concept 281
 definition of 'representation' 280–1
mental self 450, 452–3, 471, 493, 500, 510
mental states 7, 515–16, 533
 functional roles 279
 'hallmark feature' 437
 inference by analogy 515, 516
 intention to write book 279–80, 281–2, 291–3
 'knowledge-driven' versus 'process-driven' attribution 520
 location problem 354–7
 versus physical states 127–8
 'private' 354, 355
 'subjectivity' 354

supervenience on neuronal states 166–7
mental substance (Descartes) 138, 449–50, 452, 468–9, 471–2, 475–6, 500
mental world (Popper and Eccles) 96
mental-physical dichotomy 186, 189–90, 193, 197–8
'mentalizing' 520–1, 512
'mereological fallacy (Bennett and Hacker) 228
Merleau-Ponty, M. 32, 95, 103, 212–13, 438, 515
 'being in world' 235
 biography 233
 concepts of body 234
 embodiment and sensorimotor functions 234–5
 'lived body' 234
messenger ribonucleic acid (mRNA) 80
Messi, L. 255
meta-analyses 458–9, 485, 499, 507
meta-representation 38, 454, 457, 501
metaphorical comparison
 demystification of heart and brain 16
 hard problem of heart 8–9
 knowledge of heart (explanatory gap) 9–10
metaphysical assumptions 199–201, 207–8, 221–2, 271
metaphysical claims
 explanatory gap problem 376–7
metaphysical domain
 brain in (McGinn) 218–20
 inference from empirical domain 149
 mind–brain problem as 'border problem' 150
 naturalized 70
 versus phenomenal domain 36
metaphysical dualism 259
metaphysical elimination 271–2
 relevance in neurophilosophy 274–5
metaphysical and empirical domains
 brain at border between 223–5
 differences 148–9
 linkage 148
'metaphysical explanatory gap' 376
metaphysical naturalism 43, 57–62, 92, 97–8, 100
 analytic naturalism 59, 60
 definition 57–8
 same as 'naturalization of metaphysics' 61
 and neurophilosophy 60–2
 synthetic naturalism 59–60
metaphysical self 453–4

metaphysics **44**
  and different domains of philosophy (ancient Greece) 23–4
  dualism between mind and body 24–5
  versus explanation 258
  purpose 24
  Quine 52
method versus subject of investigation 258–9
methodological behaviorism 157
methodological dualism,
  introspection and observation (Descartes) 25–6
methodological monism 66, **99**, 99n, **100**, 101–2, 112
methodological naturalism 43, **62–7**, 92, 98, 100
  cooperative naturalism 63, **64–6**
  definition 62–3
  and neurophilosophy 67
  replacement naturalism (Quine) 63–4, 66
methodological pluralism 66, **101**, 102
methodological strategies 247, 329
  brain-based 298–9
  elimination of mind and 297–8
  mind-based 297
  trans-disciplinary 92, **119**, 122, 123
methodology 1–2, 9, 12–15, 17, 18, 31, 42–3, 47, 428
  philosophy versus science **44–5**
  philosophy and science (continuum) 52–3
methods 33–4
Metzinger, T. 454
  'self-models' **452–3**
midline regions (of brain) 342, 403, 405, 409, 415, 459, 485, 493, 508–9
  anterior-to-posterior connectivity 499
  see also cortical midline structures
Mill, J. S. **514**
Millikan, R. 296
mind 1, 12–14f, 43, 63, 355, 363, 452, 514, 533
  'bundle' of extrinsic stimuli (Hume) 201–2
  'central focus in philosophy' 70
  consciousness 'hallmark feature' (Dennett) 252
  'does not exist' (scientific viewpoint) 535
  embodied approach 394
  hallmarks 301, 354, 375, 447
  linguistics 27–8

mental features 29–30
metaphysical naturalism 58
metaphysics 28–9
philosophical versus scientific approach 45
'purely physical approach' 381
representational account 300
and semantic content 279
subject of cognition 227–8
mind: concept
  Aquinas 133–4
  Aristotle 132–3
  Plato 131–2, 133
mind and body,
  metaphysical dualism (Descartes) 24–5
mind and brain
  from elimination of mind to reductive neurophilosophy 4–5
  'indirectedness of studies' 316–17
  past times 2–3
  from philosophy through neuroscience to neurophilosophy 16–17, 21–124
  from presence of mind to philosophy of mind 2–3
  present times 4–5
mind, brain, science: psychology and neuroscience 16, 21, 63, 66, **69–90**, 91, 93, 95, 106, 123
  Berger and EEG 82
  chapter focus 70
  cognitive functions and consciousness 83–4
  concepts/facts: discrepancy with method of neurophilosophy 85–6
  Hebb and synapses 80–1
  neuroscience **78–88**
  neuroscience: history 73–8
  philosophy of mind to science of mind 69–70
  psychology: history 70–3
  sensorimotor functions and consciousness 76–7
  see also psychology
mind and existence (Descartes) 135–6
mind and God (Aquinas) 133–4
mind and meaning 17, 247, **278–301**
mind problem 249, 275
mind-based approach 95, 212, **214–16**, 217, 219, 224, 243, 298, 351
  top-down 278, 297, 301
  brain-based methodological strategy 297, 298–9
mind–body problem 3, 21, 28, 40, **130**, 351–2, **355–6**

Aquinas 134–5
Descartes 136
mental approaches 355
see also body
mind–brain dualism 153
  Eccles 96
  rejected 98
mind–brain identity theory 170
mind–brain models 258–9
mind–brain problem 3, 17, 125–245, 535–6
  'awaits solution' 536
  as 'border problem' 150
  'condensation of mind-body problem' 28
  'core issue in philosophy of mind' 70
  definition 128–9
  'metaphysical problem' 250
  mind-based approach 179
  nature 28
  'philosophy of mind' to 'philosophy of brain' 125–6
  reformulation 242–3
  scientific (versus philosophical) approach 67
  twentieth century 155–6
mind–brain problem: brain-based approaches 17, 212–45
  brain and body 233, 234–5
  brain at border between metaphysical and empirical domains 223–5
  brain in epistemological domain 225–7
  brain in metaphysical domain 218–20
  brain and world 236–40
  'brain-based approach' 215–17
  critical reflections 221–3, 235–6
  'domain monism' versus 'domain pluralism' 213
  'fixed domain location' versus 'flexible domain location' 214
  future research 224, 227–33, 236, 240–3
  'mind-based approach' 214–16
  neurophilosophical versus philosophical approaches 217
  physicalism 216–17
mind–brain problem: computational escapes
  behaviorism 174–5
  functionalism 172–6
  physicalism 174–5
mind–brain problem: critical reflections
  biological naturalism and brain 194–5
  brain and sensorimotor functions: relation to body during consciousness 235–6
  different forms of identity in neurophilosophy 162–3
  dualism and science 147
  'embodied and embedded brain' 236
  empirical and metaphysical domains: differences 148–9
  empirical and metaphysical domains: linkage 148
  epistemic dualism 206–7
  functionalism and meaning 178–9
  inference from empirical to metaphysical domain 149
  Kant's stance (current interpretations) 206–7
  Kripke: non-rigid designators and rejection of identity 164–5
  Kripke: rigid designators and identity 163–4
  McGinn versus Kant 221–3
  'metaphysical problem' 197–8
  mind–brain problem as 'border problem' 150
  phenomenal features (of consciousness) and physicalism 171–2
  'phenomenal-physical inference' 195–7
  'physical' (notion) 169–70
  physical and phenomenal features 170–1
  physicalism 169–70
  semantic meaning in neurophilosophy 179–80
mind–brain problem: epistemic escapes 198–206
  epistemic limitations in knowledge of mind (Kant) 198–9
  epistemic limitations versus metaphysical assumptions (Kant versus Descartes) 199–200
  epistemic limitations and physical approaches 204–5
  'epistemic-metaphysical inference' and 'empirical-metaphysical inference' (Kant) 205–6
  'epistemic-metaphysical inference' and mental approaches (Kant on Descartes) 200–1
  intrinsic features and epistemic limitations (Kant) 203–4
  intrinsic features versus extrinsic stimuli (Kant versus Hume) 202–3
  metaphorical comparison (umbrellas in Edinburgh and Koenigsberg) 203
  mind as 'bundle' of extrinsic stimuli (Hume) 201–2

mind–brain problem: ideas for future research
  'brain as functioning' 230–1, 232–3, 241–2
  'brain as observed' 229–30, 231–2
  brain as object of cognition 231–2
  brain as subject of cognition 232–3
  brain-based approach to mind-brain problem 178
  'brain-paradox' 227–8
  'degradation of brain' and brain's extrinsic features 176–7
  'elevation of brain' and brain's intrinsic features 177–8
  empirical implausibility of contingent identity 166
  'extended brain' 240–2
  mind–brain problem as domain problem 208–9
  mind–brain problem as epistemological problem 207–8
  necessary versus contingent identity in neurophilosophy 165–6
  'philosophy of brain' 242–3
  shift from 'brain paradox' and mind-brain problem 228–9
mind–brain problem: mental approaches 17, 127–54, 213
  'causal closure' of physical causes 129–30
  critical reflections 147–50
  dualism 137–42
  dualism: history 130–7
  mental escapes 150–2
  mental versus physical states 127–8
  naturalistic escapes 142–4
  neuronal escapes 145–6
mind–brain problem: metaphysical escapes 190–2
  neutral monism: problems 191–2
  neutral monism: Russell 190–1
mind–brain problem: metaphysical-empirical escapes 192–4
  biological naturalism (Searle) 192–4
  identity theory 193–4
mind–brain problem: non-mental and non-physical approaches 17, 183–211, 213
  critical reflections 194–8, 206–7
  definition of concepts 185–6
  'definition problem' as neurophilosophical rather than philosophical problem 186–7
  'definition problem' as neurophilosophical rather than neuroscientific problem 187–8

  'domain problem' 209f
  empirical and phenomenal implausibility of physical and mental approaches 184
  epistemic escapes 198–.206
  further research 195
  ideas for future research 195, 207–9
  inference from absent to present phenomenal features in mental approaches 188
  'inference problem' 209f
  'inference problem' and inferences between different domains 188–90
  Kant 198, 206–7
  metaphysical escapes 190–2
  metaphysical-empirical escapes 192–4
  'plausibility problem' 185
  'relationship problem' 187
mind–brain problem: physical and functional approaches 17, 155–82, 213
  background 155–6
  computational escapes 172–6
  critical reflections 162–5, 169–72, 178–80
  ideas for future research 165–6, 176–8
  physical escapes 166–8
  physicalism: history 157–62
'mind-dependence' (Kant) 204
mind-reading 18, 461
'mineness' 463–5, 469, 476, 497, 501–2, 504–5
minimal self 449–50, 455–6, 463, 464–5, 466, 468–9, 472, 493, 502–3, 515
minimally conscious state (MCS) 403, 480
mirror neurons 108, 512, 522, 532–4
  and different regions and functions 524–5
  and goal orientation 523–4
MisMatch Negativity (MMN) 480
modularity and modules 332–6
  cognitive functions 335–6
  definition 332–3
  sensory functions 333–4
  sensory functions as modules 334–5
modules 329, 349
  characterization 333
  'encapsulated' 333
monism 167, 168, 189
monkeys 522–4
'mosaic unity' (Craver) 312, 328
motor cortex 230, 336, 342, 461, 465, 502
Müller, J. P. 78
'multiple realizability' 155, 175, 176
music 288–9, 300, 301

formal-syntactic properties 289–90
resemblance to brain 290
music of thought (MOT) 278
and brain 290–1
'mysterianism'/'mysterians' 109, 220, 400

Nagel, T. 108–9, 212, **223**, 229–33, 244, 364, 369, 400, 411, 444
characterization of consciousness 435
versus McGinn 224–5
see also 'what it is like'
names 479–80, 483, 486, 499, 505
nativism 519
natural world 57, 353, **370**–2, 418, 420–1
naturalism 47–8, 307, 308
a priori versus *a posteriori* knowledge (continuum) 50–1
analytic versus synthetic sentences (continuum) 49–50
analytic versus synthetic sentences (dichotomy) 48–9
epistemology naturalized 52–3
metaphysical 57–62
methodological 62–7
non-reductive 101
observation sentences 53–6
philosophy and naturalism 58f
philosophy and science (continuum) 51–2
same as 'naturalization of philosophy' 56–7
see also 'philosophy and science'
naturalistic escapes 142–4
appeal of dualism (Searle) 142
predicative or conceptual dualism (Bennett and Hacker) 144
zombies and informational dualism (Chalmers) 142–4
'naturalistic property dualism' 143–4
naturalization 97, 307
naturalization of mind 58–9, 96
naturalization of philosophy 61, 91, 92, 98
same as 'naturalism' 56–7
NCC (neural correlates of consciousness) 17, 382–401
affective functions and consciousness 393
chapter focus 400
cognitive functions: working memory and attention 390–2
conscious versus unconscious contents 401
consciousness as higher- or lower-order function 394–6, 396f

content-based 402, **404**, 406, 408, 416, 424
empirical and neuroscientific mechanisms 400
gamma oscillations and neuronal synchronization 384–5
global workspace 386–7
hard problem and **419f**
investigation of consciousness 427
metaphorical comparison (floors, tables, flowers) 395–6
neurophilosophical discussion 387–90, 392–3, 397–400, 408–10
neuroscientist and philosopher (dialogue) 398–400
philosophical arguments 397–8
qualia (absent and inverted) 398–400
re-entrant processing and information integration 385–6
sensorimotor functions and consciousness 394
specificity 387
term introduced by Crick and Koch 384
'unspecificity' 400–1
NCC: background 382–3
starting point in neuroscience 383
starting point in philosophy (phenomenal features of consciousness) 382–3
NCC: level-based 402, 424
energy and consciousness 407–8
intrinsic activity and low-frequency fluctuations 404–5
low-frequency fluctuations and consciousness 405–6
metabolism and energy demand of brain 406–8
slow-wave hypothesis 405–6
NCC: neurophilosophical discussion
'brain without oscillations' and brain's intrinsic features 389–90
cognitive functions and consciousness 392–3
consciousness as basic function 397
global workspace and neuronal synchronization: specificity 388–9
NCC and qualia (absent and inverted) 398–400
NCC and philosophical arguments 397–8
're-entrant connections': neuronal and experimental specificity 387–8
specificity of NCC 387
nervous system 417
neural abnormalities 507–8

neural activity   288–9, 298, 321, 325, 332, 336, 340f, 341–3, 346, 349, 353, 388–9, 401, 406, 409, 413–14, 451, 491
  abnormal (during self-specific stimuli)   499
  biochemistry   78–9
  and consciousness   407
  encoding   468
  genes and   79–80
  'ignition'   432
  self-specificity   449
  spatial patterns during self-specific stimuli   457–8
  spatiotemporal patterns   449, 460
  structure and organization   347–8
  temporal patterns during self-specific stimuli   458–9
'neural coalition'   384
neural correlates   5–6, 20, **408–9**, 418, 419
  see also NCC
neural excitation/inhibition   319–20
neural functions   **105–6**
neural mechanisms   54, 236, 409, 434, 440, 461, 468, 538
neural networks   304, 307, 316, 338–9, 521
neural overlap   459, 462, 468, 471, 487, 524
  self and resting state activity   **484–5**
neural predisposition of self (NPS)   471, 490–1
  and brain's intrinsic activity   **488–9**
  and explanatory gap   **489–90**
neural prerequisites of consciousness   408–9, 410, 418, 419
  hard problem and   **419f**
'neuralism'   38, 61–2, 91, 100, 102, 123
Neurath, O.   156
neuroanatomy   77, 339, 534
neuroanthropology   304, 307, 309f
neuroeconomics   87, 305, 307, 309f
neuroepistemic limitations   **106–7**
'neuroepistemological constraints' (Northoff)   326
neuroepistemology   91, 92, 99f, 122–3, 422
  'bridge discipline' (Churchland)   106
  neural functions and knowledge   **105–6**
  neuroepistemic limitations   **106–7**
neuroethics   91, 99f, **110–11**, 122–4, 304, 309f
  definition   110
'neuroexperimental constraints' (Northoff)   326
neurogenetics   79, 80
neuroimaging   80, 87, 512
neurolaw   305, 307, 309f
neurology   78, 89, 447, 495

'neuron doctrine' (Ramón y Cajal)   69, 74
neuronal activity   356–7, 398, 400, 411, 415, 421, 423, 440, 453
neuronal causation   **129**
'neuronal chauvinism'   **173**
neuronal escapes
  dualism and brain (Sherrington and Penfield)   **145–6**
  psychons and brain (Eccles)   **146**
neuronal explanations   375–9, 439
neuronal features   353, 355, 357–9, 366, 374–6, 383, 397, 400–1, 418, 423–4, 444
neuronal functions   291, 330
neuronal mechanisms   39–40, 104, 231, **241**, 261, 267, 297, 307, 319, 353, 356, 367–8, 370, 373–5, 377, 382, 384, 387, 398–404, 408, 411–12, 418, 419n, 423, 427–8, 430, 434, 437, 439, 441, 443, 447, 449, 452, 464, 484, 491, 493, 496, 531–3
neuronal processes   232–3, 317, 319–23, 453, 467
  consciousness-yielding (or not)   **194–5**
  objective versus subjective   194–5
neuronal processing   399, 400
'neuronal specificity' (Northoff)   382, **387–8**, 398, 406, 449, 468–9
  definition   460
  midline regions   **460–1**
neuronal states   164–5, 173, 176–7, 181, 222, 230, 330, 354–5, 360–2, 369, 375–6, 383, 421, 423, 428, 439–40
  supervenience of mental states   **166–7**
neuronal synchronization   **384–5**, 382, 387–90, 395, 398, 400–1, 405–6, 408, 469
  specificity   **388–9**
neuronal–mental dissociation   **362–3**
neuronal–neuronal dissociation   **362–3**
neurons   4, 18, 88, 167, 255–6, 337, 414, 416, 524
  basic functional unit of brain   **73–4**
neurons and cognitive functions   **81–2**
neuroontology   91–2, 99f, 122–3
  brain-based   **109–10**
  mental characterization of brain   **108–9**
  physical characterization of brain   **107–8**
neurophenomenology   42, 77, 95, 99f, 102–3, 123, **439**
neurophilosophers   34, 319
  avant la lettre   95
neurophilosophical discussion
  definition   **460**
  empirical plausibility of qualia as context–content relationship   **363–4**
  epistemic–metaphysical inference   **367–9**

explanatory gap as brain-based problem 379
hard problem and brain's intrinsic features 377
lack in current knowledge 378–9
logical and natural worlds 371, 372
neuronal–mental versus neuronal-neuronal dissociation 362–3
zombies 371–3
neurophilosophy ix-x, 12–13, 16–17, 20, 26, 60–2, 65, 67, 276, 299, 301, 315–16, 340f, 353
avant la lettre 94
brain and philosophy 91–124
brain-based (versus mind-based philosophy) 275
brain-based versus brain-reductive 103–4, 104–5, 108–9, 113, 120, 125
concepts 97–103
core problems 351
'current sense' 1, 96
definition 5
different forms of identity 162–3
distinguished from 'neuroscience' 305
domains 105–11
explanatory levels and 266
history 94–7
mental concepts (elimination) 270–1
mental states (location problem) 355–6
method 85–6, 111–16
'necessary' versus 'contingent' identity 165–6
phenomenal and metaphysical elimination (relevance) 274–5
and philosophical and psychological functionalism 266–7
versus philosophy 121–2, 217, 260–1
reductive versus non-reductive 103–4, 247, 273–4, 316, 368
relationship with other disciplines 306f, 309f
semantic meaning in 179–80
world-based 538
see also non-reductive neurophilosophy
neurophilosophy: concept
domain and method (dissociation) 97
intertheoretic reduction and replacement naturalism 97–8
reductive neurophilosophy 98–100
neurophilosophy: concepts
neurophenomenology 102–3
non-reductive neurophilosophy 100–2
neurophilosophy: history 94–7
explicit neurophilosophy 95–7
implicit neurophilosophy 94–5

neurophilosophy: ideas for future research 117–22
Churchland versus Searle 117
'concept-fact iterativity' versus 'concept-argument iterativity' 120–1
'concept-fact iterativity' as trans-disciplinary methodological strategy 119
'conceptual feedforward input' and 'feedback output' 117–18
'conceptual re-entrant loop' and 'concept-fact iterativity' 118–19
philosophy versus neurophilosophy 121–2
neurophilosophy: method 111–16
'bootstrapping with concepts as outputs (Churchland): definition 113–14
'bootstrapping with concepts as outputs: example 114–15
concepts: unilateral adaptation to facts (Searle) 115–16
concepts: 'unilateral inference from facts' 112–13
conceptual propaedeutic as input (Searle) 116
neurophilosophy of consciousness 351–445
neurophilosophy of self 424
neuropsychiatry 79, 81, 324
neuropsychoanalysis 68, 77–8, 511
neuropsychology 81, 86, 495
neuroscience 4, 6, 16–17, 302–3
background 535–6
beginnings 495
biochemistry of neural activity 78–9
conceptualization 313–14
contextualization 313–14
domains 332
empirical investigation of self 451–2
explanations, concepts and observer 302–28
extension into other disciplines 87–8
genes and neural activity 79–80
holism in history of 337–8
'how' questions 332, 536
incorporation of other disciplines 309f
interdisciplinary foundation 86, 88, 304–6, 536
introduction as discipline 86–7
levels of explanation 304
'localizationist' versus 'holistic' approach (to brain) 339
'*mélange de tout*' 78
merger with cognitive psychology 88
merger with philosophy 78

neuroscience – *continued*
  multi-level causal explanations   312–13, 314
  NCC   383
  neural basis of cognitive functions   83
  neurons and cognitive functions   81–2
  and neurophilosophy   93
  'no homogeneous entity'   87–8
  NPC and hard problem   418–20
  phenomenal and neuronal specificity lacking   437
  present findings   341–2
  raises 'how' questions   473
  regions (of brain) and cognitive functions   82–3
  relation with psychology   303–4
  relationship with other disciplines   306f, 309f, 311f, 316
  self and consciousness   447
  theoretical and methodological issues   330
  see also philosophy of neuroscience
neuroscience: history   73–8, 340–1
  association of brain science with different disciplines   77–8
  clinical disorders and brain   74–5
  from clinical disorders of brain to consciousness   75–6
  institutionalization of neuroscience as discipline   78
  neurons as basic functional unit of brain   73–4
  see also 'mind, brain, science'
neuroscience: levels of explanation   306–12, 314–16
  brain-based   315–16
  brain-reductive   314–16
  'central issue'   305
  domain overlap   311f, 315
  explanatory context and critical neuroscience   310
  explanatory context and 'contextualization model'   310–12
  explanatory reduction and 'incorporation model'   308–9
  multi-level   312–13
  'neuronal' versus 'non-neuronal'   307, 310–15, 330
  'over-interpretation'   310
  relation between different levels   307–8
neuroscience of consciousness   382, 424
  diabetes comparison   410–11
'neuroscience of ethics'   110–11
neuroscientist   55
  dialogue with philosopher   398–400, 487–90

neurosocial structure and organization   467–8
neutral monism (Russell)   183, **190**–1, 197, 206, 210, 213, 250
  problems   191–2
Newton, Sir Isaac   44, 46
Noe, A.   234, 442
'nomological danglers' (Feigl)   141
non-conceptual content, attribution of phenomenal features   296
non-conceptual representation of content   295–6
non-derivability   256, 259
'non-naturalistic dualism'   143–4, 145
non-phenomenal states   375, 376, 377
non-reductive neurophilosopher (NN)   34, 103–5
non-reductive neurophilosophy   1–2, 100–2
  conceivability argument   372–3
  concepts of self   447
  nature   6
  NPC and hard problem   418
  see also reductive neurophilosophy
non-reductive physicalism   5, 6, 155, 180
  see also physicalism
nonconsciousness   39–40, 358t, 374–8, 380, 397–400, 404, 407, 409–10, 416–18, 419f, 424, 426–7, **432**, 434, 443–4, 483
  see also vegetative states
normative dimension   110
normative-prescriptive explanations   256–7
Northoff, G.   78, 80, 149, 152, 317, 320–1, 416, 439, 451, 454, 458, 462, 470, 479, 485–6
  radial-concentric threefold anatomical organization   320
noumenal world (Kant)   204
neural predispositions, and hard problem   409–10
NPC (neural predispositions of consciousness)   5–6, 17–18, 402–25
  chapter focus   402, 424
  current research themes   412–16
  definition   410
  disorders of consciousness   402–4
  further research   415, 424
  hard problem and   419f
  ideas for future research   416–18
  investigation of consciousness   427
  knowledge lacking   417
  metaphorical comparison (diabetes)   410–12
  neurophilosophical discussion   408–10, 418–24

NPC: current research themes 412–16
  brain's intrinsic activity: spatial
    characterization 412–13
  brain's intrinsic activity: temporal
    characterization 413–14
  resting state and stimuli: interaction
    414–15
  resting state and stimuli: non-linear
    interaction 415–16
NPC: ideas for future research 416–18
  intrinsic activity and consciousness 416
  spatiotemporal structure and
    consciousness 417–18
NPC: neurophilosophical discussion
  408–10, 418–24
  brain's intrinsic activity 421
  explanatory gap problem 418, 421–2
  neural correlates versus neural prerequisites
    408–9
  neural predispositions and hard problem
    409–10
  NPC and absent and inverted qualia 422–3
  NPC and explanatory gap problem 421–2
  NPC and 'what it is like' and 'knowledge'
    arguments 423–4
NPC and hard problem
  neuroscience and reductive
    neurophilosophy 418–20
  non-reductive neurophilosophy 418
  philosophy 420–1
Nussbaum, M. C. 132

Obama, B. H., Jr 475
'objective body' (Merleau-Ponty) 32, 394
objectivity 7, 29, 354–5, 360, 364, 383, 435,
  438–40, 501
observation 37–8, 247, 249–50, 256, 269,
  330, 371, 515, 526–8, 535
  Descartes 25–6
  versus 'experience' 435
  see also third-person perspective
observation sentences (Quine) 43
  and 'experiential similarity' 55–6
  and internal criteria 55
  receptual and perceptual similarity 53–4
  social similarity 54–5
observational-experimental methodology
  (science) 33–4, 47–8, 53, 62–7, 70–2,
    89, 92–3, 96–7, 100–3, 112–13, 118–20,
    122–3, 309, **537**, 538
observer 316–22
  neuroscience 302–28
observer-based concepts 302, 317–18,
  319–22, 327, 329–30, 347
observer-related intrusions 322–6, 346

definitions 323, **326**
  extrinsic **323**, 324, 330
  intrinsic 324, **325–6**, 330
'offline' versus 'online' methodological
  strategy 459
ontology 24, 224, 232
  first- and third-person (Searle) **193–4**
  subsumed by author under 'metaphysics'
    44
orbitofrontal cortex (OFC) 521
Owen, A. M. **391–2**
'ownership' 451, 461, **465**, 474

pain 161–2, **164–6**, 271, 295, 403, 476, 515,
  520–2, 524–5
pain versus C-fibers **59**, 60
Panksepp, J. 393, 466, 470, 511
panpsychism 127, 153, 155–6, 180, 184–5,
  536
  history of philosophy **150–1**
  strong and weak forms **152**
  twentieth-century philosophy **151–2**
PANS [Positive and Negative Syndrome]
  scale 498
Papineau, D. 296
parallel-distributed processing computer
  (PDP) 286
parallel-distributed processing **286–7**
parallelism (brain –mind) **140–2**, 150
parietal cortex 77, 336, 338, 390–2, 465,
  521, 523
  bilateral 341, 405, 413
  lateral 412, 391
  posterior 341, 523, 524
Parkinson's disease 75, 78, 486
Parnas, J. **497**, 499
patients 114, 334–5, **480–2**
  neurological 35
  psychiatric 35, 361
Pavlov, I. 158
Penfield, W. 75, **145–6**, 147
perception **130–1**, 134, 151, 207–8, 219–22,
  231n, **237–9**, 253, 269, 293, 331, 366,
  436–7, 441–2, 450, 452, 461–2, 475, 486,
  503, 506, 532
perceptual similarity (Quine) **53–4**, 55, 56
periaqueductal gray (PAG) 166, 465, 507
perigenual anterior cingulate cortex (PACC)
  415, 458, 480, 485–7, 507, 521
'personal identity' 450
  types (Gallese) **528**
personal level 299–300, 303
  'central hallmark' 278
  semantic contents 283
  'structural isomorphism' (Fodor) **284–5**

'personal levels' versus 'sub-personal levels' of explanation 17, 247, **249–77**
Pfeiffer, U. 459
'phase durations' 405
'phase-phase coupling' 414
'phase-power coupling' 414
phenomena and consciousness 30–1
'phenomenal' (concept) 272
phenomenal analysis 43, 101f, 112
phenomenal consciousness 375, **388–9**, 396f, 397, 399, **419f**, 426, 428, **429–30**, 444
  definition **429**
  relation to 'access consciousness' 430
phenomenal content/s 419f
  'central dimension of consciousness' 383
  same as 'content of consciousness' 383
phenomenal domain 41, 62, 99f, 101f, 123
  versus epistemological domain 36
  versus metaphysical domain 36
phenomenal elimination 272
  relevance in neurophilosophy **274–5**
phenomenal explanations 375–6, 378–9, 439
phenomenal features 77, 160, 278, 353, 358t, 376, 411, 423, 443, 464
  attribution to non-conceptual content 296
  and higher-order representation **294–5**
phenomenal features of consciousness 357–62, **382–3**, 412, 417, 421, 424, 426, 428
'phenomenal implausibility' **184**, 185, 189
phenomenal limits **464**
phenomenal method 47, 100, 102
phenomenal plausibility 188, 189
phenomenal self 449–50, **454**, 466, 468–9, 493, 515
phenomenal specificity 449, 460, 468, 469, 500
  of self-reference **463–4**
phenomenal states 273, 376, 377, 421
'phenomenal time' (Husserl) 32, 417, 436
phenomenal-empirical fallacy **196f**, 196n
'phenomenal-physical inference' **195–7**
phenomenal-qualitative features 366, 383, 428
'phenomenal-qualitative feel' 357, 359, 411, 435
phenomenological philosophy/ phenomenology 23, **30–41**, 70, 76, 91–2, 95, 100, 124, 437–8, 454, 463–5, 472, 475–6, 483, 487, 492, 501, 517
  body 32
  brain ('category error in non-reductive neurophilosophy') **33–4**

brain ('nothing but category error') 33
brain 'at junction between disciplines' **34–5**
  'claim' 36
  definition 31
  experience and introspection **37–8**
  phenomena and consciousness 30–1
  versus psychology **35–7**
  self and other **515–16**
  structural hallmarks 31
  subject matter 36
  subjectivity and intentionality **31–2**, 33, 35, 37, 38
  unconscious and consciousness **39–40**
Philosopher,
  dialogues with neuroscientist **398–400**, **487–90**
philosophical approach,
  versus neurophilosophical approach (mind-brain problem) 217
philosophical concepts,
  brain and self-consciousness **474–9**
philosophical functionalism 250, 266, 276
philosophy ix-xi, **16–17**, 18–20, 58f, 88–9, 311f, 513, 535
  'a priori knowledge' 51
  concept(s) of brain **10–11**
  conceptual determination of self **449–50**
  consciousness (hard problem of brain) **7–8**
  definition 23
  domain and method (adjustment) 56
  focus 24
  hallmark feature 538
  Husserl's aim 31
  identity: different forms **161–2**
  identity theory **160–1**
  intersubjectivity 533
  investigation of brain **11–13**
  and naturalism 64f
  naturalization (Quine) 48, **56–7**
  need for brain **1–2**
  versus neurophilosophy **121–2**, **260–1**
  NPC and hard problem **418**
  phenomenal features of consciousness **382–3**
  versus psychology 71
  questions asked 473, 536
  self and consciousness 447
  task and purpose (Quine) **51–2**
philosophy of brain 15f, **15–16**, 66, 212–13, 242–3, 298, 305–6, 306f, 331
  definitions 242, 305, 332
  domains 332
  'what' questions 332

philosophy of brain: characterization of
brain   11, 15, **17**, 55, 66, 76–7, 94, 247,
329–50, 405
  brain: views   340–5
  brain and function   332–9
  brain 'nothing but car'   331–2
  'central question'   349
  chapter focus   329, 332, 349
  critical reflections   345–8
  philosophy of neuroscience   330–1
  theoretical and methodological issues in
    neuroscience   330
philosophy of brain: critical reflections
  345–8
  dialogue between neuroscientist (NS) and a
    brain (BR)   345–8
  intrinsic activity (concept)   346–7
  intrinsic activity and consciousness   347–8
  'need to talk to brain itself'   345–6
philosophy of mind   2–3, 12f, **15–16**, 23,
  43, 59, 65–6, **69–70**, 91, 242, 249, 275,
  299, 301, 353, 491
  'core problem'   352
  definition   28
  linguistics of mind as linguistic analysis of
    mental features   27–8
  mental features of mind   29–30
  as metaphysics of mind   28–9
  phenomenology   30–8
  from to psychology   250–1
  relation between first- and second-person
    perspectives   38
  unconscious and consciousness   39–40
philosophy, mind, naturalism   92–3
philosophy of mind and phrenology   16,
  17, 21, **23–42**, 44, 61, 66, 69–70, 76, 88,
  91, 92, 122
  chapter focus   26
  living room and kitchen (metaphorical
    comparison)   26
  metaphysical dualism between mind and
    body   24–5
  methodological dualism between
    introspection and observation   25–6
  metaphysics and different domains of
    philosophy (ancient Greece)   23–4
  philosophy of mind   27–30
philosophy of neurophilosophy   247,
  315–16
philosophy of neuroscience   247, 298,
  304–5, 306f, **315–16**, 329, **330–1**
  critical reflections   312–14
  definition   305, 330
  distinguished from 'neurophilosophy'
    305

function   331
levels of explanation   326–7
subject matter   331
theoretical and methodological issues   330
see also cognitive neuroscience
philosophy of neuroscience: explanations,
  concepts and observer   15, **17**, 20, 109,
  247, **302–28**, 329, 348–9
  brain and observer   316–22
  Buszáki (biography)   323–4
  critical reflections   312–14
  levels of explanation   305, **306–12**,
    314–16
  philosophy of brain   305–6
  philosophy of neuroscience   304–5
  from psychology to neuroscience   302–3
  relation between psychology and
    neuroscience   303–4
philosophy of neuroscience: ideas for future
  research   322–6
  intrinsic features of brain   324–6
  observer-related intrusions   322–3
  observer-related intrusions: extrinsic   323
  observer-related intrusions:
    intrinsic   325–6
philosophy of psychology   275, 276, 302–3
  'central topic'   249
philosophy of psychology: critical
  reflections
  criteria for success and failure of matching
    (Stich)   270
  explanatory dualism versus metaphysical
    dualism   259
  explanatory levels and
    neurophilosophy   266
  explanatory models and mind-brain
    models   258–9
  explanatory relevance   260
  linguistic versus vectorial
    representation   287–8
  localized versus distributed representation
    286
  matching between folk and scientific
    psychological concepts   269–70
  mental concepts (elimination in
    neurophilosophy)   273–4
  metaphysical elimination (relevance in
    neurophilosophy)   274–5
  metaphysics versus explanation   258
  method versus subject of
    investigation   258–9
  methodological strategy   297–9
  neurophilosophy and philosophical and
    psychological functionalism   266–7
  parallel-distributed processing   286–7

philosophy of psychology: critical
  reflections – *continued*
  phenomenal elimination (relevance in
    neurophilosophy) 274–5
  philosophical versus neurophilosophical
    approach 260–1
philosophy of psychology: ideas for future
  research 288–91, 300
  brain and music (resemblance) 290
  formal-syntactic properties in language,
    music and brain 289–90
  LOT and brain 288–9, 291
  MOT and brain 290–1
philosophy of psychology: mind and
  meaning 17, 247, 278–301
  chapter focus 279
  ideas for future research 288–91, 300
  mind and semantic content 279
  representation 291–6
  semantic content 280–6
  semantic content and
    representation 279–80
philosophy of psychology: personal versus
  sub-personal levels of explanation 17,
  247, 249–77, 278, 299
  explanatory continuity 261–6
  explanatory elimination 267–73
  explanatory irreducibility 253–7
  folk versus scientific psychology 251
  interface problem 251–2
  from philosophy of mind to
    psychology 250–1
  root question 249
philosophy of psychology and
  neuroscience x, 2, 6, 17, 36, 56, 72, 76,
  88, 106, 512, 538
  from explanation of mind to explanation
    of brain 247–350
philosophy and science 44–6, 65f, 66–7,
  247
  'clear dividing border' 34
  continuum 51–3, 57
  'dichotomy undermined' 56, 57
  domains and methods (bridged) 47–8
  'principally different' 45–6
  segregation or parallelism 58f
philosophy and science: naturalism x, 6,
  13, 15, **16**, 21, 29, 34, 36, 38, **43–68**, 69,
  70, 88, 91, **92**, 105, 122
  breakdown of 'analytic-synthetic'
    distinction 50, 51
  domains versus disciplines **46–7**
  see also cooperative naturalism
philosophy–science continuum 91
photosynthesis (Searle) 193

phrenology 16, 17, **23–42**
'physical' (notion) 169–71
physical approaches 204–5
physical causation 129, 140
  'causal closure' 129–30
  see also epiphenomenalism
physical escapes 166–8
  supervenience and epiphenomenalism
    167–8
  supervenience of mental states on
    neuronal states 166–7
physical (p) versus physical (o) 169–70
'physical stance' 253
physical states 129, 221, **354**, 355, 357,
  369, 375
  versus mental states 127–8
  'objective' versus 'subjective' 193
physical world (Popper and Eccles) 96
physicalism 155, 166, **169–72**, **174–5**,
  178–80, 189, 217, 250, 360, 365
  versus 'brain-based approach' 216–17
  definition 156
  non-reductive versus reductive 155, **167**,
    168, 180, 184
  see also materialism
physicalism: history 157–62
  behaviorism: objections (Chomsky) 159
  behaviorism: rejection of inner mental
    states 157–8
  identity: different forms in philosophy
    161–2
  identity theory in philosophy 160–1
  logical behaviorism 158–9
  logical behaviorism: objections 159–60
  methodological behaviorism 157
physico-chemical synthesis (Sherrington)
  145
physics 44, 46, 56, 59–60, 89, 156, 169
physiology 78, 89, 306, 379
pineal gland 137, 146, 153, 199, 213
Place, U. T. 160
Plato 3, 24, 44, 73, 153, 431, 534
  allegory of cave 130–1, 132–3, 135
  concept of mind 131–2, 133
  perception and cognition 134
  skyscraper as modern cave (metaphorical
    comparison) 131, 134
  theory of forms 132
'plausibility problem' 183, **185**
'point of view' (Nagel) 365–6, 382–3, 411,
  423, **432**, 438
  see also subjectivity
polymorphisms 79–80
Pomarol-Clotet, E. 498
Popper, K. **95–6**, 146

positive BOLD response (PBR)   413
positron emission tomography (PET)   **82–3**, 87–8, 139, 339, 407, 412
possible consciousness   402, 409–10, 420, 422–4
  necessary conditions   418, 427
  see also   'actual consciousness'
posterior cingulate cortex (PCC)   338, 412, 458–9, 479–80, 498–9, 507, 521
posterior CMS (cortical midline structures)   341, 498, 499
pre-established harmony (Leibniz)   141
pre-reflective self-awareness   450, 454, 497
'pre-reflective self-consciousness'   454–5, 463, 471–2, **475–6**, **482–3**, 488, 493, **501–2**
  and brain's intrinsic activity   **483–4**
  philosophical definition   483
  and VS   **482–3**
'preconscious' (Searle)   431, 432
'preconscious system' (Freud)   431
precuneus   412, 458–9, 485, 498, 507
predicative dualism (Bennett and Hacker)   **144**, 153
'predispositions' (McDowell and Hornsby)   260
prefrontal cortex   297, 303, 336, 338, 342, 384, 390–1, 430, 495
  inferior   521
  lateral   413, 415
prefrontal cortical mirror system   523
prefrontal mirror mechanisms   524
prefrontal-parietal network   386–7
premotor cortex   146, 465, 523
'principle of tolerance'   **49–50**
'problem of other minds'   456, **514**
promoter polymorphism (5–HTTLPR)   80
property dualism   **138–9**, 142, 144, 145, 153
property P   212, 213, 225, 232, 244
  access to, and knowledge of   **219–20**
  and brain   **218–19**
  'cognitively closed'   220
  McGinn versus Kant   **221–3**
'propositional attitudes'   251, 278, **281–2**, 283–4, 286, 294, 300, 520
  generation   285
  versus 'proposition'   282
'propositional awareness'   471, 475, **482**, 491
propositional content   286, 300
'proto-mental features'   152
'proto-mental properties'   400
proto-self   **465–6**

Prozac   79, 433
psychiatric disorders   419n, 493
  and self   **495–6**
psychiatric disorders: abnormalities of self and brain   18, 447, **493–511**
  chapter topic   493, 532–3
  data deficiencies/further research required   494–5, 499, 501–3, 505–10
  depression   **505–8**
  schizophrenia   **496–505**
psychiatric disorders: background   **494–6**
  self and brain   **494–5**
  self and self-consciousness   **493–4**
psychiatric disorders: neurophilosophical discussion
  brain: intrinsic activity (conceptual and empirical characterization)   **508–9**
  embodied and relational concept of self   **508**
  immunity and brain's intrinsic activity   **504–5**
  see also   schizophrenia
psychiatry   75, 78, 89, 361, 306, 447
psychofunctionalism   **175**, 176
psychological functionalism   250, **266–7**, 276
psychological specificity   449, 460, **461–2**, 468, 469
psychological synthesis (Sherrington)   145
'psychologism' (Husserl)   38
psychology   59–60, 66, 77, 156, 180, 190, 233, 247, 249, 275, **302–3**, 333, 349, 451, 457, 513, 517, 533
  brain and intersubjectivity   **518–20**
  concepts   303–4
  empirical domain   250
  methodological behaviorism   **157–8**
  versus phenomenology   **35–7**
  versus philosophy   **71**
  from philosophy of mind to   **250–1**
  relation with neuroscience   **303–4**
  'science of behavior' versus 'science of mind'   157
  see also   scientific psychology
psychology: history   **70–3**
  from psychology over cognitive science to cognitive neuroscience   **72–3**
  psychology and consciousness   **71–2**
  psychology as science of mind   **70–1**
psychons   **146**
psychopathology   511
psychopharmacology   79, 80
pulvinar   485, 507
'pure experience' (James)   191
Putnam, H.   **237–9**

Qin, P. xi, 479, 485
qualia **171–2**
  content–context relationship 362–3
  elimination **272–3**
  empirical plausibility as context–content relationship **363–4**
  'extrinsic to brain' 362
  'hallmark features of consciousness' 435
  phenomenal features of consciousness **434–5**
  'phenomenal hallmark feature of consciousness' 360
  see also absent qualia
'qualitative subjective unity' (Searle) **441–2**
quanta 169–70, 191
quantum physics 151, 169–70, 186, 190
Quine, W.v.O. 67, 70, 92, 98, 106, 270
  a priori versus *a posteriori* knowledge (distinction undermined) 57
  'analytic' versus 'synthetic' sentences' **48–9**, 57
  'eliminated principal distinction between conceptual-logical and observational-experimental methodological strategies' 112–13
  metaphysics 52
  methodological naturalism **62–4**
  naturalization of philosophy **48**, 98
  observation sentences **53–6**
  'regimented theory' 52
  rejected 'principle of tolerance' **49–50**
  replacement naturalism **63–4**
  task and purpose of philosophy **51–2**
'Quining Qualia' (Dennett, 1988) 273

race/racism 313, 486, 487
Raichle, M. E. 405
Ramón y Cajal, S. 74
raphe nuclei (RN) 77
  'dorsal raphe' 507
rational-argumentative method 65f, **70**, 97, **537**, 538
rational-logical argumentation 44
re-entrant circuits 382, 385, 387, 398
're-entrant connections' 400
  neuronal and experimental specificity **387–8**
re-entrant processing **385–6**, 395, 408
reading 436–7, 455, 477, 503, 513
reality 106, 115, **130–1**
reason 133, 134
receptual similarity (Quine) **53–4**, 56
recognition (concept) **518**
reduction 249, 250, 303
reductionism 2, 190

'reductionism in practice' (Bickle) 313
reductive neurophilosopher (RN) 34, **103–5**
reductive neurophilosophy ix, 2, **4–5**, 6, 10–15, 38, 63–7, 85, 91, **98–100**, 101, 103, 122–3, 148, 268, 309, 351, 356, 536
  as brain-reductive approach 13f
  conceivability argument **371**
  concepts of self 447
  considers 'how' as detached from 'why' 536–7
  NPC and hard problem **418–20**
  see also 'brain and philosophy: neurophilosophy'
reflective self-consciousness **475–6**, 482–3
reflex arc (Sherrington) 145
regional cerebral blood flow (rCBF) 484, 485
relational concept of self **508**, 509
relational self 493, 510
religion 135, 253, 311f
replacement naturalism **63–4**, 65, 66, **96–8**, 100–2, 123, 307
representation 278, 288, **291–6**, 297, 299, 301, 521
  additional mechanism 291, **293–4**
  first-order, second-order, third-order **292–3**
  localized versus distributed **286**
  non-conceptual content **295–6**
  semantic content and **279–80**
  see also higher-order representation
representation (cognitive function) 391
representation (of world by brain) 98, 237
representation of contents 287–8
representational contents 295
representational roles 278, 280
'repressed unconscious' (Freud) 431
res cogitans 25
res extensa 25
researchers 20
rest–stimulus interaction 340f, 343–4, 345, **414–15**, 416, 425
  abnormal 505
  stimulus–rest interaction **415–16**
resting state activity 338–9, 341–3, 345–8, 384, 405–7, 413, 427, 459, 463, 468, 480, 488, 492, 498–500, 508
  abnormal 499, 501, 502–3, **510**
  hypoactive 507
  insulin and **410–12**
  neural overlap 494–5
  neural overlap between self and **484–5**
  prediction of self-specificity **486–7**, 510
  same as 'brain: intrinsic activity' 503
  spatiotemporal pattern 495
  and stimulus-induced activity **321–2**

resting state and stimuli
  interaction **414–15**
  interaction (non-linear) **415–16**
Revonsuo, A. 433
Rizzolatti, G. **522–3**
Rorty, A. O. 132
Rorty, R. 270
Rosenthal, D. M. 292
Roskies, A. 110
Rotarska-Jagiela, A. 499
Rowlands, M. 234
Ruby, P. 470
Ryle, G. **27–8**, 159

S-allele 80
s-identity (Gallese) **528**, 529–31
Sartre, J-P. 515
Sass, L. A. 497
Schilbach, L. 457, 459
schizophrenia **496–9, 509–10**
  abnormal intrinsic activity in cortical midline structures **497–9**
  abnormal neural activity during self-specific stimuli 499
  alteration of 'presence' (Parnas) 497
  basic disturbance of self **496–7**, 499–501, 503, 505, 510
  'central factors' 497
  definitions 496
  low-frequency oscillations 498–9
  pre-onset stages 497
  subjective experience of abnormal self 497
schizophrenia: neurophilosophical discussion **500–5**
  concept of empirical self **500–1**
  concept of mental self 500
  concept of minimal self **502–3**
  'immunity against error through misidentification' **503–4**
  pre-reflective self-consciousness **501–2**
  self as subject and object **503–5**
Schlicht, T. 206
Schopenhauer, A. 106, 213, **231f, 232–3**, 244
  biography 94, 225
  and brain 225–6, 229–30
  brain-paradox 95, 227–9
  cognition and knowledge **94–5**
  implicit neurophilosophy **94–5**
  'neurophilosopher *avant la lettre*' 226
  re-conceptualized **229–30**
  'shift from mind to brain' 226
science 58f, 65f, 435, **438**, 535
  '*a posteriori* knowledge' 51
  dualism and **147**
  versus philosophy **66–7**
  see also cognitive science
science of brain 66, 89
'science of experience' **439–40**
science of mind 65, 66, **69–70**, 88–9
science and philosophy,
  domains and methods (bridged) **47–8**
scientific method 36, 44, 46, 53, 67, 70
scientific psychology 249, 250, **256–9**, 266–9, 273–4, 275–6, 303, 337
  concepts (matching folk psychology) **269–70**
  definition 251
  eliminative materialism **271–2**
  interface problem (with folk psychology) **251–2**
  'reduction to neuroscience' 304
  see also cognitive psychology
Searle, J. R. 119n, 213, 431, 489
  appeal of dualism 142
  biological naturalism **192–4**
  building block approach **440–1**, 441–2
  'Chinese room argument' 179
  versus Churchland 117
  concepts: unilateral adaptation to facts **115–16**
  concepts of unconscious 432
  conceptual and empirical plausibility 117
  conceptual propaedeutic as input 116
  'conscious units' 441
  definition problem 193, 195
  empirical plausibility 117, **121–2**
  'explanatory reduction' 194
  first- and third-person ontology **193–5**
  mental–physical dichotomy 193
  neuronal processes (consciousness-yielding or not) **194–5**
  phenomenal-empirical fallacy 196f, 196n
  'phenomenal-physical inference' **195–7**
  'qualitative subjective unity' **441–2**
  'unified field model' **441–2**
second-person perspective **37–8**, 269, 457
self 18, **447–534**
  'basic disturbance' (schizophrenia) **496–7**
  and brain **494–5**
  as brain-based neurosocial structure and organization **467–8**
  'central hallmark' **532–3**
  concepts 447
  and consciousness (VS patients) **481–2**
  and decision-making 462
  definition problem **481–2**
  Descartes 135
  embodied and relational concept 508

self – *continued*
  and emotions 462
  encoded in intrinsic activity 489, 501, 505
  indirect and direct access to in consciousness 474–5
  location in brain's intrinsic activity 487–8
  neural overlap between and resting state activity 484–5
  phenomenological concept 469
  philosophical and neuroscientific approaches 471–2, 491
  philosophical concepts 449, 468
  philosophy 481
  'presence in vegetative patients' 480–1
  and psychiatric disorders 495–6
  and reward 462
  and self-consciousness 493–4
  versus 'self-reference' 466–7
  as subject and object 478–9, 482, 483
  temporal dimension 450
  see also 'brain and self'
self and other 18, 512, 529, 530f, 533
  phenomenological philosophy 515–16
  see also intersubjectivity
self as subject and object 503–5
self-consciousness 435, 438, 447, 450, 454, 471–92, 493, 509, 510
  characterization 474
  epistemological features 491
  'explanatory gap (s)' and 473–4
  'hallmark' 503
  neuroscience –philosophy linkage 494
  philosophy 481
  reflective and pre-reflective 475–6
  self and 493–4
  see also brain and self-consciousness
'self-models' (Metzinger) 452–3
self-perspectival organization 426, 443–4, 489
  'hallmark feature of consciousness' 438
  phenomenal feature of consciousness 437–8
self-reference 458, 460, 466–7, 472–4, 479–80, 484, 490–2, 494–5, 533
  phenomenal specificity 463–4
  social patterns of neural activity 459
self-reference effect (SRE) 449, 451, 457, 466, 468, 469, 472
self-referential processing 466, 530f
self-reflection 451, 506
self-relatedness 416, 466
self-relevance 463–4
'self-representation' 293–4, 449, 451, 468–9, 472, 500–1

self-specific stimuli 451, 466, 472, 479–90, 494–5, 497–8, 501–2, 507, 510
  abnormal neural activity 499
  spatial patterns of neural activity 457–8
  temporal patterns of neural activity 458–9
self-specificity 449, 460–1, 462–3, 468–9, 473–4, 476, 490–1, 505, 530f
  prediction by resting state 486–7, 510
  vegetative state 479–80
semantic contents 36, 160, 244, 278, 280–6, 291–4, 299–300
  LOT on sub-personal level (Fodor) 284–5
  mental representation 280–1
  mind and 279
  propositional attitudes 281–2
  and representation 279–80
  'structural isomorphism' (Fodor) 284–5
  sub-personal level 282–3
  syntax and semantics: 'associated with sub-personal and personal levels' 285–6
semantic externalism 212, 213, 238–41
semantic internalism 238–41
semantic meaning 179–80, 278, 300
semantic properties 267–8, 278, 300
semantics,
  'associated with sub-personal and personal levels' (Fodor) 285–6
sensible continuity (James) 72
sensorimotor cortex 166, 392
sensorimotor functions 32, 73, 75–6, 95, 102, 234–5, 244, 342, 392, 397, 407, 438, 462–5, 472, 501–3, 509–10, 517, 532–3
  brain-based 103
  and consciousness 76–7
  relation to body during consciousness 235–6
sensorimotor functions and consciousness 394, 395–6, 396f
sensorimotor states 28, 530n, 531
'sensory chains' (Plato) 131
sensory cortex 297, 334–6, 349, 391, 394, 461, 465, 502
sensory functions 333–5, 394
  as modules 334–5
serotonin 77, 79–80, 82, 415, 433
Shamay-Tsoory, S. 534
'shared manifold' (Gallese) 512, 529–32
Sherrington, Sir Charles 145, 147, 340, 341
Shulman, R. G. 406–7, 408, 411
'Simple Ego-type Life Form' (Panksepp) 466
simulation 525, 526
  concepts (Gallese) 527–8

simulation theory (ST)   512, 519, **520**, **525**, 530f, 533
Singer, W.   384
Skinner, B. F.   157–8
sleep   383, 386, 416, 424, 433
  non-REM (NREM) sleep   386, 390, 403–4
  rapid eye movement (REM) sleep   403, 404
  see also   dreams
slow cortical potentials (SCP)   413–14
'slow wave hypothesis'   **402**, **405–6**, 424
Smart, J. C.   160
smiling   518, 519
social anxiety (propositional attitude)   **282**
social functions (of brain)   337, 342
social neuroscience   93, 307
social sciences   88, 304, 311f
social self   **456–7**, 467
social similarity (observation sentences)   **54–5**, 56
socio-cultural world (Popper and Eccles)   96
Socrates   130
'software'   155, **173**, 174–7, 180–1, 186, 206, 213, 215
Solms, M.   78
soul   3, 134, 136, 147
'sparseness'   289–90
spatial characterization   **412–13**
spatial patterns,
  neural activity during self-specific stimuli **457–8**
spatiotemporal activity   **290–1**
spatiotemporal continuity   344, 442–4, 464
  phenomenal feature of consciousness **436**, 437
spatiotemporal structures   403, 421–4, 427, 531
spatiotemporal structure and consciousness **417–18**, 419f
spatiotemporal unity   440, 442–3
  phenomenal feature of consciousness **436**
  see also   unity
spatiotemporality   325, 326, 419f, 440–1, 510
specificity,
  different forms   **460**
'specious present' (James)   436
speculation   201, 202
Spinoza, B.   151
'split brain'   **114**
'standard simulation theory' (Gallese)   **527**
state of consciousness   402, 404, 419f
Stich, S.
  folk versus scientific psychology   270

'reference failure' versus 'reference success' 270
stimuli   333, 398–401, 429, 432, 452, 466, 473–4, 490
  encoded by brain into neural activity   321
  origins   **320–1**
  self-specific versus non-self-specific   **460–1**
  types   320
stimuli and resting state
  interaction   **414–15**
  interaction (non-linear)   **415–16**
stimulus-induced activity   321–2, 340f, 341, 343, 345–6, 348, 377, 404–8, 411, **412**, 414, 419f, 427, 485, 487, 498–501, 507, 510
stimulus-related effects,
  versus task-related effects   **461–2**
stimulus–response   157–9, 341
Strauss, E.   75, 95, 102, 234
'stream of consciousness' (James)   **71–2**, 191
striatum   77, 521
'structural isomorphism' (Fodor)   278, **284–5**, 292, 294, 300
students   **18–19**
subcortical regions   77, 79, 335, 342, 385, 393, 396, 465–6, 485, 507, 511
subgenual cingulate cortex   486
subjectivity   **31–2**
  abnormal self   **497**
  'core problem of philosophy'   **495**
  Nagel   223
  see also   FPP
'subliminal'   **431–2**
subpersonal level   286, 299–300, 303
  LOT (Fodor)   **284–5**
  semantic contents   **282–5**
  'structural isomorphism' (Fodor)   **284–5**
substance dualism (Descartes)   25, **137–8**, 142, 144–5, 199
substantia nigra (SN)   507
subthalamic nucleus (STN)   465, 486
superior temporal sulcus (STS)   521, 523, 525
supervenience   5, 155, **167–8**, 172, 180
  mental–neuronal   167
  of mental states on neuronal states   **166–7**
  neuronal–neuronal   167
  'non-reductive form of materialism'   168
supragenual anterior cingulate cortex (SACC) 166, 458, 479
sympathy (concept)   **518**
synapses   **80–1**, 146
syntax   179, 300
  'associated with sub-personal and personal levels' (Fodor)   **285–6**

synthetic naturalism **59–60**
  non-reductive 60, 61
  reductive 60
synthetic sentences 49, 57–9, 62

tables 37–8, **395–7**, 436, 452, 537
task-induced deactivation (TID) **498**
Taylor, S. F. 499
teachers **19–20**
temporal characterization **413–14**
temporal continuity 426
  see also spatiotemporal continuity
temporal cortex 336, 390
temporal patterns,
  neural activity during self-specific stimuli 458–9
temporal pole (TP) 521
temporo-parietal junction (TPJ) 459, 521
thalamo-cortical connectivity 386, 387, 403–4
thalamus 77, 339, 342, 385–6, 401, 485
  dorsomedial 507
Thales of Miletus 21
theoretical posits **267–8**
'theory of brain' 331
theory of forms (Plato) 132
'theory of mind' (ToM) **518**, 519, 525
  and brain imaging **520–1**
'theory-formation' **519**
theory-theory (TT) 267, 512, **518–20**, 525, 530f, 533
'third world' (Popper and Eccles) 146
third-person access **438–9**
'third-person ontology' (Searle) 116
third-person perspective (TPP) **193**
  investigation of consciousness **427–8**
  see also empirical domain
Thompson, E. 103, 234, 528
time **169**
token–token identity 161–3, 166–7, 185
Tononi, G. **385–6**, 387–9
top-down (personal to sub-personal) approach 285
'top-down modulation' **391**, 396f
trains (perception) 477, 503
trait adjectives 463, 507
transcendental method (Kant) 6
transcendental self (Kant) 489, 490
transcranial magnetic stimulation (TMS) 522
Treatise of Human Nature (Hume, 1746) 201
Treynor, W. 505
Turnbull, O. 78
'Two dogmas of empiricism' (Quine, 1951) 48

Tye, M. **295–6**
type-type identity 161–3, 166, 175, 185

unconscious 71, **291–2**, 294–5, 374, 377, 383–4, 388, 397–8, 426–7
  and consciousness **39–40**
  see also easy problems
unconsciousness 386, 409–10, 416, **430–2**, **433**, 434, 443–4
'unified field model' (Searle) 426, **441–2**, 443
  brain's intrinsic activity **442–3**
unity 272, 426, 443, 444, 464
  see also spatiotemporal unity

Varela, F. 103
vectorial representation **287–8**, 298, 300
vegetative state (VS) 30, 35, 75, 113–14, 136, 324, 386, 390, **391–2**, 393, 402–3, 406–8, 411–12, 419n, 422, 424–5, 447, 471, 490–4, 496
  definition problem 481–2
  self-specificity and consciousness **479–80**
  see also hard problem
vegetative state: brain and self-consciousness
  data deficiency 483
  presence of self **480–1**
  pre-reflective self-consciousness **482–3**
  propositional awareness **482**
  relationship between self and consciousness **481–2**
vehicle (sub-personal) **284–5**
vehicle-content distinction 278, 283, 285, 286
  see also semantic contents
Velmans, M. 439
ventral premotor cortex (F5) 522–5
ventral striatum (VS) 79, 462, 525
ventral tegmental area (VTA) 77, 507, 525
ventromedial prefrontal cortex (VMPFC) 412, 458–9, 462, 479–80, 484–5, 507
vertical explanations 249, 254, **255–6**, 276
Vesalius, A. 73–4
visual cortex 253, 297, 303, 333–5, 342–3, 349, 384, 415–16, 430, 432
  'primary visual cortex' (V1) 385, 388
'vital spirits' (Harvey) 16, 147

water 144, 161, 165, 194
  'rigid designator' 164
Watson, J. 157–8, 384
'we-centric space' 529
Wernicke, C. 75

'what it is like'   171, 272, 322, 357, 359, 361, 363, 374–6, 382, 388–9, 397–9, 426, 429–30, 444
  phenomenal feature of consciousness **434–5**
'what it is like' argument   353, **358t**, **364–5**, 367, 368, 380
  NPC and   **423–4**
  refutation   **365–6**
'What is it like to be a bat?' (Nagel, 1974)   223, 294, **364**
white matter   **74**
Whitehead, A. N.   151
Whitfield-Gabrieli, S.   498
Wicker, B.   524
Wiebking, C.   xi

Wittgenstein, L.   27, 163
Witzak, J.   511
Williord, K.   293
working memory   106, **390–1**, 416
Wundt, W.   **70–1**, 89

Zahavi, D.   454–6, 515
Zeki, S.   345
zombies   **142–4**, 357, 358t, 380, 514
  and conceivability argument   **369–70**
  definition   369
  irrelevance   **371**
  logical: transition from logical to natural worlds   **372**
  natural and logical worlds   **370–1**
  plausibility   **372–3**

Printed and bound by CPI Group (UK) Ltd, Croydon, CR0 4YY